AF773691

315
Current Topics in Microbiology and Immunology

Editors

R.W. Compans, Atlanta/Georgia
M.D. Cooper, Birmingham/Alabama
T. Honjo, Kyoto · H. Koprowski, Philadelphia/Pennsylvania
F. Melchers, Basel · M.B.A. Oldstone, La Jolla/California
S. Olsnes, Oslo · P.K. Vogt, La Jolla/California

James E. Childs John S. Mackenzie
Jürgen A. Richt (Eds.)

Wildlife and Emerging Zoonotic Diseases: The Biology, Circumstances and Consequences of Cross-Species Transmission

With 49 Figures and 21 Tables

Springer

James E. Childs, ScD
Senior Research Scientist
Department of Epidemiology
and Public Health and
Center for Eco-Epidemiology
Yale University School of Medicine
60 College St., P.O. Box 208034
New Haven, CT 06520-8034
USA
e-mail: Jamesechilds@comcast.net

John S. Mackenzie, Ph.D.
Premier's Fellow and Professor of Tropical
Infectious Diseases, and Deputy CEO
Australian Biosecurity Cooperative Research
Centre for Emerging Infectious Diseases
Curtin University of Technology
GPO Box U1987
Perth, WA
Australia 6845
e-mail: J.Mackenzie@curtin.edu.au

Jürgen A. Richt, DVM, Ph.D.
Veterinary Medical Officer, Lead Scientist
Virus and Prion Diseases of Livestock Research Unit
National Animal Disease Center
USDA, ARS
2300 Dayton Ave Ames, IA 50010
USA
e-mail: juergen.richt@ars.usda.gov

ISBN 978-3-642-08996-1 e-ISBN 978-3-540-70962-6
ISSN 0070-217X

This work is subject to copyright. All rights reserved, whether the whole or part of the material is concerned, specifically the rights of translation, reprinting, reuse of illustrations, recitation, broadcasting, reproduction on microfilm or in any other way, and storage in data banks. Duplication of this publication or parts thereof is permitted only under the provisions of the German Copyright Law of September, 9, 1965, in its current version, and permission for use must always be obtained from Springer-Verlag. Violations are liable for prosecution under the German Copyright Law.

Springer is a part of Springer Science+Business Media
springeronline.com
© Springer-Verlag Berlin Heidelberg 2007
Softcover reprint of the hardcover 1st edition 2007

The use of general descriptive names, registered names, trademarks, etc. in this publication does not imply, even in the absence of a specific statement, that such names are exempt from the relevant protective laws and regulations and therefore free for general use.
Product liability: The publisher cannot guarantee the accuracy of any information about dosage and application contained in this book. In every individual case the user must check such information by consulting the relevant literature.

Editor: Simon Rallison, Guildford
Desk editor: Anne Clauss, Heidelberg
Cover design: WMX Design, Heidelberg

List of Contents

List of Contributors

(Addresses stated at the beginning of respective chapters)

CTMI (2007) 315:1–31
© Springer-Verlag Berlin Heidelberg 2007

Introduction: Conceptualizing and Partitioning the Emergence Process of Zoonotic Viruses from Wildlife to Humans

J. E. Childs[1] · J. A. Richt[2] · J. S. Mackenzie[3] (✉)

[1]Department of Epidemiology and Public Health, Yale School of Medicine, 60 College St., P.O. Box 208034 New Haven, CT 06520-8034, USA

[2]Veterinary Medical Officer, National Animal Disease Center, 2300 Dayton Ave, Ames, IA 50010, USA

[3]Premier's Research Fellow, Australian Biosecurity CRC, Curtin University of Technology, GPO Box U1987 Perth, WA, Australia 6845
J.Mackenzie@curtin.edu.au

Abstract This introduction provides a telegraphic overview of the processes of zoonotic viral emergence, the intricacies of host–virus interactions, and the distinct role of biological transitions and modifying factors. The process of emergence is conceptualized as two transition stages which are common and required for all disease emergence, (1) human contact with the infectious agent and (2) cross-species transmission of the agent, and two transition stages which are not required for emergence and appear unavailable to many zoonotic pathogens, (3) sustained human-to-human transmission and (4) genetic adaptation to the human host. The latter two transitions are presumably prerequisites for the pandemic emergence of a pathogen. The themes introduced herein are amplified and explored in detail by the contributors to this volume. Each author explores the mechanisms and unique circumstances by which evolution, biology, history, and current context have contrived to drive the emergence of different zoonotic agents by a series of related events; although recognizable similarities exist among the events leading to emergence the details and circumstances are never repetitive.

1 Introduction

The process of zoonotic disease emergence can be understood by coupling knowledge of how zoonotic viruses have evolved and are maintained among their wildlife hosts, transmitted across a species barrier to cause productive infection in a taxonomically distinct secondary host, initiate a pathologic process causing disease, and, by repetitive infection within the secondary host species, result in incident morbidity or mortality of sufficient magnitude to be detected and characterized as a novel health concern of local, regional, or global significance (see the chapter by Childs, this volume). Obviously, we possess no such knowledge for any zoonotic virus or zoonotic disease, but casting the emergence process in this context underscores how disciplinary boundaries are blurred; advances require approaches spanning the spectrum of biological inquiry, and solutions to imminent threats require approaches unbounded by the notion of specific scientific discipline.

The emergence process involves ecological interactions at the individual, species, community, and global scale. The dynamic circumstances and relative importance of the participants reflect the evolutionary context in which zoonotic agents have become accommodated to, and been accommodated by, their reservoir hosts (H_Rs) (see the chapters by Cleaveland et al. and by Holmes and Drummond, this volume), the diversity of reservoir species involved, their geographic ranges and the local dispersion of host and pathogen populations. In turn, historical factors have modified and blurred traditional patterns of species distribution, abundance, and diversity, and are continually transforming the landscape of opportunity on which zoonotic viruses with their H_Rs mingle with novel, potentially susceptible secondary host species (H_S)s (see the chapters by Daszak et al., Field et al., Regnery, and Wang and Eaton, this volume). The current historical circumstances are unprecedented in their efficiency for continually shuffling an expanding repertoire of zoonotic viruses and hosts, introducing them in novel ecologic circumstances to a wealth of previously unavailable and unexplored niches. Within the last decade, the accelerated pace of rapid translocations of infected H_Rs or H_Ss have heralded a sea change in how we view the public health threat posed by zoonotic viruses (Childs 2004), as testified by the emergence of SARs coronavirus (SARS CoV) (Drosten et al. 2003; see the chapter by Wang and Eaton, this volume), influenza A subtype H5N1 (Peiris et al. 2004; see chapter by Webby et al., this volume), West Nile virus (WNV) (Lanciotti et al. 1999), Nipah virus (NiV) (Chua et al. 1999; see the chapter by Field et al., this volume), and Monkeypox virus (Anderson et al. 2003; see the chapter by Regnery, this volume).

1.1 Cross-Species Transmission (Spillover)

Inherent in the term "cross-species transmission" (or spillover) is the ability for a foreign virus, once introduced into an individual of a H_S population, to complete the virus infectious cycle: (1) adsorption, penetration, and uncoating, or separation of the viral nucleic acid from the capsid; (2) transcription, translation, and replication, and; (3) assembly and release (Nayak 2000). Binding and entry into permissive H_S cells is mediated by common or related cellular receptors. Additional bouts of infection following virus release from infected cells lead to the dissemination of virus throughout the host's tissue(s), precipitating, as a byproduct, pathologic alterations in the individual H_S identifiable as symptomatic disease.

1.2 Pathogenesis in the Reservoir Host and Secondary Host

The pathogenic course of infection and disease within the secondary host (H_S) may bear little correspondence to the infectious process and outcome within the reservoir host (H_R). Oral lesions caused by herpesvirus B (CeHV-1) infection among individual macaques of the H_R are transmogrified into an often fatal (~70%) meningoencephalitis in the human H_S (Huff and Barry 2003). Hantaviruses cause subclinical infections or subtle behavioral changes with limited pathology in individual rodents of species constituting the virus-specific H_Rs (Hinson et al. 2004; Llyubsky et al. 1996; see the chapter by Klein and Calisher, this volume), accompanied by no notable loss of fitness (Childs et al. 1989). However, these subtleties are lost in the severe and often fatal hemorrhagic fever with renal syndrome (HFRS) and hantavirus pulmonary syndrome (HPS), developing in the human H_S after virus spillover (Zaki et al. 1995; Tsai 1987).

No matter how different the disease course among the human H_S, the pathologic component of intra- H_R transmission is highly relevant when considering strategies to prevent human infection rather than treating post-spillover disease (see the chapter by Daniels et al., this volume). Ignoring the intricacies of zoonotic virus transmission among wildlife H_Rs guarantees that solutions springing from a traditional anthropocentric disease-treatment/vaccine-preventative approach will consider a limited universe of defensive prevention targets and generate a restricted arsenal of intervention tools.

2 The Comparative Ecology of Zoonosis Emergence and Species Invasion

2.1 Four Transition Stages to Emergence: The First Two Are Prerequisite

The ecologic process of zoonotic disease emergence can be schematized by four transition stages (Fig. 1), of which only the first two are prerequisites for emergence: (1) contact between infectious propagules originating from the wildlife H_R with individuals of a susceptible H_S and (2) cross-species transmission, a transition subsuming the complex interactions of the virus infectious cycle within the H_S (Nayak 2000; Childs 2004). These first two transitions may require a mediating host such as an arthropod vector (H_V) or an intermediary vertebrate host (H_I); these additional host populations are readily accommodated by the modular emergence schema (Fig. 1).

2.2 Two Transition Stages Are Required for Pandemic Emergence

The latter two transition stages demarcate a change in the interrelationship of host and virus (Fig. 1): (3) sustained transmission of the once zoonotic virus between members of the new H_S, subsequent to, and independent of, new spillover events, and (4) genetic adaptation and phenotypic changes accompanying sustained intra-H_S transmission. Once sustained transmission occurs within the human host, evolutionary adaptation between virus and host can transform the once zoonotic virus into a distinctive new virus with a new human H_R. The new virus associated with humans must be quantitatively and qualitatively different from ancestral strains in genetic and phenotypic characters, in order to designate the emergence of a new biological entity. With HIV and pandemic influenza subtypes, the qualities of the newly adapted viruses to humans are readily apparent in terms of host preference and host pathogenicity (Hahn et al. 2000; Claas 2000). With SARS CoV infecting humans, the specific genetic changes are less clear-cut (Song et al. 2005), most probably because the transmission of SARS CoV was curtailed early its relationship to the new human host. Support for this conclusion is based on the genetic differences accrued by SARS CoVs sustained through multiple generations of human-to-human transmission as compared with those viruses with shorter interhuman passages (Liu et al. 2005).

Some viruses are capable of sustained human-to-human transmission with minimal or no genetic change [i.e., SARS CoV; see the chapter by Wang and Eaton, this volume, although limits to genetic adaptation within humans may be imposed by the requirement for an intermediate vector or extensive prior adaptation to a specific reservoir host (Gould et al. 2003)]. The arboviruses, yellow fever virus, and the four dengue serotypes circulate in a human-to-human transmission cycle mediated by anthropophilic H_Vs after introduction by bridging H_Vs feeding on infected primate H_Rs (de Silva et al. 1999; Wolfe et al. 2001; Monath 1989; Downs 1982); these viruses appear closely related to the wild type viruses circulating in sylvatic cycles, although regional variation is apparent (Bryant and Barrett 2003).

Viral adaptation to the human H_R appears in most cases to be critical to developing a virus with pandemic potential (Mims 1991, 1995). The introduction of avian-like gene segments into preexisting, aerosol-transmitted, human influenza A viruses, or alternatively, the introduction of key genetic components into preexisting avian viruses (see the chapter by Webby et al., this volume) may be prerequisite to pandemic influenza A emergence (Claas 2000). The emergence of SARS into the human population was accompanied by strong and rapid positive selection of different subtypes of virus as indicated by comparisons of sequence data from humans and from palm civets and rhinolophid bats, putative H_Rs, or intermediate hosts (H_Is) for SARS CoV (Lau et al. 2005; Song et al. 2005; see the chapter by Wang and Eaton, this volume).

2.3
The Basic Reproductive Potential R_0 as a Measure of Viral Relative Fitness

To capture the rate at which outbreaks spread among hosts, epidemiologists have relied upon the reproduction potential, R_0, as a measure of the expected number of secondarily infected and infectious hosts produced during the infectious period of a single infected host when introduced into a freely mixing population of susceptible individuals (Halloran 1998). The relative fitness defined by R_0 is a composite of three terms *c*, the contact rate or number of contacts per unit time, *p* , the transmission probability per contact, and *d*, the duration of infectiousness (see the chapter by Real and Biek, this volume). Examples of zoonotic viruses taking alternative paths to emergence, with highly variable R_0s are discussed below.

3
Modifying Factors in the Emergence Process

The underlying feature distinguishing modifying factors (Fig. 1, right panel) from transition stages in zoonotic virus emergence (Fig. 1, left panel) is that the former requires the substrate provided by the latter on which to act. Modifying

Fig. 1 A schema for partitioning the process of zoonotic disease emergence into four transitions and modifying factors which alter the likelihood of transitions occurring. Disease emergence can occur at the local and regional level or as a pandemic depending on the nature of the pathogen and the influence of modifying factors. Modifying factors are largely responsible for driving the magnitude and geographic scope of an emergent event, but by themselves are insufficient to lead to disease emergence. Although only a single population source for a zoonotic pathogen is indicated, the reservoir host (H_R) population, the schema is modular and readily accommodates inclusion of vector (H_V) populations and intermediate vertebrate host (H_I or H_{S1}) populations antecedent to spillover to humans (see Fig. 1 in the chapter by Childs, this volume). Transition stages, with the exception of contact (*Transition 1*) and cross-species transmission (spillover; *Transition 2*), are not strictly hierarchical in the emergence process. Transition stages are shown to the *left of the population boxes* and the two transitions required for emergence (contact and spillover) are shaded *gray*. *In the center* are two population boxes, the *top shaded box* indicating a H_R population, in which a zoonotic virus or some other zoonotic pathogen circulates, and the *bottom shaded box* indicating the secondary host population (H_S) affected by pathogen spillover (assumed in most instances to be humans). The *graded shaded pyramid within the H_S population* indicates that emergence often proceeds through a gradient of human population sizes and social connectivity. Spillover and human transmission chains in remote villages (apex of pyramid) can lead to spread to urban centers (*base of pyramid*), at which point a pathogen is assumed to have access to the entire H_S population demarcated by the *H_S box*. To the left of the population boxes are

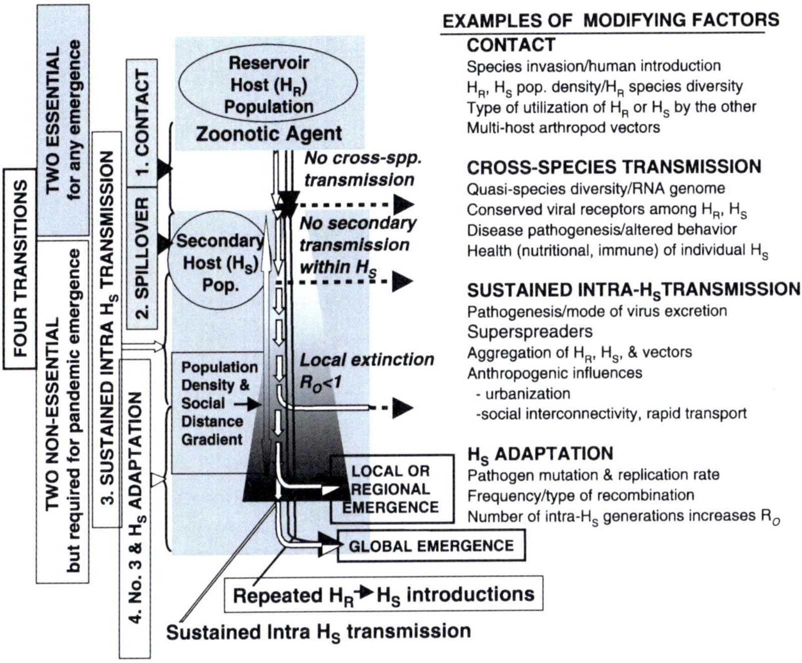

Fig. 1 *Continued* examples of modifying factors. Contact and spillover are sufficient to result in disease emergence at the local, regional, or even continental scale through reiterative introductions, as exemplified by zoonotic diseases such as rabies or West Nile fever. The two *solid black lines with arrowheads leading from the H_R then directly through the H_S to local or regional emergence* (the first emergence box) represent reiterative events as a pathway to emergence. Two other transitions not essential for emergence, but critical to pandemic disease emergence, require sustained intra-H_S transmission of the zoonotic pathogen (*Transition 3*) and, potentially, adaptation to the human host (e.g., SARs coronavirus). Sequential human-to-human transmission of a zoonotic pathogen at the local and regional scales is indicated by the series of *broken white lines in the H_S pyramid*. Evolutionary forces can transform a zoonotic agent into a genetically distinct agent adapted to a new reservoir host, establishing the former H_S as a new H_R (e.g., SIV to HIV; avian influenza to pandemic human influenza). This transition and the modifying factors associated with the geographical location and context of the initial disease outbreak (*white arrows in shaded H_S cone*) can ultimately precipitate a pandemic emergence (e.g., HIV and pandemic influenza). The process of emergence for any zoonotic pathogen can fail at a minimum of three points, indicated by the labeled dashed lines with arrowheads leading out of the H_S population box to the right. Zoonotic pathogens may fail to initiate cross-species infection following exposures (*top dashed line*), fail to generate any additional infections within the H_S (*second dashed line*), or experience epidemic fadeout when sustained human-to-human transmission fails and R_0 decreases below unity (*third dashed line*). The initial transition to sustained intra-H_S transmission is prone to failure when populations are sparse or social connectivity is limited. (Modified from Childs 2004)

factors alter the likelihood of a transition occurring and drive the geographic spread and determine the magnitude of morbidity and mortality resulting from a particular instance of emergence.

3.1 Abiotic Factors in Emergence

Abiotic factors alter the potential for contact between H_R and H_S populations, or infectious intermediaries, and modulate the potential for spillover; zoonotic diseases highly dependent on abiotic factors are often labeled environmentally driven epizootics (Allen and Cormier 1996). On a global scale, climate change has been increasingly linked to instances of zoonotic disease emergence, with El Nino Southern Oscillation (ENSO) providing the largest interannual signal of climate variation (Wang et al. 1999). One hypothesized mechanism by which ENSO triggers increased incidence of zoonotic disease among humans, is through a chain of sequentially induced events referred to as a trophic cascade (Polis et al. 2000), ultimately leading to increased numbers of individuals among H_R or H_V populations and increasing the risk of human exposure to a zoonotic pathogen (Nicholls 1986; Kelly-Hope et al. 2004; Bi and Parton 2003; Glass et al. 2002; Anyamba et al. 2001). ENSO events have been correlated with increased risk of HPS and plague in the southeastern United States (Glass et al. 2002; Parmenter et al. 1999), increased infection by Ross River virus in Australia (Lindsay and Mackenzie 1997; Kelly-Hope et al. 2004), and arthropod-vectored *Bartonella bacilliformis* and visceral leishmaniasis in South America (Chinga-Alayo et al. 2004; Franke et al. 2002).

Local weather conditions, potentially driven by global climate variation, have been repeatedly shown to influence the emergence of zoonotic and vector-borne viruses. Drought can serve to amplify enzootic transmission of St. Louis virus (Shaman et al. 2002) and possibly Japanese encephalitis (Hanna et al. 1999; Mackenzie et al. 2002) and Ebola viruses (Pinzon et al. 2004; see the chapter by Gonzalez et al., this volume), ultimately placing humans at higher risk for spillover. The converse, excessive rainfall, can increase breeding populations of H_Vs, driving enzootic transmission levels of western equine encephalomyelitis virus, Ross River virus, and Rift Valley fever virus to heightened levels, and ultimately increasing zoonotic virus spillover to humans (Lindsay et al. 1993; Wegbreit and Reisen 2000; Linthicum et al. 1999).

3.2 Evolutionary and Intrinsic Biotic Factors in Emergence

Intrinsic biotic and evolutionary factors enhance the ability of certain zoonotic viruses, notably those with RNA genomes (Cleaveland et al. 2001; Dobson and

Foufopoulos 2001; see the chapters by Cleaveland et al. and by Holmes and Drummond, this volume), to cross species barriers. Viruses with high replication rates, high mutation rates, and increased potential for recombination or reassortment may more readily adapt to new fitness landscapes and become transmitted among humans to emerge as pandemic threats (Burke 1998; Nichol et al. 2000); examples include HIV and subtypes of Influenza A (Hahn et al. 2000; Claas, 2000; see the chapter by Webby et al., this volume). The intrinsic genetic variability in susceptibility to infectious diseases within the human H_S (Segal and Hill 2003) is further modulated by an individual's cumulative life experience and history of infection by various pathogens, reflected by acquired immunological memory or, possibly, an individual's ancestry and evolutionary imprint of prior exposure to pathogens (Gillespie 1975; Lipsitch and Sousa 2002). Furthermore, immunologic function and the susceptibility of individual humans to infection and disease are dynamic and vary with factors such as nutritional status and age (Boelle et al. 2004).

Strong evolutionary forces may be in play in circumstances where zoonotically acquired viruses are intermittently maintained among small and sparsely distributed human populations where R_0 may hover close to unity. In theory, virus evolution is affected by socially structured host populations, such as where some human populations are aggregated in remote villages, with limited opportunity for social interchange. Models of virus transmission which assume homogeneous or freely mixing populations are of limited use in such circumstances. In such settings, modest increases in the number of generations of human-to-human transmission sustained by a new virus prior to fadeout (Fig. 1, second terminal dotted line) improves the likelihood of virus evolution to a higher average R_0, and hence emergence (Antia et al. 2003).

Additionally, sparsely distributed populations where contact rates, c, between infectious and susceptible individuals are low can support bistable evolutionary dynamics. One trend leads to the rapid evolution of increasingly virulent viruses. When viruses of relatively low virulence are transmitted among dispersed metapopulations of hosts, the result can be a cluster of infected individuals surrounding an index case, which is rapidly transformed into a semi-impermeable barrier of immune individuals (Boots et al. 2004), effectively terminating additional transmission. Virulent viruses causing lethal infections leave no immune survivors to block transmission and, in the course of removing infected individuals, further enhance the sparseness of the existing social structure. In situations where viruses of relatively low virulence circulate, the introduction of a highly virulent virus strain, either through an infected immigrant or from viral recombination, can alter the evolutionary trajectory of virus–host adaptation favoring selection for increasing virulence and an alternative, evolutionarily stable situation (White et al. 2002; Boots et al. 2004).

3.3 Extrinsic Biotic Interactions in Emergence

Extrinsic biotic interactions, such as natural or human-assisted translocations of infected or latently infected individuals of H_R or H_V species have played an exaggerated role in the rapid emergence of zoonotic diseases within the last few years. Monkeypox transported with African rodents destined for the US pet trade (Centers for Disease Control and Prevention 2003; see the chapter by Regnery, this volume), globe-trotting humans infected with SARS-CoV (Olsen et al. 2003; see the chapter by Wang and Eaton, this volume), domestic dogs incubating rabies accompanying human colonialists (see the chapter by Nel and Rupprecht, this volume), and the stowaway mosquito, bird, or human infected with WNV (Lanciotti et al. 2002) bear witness to the growing problems of a shrinking interconnected world (see the chapter by Daszak et al., this volume). Mosquito-borne viral diseases have resulted from the introduction of exotic viruses into indigenous local populations of mosquitoes previously not involved as vectors (Lanciotti et al. 2002), in addition to the establishment and spread of exotic mosquito species harboring viruses into new geographic locations (Lounibos 2002; Mackenzie et al. 2004).

However, not all biological invasions or disease introductions survive to cause epidemics, as was the case with SARS and monkeypox in North America. In contrast, WNV was an entirely different matter. The rapid establishment and spread of WNV in North America was nearly assured by the presences of indigenous species of competent wild bird H_Rs and mosquito H_Vs. Certain bird species sustained WNV viremias of sufficient titer and duration to infect blood-feeding "bridge vectors" (Turell et al. 2001; Komar et al. 2003), maintaining transmission to humans and spreading WNV as migrating birds followed traditional flyways (Peterson et al. 2003). The same extrinsic phenomena of a community of seemingly preadapted and widely available potential H_Vs and H_Rs in conjunction with the biogeography of avian migration aided the introduction and spread of WNV in Europe and the Middle East (Malkinson and Banet 2002). By an alternate route of introduction, wind-blown infected mosquitoes may have introduced Japanese encephalitis virus (JEV) into northern mainland Australia in 1998 (Ritchie and Rochester 2001).

3.4 Anthropogenic Influences as a Special Class of Extrinsic Factors in Emergence

As indicated above, many of the most important and widely cited factors modifying the scope and scale of zoonotic disease emergence are anthropogenic in origin; a few examples are described to highlight their importance and their distinctiveness from required transition stages.

3.4.1 Habitat Modification, Human Encroachment, and Modern Agricultural Practices

Human population growth and modern agricultural practices have enticed human settlers into clearing patches within ecosystems of maximally high biodiversity, such as tropical rain forests, converting substantial areas into cultivated fields and pastures (Patz et al. 2004; LoGiudice et al. 2003). Commercial farming operations inserted into clearings in forest habitats juxtapose and intermingle humans and livestock with native animal populations (Kock et al. 2002; Daszak et al. 2001; see the chapter by Field et al., this volume), and, coincidentally, with whatever zoonotic pathogens exist within these natural nidi (Pavlovsky1957). In many instances, the cleared land has been used for irrigated agriculture, resulting in an increase in vector-borne diseases such as JEV as mosquitoes and water bird H_Rs are brought in close proximity to domestic pigs in nearby villages (Morse 1995; Keiser et al. 2005). Dams are built to maintain water for human consumption and for use in irrigated agriculture, but they too may lead to increased zoonotic disease emergence as they provide the milieu for intermingling mosquito vectors and reservoir hosts of arboviruses as well as the spread of other diseases such as schistosomiasis.

Modern agricultural practices have also provided the mechanism by which bovine spongiform encephalopathy emerged in the United Kingdom in the early 1980s (Pattison 1998).

3.4.2 Domestic Animals Provide a Bounty of Novel Niches

Species now linked by domestication to *Homo sapiens* provide rich fodder for evolutionary forays by zoonotic viruses into potential new hosts. The emergence of zoonotic viruses among humans or domestic livestock where our species has drifted into preexisting sylvatic foci of zoonotic viruses is driven by local circumstance, history, and serendipity. The role of livestock, such as horses and pigs, can be pivotal in a transmission chain leading to human infection, as illustrated by the henipaviruses (see the chapters by Daniels et al. and Field et al., this volume). NiV and HeV first jumped the species barrier to infect pigs and horses, respectively, and only then were transmitted by these H_Is to humans (Barclay and Paton 2000; Chua et al. 1999). However, these two viruses also demonstrate the importance of transmissibility in the H_Is influencing the ultimate emergence of human disease; NiV was readily transmitted among pigs while HeV was rarely transmitted among horses or from horse to human (see the chapter by Daniels et al., this volume).

Rabies virus associated with domestic dogs incubating infection and transported with humans was the likely source of endemic cycles of rabies involving most terrestrial mammals in North and South America and in many areas of Africa (Childs et al. 2002; Smith et al. 1992; see the chapter by Nel and Rupprecht, this volume). In addition to causing an estimated 50,000 human deaths annually, rabies virus associated with domestic dogs have driven naïve indigenous populations of African wild dogs (*Lycaon pictus*) and Ethiopian wolves (*Canis simensis*) to the threshold of extinction and caused declines among other large carnivore populations (Roelke-Parker et al. 1996; Sillero-Zubiri et al. 1996; Gascoyne et al. 1993; Chapman 1978; see the chapter by Nel and Rupprecht, this volume).

Other domesticated species have become efficiently enlisted as H_Is or H_Rs, in a bridging process leading to human disease. Swine production management practices have improved the efficacy of this economically important livestock species as an amplifying H_I for JEV and NiV transmission to humans (Daniels et al. 2002; Singh and Jamaluddin 2002; Mohd Nor et al. 2000; see the chapter by Field et al., this volume). Swine may also serve as the mammalian mixing vessel for influenza A viruses of domestic and wild birds, offering the opportunity for avian viruses to obtain the complement of genes required for their sustained transmission within mammalian hosts, such as humans (Suarez et al. 2002; Gibbs et al. 2001; see the chapter by Webby et al., this volume).

3.4.3
Human Population Demographics and Urbanization

Significant changes in the demography of global human populations during the past five decades have been driven not only by population growth, but by changes in population distribution and social structuring brought about by migration, the ongoing movement of persons from rural to urban environments and the resettlement of refugees. The concentration of humans in the urban environment has given rise to mega-cities where a large proportion of persons may live in substandard conditions in marginal areas, sometimes referred to as shanty towns, surrounding the urban core. The crowded living conditions within shanty towns are further degraded by poor sanitation and lack of water; these conditions have been associated with the emergence of diseases, notably those involving vector-transmitted pathogens (Gratz 1999; Gubler 2002; Mackenzie et al. 2004).

Urban and periurban changes in land use have altered the availability and quality of habitat available to wildlife, and ecological changes in resource availability have in instances increased the potential for human–animal–vector interactions. Later chapters illustrate how ecological changes have influenced the abundance and accessibility of wildlife species serving as reservoir hosts for different pathogens, leading to the emergence of zoonotic pathogens associated with pteropid bats (see the chapter by Field et al., this volume) and white-tailed deer (see the chapter by Paddock and Yabsley, this volume).

3.4.4
The Miracle of Modern Transport

Perhaps the most influential and certainly the most infamous anthropogenic modifiers driving the emergence process have been those enhancing social connectivity through road construction (Larkin 2000), railroads, and, the crown jewel of rapid modern transport, jet plane-assisted travel (Fig. 1; Childs 2004; see the chapter by Daszak et al., this volume). Nowhere has the role of rapid transportation been more evident than with SARS CoV (Table 1), where a presumed focus of human infection in the wet markets of Guangzhou, Guangdong Province, China, where live animals or their products are available for purchase, was transformed into a global health problem affecting 27 nations on every populated continent (Heymann 2004; see the chapter by Wang and Eaton, this volume). The human SARS CoV appears to have been inadvertently transported to a wet-market, along with an infected H_I or H_R, on a journey destined to end with human consumption (Bell et al. 2004). Wildlife farming and an immense network of illegal national and international trade in wildlife has been fueled by human demands for wildlife products of unusual culinary or putative medicinal properties (Bell et al. 2004). These cultural propensities enriched the range of opportunities for novel host/zoonotic virus interchange, but alone, as with rapid transport of persons, would not have resulted in a case of SARS without the biological capabilities of the virus to readily establish spillover.

3.4.5
The Miracle of Modern Medicine

Modern medical practices requiring the widespread use of needles, increased application of immunosuppressive therapies, organ transplant, and blood transfusions have contributed substantially to the spread and emergence of zoonotic pathogens (Institute of Medicine 2003; see the chapter by Paddock and Yabsley, this volume). In certain exceptional instances, medical technology has permitted zoonotic viruses, generally limited in their capacity for human-to-human transmission, to flirt briefly with the transmission route prerequisite to pandemic emergence (Fig. 1). Illustrative of this phenomena were instances of WNV and rabies virus transmission from infected donors to susceptible recipients receiving blood transfusion (WNV) and organ and tissue transplants

Table 1 Examples of transitions and modifying factors in zoonotic disease emergence; SARS CoV and HIVs

Transition	Initial context and circumstances	Examples of modifying factors
Contact between H_R or infectious propagules from H_R and H_S	Coronaviruses extant among sylvatic species of mammals and birds Coronaviruses extant among commensal and domestic animals, and humans raising potential for coinfection or superinfection SIVs circulate in at least 40 species of nonhuman primates in West Africa, most of which are used for human consumption. High potential for retrovirus transmission through contact with primate blood as indicated by infection of humans with primate foamy viruses	• Wildlife consumed as culinary delicacies or medicinal purposes; national and international trade in wildlife • Increased connectivity between humans in remote sites to cities via public and commercial roads (logging, mining, etc.); increased availability of ground transportation • Rapid transportation of game animals to national and international markets • Concentration of diverse wildlife species, domestic animals and humans in wet markets • Exposure to animal blood and tissues during butchering by humans and during predation by animals
Cross-species transmission	SARS CoV infected several orders of mammals in wet markets of China SIVs circulating in nonhuman primates are often mosaics of several SIV subtypes, indicating contact and recombination among nonhuman primate H_Rs	• RNA genomes, most notably retroviruses, capable of high-frequency mutation and recombination. High intrahost genetic diversity provides wide range of genotypic/phenotypic viral variants for selection by H_S during spillover • SARS CoV ability to infect multiple orders of mammals suggests evolution has of a viral "generalist" preadapted for species invasion • Close taxonomic relationship between H_R and H_S may facilitate viral adaptation once SIV has been introduced into individuals of H_S • Contact with blood and other tissues containing high titers of virus ensured by use of H_Rs as food by human H_S

Sustained intra-H_S transmission	Once SARS CoV or HIV established within a human H_S, access to urban setting with high density of susceptibles increases potential for average $R_0 > 1$ Respiratory transmission of SAR CoV Long period of infectiousness without overt clinical disease contribute to high R_0 with HIV	• SARS CoV pathology in H_S results in respiratory transmission by droplet or possibly aerosols • Superspreader phenomena • Increased number of intrahuman passages may have increased R_0 of SARS CoV • Medical technology (reused nonsterile needles, blood transfusion, etc.) provides new routes of transmission for HIV
Sustained intra-H_S transmission and adaptation to H_S and global emergence	High degree of within individual diversity of HIV viral subtypes driven by immune pressure Long period of infectiousness prior to clinical disease increase number of intrahuman passages of HIV Infectiousness of SARS CoV coincides with clinical disease and peaks several days after disease onset	• Rapid international transport of infected persons with SARS or HIV • Recombination may have produced novel SARS CoV adapted to humans • Poor surveillance and reporting in China slowed control measures • SARS controlled by traditional public health measures (quarantine, isolation of exposed persons, etc.) because infectiousness and disease coincident (pandemic averted) • Reparatory transmission aided SARS CoV spread • Sexual transmission and high R_0 assured HIV global spread

(WNV and rabies virus) (Iwamoto et al. 2003; Goldrick 2003; Centers for Disease Control and Prevention 2004; Gode and Bhide 1988). These rare instances involved transient, human-to-human transmission of viruses normally requiring a mosquito vector (WNV) or direct contact (rabies virus) for their transmission. Medical interventions limited further transmission, although biologic constraints inherent to the virus and host would have self-limited any sustained human-to-human transmission.

4 Invasion Biology as a Paradigm for Disease Emergence

The schema for emerging diseases (Fig. 1) emphasizes viral interactions within the newly colonized secondary host, of which humans may be but one of several susceptible species ($H_{S\ldots n}$). The process outlined is similar to the schema developed to characterize biological invasions by nonindigenous species (Kolar and Lodge 2001). The transition states proposed for emerging diseases and those for invasive species are largely parallel: (1) contact with infectious propagules aligns with the nonindigenous species in a transport pathway to a foreign shore; (2) cross-species transmission aligns with the nonindigenous species surviving transport and being introduced into a foreign environment; (3) sustained intra-H_S transmission of a zoonotic virus aligns with the establishment (self-perpetuation) of the invasive species within the new environment; and (4) sustained intra-H_S transmission accompanied by evolutionary adaptation of the once zoonotic virus to a new H_R, prior to emergence, aligns with adaptive radiation and spread of the invasive species beyond the local site of introduction (Grant et al. 2001).

4.1 Termination Points and Pitfalls on the Route to Emergence or Invasion

The potential terminating points in the process of virus emergence or biological invasion (broken arrows leading outside of the H_S population block in Fig. 1) are consequences of similar circumstances. Failure to cross the species barrier (spillover) aligns with "fails in transport"; failure to sustain transmission, with a transmission potential, $R_0<1$, aligns with "fails to establish"; and interruption of sustained intra-H_S transmission, an average $R_0<1$, aligns with "noninvasion" by the nonindigenous species. Differences between disease emergence and biological invasion exist, as transitions leading to disease emergence are not strictly hierarchical. Reiteration of contact and spillover (Fig. 1, transitions 1 and 2) at sufficiently high levels can suffice for a disease to emerge, but if an invading species never moves and perishes at the site of its introduction, even if repeatedly introduced to the site, further establishment and spread, prerequisites of invasion, is precluded.

In addition to biological factors which establish the setting in which zoonotic pathogens may emerge (see the chapters by Cleaveland et al. and Daszak et al., this volume), the emergence of a zoonotic agent within human or animal populations must be detected by humans. Too often the presence of a zoonotic agent is first identified by the presence of disease in humans, and surveillance for disease emergence is largely restricted to identifying incident cases of disease in humans rather than monitoring infection or disease among wildlife H_Rs or H_Is (see the chapters by Childs, by Merianos, and by Stallknecht, this volume). The challenges present to designing programs aiming to disrupt transmission of a zoonotic pathogen within a wildlife reservoir host population prior to spillover and disease emergence are discussed in the chapters by Childs and by Stallknecht in this volume.

4.2
Human Invaded or Human Invader?

Altering the environmental unit being invaded produces a radically different schema. The invasion process in Fig. 1 has an organismal or medical orientation, which can be transformed to a population or community orientation by regarding humans, rather than a zoonotic pathogen, as the invasive species. Human invasion of new habitats and new environments is a frequently cited factor in the emergence process of viral zoonoses (Morse 1995; Institute of Medicine 2003). Where native H_Rs and H_Vs and their co-evolved viral pathogens exist in natural foci (Pavlovsky 1957), enhanced opportunities for novel ecologic interactions await. Initial instances of emergence have proven unpredictable as exemplified by HPS resulting from transmission of hantaviruses maintained by sigmodontine rodent H_Rs in North America (Monroe et al. 1999), HIV resulting from transmission of SIVs circulating among nonhuman primate H_Rs in West Africa (Apetrei et al. 2004), and NiV and HeV viruses from pteropid bat H_Rs in Asia and Australia (Field et al. 2001). Herein, we stress the organismal– medical orientation, humans colonized or invaded by zoonotic viruses. Human intrusion into novel environments is regarded as an anthropogenic factor which modifies the likelihood of contact and spillover transitions occurring. However, without the preexisting sylvatic zoonotic cycle, human invasion alone would not engender the first case of illness along the path to emergence.

5
Qualities of Zoonotic Viruses Emerging by Different Transition Routes

Insights as to why and how certain zoonotic viruses appear predisposed to spillover and the various paths they take in the emergence process, are to be gleaned by examining the evolutionary history and current context of where and how zoonotic viruses exist and just how they become identified as etiologic agents of human disease (see the chapter by Childs, this volume). Predisposing biological characteristics include evidence of multiple H_Rs (Dobson and Foufopoulos 2001; Cleaveland et al. 2001; see the chapters by Cleaveland and by Holmes and Drummond, this volume), high replication rates, high mutation rates, and the potential for homologous or heterologous recombination, which reach maxima in zoonotic viruses with RNA genomes (Holland et al. 1982; Arias et al. 2001).

5.1
Emergence Via Reiterative Processes of Contact and Spillover

Two zoonotic viruses with histories of reemergence are rabies virus and WNV, both of which depend solely, with the exception of rare instances mentioned above, on repetitive contact and spillover between infected H_Rs or infected H_Vs (WNV) for their transmission to the human H_S. Although the RNA genomes of these two zoonotic viruses are markedly different in terms of organization, polarity, and replication strategy, both viruses show evidence of reduced positive selection (Woelk and Holmes 2002), even where established within novel H_Rs or H_Vs (Holmes et al. 2002). The term "evolutionary generalists" has been applied to both viruses as they share, to some extent, the requirement of being able to infect and multiply within cells belonging to different species and, in the case of vector-borne WNV, the need to infect and multiply within avian, mammalian, and insect H_Rs, H_Vs, and H_Ss. The rare instances of human-to-human transmission of these viruses are epidemiologically insignificant (Dietzschold and Koprowski 2004; Iwamoto et al. 2003).

5.2
Spillover Subsequently Sustained by Human-to-Human Transmission

Although humans are, with few exceptions, incidental hosts for zoonotic viruses emerging from sylvatic transmission cycles, a few zoonotic arboviruses can be maintained by human-to-human transmission mediated by anthropophilic vectors in urban settings where large populations of humans and competent

H_Vs coexist, particularly in environments with poor sanitation and overcrowding. Yellow fever virus (Wolfe et al. 2001; de Silva et al. 1999) and dengue virus serotypes (Kuiken et al. 2003; Ksiazek et al. 2003) are arboviruses where major epidemics are associated with urban transmission cycles rather than sporadic spillover from sylvatic H_Rs and H_Vs, and dengue serotypes have become endemic among some suitably large human populations in Asia (Gubler 2002).

Rabies virus crosses mammalian orders and species and can establish sustained transmission within new H_Rs (Badrane and Tordo 2001), as has been observed on several occasions where bat-associated variants of rabies virus have achieved temporary sustained transmission among terrestrial carnivores, such as red foxes (*Vulpes vulpes*) and striped skunks (*Mephitis mephitis*) (Daoust et al. 1996; Engeman et al. 2003). The maintenance of rabies virus, considered a single species of *Lyssavirus*, serotype 1/genotype 1, is achieved as a myriad of distinct viral variants maintained within different specific mammalian H_Rs, rather than a homogeneous virus infecting multiple H_Rs; control or elimination of rabies in a specific H_R may be achieved but the diversity of host–virus dyads is a formidable buffer against any overall elimination scheme.

Epidemics of rabies virus are sustained when there are sufficient individuals of the primary H_R(s) to sustain intra-H_R transmission, with coincidental spillover to H_Ss by reiterative introductions by inoculation of infectious virus in saliva. As rabies is fatal among most mammalian species, population declines among the principal H_R generally coincide with declines in incidental rabies epizootics among H_Ss (Gordon et al. 2004; Wandeler et al. 1974). Epizootics can reemerge at periodic intervals as H_R populations recover above the critical threshold density (K_T) required to sustain virus transmission at $R_0 > 1$ (Anderson et al. 1981; Childs et al. 2000; Coyne et al. 1989).

In an analogous manner, epidemics caused by WNV involve reiterative introductions of infectious virus by any of a number of competent mosquito H_V (s). WNV readily infects at least three classes of vertebrates (Avia, Mammalia, Reptilia) and mosquito species, and some species of ticks (Gould et al. 2003; Komar et al. 2003; Lvov et al. 2004; Sardelis et al. 2002; Turell et al. 2001). Although WNV appears to be reasonably homogeneous in regions of North America over time, geographic clustering of genetically similar strains is detectable and certain epizootiologically dominant genetic clades have emerged, some with shorter extrinsic incubation periods within North American vectors (Davis et al. 2003; Ebel et al. 2004). Subtleties associated with host–vector–virus relationships are being uncovered, such as the greater frequency of *Flavivirus* recombination among mosquito H_Vs compared to tick H_Vs (Twiddy and Holmes 2003).

The temperature and humidity requirements for survival and breeding of mosquito vectors, and the demand for temperature-sensitive extrinsic viral incubation within H_Vs, drive the strong seasonal transmission dynamics of

WNV and other arboviral diseases. With the onset of cold temperatures in temperate zones, WNV transmission ceases and epidemics of human disease desist (Woodring et al. 1996).

5.3 The Road to Human Adaptation: A Still-Life with SARS CoV?

That SARS-CoV is new to science is not in question. However, the origin of SARS-CoV as a human pathogen arising from direct cross-species transmission of a preexisting, but previously unknown virus (Gibbs et al. 2004; Holmes and Rambaut 2004; see the chapters by Holmes and Drummond and by Wang and Eaton, this volume) or as a virus formed from the recombination of existing mammalian and avian coronaviruses (Rest and Mindell 2003; Zhang et al. 2004) has been the subject of debate. SARS- CoV is sufficiently distinct in genetic sequence from previously known coronaviruses (Rota et al. 2003) that a long history of preexistence with its natural H_R population is surmised (Parashar and Anderson 2004). Recently, coronaviruses related to SARS-Cov have been amplified by PCR from three communal cave-dwelling species of the genus *Rhinolophus* in the family Rhinolophidae. Genome sequence analysis indicated that SARS-like coronaviruses among these bats have an almost identical genome organization to those of SARS-CoVs isolated from humans or civets (Li et al. 2005; see the chapter by Wang and Eaton, this volume). These data suggest that bats serve as the H_R of SARS-CoV and that palm civets served as a H_{S1} in a chain of events leading to infection of humans as secondarily infected H_{S2}. SARS-CoV's global emergence may be an extraordinary example of a relatively unmodified zoonotic virus, successfully sustained by intrahuman transmission. However, increasing data suggest SARS-CoV was a virus rapidly adapting to its new human host and the rapid and effective public health response terminating its transmission halted an evolutionary dramas in the making (see the chapter by Wang and Eaton, this volume).

Genetic sequence data indicate that strong positive selection accompanied SARS-CoV's emergence and that distinctive human-associated changes in the genome distinguish virulent SARS-CoV from isolates of virus from palm civets (Song et al. 2005). Additional data indicate genetic changes were accompanying longer chains of human-to-human transmission. Heterogeneous viral sequences recovered from a single patient's samples (Liu et al. 2005) indicate the degree of viral variation available for selection within the individual human. These findings are compelling evidence of evolutionary events underway, presaging the emergence of a virus with a unique genetic signature associated with its human host.

Whatever the exact origin of SARS-CoV, the genetic endowments of this virus facilitated cross-species infection. An evolutionary history that includes viral preadaptation permitting infection to occur among a broad range of H_Rs and H_Ss is suggested by SARS CoV's ability to infect a range of mammalian orders (Ng 2003; Song et al. 2005; Bell et al. 2004). Such preadaptation can be assumed to have endowed SARS-CoV with a suite of traits readily adaptable for establishing sustained intrahuman transmission (Riley et al. 2003; Isakbaeva et al. 2004). Pathogenesis within the novel human host's respiratory tissues offered an efficient means for sustained transmission by expressed droplets, or possibly aerosol (Yu et al. 2004).

The biological properties of SARS-CoV, the human behaviors and societal practices which increased the likelihood of contact and spillover and the rapid transport of already infected individuals drove the trajectory of the emergence of this global health problem. The distinctions and critical interactions between required biological transitions and modifying factors are clearly illuminated by this emergence process (Fig. 1, Table 1). Fortunately, SARS-CoV was effectively controlled by tried and true public health measures serving to increase social distance and diminish infectious contacts, c, perhaps curtailing a rapid evolutionary path toward a R_0 sufficiently high to bypass these methods (Song et al. 2005; Antia et al. 2003; Fraser et al. 2004).

5.4 Adaptation of Zoonotic Viruses to the Human H_R and Pandemic Emergence

The zoonotic viruses leading to potentially uncontrolled, pandemic health problems have adopted unique qualities associated with their sustained transmission within the human host. Adaptation to the human host may be mediated by viral preadaptation to a genetically similar intermediate host, as is hypothesized to occur among swine for avian-adapted influenza A subtypes (see the chapter by Webby et al., this volume). Permissive cells for subtype A influenza virus replication in the pig's respiratory track have cell-surface glycoprotein receptors recognized by some avian-adapted viruses, in addition to some human-adapted influenza viruses (Basler et al. 2001). In the case of HIV, the H_R for SIVs giving rise to HIV-1 was our closest living genetic relative, the chimpanzee (*Pan troglodytes troglodytes*) (Gao et al. 1999), and the H_R for HIV-2 was a sooty mangabey (*Cerocebus atys*) (Hirsch et al. 1989), a respectfully close relative of the order Primates. Spillover was enhanced by the enormous population of candidate viruses within the genetically heterogeneous "quasispecies" of viruses present in the infected H_R at spillover. Sustained intrahuman transmission was accompanied by viral adaptations to the human host, readily detectable and quantifiable by sequence changes in the RNA genome and marked by qualitative phenotypic changes identifiable by host species preference and pathogenic interactions (Hirsch et al. 1989; Gao et al. 1999).

Evaluations of the genetic relatedness of HIV-1 and HIV-2 to SIVs circulating among nonhuman primates has led to wide acceptance that these distinctive human lentiviruses, now globally distributed in humans, originated from cross-species transmission in the not so distant past, perhaps within the first half of the twentieth century (May et al. 2001). The human lentiviruses, HIV-1 and HIV-2, have evolved and escaped from remote African settings on at least eight independent occasions to emerge as distinctive genetic subtypes responsible for regional or pandemic human disease (Apetrei et al. 2004; B. Hahn, personal communication). These recognized cases of emergence are certainly not the first instances where SIVs have successfully crossed species and evolved as distinctive HIVs of humans. Early emergences were likely restricted to local occurrences in remote locations where human contact rates, c , and population size were insufficient to support a $R_0 > 1$, even if infections were of a long duration, d; such occurrences would be highly prone to transmission fadeout (Fig. 1). Road construction, automotive transport, and, perhaps, reuse of nonsterile needles (Gisselquist 2003) were presumably key anthropogenic factors increasing the level of social connectivity by providing HIV-infected individuals access to larger aggregates of susceptible hosts in cities (Apetrei et al. 2004).

The number of SIVs described among nonhuman primates in Africa as of 2004 was approximately 40 (Zhuang et al. 2002). The biological capacity of lentiviruses includes rapid replication, high mutability, and the highest recorded rates of recombination known in virology. Knowledge of the frequency of potential opportunities for SIV spillover, based on transmission of a zoo of diverse simian foamy viruses during encounters between monkeys, apes, and human hunters in West Africa (Wolfe et al. 2004) indicate transmission of blood-borne retroviruses is not rare. These facts highlight two important features of emergence; first, emergence is a process, not an event; second, the probability of new genetic lineages of human HIVs arising approximates unity. The same lessons apply to numerous other conditions which make up the body of this volume.

References

Allen LJ, and Cormier PJ (1996) Environmentally driven epizootics. Math Biosci 131:51–80

Anderson MG, Frenkel LD, Homann S, Guffey J (2003) A case of severe monkeypox virus disease in an American child: emerging infections and changing professional values. Pediatr Infect Dis J 22:1093–1096

Anderson RM, Jackson HC, May RM, Smith AM (1981) Population dynamics of fox rabies in Europe. Nature 289:765–771

Antia R, Regoes RR, Koella JC, Bergstrom CT (2003) The role of evolution in the emergence of infectious diseases. Nature 426:658–661
Anyamba A, Linthicum KJ, Tucker CJ (2001) Climate-disease connections: Rift Valley Fever in Kenya. Cad Saude Pub 17 Suppl:133–140
Apetrei C, Robertson DL, Marx PA (2004) The history of SIVS, AIDS: epidemiology, phylogeny and biology of isolates from naturally SIV infected non-human primates (NHP) in Africa. Front Biosci 9:225–254
Arias A, Lazaro E, Escarmis C, Domingo E (2001) Molecular intermediates of fitness gain of an RNA virus: characterization of a mutant spectrum by biological and molecular cloning. J Gen Virol 82:1049–1060
Badrane H, Tordo N (2001) Host switching in Lyssavirus history from the Chiroptera to the Carnivora orders. J Virol 75:8096–8104
Barclay AJ, Paton DJ (2000) Hendra (equine morbillivirus). Vet J 160:169–176
Basler CF, Reid AH, Dybing JK, Janczewski TA, Fanning TG, Zheng H, Salvatore M, Perdue ML, Swayne DE, Garcia-Sastre A, Palese P, Taubenberger JK (2001) Sequence of the 1918 pandemic influenza virus nonstructural gene (NS) segment and characterization of recombinant viruses bearing the 1918 NS genes. Proc Natl Acad Sci U S A 98:2746–2751
Bell D, Roberton S, Hunter PR (2004) Animal origins of SARS coronavirus: possible links with the international trade in small carnivoRes Philos Trans R Soc Lond B Biol Sci 359:1107–1114
Bi P, Parton KA (2003) El Nino and incidence of hemorrhagic fever with renal syndrome in China. JAMA 289:176–177
Boelle PY, Cesbron JY, Valleron AJ (2004) Epidemiological evidence of higher susceptibility to vCJD in the young. BMC Infect Dis 4:26
Boots M, Hudson PJ, Sasaki A (2004) Large shifts in pathogen virulence relate to host population structure. Science 303:842–844
Bryant JE, Barrett AD (2003) Comparative phylogenies of yellow fever isolates from Peru and Brazil. FEMS Immunol Med Microbiol 39:103–118
Burke DS (1998) Evolvability of emerging viruses. In: Nelson AM, Horsburgh CR Jr (eds) Pathology of emerging infections 2. American Society for Microbiology, Washington, DC, pp 1–12
Centers for Disease Control and Prevention (2003) Multistate outbreak of monkeypox–Illinois, Indiana, and Wisconsin 2003. MMWR Morb Mortal Wkly Rep 52:537–540
Centers for Disease Control and Prevention (2004) Investigation of rabies infections in organ donor and transplant recipients--Alabama, Arkansas, Oklahoma, and Texas 2004. MMWR Morb Mortal Wkly Rep 53:586–589
Chapman RC (1978) Rabies: decimation of a wolf pack in artic Alaska. Science 201:365–367
Childs JE (2004) Zoonotic viruses of wildlife: hither from yon. Arch Virol Suppl1–11
Childs JE, Glass GE, Korch GW, LeDuc JW (1989) Effects of hantaviral infection on survival, growth and fertility in wild rat (*Rattus norvegicus*) populations of Baltimore Maryland. J Wildl Dis 25:469–476
Childs JE, Curns AT, Dey ME, Real LA, Feinstein L, Bjornstad ON, Krebs JW (2000) Predicting the local dynamics of epizootic rabies among raccoons in the United States. Proc Natl Acad Sci U S A 97:13666–13671

Childs JE, Krebs JW, Smith JS (2002) Public health surveillance and the molecular epidemiology of rabies. In: Leitner T (ed) The molecular epidemiology of human viruses. Kluwer Academic, Dordrecht, pp 273–312

Chinga-Alayo E, Huarcaya E, Nasarre C, del Aquilla R, Llanos-Cuentas A (2004) The influence of climate on the epidemiology of bartonellosis in Ancash Peru. Trans R Soc Trop Med Hyg 98:116–124

Chua KB, Goh KJ, Wong KT, Kamarulzaman A, Tan PS, Ksiazek TG, Zaki SR, Paul G, Lam SK, Tan CT (1999) Fatal encephalitis due to Nipah virus among pig-farmers in Malaysia. Lancet 354:1257–1259

Claas EC (2000) Pandemic influenza is a zoonosis, as it requires introduction of avian-like gene segments in the human population. Vet Microbiol 74:133–139

Cleaveland S, Laurenson MK, Taylor LH (2001) Diseases of humans and their domestic mammals: pathogen characteristics, host range and the risk of emergence. Philos Trans R Soc Lond B Biol Sci 356:991–999

Coyne MJ, Smith G, McAllister FE (1989) Mathematic model for the population biology of rabies in raccoons in the mid-Atlantic states. Am J Vet Res 50:2148–2154

Daniels TJ, Williams DT, Mackenzie JS (2002) Japanese encephalitis virus. In: Morrilla A, Yoon KJ, Zimmerman JJ (eds) Trends in emerging viral infections of swine. Iowa State Press, Ames IA, pp 249–263

Daoust PY, Wandeler AI, Casey GA (1996) Cluster of rabies cases of probable bat origin among red foxes in Prince Edward Island Canada. J Wildl Dis 32:403–406

Daszak P, Cunningham AA, Hyatt AD (2001) Anthropogenic environmental change and the emergence of infectious diseases in wildlife. Acta Trop 78:103–116

Davis CT, Beasley DW, Guzman H, Raj R, D'Anton M, Novak RJ, Unnasch TR, Tesh RB, Barrett AD (2003) Genetic variation among temporally and geographically distinct West Nile virus isolates United States, 2001, 2002. Emerg Infect Dis 9:1423–1429

De Silva AM, Dittus WP, Amerasinghe PH, Amerasinghe FP (1999) Serologic evidence for an epizootic dengue virus infecting toque macaques (*Macaca sinica*) at Polonnaruwa Sri Lanka. Am J Trop Med Hyg 60:300–306

Dietzschold B, Koprowski H (2004) Rabies transmission from organ transplants in the USA. Lancet 364:648–649

Dobson A, Foufopoulos J (2001) Emerging infectious pathogens of wildlife. Philos Trans R Soc Lond B Biol Sci 356:1001–1012

Downs WG (1982) History of epidemiological aspects of yellow fever. Yale J Biol Med 55:179–185

Drosten C, Gunther S, Preiser W, van der WS, Brodt HR, Becker S, Rabenau H, Panning M, Kolesnikova L, Fouchier RA, Berger A, Burguiere AM, Cinatl J, Eickmann M, Escriou N, Grywna K, Kramme S, Manuguerra JC, Muller S, Rickerts V, Sturmer M, Vieth S, Klenk HD, Osterhaus AD, Schmitz H, Doerr HW (2003) Identification of a novel coronavirus in patients with severe acute respiratory syndrome. N Engl J Med 348:1967–1976

Ebel GD, Carricaburu J, Young D, Bernard KA, Kramer LD (2004) Genetic and phenotypic variation of West Nile virus in New York, 2000–2003. Am J Trop Med Hyg 71:493–500

Engeman RM, Christensen KL, Pipas MJ, Bergman DL (2003) Population monitoring in support of a rabies vaccination program for skunks in Arizona. J Wildl Dis 39:746–750

Field H, Young P, Yob JM, Mills J, Hall L, Mackenzie J (2001) The natural history of Hendra and Nipah viruses. Microbes Infect 3:307–314

Franke CR, Ziller M, Staubach C, Latif M (2002) Impact of the El Nino/Southern Oscillation on visceral leishmaniasis Brazil. Emerg Infect Dis 8:914–917

Fraser C, Riley S, Anderson RM, Ferguson NM (2004) Factors that make an infectious disease outbreak controllable. Proc Natl Acad Sci U S A 101:6146–6151

Gao F, Bailes E, Robertson DL, Chen Y, Rodenburg CM, Michael SF, Cummins LB, Arthur LO, Peeters M, Shaw GM, Sharp PM, Hahn BH (1999) Origin of HIV-1 in the chimpanzee *Pan troglodytes troglodytes*. Nature 397:436–441

Gascoyne SC, Laurenson MK, Lelo S, Borner M (1993) Rabies in African wild dogs (*Lycaon pictus*) in the Serengeti region Tanzania. J Wildl Dis 29:396–402

Gibbs AJ, Gibbs MJ, Armstrong JS (2004) The phylogeny of SARS coronavirus. Arch Virol 149:621–624

Gibbs MJ, Armstrong JS, Gibbs AJ (2001) The haemagglutinin gene, but not the neuraminidase gene, of 'Spanish flu' was a recombinant. Philos Trans R Soc Lond B Biol Sci 356:1845–1855

Gillespie JH (1975) Natural selection for resistance to epidemics. Ecology 56:493–495

Gisselquist D (2003) Emergence of the HIV type 1 epidemic in the twentieth century: comparing hypotheses to evidence. AIDS Res Hum Retroviruses 19:1071–1078

Glass GE, Yates TL, Fine JB, Shields TM, Kendall JB, Hope AG, Parmenter CA, Peters CJ, Ksiazek TG, Li CS, Patz JA, Mills JN (2002) Satellite imagery characterizes local animal reservoir populations of Sin Nombre virus in the southwestern United States. Proc Natl Acad Sci U S A 99:16817–16822

Gode GR, Bhide NK (1988) Two rabies deaths after corneal grafts from one donor. Lancet 2:791

Goldrick BA (2003) West Nile virus update: a new route of transmission is found. Am J Nurs 103:27

Gordon ER, Curns AT, Krebs JW, Rupprecht CE, Real LA, Childs JE (2004) Temporal dynamics of rabies in a wildlife host and the risk of cross-species transmission. Epidemiol Infect 132:515–524

Gould EA, de Lamballerie X, Zanotto PM, Holmes EC (2003) Origins, evolution, and vector/host coadaptations within the genus *Flavivirus*. Adv. Virus Res 59:277–314

Grant PR, Grant BR, Petren K (2001) A population founded by a single pair of individuals: establishment, expansion, and evolution. Genetica 112–113:359–382

Gratz NG (1999) Emerging and resurging vector-borne diseases. Ann Rev Entomol 4451–4475

Gubler DJ (2002) The global emergence/resurgence of arboviral diseases as public health problems. Arch Med Res 33:330–342

Hahn BH, Shaw GM, De Cock KM, Sharp PM (2000) AIDS as a zoonosis: scientific and public health implications. Science 287:607–614

Halloran ME (1998) Concepts of infectious disease epidemiology. In: Rothman KJ, Greenland S (eds) Modern epidemiology. Lippencott Williams Wilkins, Philadelphia, pp 529–554

Hanna JN, Ritchie SA, Phillips DA, Lee JM, Hills SL, van den Hurk AF, Pyke AT, Johansen CA, and Mackenzie JS (1999) Japanese encephalitis in North Queensland, 1998. Med J Aust 170:533–536

Heymann DL (2004) The international response to the outbreak of SARS in (2003) Philos Trans R Soc Lond B 359:1127–1129

Hinson ER, Shone SM, Zink MC, Glass GE, Klein SL (2004) Wounding: the primary mode of Seoul virus transmission among male Norway rats. Am J Trop Med Hyg 70:310–317

Hirsch VM, Olmsted RA, Murphey-Corb M, Purcell RH, Johnson PR (1989) An African primate lentivirus (SIVsm) closely related to HIV-2. Nature 339:389–392

Holland J, Spindler K, Horodyski F, Grabau E, Nichol S, VandePol S (1982) Rapid evolution of RNA genomes. Science 215:1577–1585

Holmes EC, Rambaut A (2004) Viral evolution and the emergence of SARS coronavirus. Philos Trans R Soc Lond B Biol Sci 359:1059–1065

Holmes EC, Woelk CH, Kassis R, Bourhy H (2002) Genetic constraints and the adaptive evolution of rabies virus in nature. Virology 292:247–257

Huff JL, Barry PA (2003) B-virus (*Cercopithecine herpesvirus* 1) infection in humans and macaques: potential for zoonotic disease. Emerg Infect Dis 9:246–250

Institute of Medicine (2003) Microbial threats to health; emergence, detection, and response. National Academies Press, Washington, DC

Isakbaeva ET, Khetsuriani N, Beard RS, Peck A, Erdman D, Monroe SS, Tong S, Ksiazek TG, Lowther S, Pandya-Smith I, Anderson LJ, Lingappa J, Widdowson MA (2004) SARS-associated coronavirus transmission United States. Emerg Infect Dis 10:225–231

Iwamoto M, Jernigan DB, Guasch A, Trepka MJ, Blackmore CG, Hellinger WC, Pham SM, Zaki S, Lanciotti RS, Lance-Parker SE, DiazGranados CA, Winquist AG, Perlino CA, Wiersma S, Hillyer KL, Goodman JL, Marfin AA, Chamberland ME, Petersen LR (2003) Transmission of West Nile virus from an organ donor to four transplant recipients. N Engl J Med 348:2196–2203

Keiser J, Maltese MF, Erlanger TE, Bos R, Tanner M, Singer BH, and Utzinger J (2005) Effect of irrigated rice agriculture on Japanese encephalitis, including challenges and opportunities for integrated vector management. Acta Trop 95:40–57

Kelly-Hope LA, Purdie DM, Kay BH (2004) El Nino Southern Oscillation and Ross River virus outbreaks in Australia. Vect Borne Zoonot Dis 4:210–213

Kock R, Kebkiba B, Heinonen R, Bedane B (2002) Wildlife and pastoral society--shifting paradigms in disease control. Ann N Y Acad Sci 969:24–33

Kolar CS, Lodge DM (2001) Progress in invasion biology: predicting invaders. Trends Ecol Evol 16:199–204

Komar N, Langevin S, Hinten S, Nemeth N, Edwards E, Hettler D, Davis B, Bowen R, Bunning M (2003) Experimental infection of North American birds with the New York 1999 strain of West Nile virus. Emerg Infect Dis 9:311–322

Ksiazek TG, Erdman D, Goldsmith CS, Zaki SR, Peret T, Emery S, Tong S, Urbani C, Comer JA, Lim W, Rollin PE, Dowell SF, Ling AE, Humphrey CD, Shieh WJ, Guarner J, Paddock CD, Rota P, Fields B, DeRisi J, Yang JY, Cox N, Hughes JM, LeDuc JW,

Bellini WJ, Anderson LJ (2003) A novel coronavirus associated with severe acute respiratory syndrome. N Engl J Med 348:1953–1966

Kuiken T, Fouchier RA, Schutten M, Rimmelzwaan GF, van Amerongen G, van Riel D, Laman JD, de Jong T, van Doornum G, Lim W, Ling AE, Chan PK, Tam JS, Zambon MC, Gopal R, Drosten C, van der WS, Escriou N, Manuguerra JC, Stohr K, Peiris JS, Osterhaus AD (2003) Newly discovered coronavirus as the primary cause of severe acute respiratory syndrome. Lancet 362:263–270

Lanciotti RS, Roehrig JT, Deubel V, Smith J, Parker M, Steele K, Crise B, Volpe KE, Crabtree MB, Scherret JH, Hall RA, Mackenzie JS, Cropp CB, Panigrahy B, Ostlund E, Schmitt B, Malkinson M, Banet C, Weissman J, Komar N, Savage HM, Stone W, McNamara T, Gubler DJ (1999) Origin of the West Nile virus responsible for an outbreak of encephalitis in the northeastern United States. Science 286:2333–2337

Lanciotti RS, Ebel GD, Deubel V, Kerst AJ, Murri S, Meyer R, Bowen M, McKinney N, Morrill WE, Crabtree MB, Kramer LD, Roehrig JT (2002) Complete genome sequences and phylogenetic analysis of West Nile virus strains isolated from the United States, Europe, and the Middle East. Virology 298:96–105

Larkin M (2000) Hunting and logging linked to emerging infectious diseases. Lancet 356:1173

Lau SK, Woo PC, Li KS, Huang Y, Tsoi HW, Wong BH, Wong SS, Leung SY, Chan KH, and Yuen KY (2005) Severe acute respiratory syndrome coronavirus-like virus in Chinese horseshoe bats. Proc Natl Acad Sci U S A 102:14040–14045

Linthicum KJ, Anyamba A, Tucker CJ, Kelley PW, Myers MF, Peters CJ (1999) Climate and satellite indicators to forecast Rift Valley fever epidemics in Kenya. Science 285:397–400

Li W, Shi Z, Yu M, Ren W, Smith C, Epstein JH, Wang H, Crameri G, Hu Z, Zhang H, Zhang J, McEachern J, Field H, Daszak P, Eaton BT, Zhang S, Wang LF (2005) Bats are natural reservoirs of SARS-like coronaviruses. Science 310:676–679

Lindsay M, Mackenzie J (1997) Vector-borne viral diseases and climate change in the Australian region: major concerns and the public health response. In: Climate Change and Human Health in the Asia Pacific Region (eds) Curson P, Guest C and Jackson E. Australian Medical Association and Greenpeace International, Canberra, pp 47–62

Lindsay MD, Broom AK, Wright AE, Johansen CA, Mackenzie JS (1993) Ross River virus isolations from mosquitoes in arid regions of Western Australia: implication of vertical transmission as a means of persistence of the virus. Am J Trop Med Hyg 49:686–696

Lipsitch M, Sousa AO (2002) Historical intensity of natural selection for resistance to tuberculosis. Genetics 161:1599–1607

Liu J, Lim SL, Ruan Y, Ling AE, Ng LPF, Drosten C, Liu ET, Stanton LW, Hibberd ML (2005) SARS transmission pattern in Singapore reassessed by viral sequence variation analysis. PLoSMed 2:162–168

Llyubsky S, Gavrilovskaya I, Luft B, Mackow E (1996) Histopathology of *Peromyscus leucopus* naturally infected with pathogenic NY-1 hantaviruses: pathologic markers of HPS infection in mice. Lab Invest 74:627–633

LoGiudice K, Ostfeld RS, Schmidt KA, Keesing F (2003) The ecology of infectious disease: effects of host diversity and community composition on Lyme disease risk. Proc Natl Acad Sci U S A 100:567–571

Lounibos LP (2002)Invasions by insect vectors of human disease. Annu Rev Entomol 47:233–266

Lvov DK, Butenko AM, Gromashevsky VL, Kovtunov AI, Prilipov AG, Kinney R, Aristova VA, Dzharkenov AF, Samokhvalov EI, Savage HM, Shchelkanov MY, Galkina IV, Deryabin PG, Gubler DJ, Kulikova LN, Alkhovsky SK, Moskvina TM, Zlobina LV, Sadykova GK, Shatalov AG, Lvov DN, Usachev VE, Voronina AG (2004) West Nile virus and other zoonotic viruses in Russia: examples of ermerging-reemerging situations. Arch Virol [Suppl] 18:85–96

Mackenzie JS, Johansen CA, Ritchie SA, van den Hurk AF, Hall RA (2002) Japanese encephalitis as an emerging virus: the emergence and spread of Japanese encephalitis virus in Australasia. Curr Topics Microbiol Immunol 267:49–73

Mackenzie JS, Gubler DJ, and Petersen LR (2004) Emerging flaviviruses: the spread and resurgence of dengue Japanese encephalitis and West Nile viruses. Nature Med 10: S98–S109

Malkinson M, Banet C (2002) The role of birds in the ecology of West Nile virus in Europe and Africa. Curr Top Microbiol Immunol 267:309–322

May RM, Gupta S, Mclean AR (2001) Infectious disease dynamics: What characterizes a successful invader? Philos Trans R Soc Lond B Biol Sci 356:901–910

Mims CA (1991) The origin of major human infections and the crucial role of person-to-person spread. Epidemiol Infect 106:423–433

Mims CA (1995) Virology research and virulent human pandemics. Epidemiol Infect 115:377–386

Mohd Nor MN, Gan CH, Ong BL (2000) Nipah virus infection of pigs in peninsular Malaysia. Rev Sci Tech 19:160–165

MonathTP (1989) Yellow fever. In: Monath TP (ed) The arboviruses: epidemiology and ecolog. CRC Press, Boca Raton FL, pp 139–231

Monroe MC, Morzunov SP, Johnson AM, Bowen MD, Artsob H, Yates T, Peters CJ, Rollin PE, Ksiazek TG, Nichol ST (1999) Genetic diversity and distribution of *Peromyscus* -borne hantaviruses in North America. Emerg Infect Dis 5:75–86

Morse SS (1995) Factors in the emergence of infectious diseases. Emerg Infect Dis 1:7–15

Nayak DP (2000) Virus morphology, replication, and assembly. In: Hurst CJ (ed) Viral ecology. Academic, New York, pp 63–124

Ng SK (2003) Possible role of an animal vector in the SARS outbreak at Amoy Gardens. Lancet 362:570–572

Nichol ST, Arikawa J, Kawaoka Y (2000) Emerging viral diseases. Proc Natl Acad Sci U S A 97:12411–12412

Nicholls N (1986) A method for predicting Murray Valley encephalitis in southeast Australia using the southern oscillation. Aust J Exp Biol Med Sci 64:578–594

Olsen SJ, Chang HL, Cheung TY, Tang AF, Fisk TL, Ooi SP, Kuo HW, Jiang DD, Chen KT, Lando J, Hsu KH, Chen TJ, Dowell SF (2003) Transmission of the severe acute respiratory syndrome on aircraft. N Engl J Med 349:2416–2422

Parashar UD, Anderson LJ (2004) Severe acute respiratory syndrome: review and lessons of the 2003 outbreak. Int J Epidemiol 33:628–634
Parmenter RR, Yadav EP, Parmenter CA, Ettestad P, Gage KL (1999) Incidence of plague associated with increased winter-spring precipitation in New Mexico. Am J Trop Med Hyg 61:814–821
Pattison J (1998) The emergence of bovine spongiform encephalopathy and related diseases. Emerg Infect Dis 4:390–394
Patz JA, Daszak P, Tabor GM, Aguirre AA, Pearl M, Epstein J, Wolfe ND, Kilpatrick AM, Foufopoulos J, Molyneux D, Bradley DJ (2004) Unhealthy landscapes: Policy recommendations on land use change and infectious disease emergence. Environ Health Perspect 112:1092–1098
Pavlovsky YN (1957) Human Diseases with Natural Foci (Translated from Russian by D Rottenberg). Foreign Languages Publishing House, Moscow
Peiris JS, Yu WC, Leung CW, Cheung CY, Ng WF, Nicholls JM, Ng TK, Chan KH, Lai ST, Lim WL, Yuen KY, Guan Y (2004) Re-emergence of fatal human influenza A subtype H5N1 disease. Lancet 363:617–619
Peterson AT, Vieglais DA, Andreasen JK (2003) Migratory birds modeled as critical transport agents for West Nile virus in North America. Vect Borne Zoonot Dis 3:27–37
Pinzon JE, Wilson JM, Tucker CJ, Arthur R, Jahrling PB, Formenty P (2004) Trigger events: enviroclimatic coupling of Ebola hemorrhagic fever outbreaks. Am J Trop Med Hyg 71:664–674
Polis GA, Sears AL, Huxel GR, Strong DR, Maron J (2000) When is a trophic cascade a trophic cascade? Trends Ecol Evol 15:473–475
Rest JS, Mindell DP (2003) SARS associated coronavirus has a recombinant polymerase and coronaviruses have a history of host-shifting. Infect Genet Evol 3:219–225
Riley S, Fraser C, Donnelly CA, Ghani AC, Abu-Raddad LJ, Hedley AJ, Leung GM, Ho LM, Lam TH, Thach TQ, Chau P, Chan KP, Lo SV, Leung PY, Tsang T, Ho W, Lee KH, Lau EM, Ferguson NM, Anderson RM (2003) Transmission dynamics of the etiological agent of SARS in Hong Kong: impact of public health interventions. Science 300:1961–1966
Ritchie SA, Rochester W (2001) Wind-blown mosquitoes and introduction of Japanese encephalitis into Australia. Emerg Infect Dis 7:900–903
Roelke-Parker ME, Munson L, Packer C, Kock R, Cleaveland S, Carpenter M, O'Brien SJ, Pospischil A, Hofmann-Lehmann R, Lutz H (1996) A canine distemper virus epidemic in Serengeti lions (*Panthera leo*). Nature 379:441–445
Rota PA, Oberste MS, Monroe SS, Nix WA, Campagnoli R, Icenogle JP, Penaranda S, Bankamp B, Maher K, Chen MH, Tong S, Tamin A, Lowe L, Frace M, DeRisi JL, Chen Q, Wang D, Erdman DD, Peret TC, Burns C, Ksiazek TG, Rollin PE, Sanchez A, Liffick S, Holloway B, Limor J, McCaustland K, Olsen-Rasmussen M, Fouchier R, Gunther S, Osterhaus AD, Drosten C, Pallansch MA, Anderson LJ, Bellini WJ (2003) Characterization of a novel coronavirus associated with severe acute respiratory syndrome. Science 300:1394–1399
Sardelis MR, Turell MJ, O'Guinn ML, Andre RG, Roberts DR (2002) Vector competence of three North American strains of *Aedes albopictus* for West Nile virus. J Am Mosq Control Assoc 18:284–289

Segal S, Hill AV (2003) Genetic susceptibility to infectious disease. Trends Microbiol 11:445–448
Shaman J, Day JF, Stieglitz M (2002) Drought-induced amplification of Saint Louis encephalitis virus Florida. Emerg Infect Dis 8:575–580
Sillero-Zubiri C, King AA, Macdonald DW (1996) Rabies and mortality in Ethiopian wolves (*Canis simensis*). J Wildl Dis 32:80–86
Singh J, Jamaluddin A (2002) Nipah virus infection in swine. In: Morilla A, Yoon KJ, Zimmerman JJ (eds) Trends in emerging viral infections of swine. Iowa State Press, Ames IA, pp 105–116
Smith JS, Orciari LA, Yager PA, Seidel HD, Warner CK (1992) Epidemiologic and historical relationships among 87 rabies virus isolates as determined by limited sequence analysis. J Infect Dis 166:296–307
Song HD, Tu CC, Zhang GW, Wang SY, Zheng K, Lei LC, Chen QX, Gao YW, Zhou HQ, Xiang H, Zheng HJ, Chern SW, Cheng F, Pan CM, Xuan H, Chen SJ, Luo HM, Zhou DH, Liu YF, He JF, Qin PZ, Li LH, Ren YQ, Liang WJ, Yu YD, Anderson L, Wang M, Xu RH, Wu XW, Zheng HY, Chen JD, Liang G, Gao Y, Liao M, Fang L, Jiang LY, Li H, Chen F, Di B, He LJ, Lin JY, Tong S, Kong X, Du L, Hao P, Tang H, Bernini A, Yu XJ, Spiga O, Guo ZM, Pan HY, He WZ, Manuguerra JC, Fontanet A, Danchin A, Niccolai N, Li YX, Wu CI, Zhao GP (2005) Cross-host evolution of severe acute respiratory syndrome coronavirus in palm civet and human. Proc Natl Acad Sci U S A 102:2430–2435
Suarez DL, Woolcock PR, Bermudez AJ, Senne DA (2002) Isolation from turkey breeder hens of a reassortant H1N2 influenza virus with swine, human, and avian lineage genes. Avian Dis 46:111–121
Tsai TF (1987) Hemorrhagic fever with renal syndrome: clinical aspects. Lab Ani Sci 37:419–427
Turell MJ, O'Guinn ML, Dohm DJ, Jones JW (2001) Vector competence of North American mosquitoes (Diptera: Culicidae) for West Nile virus. J Med Entomol 38:130–134
Twiddy SS, Holmes EC (2003) The extent of homologous recombination in members of the genus Flavivirus. J Gen Virol 84:429–440
Wandeler A, Wachendorfer G, Forster U, Krekel H, Schale W, Muller J, Steck F (1974) Rabies in wild carnivores in central Europe. I. Epidemiological studies. Zentralbl Veterinar Med B 21:735–756
Wang HJ, Zhang RH, Cole J, Chavez F (1999) El Nino and the related phenomenon Southern Oscillation (ENSO): The largest signal in interannual climate variation. Proc Natl Acad Sci U S A 96:11071–11072
Wegbreit J, Reisen WK (2000) Relationships among weather, mosquito abundance, and encephalitis virus activity in California: Kern County 1990–1998. J Am Mosq Control Assoc 16:22–27
White PJ, Norman RA, Hudson PJ (2002) Epidemiological consequences of a pathogen having both virulent and avirulent modes of transmission: the case of rabbit haemorrhagic disease virus. Epidemiol Infect 129:665–677
Woelk CH, Holmes EC (2002) Reduced positive selection in vector-borne RNA viruses. Mol Biol Evol 19:2333–2336

Wolfe ND, Kilbourn AM, Karesh WB, Rahman HA, Bosi EJ, Cropp BC, Andau M, Spielman A, Gubler DJ (2001) Sylvatic transmission of arboviruses among Bornean orangutans. Am J Trop Med Hyg 64:310–316

Wolfe ND, Switzer WM, Carr JK, Bhullar VB, Shanmugam V, Tamoufe U, Prosser AT, Torimiro JN, Wright A, Mpoudi-Ngole E, McCutchan FE, Birx DL, Folks TM, Burke DS, Heneine W (2004) Naturally acquired simian retrovirus infections in central African hunters. Lancet 363:932–937

Woodring,JL, Higgs,S, Beaty,BJ (1996) Natural cycles of vector-borne pathogens. In: Beaty BJ, Marquardt WC (eds) 'The biology of disease vectors. University Press of Colorado, Niwot CO, pp 51–72

Yu IT, Li Y, Wong TW, Tam W, Chan AT, Lee JH, Leung DY, Ho T (2004) Evidence of airborne transmission of the severe acute respiratory syndrome virus. N Engl J Med 350:1731–1739

Zaki SR, Greer PW, Coffield LM, Goldsmith CS, Nolte KB, Foucar K, Feddersen RM, Zumwalt RE, Miller GL, Khan AS, Rollin PE, Ksiazek,TG, Nichol ST, Mahy BWJ, Peters CJ (1995) Hantavirus pulmonary syndrome. Pathogenesis of an emerging infectious disease. Am J Pathol 146:552–579

Zhang XW, Yap YL, Danchin A (2004) Testing the hypothesis of a recombinant origin of the SARS-associated coronavirus. Arch Virol 150:1–20

Zhuang J, Jetzt AE, Sun G, Yu H, Klarmann G, Ron Y, Preston BD, Dougherty JP (2002) Human immunodeficiency virus type 1 recombination: rate, fidelity, and putative hot spots. J Virol 76:11273–11282

CTMI (2007) 315:33–49
© Springer-Verlag Berlin Heidelberg 2007

Infectious Disease Modeling and the Dynamics of Transmission

L. A. Real (✉) · R. Biek

Department of Biology and Center for Disease Ecology, Emory University, Atlanta, GA 30322, USA
lreal@emory.edu

Abstract The dynamics of any infectious disease are heavily dependent on the rate of transmission from infectious to susceptible hosts. In many disease models, this rate is captured in a single compound parameter, the probability of transmission β. However, closer examination reveals how β can be further decomposed into a number of biologically relevant variables, including contact rates among individuals and the probability that contact events actually result in disease transmission. We start by introducing some of the basic concepts underlying the different approaches to modeling disease transmission and by laying out why a more detailed understanding of the variables involved is usually desirable. We then describe how parameter estimates of these variables can be derived from empirical data, drawing primarily from the existing literature on human diseases. Finally, we discuss how these concepts and approaches may be applied to the study of pathogen transmission in wildlife diseases. In particular, we highlight recent technical innovations that could help to overcome some the logistical challenges commonly associated with empirical disease research in wild populations.

1 Introduction

Many of the chapters in this volume have been explicitly concerned with the current increase in zoonotic disease emergence and have attempted various articulations of the causes and impediments to infectious disease transmission and spread into human populations from wildlife. An essential tool for establishing linkages between population processes of infectious disease and disease emergence is the development of mathematical models of disease processes where critical variables effecting disease dynamics can be identified and assessed. Mathematical models have a long history in infectious disease ecology starting with Bernoulli's modeling of smallpox (Bernoulli 1760) and including Ross's analysis of malaria (Ross 1911), but they have seen an expanded development over the last 25 years (Anderson et al. 1981; Anderson and May 1991). We now have models for many of the most important human emerging infectious diseases or diseases that threaten to emerge, e.g., HIV (Anderson and May 1988, 1991; Nowak and May 2000), malaria (Aron and May 1982; Macdonald 1957), SARS-coronavirus (Anderson et al. 2004; Lipsitch et al. 2003), rabies (Childs et al. 2000; Murray and Seward 1992; Murray et al. 1986; Russell et al. 2005; Smith et al. 2002), and influenza (Ferguson and Anderson 2002; Ferguson et al. 2003; Longini et al. 2005), to name a few. Mathematical models are also being used to explore wildlife disease dynamics (Grenfell and Dobson 1995; Hudson et al. 2002) and possible routes of zoonotic disease emergence. Understanding disease dynamics across hosts is an essential first step in understanding and articulating those conditions under which new diseases can emerge from wildlife reservoirs.

It is easy to recognize that the first obstacle to establishment of any infectious disease is the successful transmission from infected individuals into susceptible hosts. In the absence of sustained transmission, any infectious disease is doomed and will not spread. Most mathematical models coalesce transmission into a single phenomenological transmission rate (β) between infected and susceptible hosts, and this rate masks a great deal of information. In this chapter, we wish to examine how the transmission rate can be parameterized and decomposed into its underlying contributing variables, and how these measures can be applied to zoonotic disease dynamics.

There are three fundamental characteristics which will influence the likelihood of sustained transmission among susceptible and infected hosts: the infectiveness of the pathogen, the transmission probability, and the contact pattern and rate, which together affect the basic reproductive number (R_0) of the pathogen. In this chapter, we review some of these basic concepts with explicit attention to how these fundamental characteristics can be assessed in specific host–pathogen systems. Throughout the chapter, we will be following formulations from Halloran (1998) who has an excellent introduction to these concepts in the context of human disease dynamics.

2 Basic Concepts

Partitioning and estimating the parameters that enter into a quantitative characterization of the transmission process requires distinctions between the time course of infectiousness (i.e., that time interval over which infected individuals are capable of transmitting the pathogen to new susceptible individuals) versus the time course of disease (the expression of symptoms associated with infection). Imagine a time line beginning with a susceptible host within the population (Fig. 1). At some time point (T) the susceptible individual becomes infected by a pathogen. For the time course of infectiousness, after initial infection, the host may undergo a latent period (τ) where the pathogen can be resident in the host but not be transmitted to other hosts. The latent period is followed by an infectious period (γ) where pathogen can be transmitted. At some final time, the infectious individual loses its infectiousness and moves into a non-infectious class either through recovery or death. The time course for disease differs from infectiousness in that upon the onset of infection (T) the host moves into an incubation period (δ) where disease symptoms are absent. When symptoms appear, the host moves into the symptomatic period (σ) that lasts until the symptoms disappear and the host recovers or dies. The initiation and duration of these periods may not correspond. For example, in some diseases the latent period can be shorter than the incubation period in which case hosts are infectious before symptoms appear, e.g., ungulates infected with rinderpest virus become infectious approximately 24–48 h before the onset of symptoms (Plowright 1968). In other diseases, the latent period can be longer than the incubation period. For example, *Plasmodium falciparum* malaria has an incubation period of approximately 14 days in humans. However, the infective stages of the parasite that are infective to mosquitoes only begin to appear approximately 10 days after the onset of symptoms (Halloran 1998).

The rate of conversion of susceptible hosts into infected hosts is governed by two factors: the number of susceptibles in the host population and what has been conventionally referred to as the force of infection (Anderson and May 1991; Begon et al. 2002). The force of infection is the product of (1) the rate of contact, c, between individuals in the host population, (2) the probability, m, that an individual contact is between a susceptible individual and an infected individual that is also infectious, and (3) the transmission probability, ρ, that a contact between an infectious host and a susceptible host leads to a successful transmission event (Begon et al. 2002). Most often the infectiveness, m, is assumed to be proportional to the fraction of infectious individuals in the total population, i.e., the prevalence, P, of the disease.

3 Basic Reproductive Number

With the concepts introduced so far, it would seem that any effort to model disease transmission would require knowledge of many parameters. Often we cannot ascertain these component parts since they are exceedingly difficult to estimate. As a consequence, many disease ecologists have focused on a single index, the basic reproductive number (R_0), which captures many of the most important features of disease dynamics, especially where one is concerned with conditions leading to epidemic emergence.

R_0 is defined as the "average number of secondary infections produced when one infected individual is introduced into a host population in which every host is susceptible" (Anderson and May 1991). R_0 is defined by the following:

$$R_0 = \begin{bmatrix}\text{number of contacts}\\ \text{per unit time (c)}\end{bmatrix} \times \begin{bmatrix}\text{transmission}\\ \text{probability per contact } (\rho)\end{bmatrix} \times \begin{bmatrix}\text{duration of}\\ \text{infectiousness } (\gamma)\end{bmatrix},$$

i.e. $R_0 = c\rho\gamma$

However, there are also alternative means to estimate R_0 without knowing these components, which is certainly one reason for its popularity. For example, R_0 can be assessed phenomenologically (given its definition) as the average per capita rate of increase in infectious individuals when a pathogen emerges into a new previously unexposed population since all the individuals resident in this new population are presumed susceptible. This was the technique used by Lipsitch et al. (2003) to calculate R_0 for SARS coronavirus during its rapid emergence in 2003. Consequently, we can directly measure R_0 without necessarily knowing the details of the transmission process that generates that overall number of secondary infections in the population. The R_0 for a variety of wildlife diseases is given in Table 1. Anderson and May (1991) provide a comparable table for human infectious diseases and Dietz (1993) provides an overview of methods used to estimate R_0 from population data. Ferrari et al. (2005) have recently derived a maximum likelihood estimator for R_0 using chain binomial models as a refinement to calculating R_0 using discrete time-series data.

The capability to directly quantify R_0 can be a useful first step in predicting disease emergence. For a disease to increase in the host population, an infectious individual must at least replace itself with more than one infectious secondary case, i.e., the disease will increase if $R_0 > 1$. If $R_0 < 1$, then the disease will fade from the host population and go extinct. If $R_0 = 1$, then every infectious individual replaces itself with one and only one new infectious individual and the disease prevalence in the population will be stable, i.e., the disease will be "endemic."

However, the basic reproductive number, R_0, does have some significant limitations with regard to predicting newly emerging pathogens. R_0 is *not* a fixed property of a pathogen. Rather (as is apparent from its definition) it is only defined within a certain population of hosts governed by a specific contact pattern, duration of infectiousness, and transmission probability. One could have very different underlying biological transmission processes that generate identical basic reproductive numbers. For example, the R_0 for measles is approximately 9, which also happens to be the R_0 for HIV among intravenous drug users (Halloran 1998). However, measles has a high transmission and short duration of infection, while HIV has a low transmission and long duration of infection. This feature can make comparisons of R_0 across diseases very difficult since R_0 is a compound expression of three variables. What R_0 captures is the capacity to generate an epidemic given some (unfortunately often unknown or not assessed) transmission process. Both measles and HIV have a high capacity to generate an epidemic assuming a given transmission process, but at very different time scales and governed by very different underlying transmission components.

Ideally, for the purposes of predicting disease emergence, we would like to know the values of the underlying components of transmission that produce the overall pattern of R_0 and, ideally, how these components might change under alterations in environmental condition effecting the likelihood of disease emergence. For example, habitat fragmentation without loss of local habitat quality may generate new contact rates, c, while deterioration of habitat without changing patterns of connectivity may affect the transmission probability, ρ. Both changes might generate the same overall alteration in R_0. Yet the biological measures needed to respond to these different changes might be quite different since the pattern of emergence is driven by entirely different changes in transmission mechanism. Consequently, we should be attending to the development of methods for the direct assessment of the components of R_0 as a goal toward increasing our capacity to predict disease emergence.

4 Estimating the Transmission Probability

There are two common techniques used to estimate the likelihood that an encounter between an infected individual host and a susceptible individual will result in successful transmission of the pathogen leading to new infections. The first method, the secondary attack rate (SAR), focuses on the fate of a single infected index case (host) that comes into contact with many susceptible host individuals in the population. The second method, the binomial model of transmission probability, tracks one uninfected but susceptible host as it comes into contact with many infectious hosts. Both methods have been commonly used in human disease epidemiology but have not been used in assessing wildlife disease dynamics. Consequently, our examples will be draw from the human disease literature, but the methods should be extendable to wildlife disease dynamics.

The secondary attack rate is simply defined as the ratio of the number of hosts exposed that develop disease relative to the total number of susceptible exposed hosts, i.e.,

$$SAR = \frac{\text{total secondary cases}}{\text{total susceptible exposed}}$$

Before we can use this method, however, we must understand how one defines an exposed host and a secondary infected host. Let us observe one susceptible individual in the population. This individual host becomes infected at time *T* and will be designated the primary infected host. Primary hosts can, in general, be characterized as having (1) a maximum infectious period (I), i.e., the maximum time that individuals within the host population remain infectious, (2) a minimum incubation period (E1), i.e., the minimum time required before symptoms appear, and (3) a maximum incubation period (E2), i.e., the maximum time period before which symptoms will appear. We can arrange these time intervals along a time axis (Fig. 2) that can then be used to define secondary infections. Imagine four hosts that become symptomatic after contact with the primary infected host at time intervals specified in Fig. 2, line B. Which of these are likely to be the consequence of transmission from the primary infected host? Alternatively, which of these cases are clearly not the consequences of transmission from the primary?

Host 2 becomes symptomatic within the minimum incubation period (E1) so it could not have received its infection from the primary host. Similarly, host 5 becomes symptomatic after contact at a time greater than the sum of the maximum infectious period (I) and the maximum incubation period (E2). Consequently, it could not have been the recipient of pathogen from the

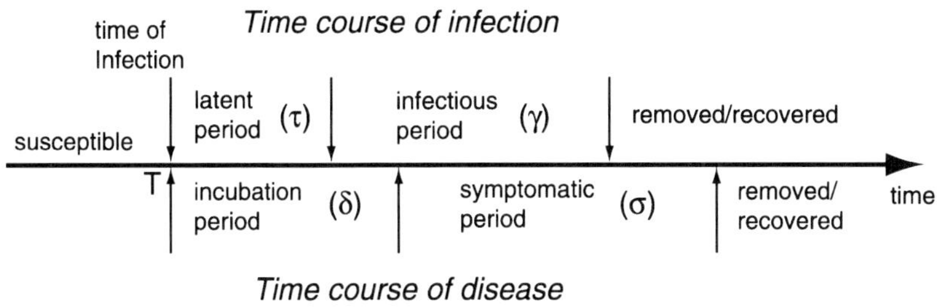

Fig. 1 Representative time intervals for the course of infection and disease used in the calculation of transmission rate

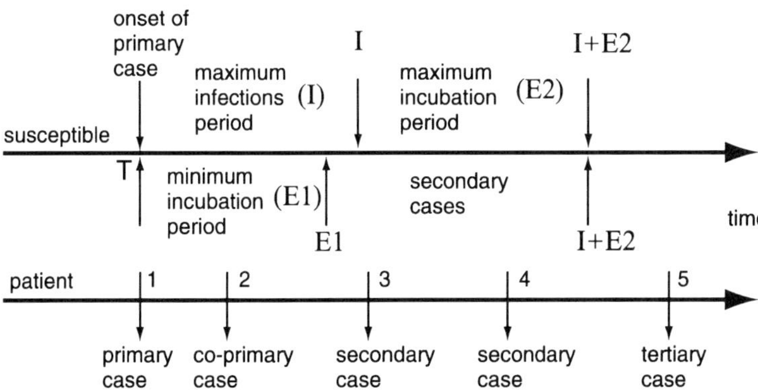

Fig. 2 Representative time intervals used for the determination of secondary cases and the calculation of the secondary attack rate

primary. The only individuals that could have become infected from the primary are those that fall within the time interval defined by E1 as the lower bound and I+E2 as the upper bound. Any individuals appearing symptomatic within this time interval after contact with the primary are considered secondary cases from the primary. Host 3 and host 4 would then be the only secondary cases from the primary.

Kendrick and Eldering (1939) used this method to calculate the SAR for pertussis. Infected individuals show positive throat cultures for 21 days after the onset of symptoms, thus I = 21 days. They ascertained through observation that the minimum incubation period was approximately 10 days and the maximum incubation period was approximately 30 days. Thus, all secondary cases were those cases of exposed individuals to the primary case who developed symptoms during the time interval 10–51 days. Then SAR equals the total

Table 1 Examples for estimation of the basic reproductive rate (R_0) for various pathogens in wildlife species

Pathogen	Host species	Scientific name	R_0	Reference
Rabies virus	Spotted hyena	*Crocuta crocuta*	1.9	East et al. 2001
Phocine distemper virus	Harbor seal	*Phoca vitula*	2.8	Swinton et al. 1998
Mycobacterium bovis	Ferret (feral)	*Mustela furo*	0.18–1.20	Caley and Hone 2005
Mycobacterium bovis	Eurasian badger	*Meles meles*	1.2	Anderson and Trewhella 1985
Classical swine fever virus	Wild boar	*Sus scrofa*	1.1–2.1	Hone et al. 1992
Heterakis gallinarum	Ring-necked pheasant	*Phasianus colchicus*	1.2	Tompkins et al. 2000

number of secondary cases across all households in the population relative to the total number of exposed susceptibles in all households. For the purposes of their study, they defined an exposure as any contact with the primary case for at least 30 min during the infectious period I. For individuals that had not received a test vaccine, the secondary attack rate was substantial, SAR = 0.685. Among individuals that had received the vaccine the secondary attack rate dropped, SAR = 0.128. The vaccine under use at the time appears to have reduced the transmission rate by approximately 82%. Calculations of secondary attack rates with and without vaccination have been traditional methods used to assess the efficacy of a particular vaccination strategy.

Similar techniques have been applied to assess the transmission probability of a variety of human infectious diseases (Table 1). For example, the Centers for Disease Control and Prevention undertook a household case study to calculate the secondary attack rate for SARS-coronavirus during the 2003 outbreak in Singapore (Goh et al. 2004). Examination of households suggests that SARS is not highly contagious among family members (secondary attack rate = 0.062), while the rate of transmission among hospital workers is strikingly higher (secondary attack rate >0.50).

For sexually transmitted disease, the transmission probability is often assessed using the binomial distribution (or its extension the chain binomial) for the following reasons. Assume that the probability of disease transmission during a single contact with an infected host is p. Then the probability of escaping infection following a contact with an infected host is $q = (1 - p)$. Suppose that a susceptible host makes n contacts with an infected host or with different infected hosts. Then the probability of escaping infection after n contacts is:

$$q^n = (1-p)^n$$

Then the probability of becoming infected after n contacts is:

$$1-q^n = 1-(1-p)^n$$

which is the description for the binomial distribution. The maximum likelihood estimate for p is given by

$$\hat{p} = \frac{\text{number of individuals who become infected}}{\text{total number of contacts with infectives}}$$

The difference between the secondary attack rate (SAR) and $\hat{p}$ is in the denominator. SAR weighs transmission relative to contacts with susceptibles while the binomial distribution weighs transmission relative to contacts with infectious hosts. The two measures are identical, i.e. SAR = $\hat{p}$, when every susceptible has contact with one and only one infectious host.

The binomial distribution method has been used quite commonly to estimate the transmission probability for HIV given the current concern over this ongoing worldwide epidemic. What is the likelihood of transmission given a sexual encounter? Halloran (1998) presents results of a transmission study in a population of 100 steady sexual couples where one partner was HIV-positive while the other partner was HIV-negative. Over the course of the study period, 25 of the 100 susceptibles became infected. The total number of sexual encounters during the study period was 1,500. From the maximum likelihood estimator then:

$$\hat{p} = {}^{25}\!/_{1500} = 0.017$$

That is, an uninfected person has a little less than a 1:50 chance of contracting HIV following a sexual encounter. The probability of infection after two encounters would be

$$1-\left(1-\hat{p}\right)^2 = 0.034$$

Table 2 provides an overview of the transmission probabilities for a variety of human diseases derived using either SAR or the binomial distribution.

Table 2 Estimates of the transmission probability (ρ) for various human pathogens

Disease	Pathogen	ρ	Method of estimation	Reference
Meningococcal disease	*Neisseria meningitidis*	0.0069	Secondary attack rate	De Wals et al. 1981
Ebola fever	Ebola virus—Sudan	Hospitals = 0.81 Community = 0.12	Secondary attack rate	Francis et al. 1978
Food poisoning	Norwalk-like virus	0.17	Secondary attack rate	Gotz et al. 2001
Cryptosporidiosis	*Cryptosporidium pavum*	0.15	Secondary attack rate	Guerrant 1997
Lower respiratory tract infection	Adenovirus	0.55	Secondary attack rate	Palomino et al. 2000
Whooping cough	*Bordetella pertussis*	Vaccinated = 0.128 Unvaccinated = 0.685	Secondary attack rate	Kendrick and Eldering 1939
AIDS	Human immunodeficiency virus	0.017	Binomial distribution	Halloran 1998
Monkeypox	Monkeypox virus	0.93	Secondary attack rate	Hutin et al. 2001

5
Estimating Transmission in Wildlife and Zoonotic Disease

As outlined above, three factors determine the rate at which new infections occur: (1) the rate of contact (c) between individuals, (2) the probability (m) that any contact is between an infectious and a susceptible individual, and (3) the probability (ρ) that such a contact actually results in a new infection. Despite their indisputable significance for disease dynamics, the estimation of these three parameters is rarely attempted for natural populations. Usually, the necessary temporal and spatial resolution at which epidemiological data have to be collected is simply not obtainable for species in the wild. However, we would argue that this may not always be true and that, especially given recent technological advancements, empirical data for the different components of transmission could be gathered in certain cases. In the following, we will therefore discuss what we feel are promising avenues of current and potential future research in this regard.

Social species probably offer the best opportunities to quantify contact rates, especially if these species are diurnal and can be observed without interfering with natural behavioral patterns. Observational studies have been used very effectively, for example, to obtain detailed information on social interactions in certain primates and ungulates (Berger 1986; Goodall 1986; Mloszewski 1983). In fact, data already collected for these species could conceivably be used to measure rates of contact between individuals. What constitutes a contact event will obviously depend on the specific infectious agent in question and its mode of transmission. Compared to within-group dynamics, determining rates of contact between social groups will usually be more difficult because such events will occur much more rarely and may require monitoring more than one group. For long-term studies, however, data on immigration of new individuals and frequency of encountering other groups may also be available. It is important to keep in mind that in many cases, such as epidemics sweeping through a population, rates of transmission between groups will be of much greater interest. This is simply because rate and course of transmission within a group is unlikely to have much effect on overall disease dynamics compared to the rate at which the disease is introduced to new groups. This is especially true for acute infections, because it is the level of host group contact relative to the length of the infectious period that will ultimately determine the rate of disease spread (Cross et al. 2005).

Contacts are less frequent and thus harder to determine for solitary species, unless opportunity for pathogen transmission is restricted to certain habitat features that can be monitored closely (e.g., water holes, bird feeders). Where use of such features cannot be determined from direct observation, it may be possible to fit animals with transmitters to record and correlate the time they spend at a common location. For example, Sutherland et al. (2005) used passive integrated transponder (PIT) tags in mice to measure their use of borrows. A hidden antenna connected to a data logger would register each time a marked individual passed the burrow entrance. Use of the same burrow by two individuals within a few minutes of each other was thereby considered an interaction. Although this study did not consider disease transmission, a similar approach could certainly be used to study contact patterns in such a context. Calisher et al. (2000) examined pairs of deer mice captured simultaneously in single-capture traps for antibody to Sin Nombre virus to infer patterns of hantaviral transmission among different demographic classes of mice (see the chapter by Klein and Calisher, this volume).

Radiotelemetry can also be used to look at simultaneous space use of individuals (see the chapter by Stallknecht, this volume), but the temporal resolution of these data is usually insufficient to infer actual encounters. This can be potentially overcome by the use of radio transmitters that note and record the nearby presence of another transmitter, a technique employed in a current study of rabies virus transmission among raccoons (L. Hungerford, personal communication). A potential problem with all approaches involving electronic tags is that they will underestimate the number of encounters unless all individuals in a population are fitted with a tag. As long as it is known what proportion of the population is tagged though, it may be possible to correct for this bias.

Assuming that contact rates can be determined with sufficient accuracy and precision, we have yet to determine whether a particular contact event involved an infectious individual and whether contact resulted in a new infection. This of course requires detailed knowledge on disease status of all individuals in a population through time. In specific cases, disease status may be inferred retrospectively. For example, mortality following rabies infection in carnivores is close to 100%, and with the help of radio transmitters, carcasses can usually be recovered quickly enough to confirm rabies as the cause of mortality, as has been demonstrated with striped skunks (Greenwood et al. 1997). Furthermore, it is known from experimental studies that animals are only infectious for a few days prior to death. With the previously mentioned technology of cross-talking radio transmitters, the number of other marked raccoons encountered during this time period can be determined along with the proportion of these individuals that subsequently develop disease.

Rabies is somewhat unusual because every infection can be considered to result in disease and ultimately death. For most pathogens, infection status and periods of infectiousness have to be established based on regular screening. Capturing and sampling individuals on a regular basis may accomplish this. The appropriate length of the interval between samples would thereby depend on the biology and epidemiology of the infectious agent. Because the humoral immune response takes several weeks to develop, individuals may be infectious even if no antibodies can be detected. Thus, screening would ideally involve serological tests as well as efforts to directly detect the infectious pathogen. Problems again arise if not all individuals can be resampled regularly, as will often be the case in wildlife populations. Furthermore, capturing and collecting blood samples may be considered too traumatic to be carried out frequently. Fortunately, considerable progress has been made in recent years regarding the use of noninvasive sampling of wildlife species, including techniques for disease screening. For example, Santiago et al. (2003a, 2003b) were able to determine infection with simian immunodeficiency virus (SIV) in wild chimpanzees using fresh fecal and urine samples that yielded both antibodies and virus RNA. Similarly successful results were obtained for simian foamy virus (SFV; B. Hahn, personal communication), suggesting that these techniques are more widely applicable. A very promising research study would therefore be to combine behavioral data on contact rates within a group with the collection of fecal samples to monitor the infection status of individuals.

The use of modern molecular techniques may even allow us to go one step further and not only determine infection status of an individual but to document the source of that infection. In rapidly evolving RNA viruses, for example, spatial spread among different host populations or geographic areas can frequently be discerned from genetic sequence data (Real et al. 2005; Walsh et al. 2005). By extension, similar methods could be used to identify the most probable donor individual for a new infection using genetic evidence. Probably the most famous application of forensic phylogenetics to date has been that of a doctor who allegedly had used blood from an HIV-infected patient to infect his ex-girlfriend. Phylogenetic analysis showed that the victim's virus sequences were nested within those of the suspected donor but were clearly distinct from other viruses circulating in the larger geographic area. This result was consistent with the proposed direction of transmission from the donor patient to the victim and held up as evidence in court (Metzker et al. 2002). Especially in situations where all possible donors are known (such as in animal social groups) and for pathogens with high standing genetic diversity (increasing the chances of pathogens in different individuals being distinct), molecular epidemiology could become a powerful tool for elucidating actual transmission histories in wildlife populations.

6 Conclusions

In this chapter, we have been concerned with the models and parameters used to describe the process of pathogen transmission. Although most of our case studies came from human diseases, the day when similar studies are being conducted in wild animal species may not be too far off. New and more sophisticated methods for tracing contact patterns and pathogen surveillance are constantly being developed, and we would expect that many of these methods will eventually also find use in the study of wildlife diseases. Better empirical data at hand will undoubtedly facilitate the development of more powerful models, improving our ability to predict, prevent, and control the future emergence and spread of zoonotic diseases.

References

Anderson RM, May RM (1988) Epidemiological parameters of HIV transmission. Nature 333:514–522

Anderson RM, May RM (1991) Infectious disease of humans: dynamics and control. Oxford University Press, Oxford

Anderson RM, Trewhella W (1985) Population dynamics of the badger (*Meles meles*) and the epidemiology of bovine tuberculosis (*Mycobacterium bovis*). Phil Trans Roy Soc London B 310:227–381

Anderson RM, Jackson HC, May RM, Smith AM (1981) Population dynamics of fox rabies in Europe. Nature 289:765–771

Anderson RM, Fraser C, Ghani AC, Donnelly CA, Riley S, Ferguson NM, Leung GM, Lam TH, and Hedley AJ (2004) Epidemiology, transmission dynamics and control of SARS: the 2002–2003 epidemic. Phil Trans Biol Sci 359:1091–1105

Aron JL, May RM (1982) The population dynamics of malaria. In: Anderson RM (ed) Population dynamics of infectious diseases. Chapman and Hall, London, pp 139–179

Begon M, Bennett M, Bowers RG, French NP, Hazel SM, Turner J (2002) A clarification of transmission terms in host-microparasite models: numbers, densities, and areas. Epidemiol Infect 129:147–153

Berger J (1986) Wild horses of the great basin: social competition and population size. University of Chicago Press, Chicago

Bernoulli D (1760) Essai d'une nouvelle analyse de la mortalité causée par la petite vérole et des avantages de l'inoculation pour la prévenir. Mem Math Phys Acad Royal Soc Paris, 1–45

Caley P, Hone J (2005) Assessing the host disease status of wildlife and the implications for disease control: *Mycobacterium bovis* infection in feral ferrets. J Appl Ecol 42:708–719

Calisher CH, Childs JE, Seney WP, Canestorp KM, Beaty BJ (2000) Dual captures of Colorado rodents: implications for transmission of hantaviruses. Emerg Infect Dis 6:363–369
Childs JE, Curns AT, Dey ME, Real LA, Feinstein L, Bjornstad ON, Krebs JW (2000) Predicting the local dynamics of epizootic rabies among raccoons in the United States. Proc Natl Acad Sci U S A 97:13666–13671
Cross PC, Lloyd-Smith JO, Johnson PLF, Getz WM (2005) Dueling timescales of host movement and disease recovery determine invasion of disease in structured populations. Ecol Lett 8:587–595
De Wals P, Hertoghe L, Borlee-Grimee I, De Maeyer-Cleempoels S, Reginster-Haneuse G, Dachy A, Bouckaert A, Lechat MF (1981) Meningococcal disease in Belgium. Secondary attack rate among households, day-care nursery and pre-elementary school contacts. J Infect 3 [Suppl]:53–61
Dietz K (1993) The estimation of the basic reproduction number for infectious diseases. Stat Meth Med Res 2:23–41
East ML, Hofer H, Cox JH, Wulle U, Wiik H, Pitra C (2001) Regular exposure to rabies virus and lack of symptomatic disease in Serengeti spotted hyenas. Proc Natl Acad Sci U S A 98:15026–15031
Ferguson NM, Anderson RM (2002) Predicting evolutionary change in the influenza A virus. Nature Med 8:562–563
Ferguson NM, Galvani AP, Bush RM (2003) Ecological and immunological determinants of influenza evolution. Nature 422:428–433
Ferrari MJ, Bjornstad ON, Dobson AP (2005) Estimation and inference of R_0 of an infectious pathogen by removal method. Math Biosci 198:14–26
Francis DP, Smith DH, Highton RB, Simpson DIH, Lolik P, Deng IM, and Gillo AL (1978) Ebola fever in the Sudan, 1976: epidemiological aspects of the disease. In: Pattyn SR (ed) Ebola virus haemorrhagic fever. Elsevier, Amsterdam, pp 1–7
Goh DL, Lee BW, Chia KS, Heng BH, Chen M, Ma S, Tan CC (2004) Secondary household transmission of SARS, Singapore. Emerg Infect Dis 10:232–234
Goodall J (1986) The chimpanzees of Gombe: patterns of behavior. Belknap Press of Harvard University Press, Cambridge MA
Gotz H, Ekdahl K, Lindback J, de Jong B, Hedlund KO, Giescke J (2001) Clinical spectrum and transmission characteristics of infection with Norwalk-like virus: findings from a large community outbreak in Sweden. Clin Infect Dis 33:622–628
Greenwood RJ, Newton WE, Pearson GL, Schamber GJ (1997) Population and movement characteristics of radio-collared striped skunks in North Dakota during an epizootic of rabies. J Wild Dis 33:226–241
Grenfell BT, Dobson AP (1995) The ecology of infectious diseases in natural populations. Cambridge University Press, London
Guerrant RL (1997) Cryptosporidiosis: an emerging, highly infectious threat. Emerg Infect Dis 3:51–57
Halloran ME (1998) Concepts of infectious disease epidemiology. In: Rothman KJ, Greenland S (eds) Modern epidemiology. Lippincott Williams and Wilkins, Philadelphia, pp 529–554

Hone J, Pech R, Yip P (1992) Estimation of the dynamics and rate of transmission of classical swine fever (hog cholera) in wild pigs. Epidemiol Infect 108:377–386

Hudson PJ, Rizzoli A, Grenfell BT, Heesterbeek H, Dobson AP (2002) The ecology of wildlife diseases. Oxford University Press, London

Hutin YJF, Williams J, Malfait P, Pebody R, Loparev VN, Ropp SL, Rodriguez M Knight JC, Tshioko FK, Khan AS, Szczeniowski MV and Esposito JJ et al (2001) Outbreak of human monkeypox Democratic Republic of Congo, 1996–1997. Emerg Infect Dis 7:434–438

Kendrick P, Eldering G (1939) A study in active immunization against pertussis. Am J Hyg B 38:133

Lipsitch M, Cohen T, Cooper B, Robins JM, Ma S, James L, Gopalakrishna G, Chew SK, Tan CC, Samore MH, Fisman D, Murray M (2003) Transmission dynamics and control of severe acute respiratory syndrome. Science 300:1966–1970

Longini IM, Nizam A, Xu S, Ungchusak K, Hanshaoworakul W, Cummings DAT, Halloran ME (2005) Containing pandemic influenza at the source. Science 309:1083–1087

Macdonald G (1957) The epidemiology and control of malaria. Oxford University Press, London

Metzker ML, Mindell DP, Liu XM, Ptak RG, Gibbs RA, Hillis DM (2002) Molecular evidence of HIV-1 transmission in a criminal case. Proc Natl Acad Sci U S A 99:14292–14297

Mloszewski MJ (1983) The behavior and ecology of the African buffalo. Cambridge University Press, Cambridge

Murray JD, Seward WL (1992) On the spatial spread of rabies among foxes with immunity. J Theor Biol 156:327–348

Murray JD, Stanley EA, Brown DL (1986) On the spatial spread of rabies among foxes. Proc R Soc Lond Biol 229:111–150

Nowak MA, May RM (2000) Virus dynamics. Oxford University Press, London

Palomino MA, Larranaga C, Avendano. LF (2000) Hospital-acquired adenovirus 7 h infantile respiratory infection in Chile. Ped Infect Dis J 19:527–531

Plowright W (1968) Rinderpest virus. Monog Virol 3:25–110

Real LA, Henderson JC, Biek R, Snaman J, Jack TL, Childs JE, Stahl E, Waller L, Tinline R, Nadin-Davis SA (2005) Unifying the spatial population dynamics and molecular evolution of epidemic rabies virus. Proc Natl Acad Sci U S A 102:12107–12111

Ross R (1911) The prevention of malaria. Murray, London

Russell CA, Smith DL, Childs JE, Real LA (2005) Predictive spatial dynamics and strategic planning for raccoon rabies emergence in Ohio. PLoS 3:1–7

Santiago ML, Bibollet-Ruche F, Bailes E, Kamenya S, Muller MN, Lukasik M, Pusey AE, Collins DA, Wrangham RW, Goodall J, Shaw GM, Sharp PM, Hahn BH (2003a) Amplification of a complete simian immunodeficiency virus genome from fecal RNA of a wild chimpanzee. J Virol 77:2233–2242

Santiago ML, Lukasik M, Kamenya S, Li Y, Bibollet-Ruche F, Bailes E, Muller MN, Emery M, Goldenberg DA, Lwanga JS, Ayouba A, Nerrienet E, McClure HM, Heeney JL, Watts DP, Pusey AE, Collins DA, Wrangham RW, Goodall J, Brookfield JF, Sharp PM, Shaw GM, Hahn BH (2003b) Foci of endemic simian immunodeficiency virus infection in wild-living eastern chimpanzees (*Pan troglodytes schweinfurthii*). J Virol 77:7545–7562

Smith DL, Lucey B, Waller LA, Childs JE, Real LA (2002) Predicting the spatial dynamics of rabies epidemics on heterogeneous landscapes. Proc Natl Acad Sci U S A 99:3668–3672

Sutherland DR, Spencer PBS, Singleton GR, Taylor AC (2005) Kin interactions and changing social structure during a population outbreak of feral house mice. Mol Ecol 14:2803–2814

Swinton J, Harwood J, Grenfell BT, Gilligan CA (1998) Persistence thresholds for phocine distemper virus infection in harbour seal *Phoca vitulina* metapopulations. J Anim Ecol 67:54–68

Tompkins DM, Greenman JV, Robertson PA, Hudson PJ (2000) The role of shared parasites in the exclusion of wildlife hosts: *Heterakis gallinarum* in the ring-necked pheasant and the grey partridge. J Anim Ecol 69:829–840

Walsh P, Biek R, Real LA (2005) Wave-like spread of Ebola Zaire. PLoS 3:e71

CTMI (2007) 315:51–66
© Springer-Verlag Berlin Heidelberg 2007

The Evolutionary Genetics of Viral Emergence

E. C. Holmes[1] (✉) · A. J. Drummond[2] (✉)

[1]Center for Infectious Disease Dynamics, Department of Biology, Mueller Laboratory, The Pennsylvania State University, University Park, PA 16802 USA
ech15@psu.edu
[2]Department of Computer Science, University of Auckland, Private Bag 92019 Auckland New Zealand
alexei@cs.auckland.ac.nz

Abstract Despite the wealth of data describing the ecological factors that underpin viral emergence, little is known about the evolutionary processes that allow viruses to jump species barriers and establish productive infections in new hosts. Understanding the evolutionary basis to virus emergence is therefore a key research goal and many of the debates in this area can be considered within the rigorous theoretical framework established by evolutionary genetics. In particular, the respective roles played by natural selection and genetic drift in shaping genetic diversity are also of fundamental importance for understanding the nature of viral emergence. Herein, we discuss whether there are evolutionary rules to viral emergence, and especially whether certain types of virus, or those that infect a particular type of host species, are more likely to emerge than others. We stress the complex interplay between rates of viral evolution and the ability to recognize cell receptors from phylogenetically divergent host species. We also emphasize the current lack of convincing data as to whether viral emergence requires adaptation to the new host species during the early stages of infection, or whether it is largely a chance process involving the transmission of a viral strain with the necessary genetic characteristics.

1 Introduction

Until recently, studies of emerging viruses frequently involved compiling lists of pathogens that were considered new to human populations or that had increased in frequency and geographical range, and describing the ecological factors responsible for their appearance. Such studies gave particular emphasis to how changes in human ecology, notably increases in population size, modifications in land use, global travel, and political upheavals, had been responsible for an elevated burden of infectious disease, often by increasing the proximity and/or density of possible reservoir populations (Morse 1995). What was largely absent from these studies was a consideration of the evolutionary processes that underlie viral emergence (see the chapter by Childs et al., this volume). Indeed, the main role played by evolutionary biology in the first studies of emerging viruses was to reconstruct the origin and spread of new pathogens, largely through phylogenetic analysis (for example, Nichol et al. 1993). Although the focus on changing human ecology and phylogenetic history was an important and necessary first step, and it will often be impossible to disentangle ecology and genetics when explaining the emergence of a specific pathogen, it is also essential to ask what evolutionary processes are responsible for the appearance and spread of pathogens? Indeed, for the study of emerging viruses to come of age, it is critical to determine whether there are any general evolutionary rules governing the process of emergence.

Evolutionary genetics aims to understand the processes responsible for the origin and maintenance of genetic variation in populations. The great obsession of evolutionary genetics has been to reveal the respective roles of random fluctuations in allele frequencies—genetic drift—and the natural selection of advantageous mutations in shaping genetic diversity (Gillespie 1998). Although, at face value, evolutionary genetics may seem of little relevance to the problem of viral emergence, we suggest that it in fact provides an essential theoretical framework. For example, it might be that emergence simply requires the chance exposure of a virus to a new susceptible population, with little involvement of natural selection. Alternatively, it may be that viruses have to adapt to successfully spread in a new species. In this scenario, different species might represent different fitness peaks on an "adaptive landscape" and that traversing between these peaks is difficult because they are connected by valleys of low fitness. Moreover, because an emergent virus will only infect a small number of individuals when it first enters a new population, genetic drift is expected to play a major role in determining what viral mutations get fixed, because drift is more potent in small populations. Finally, the respective influences of drift and selection will also vary according to the size of the population bottleneck that accompanies viral transmission among hosts, which will itself be a function of the mode of transmission.

In this paper we address some of the key questions surrounding the evolutionary genetics of viral emergence. Although evolutionary genetics relates equally well to hosts (such as differences in susceptibility and immune responses) as well as pathogens, we will concentrate on the latter where data is more abundant and their evolution can be tracked more readily. To focus our paper, we will consider four key questions:

1. Are certain types of virus more likely to emerge than others?
2. Are viruses from phylogenetically related host species more likely to undergo cross-species transmission than those viruses from distantly related host species?
3. Does viral emergence require adaptation to the new host species?
4. Is recombination a prerequisite for viral emergence?

Although definitive answers will not be forthcoming in all cases, we discuss the data required to resolve these issues. Finally, as well as shedding new light on the process of emergence in particular, answering these questions will provide more general insights into the nature of viral evolution.

2
Are Certain Types of Virus More Likely to Emerge than Others?

The broadest division in virus classification is between those viruses in which the genome is composed of DNA (DNA viruses) and those where the genomic nucleic acid comprises RNA (RNA viruses), with the latter also usually considered to include retroviruses that make a DNA copy of the RNA genome through reverse transcription. Although most known viruses have RNA genomes, even accounting for this bias it is clear that RNA viruses are more often associated with emerging diseases than DNA viruses (Cleaveland et al. 2001; Woolhouse 2002; see the chapter by Cleaveland et al., this volume). In contrast, DNA viruses are often associated with a process of virus–host co-speciation that can extend many millions of years. This is perhaps because DNA viruses more often establish persistent infections than RNA viruses and so can more easily track host evolution (Holmes 2004).

That RNA viruses seem to possess inherently more "emergibility" than DNA viruses has usually been put down to their very rapid rates of evolutionary change. RNA viruses are thought to evolve many logs faster than DNA viruses because of a combination of highly error-prone replication with RNA polymerase or reverse transcriptase, large population sizes, and rapid replication rates (Domingo and Holland 1997; Moya et al. 2004). In turn, a rapid rate of evolutionary change allows RNA viruses to quickly generate the mutations that might be required to adapt them to new environments, including new host species. Although this effect is broadly true, there is still substantial variation among RNA viruses in their ability to cause emergent diseases. Understanding the basis of this variation is critical to the development of an evolutionary model of viral emergence and for understanding the constraints on RNA virus evolution in general.

If the rate of evolutionary change is driven by the rate at which neutral variants are generated, then the most important factors determining rates of evolutionary change are the replication error rate and the generation time. The variation in error rates among viruses is a subject of considerable research activity (Malpica et al. 2002; Pugachev et al. 2004) and there is a growing body of data on how generation times vary within and among viruses (Markowitz et al. 2003; Whalley et al. 2001). However, despite a wealth of population genetic theory about the interplay between these factors and natural selection, we are still some way from producing an all-encompassing picture of what determines rates of evolutionary change in viruses. This is in part due to a lack of detailed and comprehensive experimental measurements of the critical parameters and partly because of the limited complexity of available analytical models when compared to the reality of viral evolution.

Moreover, not all RNA and DNA viruses neatly fit the picture of rapid and slow evolution, respectively. The most dramatic example is that of simian foamy virus (SFV), which has co-diverged with its primate hosts over many million of years and evolved at the remarkably low rate of approximately 10^{-8} substitutions per site, per year, similar to that seen in that in host mitochondrial DNA (Switzer et al. 2005). The most likely explanation for this low rate of change is a combination of greatly reduced rates of replication and strong purifying selection, so that the vast majority of mutations that arise are deleterious and selectively eliminated. At the other end of the evolutionary spectrum, there is growing evidence that single-stranded (ss) DNA viruses evolve at rates approaching those seen in RNA viruses, most likely because of high intrinsic error rates (Sanz et al. 1999; Shackleton et al. 2005).

Although mutation provides the raw materials for evolutionary change, it is the dual processes of genetic drift and natural selection that are the ultimate arbiters over what genetic variation remains in the long term. Because all viruses have an absolute dependence on host cellular machinery for a productive life cycle, the interaction between viral proteins and the cellular receptors of host cells makes up a large part of the viruses' fitness landscape. We would therefore expect the interaction between specific viral sequences and cellular receptors to be of particular importance in determining why some RNA viruses are more often associated with cross-species transmission than others (Baranowski et al. 2001). For example, avian influenza A viruses are usually unable to evolve human-to-human transmission because they lack the correct amino acids in a number of viral proteins (see the chapter by Webby et al., this volume). Most attention has been directed toward the viral haemagluttinin (HA), which requires specific amino acids to interact with the sialic receptors on human cells in the correct configuration (Scholtissek et al. 1993; see Sect. 4). The intimacy of the relationship between virus and cell receptor also predicts that generalist viruses, which infect a broad range of cellular receptors, are more able to cross species boundaries than specialist viruses that have a narrower tropism. Strikingly, provisional analyses suggest that this is indeed the case, as viruses that utilize conserved cell receptors are more able to jump species boundaries than viruses that use divergent cell receptors (Woolhouse 2002). If validated with more data, this result is of great importance because it allows predictions to be made as to the types of virus that are most likely to emerge in the future.

Another factor that might influence the ability of viruses to cross species boundaries is the mode of transmission. It is easy to imagine how certain types of transmission mechanism—particularly respiratory and vector-borne transmission—might more readily facilitate viral emergence than others. In both these cases, the probability of exposure to an emergent virus is relatively high compared to viruses that are blood-borne or sexually transmitted. Indeed, it is striking that many of the viruses described in the earliest lists of emerging viruses were transmitted by mosquito vectors (the arboviruses), which could potentially take blood meals from a range of mammalian hosts. However, although an increased probability of exposure will generally translate into an increased likelihood of emergence (see the chapters by Childs et al. and Real and Biek, this volume), there are a number of other factors that this simple calculus omits. For example, there appears to be an association between mode of transmission and the ability of a virus to successfully replicate in the cells of a new host species. This is best documented with the arboviruses where there is strong evidence from both comparative and in vitro studies that the necessity of replicating in very different host species, in this case arthropods and mammals, imposes strong constraints against sequence change (Holmes 2003b; Woelk and Holmes 2002; Zárate and Novella 2004). This effect is most likely attributable to an antagonistic fitness trade-off, such that mutations that increase fitness in one host species reduce it in another. Hence, the majority of amino acid changes that arise in either host are deleterious (or slightly deleterious) and eventually removed by purifying selection. In particular, it seems especially difficult to establish productive infections in insect cells (Zárate and Novella 2004), which perhaps explains why vector species seems to be a key correlate of evolution in some animal and plant RNA viruses (Gaunt et al. 2001; Chare and Holmes 2004). However, perhaps the most striking of all observations in this context is that although arboviruses are frequently associated with sporadic disease in humans, few are able to sustain long-term transmission networks and dead-end infections are commonplace (M.E.J. Woolhouse, personal communication). Hence, the intricate adaptations to replicate in divergent hosts may act to prevent many arboviruses from successful emergence in new host species. Whether similar constraints apply to viruses transmitted by other mechanisms is unknown.

3 Are Viruses from Phylogenetically Related Host Species More Likely to Experience Cross-Species Transmission?

Central to our discussion so far has been the assumption that nearly all emerging viruses have jumped to humans from another animal species, in a process of cross-species transmission. Indeed, one of the great successes of molecular epidemiology has been the identification, often very rapidly, of the reservoir species for a myriad of human viruses. The rapid discoveries of the nonhuman primate ancestry of the human immunodeficiency virus (HIV) (Huet et al. 1990), and of some bat species as the ultimate reservoir of SARS coronavirus (SARS-CoV) (Lau et al. 2005; Li et al. 2005), serve as important illustrations. However, there are exceptions. Perhaps the most notable are the Ebola virus, where many thousands of animal specimens have been surveyed in Africa without certain identification of the reservoir species (Breman et al. 1999; Peterson et al. 2004), although bats have been recently implicated (see the chapter by Gonzalez et al., this volume) and hepatitis C virus, cause of one of the most prevalent somewhat new diseases to be identified in humans but where no close relatives have been discovered (Simmonds 2004). However, it is likely that with an increased sampling of taxa the species reservoirs for these viruses will also be determined, if they still exist today.

The next question that arises is whether some viruses are more able to jump species boundaries than others. The most compelling idea in this context is that there are phylogenetic constraints to this process, such that the more closely related the host species in question, the greater the chance of successful cross-species transmission (DeFilippis and Villarreal 2000). This theory is supported by some broad-brush observations. In particular, there is no evidence that the viruses that infect humans come from organisms as divergent as plants, fish, reptiles, or amphibians (Holmes and Rambaut 2004), even though in some cases, such as plant viruses, exposure might occur on a regular basis through the consumption of infected food. Rather, the majority of human viruses are of mammalian origin, with an occasional few coming from birds. Moreover, although insect viruses often infect human populations (that is, the arboviruses), these always jump from another mammalian species rather than directly from insects and, as described above, tend to cause dead-end infections in new hosts (see the chapter by Nel and Rupprecht, this volume, for discussion of the origin of rhabdoviruses infecting mammals from a potential insect source).

A more revealing question is whether there are any phylogenetic trends with respect to the mammalian viruses that also affect humans. Specifically, are those viruses from our closest relatives, the simian primates, more able to infect us than those from other mammalian orders? At present there is insufficient data to fully test this hypothesis, although it is clearly a research priority for the future. Additionally, it is difficult to fully disentangle probability of transmission from probability of exposure; for example, although we are clearly more closely related to other primates than to rodents, the global human population is more often exposed to the latter. There are, however, some tentative pieces of evidence to suggest that primate viruses are especially able to infect us as predicted from our close evolutionary relationship. As well as the obvious cases of HIV-1 and HIV-2, whose ultimate origins lie with chimpanzees (*Pan troglodytes troglodytes*) and sooty mangabey monkeys (*Cercocebus torquatus atys*), respectively, a variety of other major human viruses seem to have their origins in nonhuman primates. These include dengue virus, yellow fever virus, GB viruses A and C, hepatitis B virus, and HTLV-I and II. That some of these viruses seem to have appeared relatively recently in humans may be a consequence of changing ecological factors, most notably deforestation and linked activities, that have increased the rate of contact between humans and other primates (although an absence of retrospective diagnoses makes it difficult to determine whether emerging viruses are more common now than at previous times in our evolutionary history). Turning the tables, there are also examples of humans transmitting their viruses to other primates (often with serious consequences), as appears to be the case for measles (Ferber 2000) and TTV (Okamoto et al. 2000). This will doubtless be a continuing problem, as perhaps will be the movement of viruses from the populations of industrialized nations to indigenous peoples with naïve genetic backgrounds.

There are also good mechanistic reasons for believing that there is a relationship between phylogenetic distance and the likelihood of viral emergence. In particular, if, as argued above, the ability to recognize and infect host cells is a key component of cross-species transmission, then phylogenetically related host species are more likely to share related cell receptors. Given the pace at which RNA viruses evolve, it is easy to see that highly dependent relationships between viruses and cell receptors will be quickly established, so that the probability of successful cross-species transmission will decrease with increasing phylogenetic distance. If true, this theory further predicts that the more slowly evolving DNA viruses should initially be able to jump wider phylogenetic boundaries but, when they do adapt to their host, will eventually find it much more difficult to make subsequent species jumps. This pattern may partly explain the tendency for slower-evolving DNA viruses to co-diverge rather than move horizontally among species (Holmes 2004).

There are, however, some complicating factors to this simple phylogenetic rule. The issue of exposure vs transmissibility has been discussed above. There are also numerous exceptions to the phylogenetic trend. In particular, a large number of the emerging viruses of humans appear to have arisen from rodents rather than primates (M.E.J. Woolhouse, personal communication). This implies that the high density of many rodent populations allows them to carry a greater diversity of pathogens and/or that rodents often live in close proximity to humans which increases the probability of exposure (see the chapters by Gonzalez et al. and Klein and Calisher, this volume). Another factor of potential importance is phylogenetically related immune responses. Specifically, closely related host species, such as humans and other primates, are also likely to share the alleles that determine immune responses to specific pathogens. This has been particularly well documented for the major histocompatibility complex (MHC) group of loci, in which certain allelic lineages have persisted for millions of years of evolutionary history (Figueroa et al. 1988). Consequently, although a species might be exposed to a novel pathogen, they might, through a combination of shared common ancestry and good fortune, already possess a sufficient immune response to prevent the infection from being established.

4
Does Emergence Require Adaptation Within the New Host Species?

Understandably, most definitions of emerging viruses focus on the issue of disease. This means that no distinction is drawn between those viruses that spread efficiently among us, and those that only cause sporadic disease, often with no human-to-human transmission. Indeed, it seems that many, if not the majority, of the emerging diseases of humans represent dead-end infections. This may represent the natural background dynamics of cross-species transmission.

For example, almost all avian-to-human transmissions of influenza A virus result in dead-end infections, yet occasional avian-to-human transmission can cause pandemics. Each time a cross-species transmission occurs, there is a small chance that it will take hold. The problem is identifying and quantifying the key factors that determine whether a particular initial infection will survive and grow into a full-fledged epidemic. To understand emergence, it is therefore crucial to understand why only some viruses are able to regularly establish long-term transmission networks (see the chapter by Childs et al., this volume).

Perhaps the central question in this respect is whether, following cross-species transmission, emergent viruses must adapt to replicate in their new species, or whether the process of emergence is essentially blind to natural selection? Arguments can be advanced on both sides and there is currently little good data to choose among them. For example, one model of viral emergence posits that adaptation to a new host species during the early period of an epidemic is of fundamental importance, because this raises the basic reproductive rate of the virus, R_0, to greater than 1, so that sustained transmission networks can be established (Anita et al. 2003). This adaptive process is thought to occur during the "stuttering chains of transmission" that might characterize the early stages of an epidemic (Anita et al. 2003). Hence, those viruses that have not evolved human-to-human transmission are simply those that have not yet fully adapted to our species as $R_0 <1$ (and human-to-human transmission would surely be favored by natural selection because it increases the number of secondary infections). Empirical evidence for this theory comes from one of the best-documented cases of emergence, that of the carnivore parvoviruses (ssDNA viruses). In this case, the feline parvoviruses that infected cats jumped to dogs in the early 1970s, therein giving rise to the canine parvoviruses, an event that was accompanied by strong positive selection and an extremely high rate of nucleotide substitution (Shackelton et al. 2005). Direct adaptation to a new host species also seems to have been central to the emergence of the Venezuelan equine encephalitis virus (Brault et al. 2002). Further, although there is no strong evidence to date that the cross-species transmission event from dengue virus in monkeys to dengue virus in humans involved adaptive evolution in the latter (Twiddy et al. 2002), experimental studies imply that adaptation to the principal vector of dengue virus in an urban human setting, the *Aedes aegypti* mosquito, is a crucial prerequisite for sustained human transmission (Moncayo et al. 2004).

An alternative model for viral emergence is that rather than the emergent virus adapting to the new host species following exposure, successful emergence will only occur if a virus that already possesses the necessary mutations (such as those for receptor-binding) is exposed to the recipient host. In other words, successful emergent strains are those that are in some sense *preadapted* to establish productive infections in the new host species (see the chapter by Childs et al., this volume), so that the probability of emergence then becomes a function of the frequency of exposure. Indeed, that the majority of emerging infections (in humans at least) result in dead-end infections implies that even short-term transmission chains are difficult to establish for most viruses because they lack the necessary mutations. Moreover, for the majority of emergent viruses it has been difficult to show that cross-species transmission is associated with adaptation in the recipient host. To take two high-profile examples, although some sequence analyses suggest that SARS-CoV was subject to adaptive evolution during its early spread through humans (Yeh et al. 2004), it is unclear whether this was adaptation to the new host or selection for immune escape. Similarly, while the transition from SIVcpz in chimpanzees to HIV in humans seems to have been associated with a change in selection pressure (Sharp et al. 2001), it is unclear whether this reflects adaptive evolution or a relaxation of selective constraints. Finally, viral exposure to hosts of the right genetic configuration may also be of critical importance in the establishment of new infections. For example, it might be that a particular host HLA type is a more willing recipient of an emergent virus than another. In these circumstances, it is the particular combination of viral sequence and host immune system that is necessary to start a successful infection.

The case of influenza A virus again provides a highly illustrative example (see the chapter by Webby et al., this volume). Central (although not sufficient) to whether this virus is able to productively infect hosts are the sialic acid cell receptors found on cell-surface oligosaccharides. All avian influenza viruses replicate in the gastrointestinal tract and bind to sialic acid in a α 2,3-linkage to galactose. In contrast, human influenza viruses replicate in the respiratory tract, producing the distinctive disease symptoms, and bind to sialic acid in a α 2,6-linkage. Hence, the shift from α 2,3- to α 2,6-linkage is critical in enabling the switch from birds to humans and often involves changes at two amino acid residues, although mutations in other genes also play key roles (Matrosovich et al. 1997; Taubenberger et al. 2005). The key question, therefore, is whether these mutations appear de novo in humans, in the short transmission networks of people who initially suffer avian influenza, or whether they preexist in the avian population, and if the appropriate strain is transmitted, then will emergence follow? Again, there is little data to determine the relative importance of these two aspects of cross-species transmission dynamics, although the population size of influenza virus in avian species must be orders of magnitude greater than that in the handful of human cases during most outbreaks.

We suggest that three important advances are required to fully elucidate the role of adaptive evolution in viral emergence. First, improvements are needed in the analytical methods available to measure selection pressures acting on genes. The methods most commonly used at present involve comparisons of the numbers of synonymous (d_S) and nonsynonymous (d_N) substitutions per site. Although informative, these methods are highly conservative and are greatly limited when detecting positive selection at sporadic amino acid sites along a single lineage, which may be the form of adaptive evolution most often associated with viral emergence. Second, despite ever-expanding sequence databases, there are surprisingly few examples where viral sequence data is available from both donor and recipient species. As a case in point, although dengue is one of the most important emerging viruses of humans and a multitude of sequence data from humans are readily available, only a handful of samples have been collected from the most likely donor species, Old World monkeys (Wang 2000).

Third, models of evolution need to be developed that take into account not only the varying selective environment in which most viral pathogens exist but also that accurately reflect the often complex life cycle of emerging viruses. These models will probably not be tractable by analytical techniques and this will mean that computationally intensive simulation studies will become an increasingly important component of research into the evolutionary genetics of emerging disease.

5
Is Recombination a Prerequisite for Viral Emergence?

Although the engine of RNA virus evolution is undoubtedly their high mutation rate, there is mounting evidence that the genetic variability observed in RNA virus populations can be shaped, in part, by recombination. Further, because recombination is a process that potentially increases fitness by creating advantageous genotypes and removing deleterious mutations, it might also be supposed that it can assist the process of emergence. This is perhaps best shown in the case of the primate lentiviruses, such as HIV, which not only experience extremely high rates of recombination, with multiple template-switching events occurring during each replication cycle (Jung et al. 2002), but where recombinant viruses seem to be associated with many cases of cross-species transmission (Bailes et al. 2003). Similarly, the cross-species transmission of influenza A virus from birds to human is often associated with reassortment among hemagglutinin (HA) and neuraminidase (NA) subtypes (Webby and Webster 2001; see the chapter by Webby et al., this volume).

A more recent case in point concerns the emergence of SARS-CoV. In this case, it has been argued that human SARS-CoV is a recombinant of avian and other mammalian coronaviruses (Stavrinides and Guttman 2004), and that recombination may even have allowed the virus to acquire the critical suite of amino acid changes required to cause infection in humans (Stanhope et al. 2004). However, a closer inspection of the pertinent sequence data casts serious doubt on this hypothesis (see the chapter by Wang and Eaton, this volume). First, the evidence for recombination in SARS coronavirus is weak at best, and it seems equally likely that the phylogenetic signal said to support recombination results from variation in substitution rate among lineages, such that different genes produce slightly different trees (Gibbs et al. 2004; Holmes and Rambaut 2004). Second, the proposed recombination in SARS-CoV would have involved such distantly related virus strains that it cannot be responsible for the very recent emergence of the virus in humans (Holmes and Rambaut 2004).

Another reason to doubt the role played by recombination in emergence in general is that, other than in the retroviruses, recombination is not a particularly common process in RNA viruses and there is no reason to suppose that it is any more than a mechanistic by-product. For example, recombination appears to be extremely rare in negative-sense RNA viruses (Chare et al. 2003), most likely because their RNA is always encapsidated, thereby greatly limiting the template-switching thought to be central to RNA recombination. As a number of emerging viruses have negative-sense RNA genomes, this automatically argues against recombination as a general process in viral emergence. Similarly, although recombination is more common in positive-sense RNA viruses, in most cases it appears to be a sporadic event that does not occur at a high enough frequency to make it a key evolutionary strategy, although, of course, rare events like recombination may sometimes be critical in kick-starting the process of viral emergence.

Most of the available evidence suggests that recombination rates in RNA viruses are controlled by two factors; the ability of the virus in question to undergo template switching and the frequency with which multiple infections occur. It is these factors that explain why HIV has such a high rate of recombination; the virus possesses two copies of the RNA genome, which means that template switching will occur readily, and the ease at which the virus has spread worldwide means that multiple infections are abundant. However, HIV appears to be the exception rather than the rule. For example, HTLV is a retrovirus like HIV and is also reported to have jumped species boundaries (between humans and other primates) during its evolutionary history. However, there is no evidence for recombination in HTLV. In fact, even in the case of HIV it is not certain that recombination has been critical to emergence, despite its frequency. Overall, as the rate of recombination, per base, will be very much lower than that of mutation for most RNA viruses, it seems logical to conclude that recombination is not a key requirement for emergence, but rather a happy coincidence.

6
Conclusions: Evolution and Emergence in RNA Viruses

Much of this chapter has explored the issue of the role played by viral evolution in the process of emergence. The result, contrary to many perceptions of the inherent adaptability of RNA viruses, is that successful emergence, characterized by sustained human-to-human transmission may be a far more difficult process than might be expected given the remarkable rates at which RNA viruses mutate. Why might this be so? One probable answer lies in the theory that the genomes of RNA viruses are so small (usually less than 15 kb in length), and so multifunctional, that most mutations are likely to affect some critical aspect of virus biology (Holmes 2003a). As such, although mutations are abundant in RNA viruses, the vast majority are deleterious, or slightly deleterious, and will reduce viral fitness in the long term. Much of the genetic variation sampled within RNA virus populations, including those of emerging viruses in new host species, will therefore consist of short-lived deleterious mutations, which can be thought of as a form of mutational straightjacket. Unlike many other evolutionary systems that are dominated by a drift–selection balance, viral evolutionary genetics may be dominated by a mutation–selection balance (Domingo and Holland 1997). The requirements of a compact and highly pleiotropic genome, coupled with a high error rate, lead to a high mutational load (Elena and Moya 1999) that leaves little room for the limitless adaptability some have attributed to RNA viruses. Indeed, mutation rates are so high that it is possible that even highly beneficial mutations will not be able to spread through a viral population because they are, by chance, linked to deleterious mutations that arise in the same genome. This will evidently hinder their ability to emerge in new hosts and does much to explain why some RNA viruses are better able to cross species barriers than others. As such, we propose that the true importance of the nearly neutral (or slightly deleterious) theory of molecular evolution (Ohta 1992, 1998) to the study of RNA viruses has not been fully appreciated, yet may be crucial to fully understanding emergence.

Acknowledgements We thank The Wellcome Trust for financial support.

References

Antia R, Regoes RR, Koella JC, Bergstrom CT (2003) The role of evolution in the emergence of infectious diseases. Nature 426:658–610

Bailes E, Gao F, Bibollet-Ruche F, Courgnaud V, Peeters M, Marx PA, Hahn BH, Sharp PM (2003) Hybrid origin of SIV in chimpanzees. Science 300:1713

Baranowski E, Ruiz-Jarabo CM, Domingo E (2001) Evolution of cell recognition by viruses. Science 292:1102–1105

Brault AC, Powers AM, Holmes EC, Woelk CH, Weaver SC (2002) Positively charged amino acid substitutions in the E2 envelope glycoprotein are associated with the emergence of Venezuelan equine encephalitis virus. J Virol 76:1718–1730

Breman JG, Johnson KM, van der Groen G, Robbins CB, Szczeniowski MV, Ruti K, Webb PA, Meier F, Heymann DL (1999) A search for Ebola virus in animals in the Democratic Republic of the Congo and Cameroon: ecologic, virologic, and serologic surveys, 1979–1980. J Infect Dis 179:S139–S147

Chare ER, Holmes EC (2004) Selection pressures in the capsid genes of plant RNA viruses reflect mode of transmission. J Gen Virol 85:3149–3157

Chare ER, Gould EA, Holmes EC (2003) Phylogenetic analysis reveals a low rate of homologous recombination in negative-sense RNA viruses. J Gen Virol 84:2691–2703

Cleaveland S, Laurenson MK, Taylor LH (2001) Diseases of humans and their domestic mammals: pathogen characteristics, host range and the risk of emergence. Phil Trans R Soc Lond B 356:991–999

DeFilippis VR, Villarreal LP (2000) An introduction to the evolutionary ecology of viruses. In: Hurst CJ (ed) Viral ecology. Academic Press, New York, pp 126–208

Domingo E, Holland JJ (1997) RNA virus mutations for fitness and survival. Ann Rev Microbiol 51:151–178

Elena SF, Moya A (1999) Rate of deleterious mutation and the distribution of its effects on fitness in vesicular stomatitis virus. J Evol Biol 12:1078–1088

Ferber D (2000) Human diseases threaten great apes. Science 289:1277–1278

Figueroa F, Günther E, Klein J (1988) MHC polymorphism pre-dating speciation. Nature 355:265–267

Gaunt MW, Sall AA, de Lamballerie X, Falconar AKI, Dzhivanian TI, Gould EA (2001) Phylogenetic relationships of flaviviruses correlate with their epidemiology, disease association and biogeography. J Gen Virol 82:1867–1876

Gibbs AJ, Gibbs MJ, Armstrong JS (2004) The phylogeny of SARS coronavirus. Arch Virol 149:621–624

Gillespie JH (1998) Population genetics: a concise course. Johns Hopkins University Press, Baltimore

Holmes EC (2003a) Error thresholds and the constraints to RNA virus evolution. Trends Microbiol 11:543–546

Holmes EC (2003b) Patterns of intra- and inter-host nonsynonymous variation reveal strong purifying selection in dengue virus. J Virol 77:11296–11298

Holmes EC (2004) The phylogeography of human viruses. Mol Ecol 13:745–756

Holmes EC, Rambaut A (2004) Viral evolution and the emergence of SARS coronavirus. Phil Trans R Soc Lond B 359:1059–1065

Huet T, Cheynier R, Meyerhans A, Roelants G, Wain-Hobson S (1990) Genetic organisation of a chimpanzee lentivirus related to HIV-1. Nature 345:356–359

Jung A, Maier R, Vartanian JP, Bocharov G, Jung V, Fischer U, Meese E, Wain-Hobson S, Meyerhans A (2002) Recombination: multiply infected spleen cells in HIV patients. Nature 418:144

Lau SK, Woo PC, Li KS, Huang Y, Tsoi HW, Wong BH, Wong SS, Leung SY, Chan KH, Yuen KY (2005) Severe acute respiratory syndrome coronavirus-like virus in Chinese horseshoe bats. Proc Natl Acad Sci U S A 102:14040–14045

Li W, Shi Z, Yu M, Ren W, Smith C, Epstein JH, Wang H, Crameri G, Hu Z, Zhang H, Zhang J, McEachern J, Field H, Daszak P, Eaton BT, Zhang S, Wang LF (2005) Bats are natural reservoirs of SARS-like coronaviruses. Science 310:676–679

Malpica MJ, Fraile A, Moreno I, Obies CI, Drake JW, García-Arenal F (2002) The rate and character of spontaneous mutation in an RNA virus. Genetics 162:1505–1511

Markowitz M, Louie M, Hurley A, Sun E, Di Mascio M, Perelson AS, Ho DD (2003) A novel antiviral intervention results in more accurate assessment of human immunodeficiency virus type 1 replication dynamics and T-cell decay in vivo. J Virol 77:5037–5038

Matrosovich MN, Gambaryan AS, Teneberg S, Piskarev VE, Yamnikova SS, Lvov DK, Robertson JS, Karlsson KA (1997) Avian influenza A viruses differ from human viruses by recognition of sialyloligosaccharides and gangliosides and by a higher conservation of the HA receptor-binding site. Virology 233:224–234

Moncayo AC, Fernandez Z, Ortiz D, Diallo M, Sall A, Hartman S, Davis CT, Coffey L, Mathiot CC, Tesh RB, Weaver SC (2004) Dengue emergence and adaptation to peridomestic mosquitoes. Emerg Infect Dis 10:1790–1796

Morse SS (1995) Factors in the emergence of infectious diseases. Emerg Infect Dis 1:7–15

Moya A, Holmes EC, González-Candelas F (2004) The population genetics and evolutionary epidemiology of RNA viruses. Nat Rev Microbiol 2:279–287

Nichol ST, Spiropoulou CF, Morzunov S, Rollin PE, Ksiazek TG, Feldmann H, Sanchez A, Childs J, Zaki S, Peters CJ (1993) Genetic identification of a hantavirus associated with an outbreak of acute respiratory illness. Science 262:914–917

Ohta T (1992) The nearly neutral theory of molecular evolution. Ann Rev Ecol Syst 23:263–286

Ohta T (1998) Evolution by nearly-neutral mutations. Genetica 102/103:83–90

Okamoto H, Fukuda M, Tawara A, Nishizawa T, Itoh Y, Hayasaka I, Tsuda F, Tanaka T, Miyakawa Y, Mayumi M (2000) Species-specific TT viruses and cross-species infection in nonhuman primates. J Virol 74:1132–1139

Peterson AT, Carroll DS, Mills JN, Johnson KM (2004) Potential mammalian filovirus reservoirs. Emerg Infect Dis 10:2073–2081

Pugachev KV, Guirakhoo F, Ocran SW, Mitchell F, Parsons M, Penal C, Girakhoo S, Pougatcheva SO, Arroyo J, Trent DW, Monath TP (2004) High fidelity of yellow fever virus RNA polymerase. J Virol 78:1032–1038

Sanz AI, Fraile A, Gallego JM, Malpica JM, García-Arenal F (1999) Genetic variability of natural populations of cotton leaf curl geminivirus, a single-stranded DNA virus. J Mol Evol 49:672–681

Scholtissek C, Ludwig S, Fitch WM (1993) Analysis of influenza A virus nucleoproteins for the assessment of molecular genetic mechanisms leading to new phylogenetic virus lineages. Arch Virol 131:237–250

Shackelton LA, Parrish CR, Truyen U, Holmes EC (2005) High rate of viral evolution associated with the emergence of canine parvoviruses. Proc Natl Acad Sci U S A 102:379–384

Sharp PM, Bailes E, Chaudhuri RR, Rodenburg CM, Santiago MO, Hahn BH (2001) The origins of acquired immune deficiency syndrome viruses: where and when? Phil Trans R Lond B 356:867–876

Stanhope MJ, Brown JR, Amrine-Madsen H (2004) Evidence from the evolutionary analysis of nucleotide sequences for a recombinant history of SARS-CoV Infect. Genet Evol 4:15–19

Stavrinides J, Guttman DS (2004) Mosaic evolution of the severe acute respiratory syndrome coronavirus. J Virol 78:76–82

Switzer WM, Salemi M, Shanmugam V, Gao F, Cong M-E, Kuiken C, Bhullar V, Beer B, Vallet D, Gautier-Hion A, Tooze A, Villinger F, Holmes EC, Heneine W (2005) Ancient co-speciation of simian foamy viruses and primates. Nature 434:376–380

Taubenberger JK, Reid AH, Lourens RM, Wang R, Jin G, Fanning TG (2005) Characterization of the 1918 influenza virus polymerase genes. Nature 437:889–893

Twiddy SS, Farrar JF, Chau NV, Wills B, Gould EA, Gritsun T, Lloyd G, Holmes EC (2002) Phylogenetic relationships and differential selection pressures among genotypes of dengue-2 virus. Virology 298:63–72

Wang E, Ni H, Xu R, Barrett ADT, Watowich SJ, Gubler DJ, Weaver SC (2000) Evolutionary relationships of endemic/epidemic and sylvatic dengue viruses. J Virol 74:3227–3234

Whalley SA, Murray JM, Brown D, Webster GJ, Emery VC, Dusheiko GM, Perelson AS (2001) Kinetics of acute hepatitis B virus infection in humans. J Exp Med 193:847–854

Webby RJ, Webster RG (2001) Emergence of influenza A viruses. Phil Trans R Soc Lond B 356:1817–1828

Woelk CH, Holmes EC (2002) Reduced positive selection in vector-borne RNA viruses. Mol Biol Evol 19:2333–2336

Woolhouse MEJ, Taylor LH, Haydon DT (2001) Population biology of multihost pathogens. Science 292:1109–1112

Woolhouse MEJ (2002) Population biology of emerging and re-emerging pathogens. Trends Microbiol 10:S3–S7

Yeh S-H, Wang H-Y, Tsai C-Y, Kao C-L, Yang J-Y, Liu H-W, Su I-J, Tsai S-F, Chen D-S, Chen P-J, and the National Taiwan University SARS Research Team (2004) Characterization of severe acute respiratory syndrome coronavirus genomes in Taiwan: Molecular epidemiology and genome evolution. Proc Natl Acad Sci U S A 101:2542–2547

Zárate S, Novella IS (2004) Vesicular stomatitis virus evolution during alternation between persistent Infection in insect cells and acute infection in mammalian cells Is dominated by the persistence phase. J Virol 78:12236–12242

CTMI (2007) 315:67–83
© Springer-Verlag Berlin Heidelberg 2007

Influenza Viruses in Animal Wildlife Populations

R. J. Webby[1](✉) · R. G. Webster[2] · J. A. Richt[3]

[1]Department of Infectious Diseases, St. Jude Children's Research Hospital, 332 N. Lauderdale, Memphis, TN 38105-2794, USA
richard.webby@stjude.org

[2]Division of Virology, St. Jude Children's Research Hospital, Memphis, TN 38105-2794, USA

[3]Virus and Prion Diseases of Livestock Research Unit, National Animal Disease Center, USDA, ARS, Ames, IA 50010, USA

Disclaimer: Mention of trade names or commercial products in this article is solely for the purpose of providing specific information and does not imply recommendation or endorsement by the US Department of Agriculture.

Abstract Influenza viruses belong to the family *Orthomyxoviridae*. Genus Influenza A viruses are true zoonotic agents with many animal reservoirs, whereas genus Influenza B viruses are generally considered to be a virus of humans. The genome of influenza A viruses consists of eight unique segments of single-stranded RNA of negative polarity; they are typed according to their surface proteins, hemagglutinin (HA) and neuraminidase (NA). HA and NA, the major antigenic determinants of influenza A viruses, are present in 16 and nine serologic subtypes, respectively. Annual epidemics and occasional pandemics of influenza in humans depend on the continued evolution of influenza viruses. Although they have numerous potential host populations, most of our genetic and biologic data are obtained from studies of domestic populations of species such as chickens, turkeys, swine, and horses. Concerning wildlife populations, including wild populations of these domesticated species, much less is known. The purpose of this review is to establish what role wildlife populations play in the continued evolution of influenza viruses. Future work needs to determine what chain of events makes it possible for an influenza virus to be successfully transmitted *to*, and more importantly *within*, an alternative host population. Even questions as fundamental as which hosts can transmit viruses to humans remain unanswered so far.

1 Introduction

Annual epidemics and occasional pandemics of influenza in humans depend on the continued evolution of influenza viruses. Although they have numerous potential host populations, we obtain most of our detailed genetic and biologic data from study of domestic populations of species such as chickens, turkeys, swine, and horses. Concerning wildlife populations, including wild populations of these domesticated species, much less is known. The purpose of this review is to establish what role wildlife populations play in the continued evolution of influenza virus, which viruses are present in wild animal populations, and the methods by which they are perpetuated.

2 The Virus

Four influenza virus genera belong to the family Orthomyxoviridae: Influenza A, B, and C viruses, and Thogotovirus. Of these, only influenza A and B viruses cause appreciable amounts of disease in humans. Influenza B, despite having been found in seals (Osterhaus et al. 2000), is generally considered to be a virus of humans. In contrast, influenza A viruses are true zoonotic agents with many animal reservoirs. The genome of influenza A consists of segmented single-stranded RNA of negative polarity. The eight unique RNA segments within the viral genome encode ten (or in some cases 11) proteins (for review see Lamb and Krug 2001): segments 7 (M) and 8 (NS) encode two proteins due to differentially spliced transcripts, and in some strains segment 2 (PB1) encodes a second short protein from an additional open-reading frame. Influenza A viruses are typed according to their surface glycoproteins, hemagglutinin (HA) and neuraminidase (NA). HA and NA, the major antigenic determinants of influenza A viruses, are present in 16 and nine serologic subtypes, respectively.

Influenza viruses have developed two distinguishable mechanisms that allow them to change antigenically. Antigenic drift, which has been observed in influenza A and B viruses, results from the low fidelity of the virally encoded RNA polymerase. The HA molecule can maintain its functionality while allowing numerous amino acid substitutions at its antigenic sites. The humoral immunity built up within an exposed population selects for variant viruses that can evade neutralizing antibodies. This continual selection for antigenically novel virus variants allows influenza to reappear seasonally within populations that have been exposed in previous years. Antigenic shift, less frequent than antigenic drift, causes greater concern because it presents the greatest threat to human health. The term describes the emergence of a novel HA subtype within a population, either from the interspecies transfer of a whole virus from animal reservoirs (see Sect. 4) or through the process of genetic reassortment. The segmented nature of the influenza genome allows doubly infected cells to produce reassortant viruses. The progeny viruses can inherit a mix of genomic segments from both parental viruses; as a result, they can have an unpredictable phenotype and antigenicity. For this reason, animal populations are of vital importance to the evolution and emergence of human influenza. Because the reservoirs of virus in animals are an essential component of antigenic shift, an understanding of what viruses are present in different animal populations is crucial if we wish to find ways to reduce the threat to human health and life from influenza epidemics and potential pandemics.

3
Hosts of Influenza Virus

Although influenza A viruses infect numerous host species, including humans, swine, horses, dogs, felines, whales, seals, and various avian species, the aquatic bird populations of the world comprise their major reservoir. Influenza viruses have generally established stable host–pathogen relationships in their aquatic bird hosts and are assumed to be in evolutionary stasis (Gammelin et al. 1990; Gorman et al. 1990, 1991, 1992; Kida et al. 1987; Suarez 2000; Webster et al. 1992). The host range of influenza A virus is tightly restricted; nevertheless, interspecies transfer does occur. Phylogenetic analysis indicates that all influenza viruses are linked ancestrally to viruses in the aquatic bird reservoir (Gammelin et al. 1990; Gorman et al. 1990) and that viruses from this reservoir sporadically cross species barriers and establish lineages in alternative species

(Fig. 1). As soon as the viruses are established in a new host, they evolve rapidly as they adapt to their new host. The incidence of interspecies transmissions from aquatic birds to alternative hosts is likely to be higher than documented but difficult to detect because of the lack of transmission within the new host. Indeed, studies by Shortridge and colleagues in China showed that members of a rural community had serologic evidence of exposure to a number of avian influenza strains (Shortridge 1992).

Influenza viruses have numerous possible host populations. However, it is from domestic species that we obtain most of the detailed genetic and biologic data concerning animal influenza virus. In contrast to the well documented

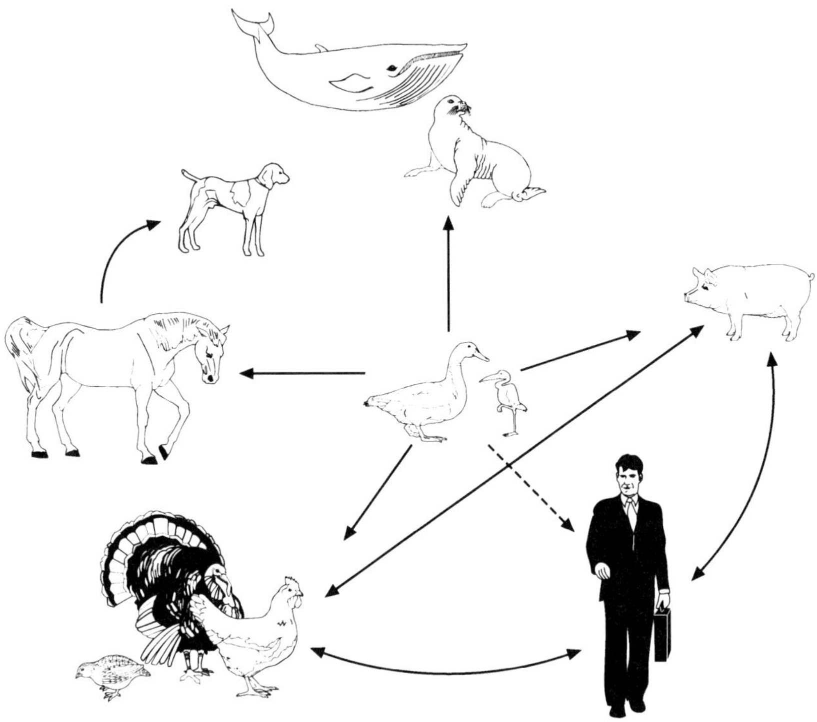

Fig. 1 Interspecies transmission of Influenza A viruses. Wild aquatic birds are the natural reservoir of influenza A viruses. It is from this reservoir that viruses sporadically transmit to other hosts and occasionally from that host to another. Although it is likely that we do not understand the full spectrum of interspecies virus flow, the *arrows* demonstrate the transmissions that have been documented

reports of influenza viruses in poultry populations, such as chicken and turkey, and in domestic swine and equine populations, information on influenza viruses in wildlife (including wild populations of the aforementioned) is scant (see chapters by Childs and by Stallknecht, this volume).

4
Influenza Virus in Wild Aquatic Birds

4.1
Ducks

The first conclusive pieces of evidence that migratory waterfowl were important hosts of influenza were presented in publications in 1967 and 1968 reporting that influenza virus antibodies were detected in wild birds (Easterday et al. 1968; Pereira et al. 1967). Shortly after, influenza viruses were isolated from a shearwater, a migratory shore bird (Downie et al. 1973), and from healthy wild ducks in California (Slemons et al. 1974). Of all wildlife hosts of influenza, aquatic birds, particularly ducks, are the most frequently studied. Duck populations represent one of the largest and most diverse reservoirs of influenza A viruses (Stallknecht 1999). Although different duck species host different influenza A viruses, we refer to ducks generically in this review. Nevertheless, it is important to note that duck species, differences in enzootic viruses aside, have different zoonotic potentials because variations in behavior, niche preferences, and flyways affect their interactions with the human population (Stallknecht and Shane 1988).

Typically, influenza virus produces asymptomatic infections in ducks, in which replication occurs primarily in the gastrointestinal tract (Webster et al. 1978). In this host, influenza virus produces an acute infection, and no available evidence suggests carrier states (Wang and Webster 1990), although experimentally infected ducks can shed virus for at least 22 days (Alexander et al. 1978). Several longitudinal studies of global duck populations have shown that most HA and NA subtypes circulate in this population, although their frequencies differ. In a 26-year study of ducks in Alberta, 12 of the 15 HA subtypes then known (Fouchier et al. 2005) were identified (Krauss et al. 2004). Of these, H3, H4, and H6 accounted for 76.8% of isolates. In other studies spanning more than two seasons, the following HA subtypes have dominated: H3 and H11 in 1986–1988 in Ohio (Slemons et al. 1991) and H3, H4, and H6 in 1998–2000 in Minnesota (Hanson et al. 2003). Studies in other regions of the world attest to the diversity of virus subtypes within the wild bird reservoirs and the variations in subtypes isolated in different flyways (Fouchier et al. 2003; De Marco et al. 2003; Ito et al. 1995; Shengqing et al. 2002; Suss et al. 1994).

The overall frequency of influenza virus isolations within different flyways also varies. In one study of waterfowl sampled in the marshalling areas of Alberta during August and September, 20% of juvenile birds were shedding influenza virus (Hinshaw et al. 1980). The percentage of those shedding virus drops to approximately 2% when birds are sampled in the lower Mississippi during November (Stallknecht et al. 1990; Webster et al. 1976) and to 0.45% when birds are sampled in Louisiana during January (Stallknecht et al. 1990). Although different scenarios have been suggested, the actual mechanisms of viral perpetuation and maintenance within these migratory populations remain largely unknown.

Wild waterfowl are generally thought to be reservoirs and donors of influenza virus rather than disseminators of virus, i.e., aquatic bird populations are the source of the viruses responsible for epidemics and pandemics in human and domestic animal populations, but their role in maintenance or spread of epidemic agents is thought to be limited, with one possible exception. In late 2003, scientists reported outbreaks of highly pathogenic H5N1 virus among avian species in several Southeast Asian countries (Sims et al. 2005; World Health Organization Global Influenza Program Surveillance Network 2005). Cases of human infection were subsequently reported in Vietnam, Thailand, Indonesia, China, Cambodia, Iraq, Turkey, Egypt, and Azerbaijan (for latest information see http://www.who.int/csr/disease/avian_influenza/en/). Migratory birds have been implicated in the regional dissemination of this virus, and although the evidence is convincing, their contribution has not yet been conclusively proven. H5N1 virus subtypes were isolated from diseased and dying wild waterfowl (geese, ducks, and swans), little egrets, grey herons, and captive greater flamingos in Hong Kong in 2002 (Ellis et al. 2004; Sturm-Ramirez et al. 2004). Many of these birds were overwintering in the public parks and not in direct contact with infected domesticated poultry. The presence of H5N1 subtypes in these species is intriguing because highly pathogenic avian influenza (HPAI) viruses are thought to emerge in domestic poultry species such as chickens and turkey and not in aquatic species. The low-pathogenic precursor avian influenza strains originate in the aquatic bird reservoirs and evolve into the highly pathogenic forms after interspecies transfer. Therefore, the deaths of ducks and other aquatic birds in Hong Kong during 2002 appear to have resulted from an interspecies transfer from domestic poultry back to wild waterfowl. HPAI infections in waterfowl have been previously documented, but only when ducks were near infected turkey and chicken populations (Mutinelli et al. 2003). Although the extent to which the 2002 H5N1 virus may have circulated among wild waterfowl is unclear, another outbreak occurred, this time among migratory waterfowl (mostly bar-headed geese) in western China's Qinghai Lake in 2005 (Chen et al. 2005). The lake is on a protected nature reserve, and no poultry were in the vicinity (Chen et al. 2005).

Although the exact origin of the virus from Qinghai Lake remains contentious, the presence of HPAI H5N1 in migratory birds concerns the scientific community because migratory birds may be able to disperse the virus over large geographic areas, indicating that the role of aquatic birds in the ecology of influenza may be changing. At the time of this writing, the role of migratory birds in the dispersal of HPAI H5N1 in Asia is still hypothetical, although scientists are surveying the area to determine the prevalence of this virus in these populations. The more recent spread of the H5N1 virus from Asia into Europe, Africa, and the Middle East adds further support to the role of migratory birds as disseminators. It has been correctly argued that dead ducks are poor disseminators of virus, but these arguments ignore the fact that not all of the contemporary H5N1 viruses are highly lethal in ducks (yet they retain their pathogenicity for chickens) (Sturm-Ramirez et al. 2004, 2005). It should also be considered that a virus that kills one species of migratory bird may not kill another.

Other evidence that influenza viruses can jump from domestic species back into wild aquatic birds comes from wild ducks shot in the United States in 2000. In 2003, Olsen et al. reported that an influenza virus had been isolated from pooled swab material from dead mallard and wood ducks (Olsen et al. 2003). The isolate, an H1N2 virus, was genetically related to a lineage of virus that had been circulating in the US swine population since at least 1999 (Karasin et al. 2000b). This virus lineage was formed from the reassortment of two different viruses endemic in swine (Karasin et al. 2000b, 2002), strongly suggesting that the virus was transmitted from swine back to ducks.

4.2 Shorebirds

Although less studied, shorebirds are also a major avian reservoir of influenza A virus. In a 16-year study of the US Eastern seaboard's Delmarva Peninsula, the recovery of influenza virus isolates from northward migrating shorebirds and gulls ranged from approximately 2% to 35% (Krauss et al. 2004). Species studied included red knots, ruddy turnstones, sanderlings, sandpipers, laughing gulls, and herring gulls. H3 and H11 subtypes predominated in this avian population, in which a greater variety of HA types was present than is usually the case in waterfowl. In contrast, Fouchier and colleagues (2003) failed to identify any influenza infections in Northern European shorebirds by using polymerase chain reaction detection or isolation techniques. They isolated strains from waterfowl and gulls. It is unclear whether the difference in isolation rates of North American and European shorebird populations represent a difference in species, susceptibilities to viruses, timing of sampling, or perhaps even differences in the viral populations themselves.

Considering a growing body of knowledge concerning the nature of the wild aquatic bird reservoirs of influenza [including ever-increasing amounts of genetic information (Hatchette et al. 2004; Spackman et al. 2005; Widjaja et al. 2004)] that we cannot predict those with zoonotic potential is somewhat sobering. What genetic features of a virus may enhance its ability to cross species barriers is unknown. Certainly differences between the receptor preference of avian and human viruses is likely a key feature (see below), but it is not the only one. The H5N1 viruses that infected limited numbers of humans in 1997 (Matrosovich et al. 1999) and in 2004–2005 (Hoffmann et al. 2005) retain the characteristic binding preferences of avian viruses. From cases of H5N1 infection in humans, we have learned that receptor preference does not absolutely preclude infection (although it might prevent efficient transmission). In contrast, H9N2 viruses that have been circulating widely in Asian poultry during the last decade have displayed receptor specificity like that of human viruses (Matrosovich et al. 2001), yet they have failed to successfully establish themselves in mammalian hosts despite sporadically infecting human and swine populations (Gou et al. 2000; Lin et al. 2000; Peiris et al. 2001; Shaw et al. 2002; Uyeki et al. 2002). Likewise, the H5N1 virus isolated from humans in 2003 (Peiris et al. 2004) displayed receptor specificities like those of human and avian viruses (Shinya et al. 2005) but was no more able to establish itself in humans than were other H5N1 viruses. More recent studies looking at the more defined specificities may help us understand these conundrums (Gambaryan et al. 2004; Iwatsuki-Horimoto et al. 2004).

5
Influenza Virus in Wild Swine

Although much is known about the components of influenza viruses in domestic swine populations, there is a paucity of information concerning the affect of the virus on feral swine. The ecology of influenza virus in domestic swine resembles that in humans, that is, they both have limited lineages of endemic viruses, and they have similar disease spectrums. Wild swine herds might be important viral reservoirs because of their potential for increased contact with wild avian species. Swine have been designated the "mixing vessel" (Scholtissek 1990) for reassortment of influenza viruses of avian and human origin due to their relatively high susceptibility to viruses from both sources (Brown 2000; Chambers et al. 1991; Kida et al. 1994). The receptor specificity of viruses from each host presents a major barrier to efficient transmission of avian influenza viruses to humans. All influenza viruses attach to oligosaccharides present on cell surface sialic acid, but avian and human viruses preferentially recognize different forms of these molecules. Although recent work is revealing more refined details (Gambaryan et al. 2005), influenza viruses isolated from aquatic birds preferentially bind to NeuAcα 2,3Gal-terminated receptors, whereas human viruses favor receptors terminated with NeuAcα 2,6Gal linkages. The relative abundance of these two sialic acid moieties in the human respiratory tract and the gastrointestinal tract of ducks is likely a key barrier to frequent interspecies transfer of influenza viruses between these hosts (Ito et al. 1998; Murphy et al. 1982; Couceiro et al. 1993). In contrast, swine contain both receptor types, a characteristic consistent with their susceptibility to avian and human influenza viruses (Ito et al. 1998).

In one study, domestic pigs experimentally inoculated with a high dose of virus were able to support the replication of at least one avian virus from subtypes H4–H13 (Kida et al. 1994). Despite their susceptibility to viruses of many subtypes, endemic influenza viruses in domestic swine herds are limited to the H1N1, H3N2, and H1N2 subtypes, although which viruses are isolated in swine herds varies on the basis of geographic area (for review see Brown 2000). Other subtypes sporadically appear (Choi et al. 2005; Karasin et al. 2000a; Peiris et al. 2001), although the extent to which they circulate among domestic swine is uncertain, and the infections appear to be transient. Although wild swine populations show evidence of seroconversion to influenza viruses, we do not know the effects of the disease and its extent in these animals. Reports of studies on wild swine in the United States and Europe suggest that these animal populations carry significant amounts of influenza virus. On the basis of hemagglutinin inhibition (HI) assays, 15 of 20 animals in a 1993–1994 study of feral swine in Kansas showed evidence of exposure to H1N1 viruses (Gipson et al. 1999). In a similar study of Oklahoma herds in 1996, seropositivity to H1N1 was 11% (Saliki et al. 1998). H1N1 seroprevalences of 4% and 24% have also been described in Spain (Vicente et al. 2002) and Poland (Markowska-Daniel and Pejsak 1999), respectively. Recently, our laboratory conducted a study on feral pigs from an isolated peninsula adjacent to the Atlantic Ocean in South Carolina. The population of feral swine in this area was estimated to be between 2,000 and 2,500 animals; no domestic swine operations are present on this peninsula. Sera from these animals (n = 178) were analyzed for evidence of influenza virus infections of the H3 and H1 subtypes. HI titers of ≤1:160 specific for the H1 subtype of swine influenza virus were found in 12% of the animals, whereas no positive HI reaction was found for the H3 subtype (J.A. Richt et al., unpublished results). These results indicate that feral pigs are a reservoir for influenza viruses significant to the US swine industry. Although valuable, these studies provide no information on the frequency of disease and, perhaps more importantly, no evidence for infection with avian strains of influenza. Certainly the former is difficult within such a study population, but the latter can be achieved.

6 Influenza Virus in Wild Turkeys

Turkeys are perhaps one of the most permissive domesticated hosts of influenza virus. Domestic turkeys are susceptible to infection with avian and swine viruses, and there are numerous reports of transmission of influenza viruses from wild aquatic birds to this host (for example, Campitelli et al. 2004; Halvorson et al. 1983; Sivanandan et al. 1991). If wild flocks of turkey are similarly susceptible, they may also be important wildlife hosts of influenza virus. Unfortunately, few investigators are studying this possibility. In a small number of studies conducted in the United States, little to no indication of influenza infection has been identified. In a study of 70 wild Rio Grande turkeys captured in 2001, investigators in Texas were unable to identify antibodies to influenza virus (Peterson et al. 2002). Likewise, none of the birds tested in two other studies showed evidence of previous exposure to influenza virus [44 in Arkansas (Hopkins et al. 1990); 210 in Eastern states in 1981–1986 (Davidson et al. 1988)]. In a Californian study of sera from 383 wild turkeys captured over a period of a decade (1986–1996), only one sample was positive for influenza virus (Charlton 2000). Taken together, these limited studies indicate that wild turkeys are not significant sources of influenza viruses. However, if wild turkeys are like their domestic counterparts and highly susceptible to transient epizootics of influenza, the true influenza burden may not be detectable by studies such as the ones described above, because of the small sample size and the time at which sampling is conducted.

7 Influenza Virus in Marine Mammals

Through virus isolation or serologic testing, influenza A virus subtypes have also been isolated from marine mammals: harbor (Geraci et al. 1982), harp (Stuen et al. 1994), hooded (Stuen et al. 1994), Baikal (Ohishi et al. 2004), ringed (Nielsen et al. 2001; Ohishi et al. 2004), and Caspian (Ohishi et al. 2002) seals, as well as walruses (Nielsen et al. 2001) and beluga whales (Nielsen et al. 2001). The isolated viruses—H7N7 (Geraci et al. 1982; Webster et al. 1981b), H4N5 (Hinshaw et al. 1984), H4N6 (Callan et al. 1995), and H3N3 (Callan et al. 1995)—have been of avian origin, although serologic evidence suggests that human viruses may also be able to infect seals (Ohishi et al. 2002, 2004).

Whales can also act as hosts of influenza. An H1N3 virus has been isolated from a striped whale (Lvov et al. 1978) and H13N2 and H13N9 viruses from a pilot whale (Hinshaw et al. 1986). Again, all isolates were closely related to viruses from avian species, suggesting that whales are susceptible to at least a subset of viruses from this reservoir. In this particular case, the viruses isolated from pilot whales were similar to those preferentially isolated from gulls, suggesting waterborne transmission in locations where these two species co-habitate. Alternatively, transmission by whales (and seals) preying on seabirds has also been touted as possible (Nielsen et al. 2001).

Although some infections in marine mammals have caused substantial disease, most investigators conclude that the epizootics have been relatively limited in extent, and that they are perpetuated by continual reinfection from avian and human sources rather than by continual circulation within the marine mammal population. The variability of influenza infection rates from study to study also supports this conclusion. The limited interaction between infected marine mammals and humans suggests that the pandemic threat posed by this reservoir is more limited than that of other wildlife. However, evidence of autopsy personnel contracting H7N7 conjunctivitis from infected seals provides a precedence for marine-mammal-to-human infection (Webster et al. 1981a).

8 Conclusion

From our examination of wildlife reservoirs of influenza A viruses, we can firmly conclude one thing: the ecology of the disease is complex. Distinct lineages of virus have been established in different animal reservoirs, and these lineages are occasionally transmitted between species. The challenge for the future is to determine what chain of events makes it possible for a virus to be successfully transmitted *to*, and more importantly *within*, an alternative host population. To claim that we know even a fraction of the mechanisms of interspecies transfer is optimistic. Even questions as fundamental as which hosts can transmit viruses to humans remain unanswered. That avian-to-human transmission of influenza viruses is limited had been interpreted to mean that aquatic bird viruses must spend some time in an intermediate mammalian host to gather the properties required for success in the human population. However, the direct avian-to-human transmission of H5N1 viruses in 1997 demonstrated that this is not the case and that domestic poultry species can also act as the link between aquatic birds and humans. Although results of studying the 2004–2006 H5N1 outbreaks now hint that intermediate hosts are not required and that viruses with the correct constellation of genes can pass directly from aquatic birds to humans, the studies are not complete, and results are inconclusive. Many data gaps exist, and they can only be addressed by further studies of influenza viruses of wildlife.

References

Alexander DJ, Allan WH, Parsons DG, Parsons G (1978) The pathogenicity of four avian influenza viruses for fowls, turkeys and ducks. Res Vet Sci 24:242–247

Brown IH (2000) The epidemiology and evolution of influenza viruses in pigs. Vet. Microbiol 74:29–46

Callan RJ, Early G, Kida H, Hinshaw VS (1995) The appearance of H3 influenza viruses in seals. J Gen Virol 76:199–203

Campitelli L, Mogavero E, De Marco MA, Delogu M, Puzelli S, Frezza F, Facchini M, Chiapponi C, Foni E, Cordioli P, Webby R, Barigazzi G, Webster RG, Donatelli I (2004) Interspecies transmission of an H7N3 influenza virus from wild birds to intensively reared domestic poultry in Italy. Virology 323:24–36

Chambers TM, Hinshaw VS, Kawaoka Y, Easterday BC, Webster RG (1991) Influenza viral infection of swine in the United States 1988–1989 Arch Virol 116:261–265

Charlton KG (2000) Antibodies to selected disease agents in translocated wild turkeys in California. J Wildl Dis 36:161–164

Chen H, Smith GJ, Zhang SY, Qin K, Wang J, Li KS, Webster RG, Peiris JS, Guan Y (2005) Avian flu: H5N1 virus outbreak in migratory waterfowl. Nature 436:191–192

Choi YK, Nguyen TD, Ozaki H, Webby RJ, Puthavathana P, Buranathal C, Chaisingh A, Auewarakul P, Hanh NT, Ma SK, Hui PY, Guan Y, Peiris JS, Webster RG (2005) Studies of H5N1 Influenza virus infection of pigs by using viruses isolated in Vietnam and Thailand in 2004. J Virol 79:10821–10825

Couceiro JN, Paulson JC, Baum LG (1993) Influenza virus strains selectively recognize sialyloligosaccharides on human respiratory epithelium; the role of the host cell in selection of hemagglutinin receptor specificity. Virus Res 29:155–165

Davidson WR, Yoder HW, Brugh M, Nettles VF (1988) Serological monitoring of eastern wild turkeys for antibodies to *Mycoplasma* spp. and avian influenza viruses. J Wildl Dis 24:348–351

De Marco MA, Foni GE, Campitelli L, Raffini E, Di TL, Delogu M, Guberti V, Barigazzi G, Donatelli I (2003) Circulation of influenza viruses in wild waterfowl wintering in Italy during the 1993–99 period: evidence of virus shedding and seroconversion in wild ducks. Avian Dis 47:861–866

Downie JC, Webster RG, Schild GC, Dowdle WR, Laver WG (1973) Characterization and ecology of a type A influenza virus isolated from a shearwater. Bull World Health Organ 49:559–566

Easterday BC, Trainer DO, Tumova B, Pereira HG (1968) Evidence of infection with influenza viruses in migratory waterfowl. Nature 219:523–524

Ellis TM, Bousfield RB, Bissett LA, Dyrting KC, Luk GS, Tsim ST, Sturm-Ramirez K, Webster RG, Guan Y, Malik Peiris JS (2004) Investigation of outbreaks of highly pathogenic H5N1 avian influenza in waterfowl and wild birds in Hong Kong in late (2002) Avian Pathol 33:492–505

Fouchier RA, Olsen B, Bestebroer TM, Herfst S, van der K L, Rimmelzwaan GF, Osterhaus AD (2003) Influenza A virus surveillance in wild birds in Northern Europe in 1999 and (2000) Avian Dis 47:857–860

Fouchier RA, Munster V, Wallensten A, Bestebroer TM, Herfst S, Smith D, Rimmelzwaan GF, Olsen B, Osterhaus AD (2005) Characterization of a novel influenza A virus hemagglutinin subtype (H16) obtained from black-headed gulls. J Virol 79:2814–2822

Gambaryan AS, Tuzikov AB, Pazynina GV, Webster RG, Matrosovich MN, Bovin NV (2004) H5N1 chicken influenza viruses display a high binding affinity for Neu5Acalpha2-3Galbeta1-4(6-HSO3)GlcNAc-containing receptors. Virology 326:310–316

Gambaryan A, Yamnikova S, Lvov D, Tuzikov A, Chinarev A, Pazynina G, Webster R, Matrosovich M, Bovin N (2005) Receptor specificity of influenza viruses from birds and mammals: new data on involvement of the inner fragments of the carbohydrate chain. Virology 334:276–283

Gammelin M, Altmuller A, Reinhardt U, Mandler J, Harley VR, Hudson PJ, Fitch WM, Scholtissek C (1990) Phylogenetic analysis of nucleoproteins suggests that human influenza A viruses emerged from a 19th-century avian ancestor. Mol Biol Evol 7:194–200

Geraci JR, St Aubin DJ, Barker IK, Webster RG, Hinshaw VS, Bean WJ, Ruhnke HL, Prescott JH, Early G, Baker AS, Madoff S, Schooley RT (1982) Mass mortality of harbor seals: pneumonia associated with influenza A virus. Science 215:1129–1131

Gipson PS, Veatch JK, Matlack RS, Jones DP (1999) Health status of a recently discovered population of feral swine in Kansas. J Wildl Dis 35:624–627

Gorman OT, Bean WJ, Kawaoka Y, Webster RG (1990) Evolution of the nucleoprotein gene of influenza A virus. J Virol 64:1487–1497

Gorman OT, Bean WJ, Kawaoka Y, Donatelli I, Guo YJ, Webster RG (1991) Evolution of influenza A virus nucleoprotein genes: implications for the origins of H1N1 human and classical swine viruses. J Virol 65:3704–3714

Gorman OT, Bean WJ, Webster RG (1992) Evolutionary processes in influenza viruses: divergence, rapid evolution, and stasis. Curr Top Microbiol Immunol 176:75–97

Gou Y, Xie J, Wang M (2000) A strain of influenza A H9N2 virus repeatedly isolated from human population in China. Zhonghua Shi Yan He Lin Chuang Bing Du Xue Za Zhi 14:209–212

Halvorson D, Karunakaran D, Senne D, Kelleher C, Bailey C, Abraham A, Hinshaw V, Newman J (1983) Epizootiology of avian influenza–simultaneous monitoring of sentinel ducks and turkeys in Minnesota. Avian Dis 27:77–85

Hanson BA, Stallknecht DE, Swayne DE, Lewis LA, Senne DA (2003) Avian influenza viruses in Minnesota ducks during 1998–2000. Avian Dis 47:867–871

Hatchette TF, Walker D, Johnson C, Baker A, Pryor SP, Webster RG (2004) Influenza A viruses in feral Canadian ducks: extensive reassortment in nature. J Gen Virol 85:2327–2337

Hinshaw V, Webster R, Turner B (1980) The perpetuation of orthomyxoviruses and paramyxoviruses in Canadian waterfowl. Can J Microbiol 26:622–629

Hinshaw VS, Bean WJ, Webster RG, Rehg JE, Fiorelli P, Early G, Geraci JR, St Aubin DJ (1984) Are seals frequently infected with avian influenza viruses? J Virol 51:863–865

Hinshaw VS, Bean WJ, Geraci J, Fiorelli P, Early G, Webster RG (1986) Characterization of two influenza A viruses from a pilot whale. J Virol 58:655–656

Hoffmann E, Lipatov AS, Webby RJ, Govorkova EA, Webster RG (2005) Role of specific hemagglutinin amino acids in the immunogenicity and protection of H5N1 influenza virus vaccines. Proc Natl Acad Sci U S A 102:12915–12920

Hopkins BA, Skeeles JK, Houghten GE, Slagle D, Gardner K (1990) A survey of infectious diseases in wild turkeys (*Meleagridis gallopavo silvestris*) from Arkansas. J Wildl Dis 26:468–472

Ito T, Okazaki K, Kawaoka Y, Takada A, Webster RG, Kida H (1995) Perpetuation of influenza A viruses in Alaskan waterfowl reservoirs. Arch Virol 140:1163–1172

Ito T, Couceiro JN, Kelm S, Baum LG, Krauss S, Castrucci MR, Donatelli I, Kida H, Paulson JC, Webster RG, Kawaoka Y (1998) Molecular basis for the generation in pigs of influenza A viruses with pandemic potential. J Virol 72:7367–7373

Iwatsuki-Horimoto K, Kanazawa R, Sugii S, Kawaoka Y, Horimoto T (2004) The index influenza A virus subtype H5N1 isolated from a human in 1997 differs in its receptor-binding properties from a virulent avian influenza virus. J Gen Virol 85:1001–1005

Karasin AI, Brown IH, Carman S, Olsen CW (2000a) Isolation and characterization of H4N6 avian influenza viruses from pigs with pneumonia in Canada. J Virol 74:9322–9327

Karasin AI, Olsen CW, Anderson GA (2000b) Genetic characterization of an H1N2 influenza virus isolated from a pig in Indiana. J Clin Microbiol 38:2453–2456

Karasin AI, Landgraf J, Swenson S, Erickson G, Goyal S, Woodruff M, Scherba G, Anderson G, Olsen CW (2002) Genetic characterization of H1N2 influenza A viruses isolated from pigs throughout the United States. J Clin Microbiol 40:1073–1079

Kida H, Ito T, Yasuda J, Shimizu Y, Itakura C, Shortridge KF, Kawaoka Y, Webster RG (1994) Potential for transmission of avian influenza viruses to pigs. J Gen Virol 75:2183–2188

Kida H, Kawaoka Y, Naeve CW, Webster RG (1987) Antigenic and genetic conservation of H3 influenza virus in wild ducks. Virology 159:109–119

Krauss S, Walker D, Pryor SP, Niles L, Chenghong L, Hinshaw VS, Webster RG (2004) Influenza A viruses of migrating wild aquatic birds in North America. Vector Borne. Zoonotic Dis 4:177–189

Lamb RA, Krug RM (2001) Orthomyxoviruses: the viruses and their replication. In: Knipe DM, Howley PM (eds) Fields virology. Lippincott Williams and Wilkins, Philadelphia, pp 1487–1532

Lin YP, Shaw M, Gregory V, Cameron K, Lim W, Klimov A, Subbarao K, Guan Y, Krauss S, Shortridge K, Webster R, Cox N, Hay A (2000) Avian-to-human transmission of

H9N2 subtype influenza A viruses: relationship between H9N2 and H5N1 human isolates. Proc Natl Acad Sci U S A 97:9654–9658
Lvov DK, Zdanov VM, Sazonov AA, Braude NA, Vladimirtceva EA, Agafonova LV, Skljanskaja EI, Kaverin NV, Reznik VI, Pysina TV, Oserovic AM, Berzin AA, Mjasnikova IA, Podcernjaeva RY, Klimenko SM, Andrejev VP, Yakhno MA (1978) Comparison of influenza viruses isolated from man and from whales. Bull World Health Organ 56:923–930
Markowska-Daniel I, Pejsak Z (1999) Serological prevalence of influenza virus in pigs and wild boar in Poland. Medycyna Weterynaryjna 55:302–305
Matrosovich M, Zhou N, Kawaoka Y, Webster R (1999) The surface glycoproteins of H5 influenza viruses isolated from humans, chickens, and wild aquatic birds have distinguishable properties. J Virol 73:1146–1155
Matrosovich MN, Krauss S, Webster RG (2001) H9N2 influenza A viruses from poultry in Asia have human virus-like receptor specificity. Virology 281:156–162
Murphy BR, Hinshaw VS, Sly DL, London WT, Hosier NT, Wood FT, Webster RG, Chanock RM (1982) Virulence of avian influenza A viruses for squirrel monkeys. Infect Immun 37:1119–1126
Mutinelli F, Capua I, Terregino C, Cattoli G (2003) Clinical, gross, and microscopic findings in different avian species naturally infected during the H7N1 low- and high-pathogenicity avian influenza epidemics in Italy during 1999 and 2000. Avian Dis 47:844–848
Nielsen O, Clavijo A, Boughen JA (2001) Serologic evidence of influenza A infection in marine mammals of arctic Canada. J Wildl Dis 37:820–825
Ohishi K, Ninomiya A, Kida H, Park CH, Maruyama T, Arai T, Katsumata E, Tobayama T, Boltunov AN, Khuraskin LS, Miyazaki N (2002) Serological evidence of transmission of human influenza A, B viruses to Caspian seals (*Phoca caspica*). Microbiol Immunol 46:639–644
Ohishi K, Kishida N, Ninomiya A, Kida H, Takada Y, Miyazaki N, Boltunov AN, Maruyama T (2004) Antibodies to human-related H3 influenza A virus in Baikal seals (*Phoca sibirica*) and ringed seals (*Phoca hispida*) in Russia. Microbiol Immunol 48:905–909
Olsen CW, Karasin A, Erickson G (2003) Characterization of a swine-like reassortant H1N2 influenza virus isolated from a wild duck in the United States. Virus Res 93:115–121
Osterhaus AD, Rimmelzwaan GF, Martina BE, Bestebroer TM, Fouchier RA (2000) Influenza B virus in seals. Science 288:1051–1053
Peiris JS, Guan Y, Markwell D, Ghose P, Webster RG, Shortridge KF (2001) Cocirculation of avian H9N2 and contemporary "human" H3N2 influenza A viruses in pigs in southeastern China: potential for genetic reassortment? J Virol 75:9679–9686
Peiris JS, Yu WC, Leung CW, Cheung CY, Ng WF, Nicholls JM, Ng TK, Chan KH, Lai ST, Lim WL, Yuen KY, Guan Y (2004) Re-emergence of fatal human influenza A subtype H5N1 disease. Lancet 363:617–619
Pereira HG, Tumova B, Webster RG (1967) Antigenic relationship between influenza A viruses of human and avian origins. Nature 215:982–983

Peterson MJ, Aguirre R, Ferro PJ, Jones DA, Lawyer TA, Peterson MN, Silvy NJ (2002) Infectious disease survey of Rio Grande wild turkeys in the Edwards Plateau of Texas. J Wildl Dis 38:826–833

Saliki JT, Rodgers SJ, Eskew G (1998) Serosurvey of selected viral and bacterial diseases in wild swine from Oklahoma. J Wildl Dis 34:834–838

Scholtissek C (1990) Pigs as the "mixing vessel" for the creation of new pandemic influenza A viruses. Med Principles Pract 2:65–71

Shaw M, Cooper L, Xu X, Thompson W, Krauss S, Guan Y, Zhou N, Klimov A, Cox N, Webster R, Lim W, Shortridge K, Subbarao K (2002) Molecular changes associated with the transmission of avian influenza a H5N1 and H9N2 viruses to humans. J Med Virol 66:107–114

Shengqing Y, Shinya K, Otsuki K, Ito H, Ito T (2002) Isolation of myxoviruses from migratory waterfowls in San-in district, western Japan in winters of 1997–2000. J Vet Med Sci 64:1049–1052

Shinya K, Hatta M, Yamada S, Takada A, Watanabe S, Halfmann P, Horimoto T, Neumann G, Kim JH, Lim W, Guan Y, Peiris M, Kiso M, Suzuki T, Suzuki Y, Kawaoka Y (2005) Characterization of a human H5N1 influenza A virus isolated in 2003. J Virol 79:9926–9932

Shortridge KF (1992) Pandemic influenza- a zoonosis? Semi Respir Infect 7:11–25

Sims LD, Domenech J, Benigno C, Kahn S, Kamata A, Lubroth J, Martin V, Roeder P (2005) Origin and evolution of highly pathogenic H5N1 avian influenza in Asia. Vet Rec 157:159–164

Sivanandan V, Halvorson DA, Laudert E, Senne DA, Kumar MC (1991) Isolation of H13N2 influenza A virus from turkeys and surface water. Avian Dis 35:974–977

Slemons RD, Johnson DC, Osborn JS, Hayes F (1974) Type-A influenza viruses isolated from wild free-flying ducks in California. Avian Dis 18:119–124

Slemons RD, Shieldcastle MC, Heyman LD, Bednarik KE, Senne DA (1991) Type A influenza viruses in waterfowl in Ohio and implications for domestic turkeys. Avian Dis 35:165–173

Spackman E, Stallknecht DE, Slemons RD, Winker K, Suarez DL, Scott M, Swayne DE (2005) Phylogenetic analyses of type A influenza genes in natural reservoir species in North America reveals genetic variation. Virus Res 114:89–100

Stallknecht DE (1999) Ecology and epidemiology of avian influenza viruses in wild bird populations. Proceedings of the Fourth International Symposium on Avian Influenza Athens GA. US Animal Health Association, pp 61–69

Stallknecht DE, Shane SM (1988) Host range of avian influenza virus in free-living birds. Vet Res Commun 12:125–141

Stallknecht DE, Shane SM, Zwank PJ, Senne DA, Kearney MT (1990) Avian influenza viruses from migratory and resident ducks of coastal Louisiana. Avian Dis 34:398–405

Stuen S, Have P, Osterhaus AD, Arnemo JM, Moustgaard A (1994) Serological investigation of virus infections in harp seals (*Phoca groenlandica*) and hooded seals (*Cystophora cristata*). Vet Rec 134:502–503

Sturm-Ramirez KM, Ellis T, Bousfield B, Bissett L, Dyrting K, Rehg JE, Poon L, Guan Y, Peiris M, Webster RG (2004) Reemerging H5N1 influenza viruses in Hong Kong in 2002 are highly pathogenic to ducks. J Virol 78:4892–4901

Sturm-Ramirez KM, Hulse-Post DJ, Govorkova EA, Humberd J, Seiler P, Puthavathana P, Buranathai C, Nguyen TD, Chaisingh A, Long HT, Naipospos TS, Chen H, Ellis TM, Guan Y, Peiris JS, Webster RG (2005) Are ducks contributing to the endemicity of highly pathogenic H5N1 influenza virus in Asia? J Virol 79:11269–11279

Suarez DL (2000) Evolution of avian influenza viruses. Vet Microbiol 74:15–27

Suss J, Schafer J, Sinnecker H, Webster RG (1994) Influenza virus subtypes in aquatic birds of eastern Germany. Arch Virol 135:101–114

Uyeki TM, Chong YH, Katz JM, Lim W, Ho YY, Wang SS, Tsang TH, Au WW, Chan SC, Rowe T, Hu-Primmer J, Bell JC, Thompson WW, Bridges CB, Cox NJ, Mak KH, Fukuda K (2002) Lack of evidence for human-to-human transmission of avian influenza A (H9N2) viruses in Hong Kong, China (1999) Emerg Infect Dis 8:154–159

Vicente J, Leon-Vizcaino L, Gortazar C, Jose CM, Gonzalez M, Martin-Atance P (2002) Antibodies to selected viral and bacterial pathogens in European wild boars from south cenral Spain. J Wildl Dis 38:649–652

Wang M, Webster RG (1990) Lack of persistence of influenza virus genetic information in ducks. Arch Virol 111:263–267

Webster RG, Morita M, Pridgen C, Tumova B (1976) Ortho-and paramyxoviruses from migrating feral ducks: characterization of a new group of influenza A viruses. J Gen Virol 32:217–225

Webster RG, Yakhno M, Hinshaw VS, Bean WJ, Murti KG (1978) Intestinal influenza: replication and characterization of influenza viruses in ducks. Virology 84:268–278

Webster RG, Geraci J, Petursson G, Skirnisson K (1981a) Conjunctivitis in human beings caused by influenza A virus of seals. N Engl J Med 304:911

Webster RG, Hinshaw VS, Bean WJ, Van Wyke KL, Geraci JR, St Aubin DJ, Petursson G (1981b) Characterization of an influenza A virus from seals. Virology 113:712–724

Webster RG, Bean WJ, Gorman OT, Chambers TM, Kawaoka Y (1992) Evolution and ecology of influenza A viruses. Microbiol Rev 56:152–179

Widjaja L, Krauss SL, Webby RJ, Xie T, Webster RG (2004) Matrix gene of influenza A viruses isolated from wild aquatic birds: ecology and emergence of influenza A viruses. J Virol 78:8771–8779

World Health Organization Global Influenza Program Surveillance Network (2005) Evolution of H5N1 avian influenza viruses in Asia. Emerg Infect Dis 11:1515–1521

CTMI (2007) 315:85–111
© Springer-Verlag Berlin Heidelberg 2007

Overviews of Pathogen Emergence: Which Pathogens Emerge, When and Why?

S. Cleaveland[1] (✉) · D. T. Haydon[2] · L. Taylor[3]

[1]Centre for Tropical Veterinary Medicine, Royal (Dick) School of Veterinary Studies, University of Edinburgh, Easter Bush, Roslin, Midlothian UK EH25 9RG
sarah.cleaveland@ed.ac.uk
[2]Division of Environmental and Evolutionary Biology, University of Glasgow, Glasgow, UK G12 8QQ
[3]Institute for Stem Cell Research, University of Edinburgh, Roger Land Building, West Mains Road, Edinburgh, UK EH9 3JQ

Abstract An emerging pathogen has been defined as the causative agent of an infectious disease whose incidence is increasing following its appearance in a new host population or whose incidence is increasing in an existing population as a result of long-term changes in its underlying epidemiology (Woolhouse and Dye 2001). Although we appear to be in a period where novel diseases are appearing and old diseases are spreading at an unprecedented rate, disease emergence *per se* is not a new phenomenon. It is almost certain that disease emergence is a routine event in the evolutionary ecology of pathogens, and part of a ubiquitous response of pathogen populations to shifting arrays of host species. While our knowledge of emerging diseases is, for the most part, limited to the time span of the human lineage, this history provides us with a modern reflection of these deeper evolutionary processes, and it is clear from this record that at many times throughout human history, demographic and behavioural changes in society have provided opportunities for pathogens to emerge.

1 Introduction

Over recent years, many reviews have been undertaken that survey emergence events and factors associated with recently emerging diseases. In this chapter, we discuss whether, and how, these surveys can improve our understanding of the mechanisms of emergence and influence our ability to predict, detect or control emerging diseases. To address the question of which pathogens emerge, we first review what can be learnt from the history of emerging diseases. We then consider surveys that characterise emerging pathogens in terms of taxonomy, host range and transmission routes, drawing on examples from both emerging human and emerging animal diseases to illustrate general patterns in disease emergence. Finally, we present an alternative framework for analysing why different pathogens emerge, attempting to identify high-risk situations and environments that might be of practical relevance for targeting disease surveillance and control measures.

2 Emerging Zoonoses and Human Population History: When Have Human Pathogens Emerged in the Past?

Pathogens can persist in host populations only if each infected host, on average, infects one or more susceptible hosts. If the average number of new hosts infected per case (which in the event that the rest of the population is entirely susceptible is the basic reproduction number, R_0,) falls below 1, then the pathogen will ultimately die out (Anderson and May 1991). Pathogen persistence requires a supply of susceptible hosts, generated through birth, immigration or loss of immunity. If a pathogen with an $R_0 < 1$ is introduced into a naïve population, there may be a small trickle of cases, but the introduction will ultimately fail. If $R_0 > 1$, then there remains a probability that simply by chance the outbreak may only number a handful of cases, but the probability of a major outbreak is much larger. As the epidemic spreads through the host population, the pool of remaining susceptibles will diminish (as more of the population becomes immune or infected individuals die) and the rate of spread will slow. If the population is smaller than an identifiable critical community size (Bartlett 1966; Keeling and Grenfell 1997), the pathogen is unlikely to persist and the outbreak will fade-out. This is particularly true for infections with short infectious periods and those that either cause high mortality or generate prolonged host immunity.

From a historical perspective, early hunter-gatherer communities would have been too small to generate sufficient susceptible hosts to maintain species-specific pathogens. At this stage of human history, outbreaks of infectious diseases would have required repeated introduction of the pathogen from other host populations and most were likely to have been zoonotic. Human-specific pathogens probably comprised only the heirloom species, such as pinworm, that were carried over from hominid ancestors (Sprent 1969).

The history of human emerging infectious diseases (EIDs) has been described with reference to key transitions (Barrett et al. 1998; McMichael 2004). The first key transition in human societies is likely to have been the domestication of livestock 10,000–15,000 years ago, which provided multiple opportunities for disease emergence, first by facilitating cross-species (zoonotic) transmission and, second by allowing the expansion of human settlements large enough for virulent pathogens, such as measles and smallpox, to persist (Diamond 2002). As settlements became cities, a second transition point was reached: the problems of sanitation and pest control increased, allowing huge epidemics of infections, such as the black death and cholera. Migration, trade, exploration and conquest gave rise to the third major transition during which human infections established in one area were brought to highly susceptible populations in another, often with catastrophic consequences. The Age of Discovery, starting in the fifteenth century, with an estimated 10–15 million deaths in 1520–1521, and other Amerindian and Pacific civilisations were destroyed by imported smallpox and measles. In return, treponemal infections were introduced into Europe.

The past history of human infectious diseases can therefore be described by major epidemiological transitions that have been associated with large-scale changes in human demography, behaviour and technology (Barrett et al. 1998; McMichael 2004). Anthropogenic factors have always been the driving force behind human epidemiological change and this situation still applies today. What makes the recent emerging and re-emerging disease trends different to those over the rest of human history is the number of diseases which are increasing and the potential scale of outbreaks (McNeill 1976, Barrett et al. 1998). New diseases are currently being detected at a rate of about one new disease per year, with more than 30 new pathogens identified over the past 30 years (CDC http://www.cdc.gov; WHO http://www.who.int; Woolhouse 2002). Given that a total of only 1,415 human pathogens have been identified (Taylor et al. 2001), it is possible that the current rate at which humans are acquiring new infections is unprecedented, although data from other major transitions are not available for comparison. Although some new pathogens, such as *Helicobacter pylori* and *Legionella pneumophila* have turned out to be newly recognised causes of old diseases, the global impact of entirely new human diseases (such as HIV/AIDS and SARS), and the increasing incidence and spread of pre-existing infectious diseases (such as tuberculosis) cannot be denied.

The recurring theme throughout reviews of historical and recent disease emergence is the importance of changes in host ecology and contact patterns. Anthropogenic impacts that have affected human demographics and contact patterns between different host populations have almost invariably resulted in disease emergence. The current rate of increase in the human population, the scale of human and animal movements and the rate of environmental change creates a situation of unprecedented global contact between people and between different human and animal populations, a clear harbinger of future risk. As we look into the future, the lessons of the past become increasingly resonant.

3 Zoonotic Origins of Human Diseases

Zoonoses have been defined as "diseases and infections that are naturally transmitted between vertebrate hosts and man" (WHO 1959; Palmer et al. 1998). Zoonotic infections have long been considered an important category of emerging diseases, with animal reservoirs providing a source of new infections for humans throughout evolutionary history.

In the past, as today, two distinct mechanisms of zoonotic disease emergence can be recognised. Some pathogens have their origins as zoonoses but appear to have evolved as predominantly or exclusively human infections, having adapted to human-to-human transmission after jumping from animals to humans (R_0 in humans >1). Others require continued re-introduction from animal reservoirs (obligate zoonoses) and have never taken off in the human population as self-sustaining epidemics (R_0 in humans < 1).

Hart et al. (1999) proposed a system of classifying zoonoses based on these distinct mechanisms and the time-scale of emergence events. In the former category, human-specific infections that have their origins in an animal host were defined as either old or recent. Many of these old diseases are thought to have originated from domestic animal pathogens at the time of animal domestication (Bennet and Begon 1997; Diamond 2002). It is suggested, for example, that measles originated from closely-related morbilliviruses of cattle (rinderpest), and smallpox from poxviruses of either camels or cattle. Examples of recent zoonoses include HIV-1 and HIV-2, which have appeared as new human diseases after jumping the species barrier from primates to humans (Gao et al. 1999; Hahn et al. 2000) and SARS, which is thought have had its origins as a zoonosis (Song et al. 2005) but has now adapted to human-to-human transmission.

While genetic analyses have provided important evidence for these recent animal-to-human species jumps, they have also cast doubt on the historic zoonotic origins of other human pathogens. For example, the conventional wisdom that *Mycobacterium tuberculosis* (the cause of human tuberculosis) originated as a zoonosis from *M. bovis* (the cause of bovine tuberculosis) now appears unlikely in the light of sequence data analysis, which shows that the genome of *M. bovis* has lost a number of genes that are present in *M. tuberculosis* and that *M. tuberculosis* evolved from the common progenitor of the tuberculosis complex earlier than *M. bovis* (Garnier et al. 2003).

Within the second broad category of zoonoses are the obligate zoonoses, which include those that are established (e.g. Q-fever, brucellosis) and those that are newly recognised (e.g. Nipah and Hendra viruses) (Hart et al. 1999). These pathogens can only be sustained in human populations by continued re-introduction from animal reservoirs

4 Which Pathogens Have Recently Emerged?

The literature on recent emerging human diseases contains accounts of many pathogens that are zoonotic (e.g. vCJD, *Escherichia coli* 0157) and many that involve wildlife hosts (e.g. Ebola virus, West Nile virus), suggesting that transmission from an animal host to humans is an important component of human disease emergence. However, most of these accounts have been largely descriptive (e.g. Morse 1995; Osburn 1996, Murphy 1998; Palmer et al. 1998; Chomel 1998; Daszak et al. 2000) and quantifying risk factors has only been possible with the construction of a database that contains all known human pathogens and thus allows the characteristics of emerging and nonemerging human pathogens to be compared (Taylor et al. 2001). A similar database has been constructed for domestic carnivore and livestock pathogens allowing features of both human and animal emerging pathogens to be identified (Cleaveland et al. 2001). The most important finding of these quantitative analyses is that emerging pathogens are not a random selection of all pathogens, but that host range and pathogen taxonomy are important risk factors for disease emergence.

5
Host Range

Links between human emerging diseases and animal hosts have been noted in several emerging infectious disease (EID) reviews (Morse 1995; Murphy 1998; Osburn 1996; Palmer et al. 1998, Chomel 1998; Daszak et al. 2000). Of the 1,415 pathogens identified in the human pathogen database, 61% pathogens from 313 different genera are known to be zoonotic and therefore infect multiple hosts (Taylor et al. 2001). Overall, 175 (12.4%) human pathogens from 96 genera were identified as the cause of emerging diseases and, of these, 133 (76%) were zoonotic (Taylor et al. 2001). In this study, zoonoses did not include those which are known to have their origin in animal hosts, but for which infection now occurs exclusively through human-to-human transmission (e.g. HIV-1 and HIV-2). Multi-host pathogens also predominate among animal EIDs, with 90% of emerging livestock diseases and 100% of emerging domestic carnivore diseases caused by multi-host pathogens (Cleaveland et al. 2001).

From these surveys, it is clear that generalist pathogens are over-represented in both human and animal emerging diseases. Thus, pathogens that have the ability to infect more than one host (which, for human diseases, includes all zoonoses), pathogens that have the ability to infect more than one taxonomic order (Fig. 1), and pathogens infecting wildlife hosts all have a higher relative risk for emergence than pathogens with more restricted host ranges (Cleaveland et al. 2001) (Table 1; Fig. 1). A broad host range is also a feature of many recent disease outbreaks in wildlife hosts, particularly endangered populations (Cleaveland et al. 2002).

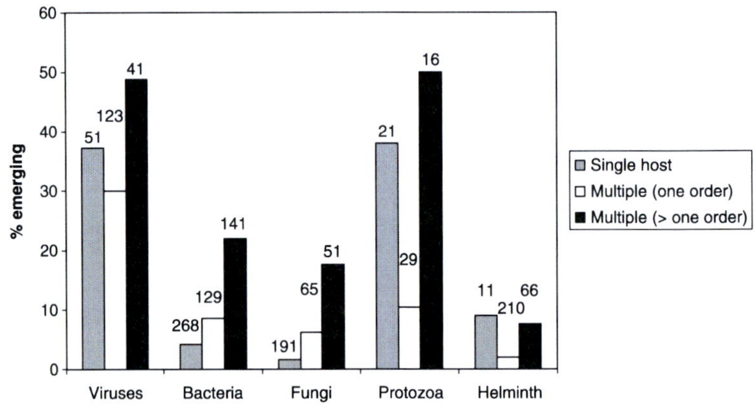

Fig. 1 The proportion of emerging pathogens in different taxonomic groups in relation to host range. The numbers shown above each column indicate the total number of human pathogens in each category

Table 1 The relative risk of emergence for different categories of pathogen in relation to host range of pathogens. Diseases for which the identity of animal hosts was unknown were excluded, hence the number of zoonoses given here (n = 800) is lower than the total number of human pathogens identified as zoonoses (n = 872)

Categories of host infected by pathogen	Number of zoonotic diseases (n = 800)	Number of emerging zoonotic diseases (n = 125)	Relative risk
Wildlife	619 (77.4%)	113 (90.4%)	2.75
Birds	82 (10.3%)	23 (18.4%)	1.97
Nonmammalian hosts	109 (13.6%)	30 (24.0%)	2.0
Ungulates	315 (39.3%)	72 (57.6%)	2.09
Carnivores	344 (43.0%)	64 (51.2%)	1.39
Primates	103 (12.9%)	31 (24.8%)	2.23
Rodents	180 (22.5%)	43 (34.4%)	1.81
Marine mammals	41 (5.1%)	6 (4.8%)	0.93
Bats	15 (1.9%)	6 (4.8)	2.64

Here, parallels can be drawn with early human communities; endangered wildlife populations are too small to maintain species-specific pathogens and the risk of emergence invariably arises as a result of cross-species transmission.

6 Pathogen Taxonomy

Although all taxonomic groups are represented within the group of human emerging pathogens, viruses appear disproportionately among emerging pathogens. For viruses, the proportion of pathogens that are emerging is four times higher than other taxonomic groups (relative risk of emergence [RR] = 4.3), with viruses comprising 15% (n = 215) of all human pathogens and 35% (n = 76) of emerging pathogens. This applies also to emerging pathogens of domestic animals (Cleaveland et al. 2001). Conversely, parasitic helminths are underrepresented in the emerging disease category (RR < 0.25). Although quantitative baseline data are lacking for wildlife diseases (and hence RR cannot be calculated), viral pathogens have also been the cause of most recent wildlife disease outbreaks (Murray et al. 1999; Dobson and Foufopoulos 2001; Funk et al. 2001). Among the viruses, RNA viruses have only a slightly higher RR of emergence (RR = 2.8) than DNA viruses (RR = 2.5), but are disproportionately represented among those pathogens that have emerged as new human and animal diseases after jumping from other host species (Table 2).

Table 2 Examples of viruses thought to have emerged as a result of species jumps A. in the historical past (5000–10,000 years ago) and B. in the recent past

A. Historical past

Disease/pathogen	Proposed original host	New host	Reference
Measles	Cattle/dogs?	Human	Bennett and Begon 1997; Diamond 2002
Smallpox	Cattle/camels?	Human	Bennett and Begon 1997; Diamond 2002
Common cold	Cattle?	Human	Bennett and Begon 1997; Diamond 2002

B. Recent past

Disease/pathogen	Original Host	New host	Year first observed in new host	
FPLV/CPV	Cats	Dogs	1978	Parrish 1994
SIVcpz/HIV-1	Chimpanzee	Human	1983	Gao et al. 1999; Hahn et al. 2000
SIVmac/HIV-2	Macaques	Human	1986	Hahn et al. 2000
Canine/Phocine	Canids	Harp seals	Harp seals?	
distemper virus	Harp seals	Harbour seals	Harbour seals 1988	Mahy 1993
Hendra virus	Fruit bats	Humans/horses	1994	Mackenzie et al. 2001
Australian bat lyssavirus	Fruit bats	Humans	1996	Mackenzie et al. 2001
Menangle virus	Fruit bats	Pigs, human	1997	Mackenzie et al. 2001
Nipah virus	Fruit bats	Humans/pigs	1999	Mackenzie et al. 2001
Canine distemper virus	Dogs	Lions	1994	Roelke-Parker et al. 1996
	Sledge dogs?	Crab-eating seals	1955	Bengston et al. 1991
	Dogs/wild canids?	Lake Baikal seals	1987/1988	Grachev et al. 1989
	Dogs/wild canids?	Caspian sea seals	2000 (disease)	Kennedy et al. 2000
H5N1 Influenza A	Chickens	Humans	1997	Li et al. 2004
Hepatitis E virus	Deer	Humans	2003	Tei et al. 2003
SARS coronavirus	Palm civets?	Human	2003	He et al. 2004; Song et al. 2005

7 Does Knowing Which Pathogens Emerge Help Us Understand How Diseases Emerge?

What does the preponderance of viral pathogens among emerging diseases tell us about mechanisms of disease emergence? Several factors has been proposed to explain this observation, such as the relative difficulty of treating viral diseases, improved detection rates, short generation and higher mutation rates (Domingo and Holland 1994). That RNA viruses are over-represented in instances of pathogens jumping into new host species is consistent with the view that mutation rates may play a role in emergence. High mutation rates in RNA viruses (Drake 1993; Domingo and Holland 1994), and the existence of multiple variants within strains of RNA viruses, provide an enormous capacity for RNA viruses to adapt to changing host environments and to overcome barriers to spread of virus both within hosts and between species. For example, it has been suggested that the spread of rabies virus within different host tissues and between host species may only be possible as a result of the combined action of virus variants with diverse tissue tropism, with multiple strain variants compensating for the simplicity and lack of regulatory elements within the rabies virus genome (Morimoto et al. 1998).

However, it has also been argued that the limitations of a very small genome act as an important constraint to the adaptability and evolution of RNA viruses. As specific sequences are required to encode multiple functions, there may be little flexibility for mutations to confer any adaptive advantage (Holmes 2003; see the chapter by Holmes and Drummond, this volume). Understanding the mechanistic basis of genomic constraints to RNA virus evolution may help explain why some RNA viruses are more able to cross species boundaries than others (Holmes and Rambaut 2004; see chapter by Holmes and Drummond, this volume).

The ability to undergo recombination, which is seen in a wide range of RNA viruses (Worobey and Holmes 1999), may also be a factor. Recombination plays a key role in the emergence of highly pathogenic strains of influenza A (Shu et al. 1996), and may contribute to the burgeoning diversity and emergence of Dengue viruses (Holmes and Burch 2000). If recombination is an important mechanism in emergence, then understanding how the genetic organisation of viral genomes influences recombination rates is an important question. For example, rates in segmented viral genomes, like influenza A, may be higher than in nonsegmented genomes, while in negative stranded RNA viruses, such as rhabdoviruses, recombination rates are likely to be lower than in positive stranded viruses.

In terms of mechanisms of disease emergence, most attention has focussed on the question of host-switching and the appearance of new pathogens, such as HIV and SARS, in the human populations. Parallels are also seen in animal EIDs, with host-switching an important feature of several new disease outbreaks, such as canine distemper virus (CDV) jumping from domestic dogs to lions (Roelke-Parker et al. 1996), Lake Baikal seals (Grachev et al. 1986; Mamaev et al. 1996) and Caspian seals (Kennedy et al. 2000), phocine distemper virus (PDV) jumping from harp seals to common seals (Goodhart 1988; Barrett 1999) and feline panleucopaenia virus in cats evolving into canine parvovirus in dogs (Hueffer et al. 2003).

Evolutionary ecologists studying adaptive radiations have long suggested that they arise from generalist ancestors, and develop through adaptive diversification into ever more specialised niches (Simpson 1953; Mayr 1942; Thompson 1994; Schluter 2000). Many taxonomic groups of pathogen fit comfortably into the paradigm of adaptive radiation, but it is not clear whether the phenomenon of emergence corresponds to a process of increasing ecological generalism or simply host-switching followed by subsequent further specialism (for example, HIV). Host-switching events are indicated throughout the evolutionary record by the frequent topological discordancies in paired host-pathogen phylogenies (Jackson and Charleston 2004), and it is reasonable to suppose that these switches corresponded to periods of pathogen emergence. But following a host switch, the outcome of opposing selective forces for further adaptation to the new host, or maintaining a broader spectrum of host species use remains unclear.

A broad host range may be a more important predictor of the potential for novel host use than close taxonomic relatedness, which is not invariably required for either pathogens that undergo species jumps (Table 2; Woolhouse et al. 2005) or for established zoonotic pathogens. Emerging zoonoses originate from a broad spectrum of different animal hosts (Table 1), with the greatest number of emerging zoonoses caused by ungulates (58%), followed by carnivores (51%) and rodents (34%). However, only relatively few zoonoses overall (13%) are known to infect primates under natural conditions, so this may simply reflect a lack of data on natural populations (Wolfe et al. 1998). Perhaps a better measure of the potential for emergence is given by the relative risk, which is greater in primates and bats than ungulates and carnivores (Table 1).

The determinants of a broad host range are poorly understood. It has been suggested that the use of host-cell receptors that are highly conserved across host species may facilitate infection in a wide range of hosts (Woolhouse 2002). For example, the rabies virus, which has the potential to infect all mammal species, gains entry to peripheral nerves via the highly conserved nicotinic acetylcholine receptor, and the foot-and-mouth disease virus (FMDV) uses the conserved vitronectin receptor (Baranowski et al. 2001). While appropriate receptors are clearly a prerequisite for entry in to the cell lines of any potentially permissive host, it is becoming increasingly clear that downstream intracellular events can also restrict host range (McFadden 2005), and much remains to be learned about these processes.

8 What Practical Lessons Can Be Learnt from Emerging Disease Surveys?

Characterising the features of emerging pathogens highlights several key issues in the approach towards human and animal EIDs. First, these surveys all demonstrate the importance of zoonotic transmission in past and current emerging human diseases, emphasising the need to understand the infection dynamics of zoonotic pathogens in both animal and human populations and to broaden the single-species focus of human medicine to incorporate knowledge available within veterinary and wildlife disciplines. It was notable that during the construction of the human infectious disease database, a substantial number of human pathogens were identified as zoonoses from veterinary reference texts, but not from medical texts. Many emerging zoonotic diseases of the future may be infections that are currently recognised by veterinarians or wildlife biologists, and their involvement is likely to be an important element in the early detection of emerging zoonoses. For example, veterinary pathologists at the Bronx zoo played a major role in the detection and identification of West Nile Virus (McNamara 2002). A granulocytic *Ehrlichia* described from meadow voles on Martha's Vineyard in the 1930s (Tyzzer 1938) is now believed to be the agent causing human granulocytic ehrlichiosis (Telford 2002), and archived veterinary material is likely to provide an important source of data for identifying potential reservoirs of new or emerging zoonotic infections.

The predominance of viral pathogens among human and animal EIDs highlights the need for maintaining expertise in virological techniques, for improved anti-viral treatments and for enhanced collaboration between medical and veterinary virologists. Prior to the emergence of SARS, human coronaviruses had been of little interest in medical virology and much of the knowledge about coronavirus biology was available only from studies of animal coronaviruses in the context of important veterinary diseases (Cavanagh 2000). This expertise was effectively harnessed in the rapid international response to SARS, contributing to the rapid isolation, diagnosis and characterisation of the SARS virus and to an understanding of aspects of pathogenesis and immune response (Cavanagh 2003; Berger et al. 2004). Similarly, insights from research on coronavirus vaccines for animals are likely to assist the development of a SARS vaccine.

In general, relatively little is still known about the infection dynamics of emerging zoonoses in animal host populations, and this is particularly true when wildlife hosts are involved. The epidemiology of generalist pathogens in multi-host populations is often complex and identifying reservoirs of infection invariably a challenging task (Haydon et al. 2001). The enduring uncertainties about the role of badgers as reservoirs and/or sources of bovine tuberculosis for cattle in the UK typify these difficulties (Krebs et al. 1998). For zoonotic diseases, integration and collaboration between disciplines is clearly important. Public health researchers and veterinarians require some understanding of ecological processes and the links between the environment, ecology and disease. Conversely, ecologists and population biologists need to understand the dynamics of pathogens at individual, population and community levels (Daszak and Cunningham 2002).

From these surveys, we know which pathogens have emerged and we are beginning to understand how they are able to do so. An important lesson is the breadth of pathogens that *can* emerge. The fact that many recent emergence events have taken us by surprise is, in itself, surprising, given the historical patterns of disease emergence and the evidence that many pathogens have the potential to emerge under favourable ecological and environmental conditions. In the next section, we therefore explore the question of why certain pathogens emerge, attempting to identify circumstances and situations where disease emergence might be expected, so that surveillance and control measures can be targeted to high-risk settings. We consider whether an appraisal of risk factors provides a useful way of reviewing past emergence events and attempt to address the question of which pathogens emerge with reference to particular environmental or demographic settings rather than a particular pathogen type.

9
Which Pathogens Emerge: Where and Why?

Many reviews have emphasised the importance of anthropogenic social and environmental factors in disease emergence (e.g. Institute of Medicine 1992; Schrag and Wiener 1995; Kuiken et al. 2003). Indeed all the six factors identified by the Institute of Medicine (1992) as contributing to EIDs are considered anthropogenic (i.e. human demographics and behaviour, technology and industry, economic development and land use, international trade and commerce, microbial adaptation and change, breakdown of public health measures). Recognition of the importance of human-related impacts has dispelled some of the early complacency about infectious diseases and suggests that the EIDs are likely to increase as the human ecological footprint continues to grow. In theory, it also suggests that counter measures to mitigate the effects of anthropogenic change might be possible. However, risk factors are often cited only in terms of broad categories, such as climate change, human population increase, urbanisation or habitat destruction. Unless we can link these factors to specific effects on the underlying dynamics of a disease, it will be difficult to design effective control measures or target surveillance to the appropriate steps of different transmission pathways.

As an example, land-use change is often suggested as a risk factor for emerging zoonoses, but there are multiple ways in which changes in land use and habitat might affect the infection dynamics of zoonotic pathogens, including (1) an increase in the number of reservoir hosts, (2) an increase in the incidence of infection in reservoir hosts, or (3) a change in the pattern, rate or frequency of contact between reservoir and human hosts. Understanding which of these factors are operating will determine how and where control measures can be targeted for optimum effect. However, identifying critical pathways may not be simple; the ecological processes that can lead to changes in zoonotic infection dynamics are often very specific (Box 1), requiring a detailed understanding of host population ecology.

Box 1: Potential mechanisms by which land-use changes can affect pathogen dynamics and emergence

The complexity of mechanisms by which changes in host-pathogen dynamics can result in emergence is illustrated here with respect to land-use change as a risk factor for disease emergence. Land-use change could result in pathogen emergence by any of the following factors, which may affect reservoir dynamics or host-reservoir contact patterns: (1) demographic host release arising from reduction of competitor and/or predator species, resulting in competitive release and an increased density of the most competent host for a pathogen (Rosenblatt et al. 1999), (2) the fence effect, whereby habitat fragmentation restricts dispersal and leads to unnaturally high densities and hence infection rates (Dobson and May 1986), (3) reduction of species diversity leading to a relative increase in alternative, more competent hosts (Ostfeld and Keesing 2000), (4) a reduction in the genetic diversity, which may increase opportunities for EIDs (Acevedo-Whitehouse et al. 2003; Keller et al. 2002) with knock-on effects on the equitability of higher trophic levels, (5) enrichment of nutrient status (by pollution or agricultural crop presence, or fertiliser), which may favour certain species that specialise on such resources, (6) elimination of biodiversity creating vacant niches for invasive species, which has been suggested as a factor in the emergence of non-polio enteroviruses following elimination of polio (Delpeyroux et al. 2000) and (7) the establishment of secondary contact zones, in which pathogens introduced into novel environments have the opportunity to come into contact with closely related but previously geographically isolated pathogens. This concept has been explored mainly in the context of plant EIDs and identified as the cause of emergence of diseases, such as Dutch elm disease (*Ophiostome novo-ulmi*) (Brasier 2001) and novel fungal diseases of alder trees (*Alnus spp.*) (Brasier et al. 1999). However, pathogen recombination in secondary animal contact zones may prove to be a rich source of novel zoonotic pathogens (e.g. Waterfield et al. 2004).

We focus on ecological risk factors for zoonotic disease emergence and propose a framework that identifies three steps for zoonotic disease emergence: (1) transmission from animal host to human samplers (individuals with a high risk of acquiring novel infections), (2) transmission from samplers to spreaders (individuals with a high potential for transmitting novel infection onwards within the new host population) and (3) transmission from spreaders to the general population. The risk of transmission at each of these steps is a function of the number of infections in the source population (I), the per capita rate of contact between populations (C), the number of individuals engaging in this type of contact behaviour (N) and the susceptibility of the host population (S) (Fig. 2). The number of cases in the source population, I, reflects both the number of hosts and the incidence/prevalence of infection in the population, and may therefore incorporate enormous complexity (as illustrated by the multiple factors outlined in Box 1 that can affect reservoir infection dynamics). Transmission between host populations is encapsulated by the terms describing both contact and host susceptibility. This framework does not specifically consider the genetic mechanisms by which pathogens acquire or increase their ability to infect humans, but assumes that a pathogen is competent to infect humans.

As in the earlier discussion of zoonotic disease classification, this framework also needs to distinguish obligate zoonoses, which can only be transmitted to humans from animals (i.e. there is no or virtually no human-to-human transmission), from human diseases that originate in animal hosts but have the potential to spread within the human population. For obligate zoonoses, such as rabies, brucellosis, and West Nile virus, there is no human spreader population and all victims are essentially samplers. Mechanisms and risk factors for disease emergence in this group are therefore concerned only with the transmission step between animal host and human samplers (Fig. 2b). To explore the value of this framework for providing insights into the mechanisms of zoonotic disease emergence and identifying key gaps in current knowledge, we examined several well-studied emerging diseases, attempting to allocate risk factors to specific components of the emergence pathway.

We chose ten relatively well-studied pathogens, or pathogen groups, in order to attempt a preliminary analysis of the epidemiological relevance of factors suggested for their emergence. These were *Borrelia burgdorferi*, Ebola virus, Hantaviruses, human immunodeficiency viruses, influenza virus, *Mycobacterium tuberculosis*, Nipah virus, severe acute respiratory syndrome coronavirus, variant CJD and *Yersinia pestis*.

A literature search using the terms "factor", "emergence" and the pathogen name identified 18 references that listed 157 risk factors, many being repeated across different references (Table 3). These were summarised into both the conventional categories such as land-use changes and urbanisation effects, and the

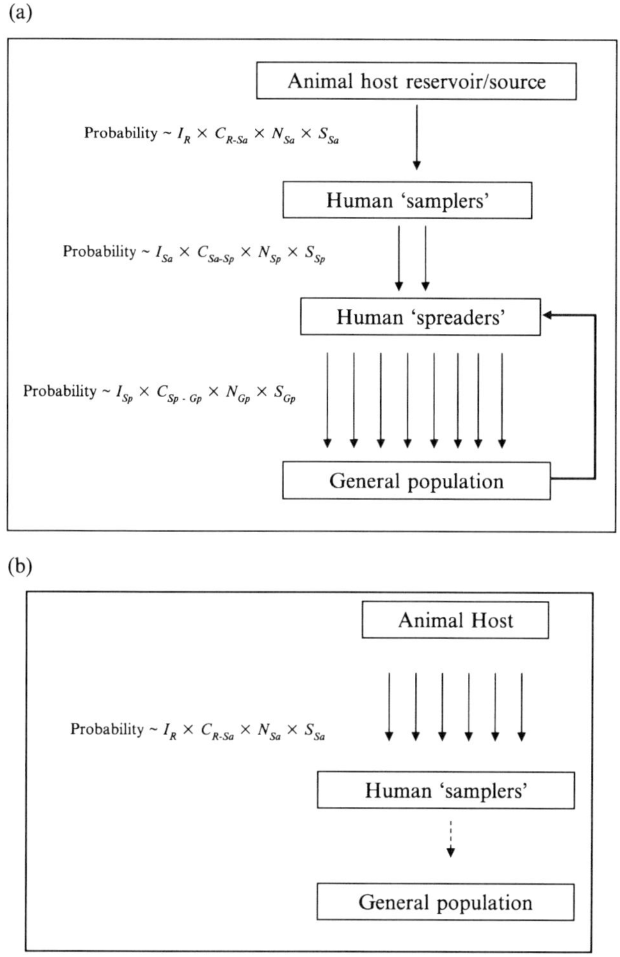

Fig. 2a, b Steps in the emergence of a zoonotic pathogen with the associated risk function. *I* is the number of infections in the source population, *C* a function of per capita contact rate between populations, *N* the number of hosts engaging in that contact activity and *S* the susceptibility of the host population. The subscripts refer to the following populations: *R*, animal reservoir or source population, *Sa*, human samplers, *Sp*, human spreaders, *Gp*, the general human population. **a** Zoonotic pathogens which have the capacity for human-to-human transmission. **b** Obligate zoonotic pathogens for which human-to-human transmission is limited

Table 3 Risk factors associated with the emergence of selected zoonotic diseases showing the elements of emergence pathways potentially affected by each factor. Key: Green = Land-use changes, Red = Bush meat trade, Blue = Hospital/medical practices, Yellow = Farming, food production and marketing, Grey = Urbanisation, Pale Blue = Local and long-distance travel

	Infection			Contact			Number			Susceptibility		
Pathogen	Reservoir	Samplers	Spreaders	Reservoir-Samplers	Samplers-Spreaders	Spreaders-Population	Samplers	Spreaders	General Population	Samplers	Spreaders	General Population
Borrelia burgdorferi	Forest fragmentation Reforestation Land-use changes			Intrusion into country Reforestation			Intrusion into country					
Ebola virus			Poor hospital hygiene	Bush-meat trade Intrusion into countryside Logging	Contaminated needles	Long-distance travel	Bush-meat trade					
Hantaviruses	Increase in rodent density Land-use changes Weather changes			Land-use changes Importation of exotic animals Intrusion into countryside Transport of rats with trade	Long-distance travel	Long-distance travel	Land-use changes					

Pathogen	Infection			Contact			Number			Susceptibility		
	Reservoir	Samplers	Spreaders	Reservoir-Samplers	Samplers-Spreaders	Spreaders-Population	Samplers	Spreaders	General Population	Samplers	Spreaders	General Population
HIV				Bush-meat trade	Promiscuous sex	Medical technology	Bush-meat trade, poverty	Poverty				
				Intrusion in country	Migration	Long-distance travel	Intrusion in country					
					Lack of control measures	Promiscuous sex						
						Medical technology						
						Lack of control measures						
Influenza A virus	Intensive farming Livestock movement Mixed livestock farming/markets			Livestock movement	Long-distance travel	Long-distance travel						
Mycobacterium tuberculosis			Public health breakdown		Migration	Public health breakdown		Public health breakdown			Co-infection with HIV Poverty	Co-infection with HIV Poverty
						High population densities						
						Long-distance travel						

(*Continued*)

Table 3 —cont'd.

Pathogen	Infection			Contact			Number			Susceptibility		
	Reservoir	Samplers	Spreaders	Reservoir-Samplers	Samplers-Spreaders	Spreaders-Population	Samplers	Spreaders	General Population	Samplers	Spreaders	General Population
Nipah virus	Intensive farming			Livestock movements								
	Livestock movements											
	Weather changes											
	Habitat destruction											
	Farming practices											
SARS-Coronavirus			Poor hygiene	Bush-meat trade	Local travel	Dense populations	Bush-meat trade			Chronic health problems		
			Poor hospital hygiene	Lab-acquired infections	Lab-acquired infections	Long-distance travel				Other health problems		
				Mixed live-stock markets	Eating out							
				Butchering practices	Blood transfusion							
Variant CJD	Cattle-feeding practices											
	BSE epidemic											
Yersinia pestis	Weather			Intrusion into country	Travel, trade	Travel, migration, trade						
	Domestic animal infections rising				Migration	Poor hygiene						

model parameters from our epidemiological framework that would be affected by those factors (Table 3). Obviously in an emerging disease context, there is a knock-on effect between our epidemiological parameters. For example, if contact between a reservoir and sampler population increases, so too does the infection level in the sampler population. We attempted to identify the root effect of the factor in question as the earliest point in the transmission pathway at which the factor operates. For example, we could consider that poor hospital hygiene, which increases transmission of Ebola virus through contact with infected bodily fluids, might influence contact between spreaders and samplers as well as increase infection rates in spreaders. Clearly there would be many ways of organizing this information and variability in the way that categories are assigned but we suggest that the broad patterns are robust to these alternative arrangements.

As a first step, we have adopted this simple approach for qualitative exploration of emergence risk factors, but the general methodology could be developed in more detail for further quantitative analyses. Using our selected examples, several issues come to light. For all the diseases selected here, emergence has been associated with multiple risk factors, which need to be operating simultaneously or sequentially for a disease to emerge or re-emerge. None of the major categories of risk factor, as they are generally summarised in the literature, operate at a single specific step in the epidemiological framework but have the potential for multiple impacts on infection dynamics. Thus, changes in farming practices can affect zoonotic disease emergence through changing infection rates in animal reservoirs, and/or by increasing contact between reservoirs and samplers.

A striking feature of Table 3 is the predominance of risk factors that affect the contact rate, with local and long-distance movements acting to increase both human-to-human and animal-to-human contact. This is perhaps unsurprising given the unprecedented speed, volume and extent of travel and international trade today. More than 1.4 million people cross international borders on flights everyday and cruise ships now have the capacity to carry 47 million passengers per year (Wilson 2003). Although long-distance movements tend to be associated with transmission by spreaders to the general population, some types of long-distance travel, such as tourism, provide travellers with the potential to act as both samplers and spreaders, and some long-distance trade movements have been associated with increased contact between animal reservoirs (e.g. rats) and humans. Wildlife and livestock movements clearly also play an important role in the emergence of zoonotic diseases, with the potential both to increase the incidence of disease in reservoir or source populations and increase reservoir-to-human contact. While limiting human contacts is often difficult, particularly with ease of travel, restricting the scale of legal and illegal movements of domestic animals (livestock and pet animals) and wildlife presents a real opportunity to minimise emergence risks.

With respect to sampler-to-spreader transmission, or dissemination from spreaders to the general population, relatively few mechanisms may be involved (e.g. international airline travel, food contamination, hospital care). Some of these contact networks, and therefore emergence risks, may be relatively simple to predict. For example, a simulation model of SARS that used global aviation routes to predict contact and transmission networks was able to provide close estimates of the number of cases occurring in different countries (Hufnagel et al. 2004). In practice, SARS cases were contained more effectively by simple hygienic precautions, such as wearing masks, than airport surveillance and detection strategies (e.g. thermal imagery) (Bell 2004; John et al. 2005). However, the benefits of targeting control and surveillance efforts to high-risk travellers may still be considerable.

While compiling the table, the difficulty of pinning down the exact epidemiological components affected by the risk factor became clear and drew attention to gaps in our knowledge. For example, urbanisation actually summarises a large number of different factors, each of which affects the underlying epidemiology of a particular pathogen in different ways. Urbanisation could lead to disease emergence as a result of poverty (which could increase susceptibility of human populations), high population densities, crowded housing, poor sanitation (which could all affect contact rates and the number of spreaders), and/or a breakdown in social values and public health (which could affect both infection rates and contact rates). The table also highlights the complete absence of information about reservoir infection dynamics for several zoonoses (Ebola virus and SARS).

10 Prediction and Surveillance of Emerging Zoonoses

It is often stated that it is impossible to predict where and when the next emerging zoonosis will appear (e.g. Murphy 1998). If the exact timing of species jumps is likely to be difficult, if not impossible to predict, early detection of emergence events is likely to be the best hope of controlling outbreaks and minimising the impact of disease.

Given the amplification effect of spreaders in the population, both the probability of transmission and the consequences (costs) of an emergence event increase with progression down the pathway from animal-to-human to human-to-human transmission. An important question is therefore to determine the point in the transmission chain at which resources are best directed. Is it better to try to detect transmission events from animal reservoirs to samplers, which may occur relatively rarely but may allow large and costly outbreaks to be prevented, or to focus surveillance on transmission events in spreader populations, which may occur more frequently, but may result in delayed control of epidemics. How important is it to understand infection dynamics in animal reservoirs, which may be costly and demanding to achieve, particularly for wildlife populations (Haydon et al. 2000)? For prevention of Nipah virus, for example, is it better to focus efforts on surveillance of fruit-bat reservoir populations, on enhancing the capacity for disease detection in high-risk pig farms (e.g. in areas with recent conversion from mangrove to oil palm plantations), on improving case surveillance in hospitals in these areas, or some combination of these three?

Factors that increase I_R or $C_{R\text{-}Sa}$ and therefore increase the probability of spill-over from animal populations into humans may be easier to predict than rare cases of species jumps. At this stage in the emergence of zoonoses, it may indeed be possible to identify sentinel (i.e. sampler) populations associated with high-risk situations (Table 4). For example, increased bushmeat consumption is cited as a risk factor for the emergence of several zoonoses and novel pathogens, and could be caused by an increase in the number of hunters/consumers (increased N_{Sa}) and/or increased contact between hunters/consumers and wildlife reservoirs (increased $C_{R\text{-}Sa}$). Improved surveillance of a sampler population of bushmeat hunters or butchers may detect host-switching and emergence events, possibly before dissemination to the general population. Similarly, farmers and market traders may be suitable sentinels for diseases

Table 4 Suggested high-risk environments and human sentinels that could be targeted for surveillance of emerging zoonoses

Risky environments/situations	Potential human sentinel population
Travel hubs	Airline crews, airport staff, frequent flyers, cruise ship staff, international conferences
Urban shanty towns	Impoverished communities, urban livestock keepers, prostitutes
Hospitals	Nurses, doctors, immunosuppressed and elderly patients
Farms and markets	Farmers, market traders, abattoir workers, vets, peri-urban livestock keepers
Interface habitats	Bush-meat hunters and butchers, market traders, consumers
Changing habitats, e.g. dam construction, logging, reforestation	New settler communities
New technologies	Transplant and blood transfusion recipients

affected by changes in farming practice, and may allow early detection of new outbreaks of SARS or Nipah virus. The potential value of high-risk human sentinels has been demonstrated, with detection of simian foamy viruses (retroviruses) in villagers in Cameroon that have direct contact with blood and body fluids of non-human primates (Wolfe et al. 2004). However, host-switching events appear to be occurring frequently and, since most outbreaks are small and may never take off (Woolhouse et al. 2001), the appropriate response to detecting new microbial agents in human populations remains very uncertain. An alternative approach would be the improvement of medical diagnostics and communication in remote communities (such as at the tropical forest interface), which might provide a more cost-effective approach to preventing large outbreaks of emerging pathogens (Shears 2000a, 2000b).

Land-use changes that affect reservoir infection rates are often associated with emerging zoonoses transmitted from wildlife reservoirs (e.g. *Borrelia burgdorferi*, Hantaviruses, Nipah viruses). While the emergence of a specific pathogen may be hard to predict, it is certainly predictable that changes in land use carry a risk of zoonotic disease emergence. Can we identify high-risk environments in which accelerating land-use changes raise particular concerns?

In summary, pathogen emergence is not an ecologically novel phenomenon, rather an inevitable consequence of changes in the abundance of host populations and the contact networks that exist between them. Throughout human history, pathogens have always exploited ecological change. Some pathogens, such as viruses and generalists, may be better at doing this than others, but many different pathogens have emerged. While there are several features that characterise many emerging pathogens and these may be combined into a profile of an emerging disease, most emerging pathogens will not fit this profile exactly. The objective perhaps should not be to predict which pathogens emerge but to plan for the inevitability of emergence events and prepare to detect and deal with them quickly. However, planning an effective response requires an understanding of their epidemiology, and once an emergence event is detected, efforts must be directed to the rapid acquisition of this information. The response to SARS provides encouragement that a flexible, integrated global strategy can be developed. SARS also highlights our inability to predict where and when the next outbreak might occur. Increasing our knowledge about the identity or infection dynamics of animal reservoirs must be a key priority that requires contributions from many different disciplines.

Over the past decade, there has been clear recognition of the problems that emerging infectious diseases pose to health care professionals throughout the world. The next decade will reveal whether solutions to these problems lie in the development of a general theory of infectious disease emergence or whether they will require case-specific approaches.

Acknowledgements This work was carried out with support of the Wellcome Trust and Department for International Development Animal Health Programme. We would like to thank Mark Woolhouse, Andy Dobson, Eric Fèvre, Louise Matthews and Sonya Gowtage-Sequeira for constructive comments and discussions.

References

Acevedo-Whitehouse K, Gulland F, Greig D, Amos W (2003) Disease susceptibility in California sea lions. Nature 422:35–35

Anderson RM, May, RM (1991) Infectious diseases of humans: dynamics and control. Oxford University Press, Oxford

Baranowski E, Ruiz-Jarabo CM, Domingo E (2001) Evolution of cell recognition by viruses. Science 292:1102–1105

Barrett R, Kuzawa CW, McDade T, Armelagos GJ (1998) Emerging and re-emerging infectious diseases: the third epidemiologic transition. Ann Rev Anthropol 27:247–271

Barrett T (1999) Morbillivirus infections, with special emphasis on morbilliviruses of carnivores. Vet Microbiol 69:3–13

Bartlett MS (1966) The critical community size for measles in the United States. J R Stat Soc 123:37–44

Bell DM (2004) Public health interventions and SARS spread (2003) Emerg Infect Dis 10:1900–1906

Bengston JL, Boveng P, Frankzen U, Have P, Heide-Jorgensen MP, Harkonen R (1991) Antibodies to canine distemper virus in Antarctic seals. Marine Mam Sci 7(1): 85–87

Bennett M, Begon ME (1997) Virus zoonoses—a long-term overview. Comp Immuno Microbiol Infect Dis 20:101–109

Berger A, Drosten C, Doerr HW, Sturmer M, Preiser W (2004) Severe acute respiratory syndrome (SARS) - paradigm of an emerging viral infection. J Clin Virol 29:13–22

Brasier CM (2001) Rapid evolution of introduced plant pathogens via interspecific hybridization. Bioscience 51:123–133

Brasier CM, Cooke DEL, Duncan JM (1999) Origin of a new *Phytophthora* pathogen through interspecific hybridization. Proc Natl Acad Sci U S A 96:5878–5883

Cavanagh D (2000) Coronaviruses and Toroviruses. Wiley, Chichester

Cavanagh D (2003) Severe acute respiratory syndrome vaccine development: experiences of vaccination against avian infectious bronchitis coronavirus. Av Pathol 32:567–582

Chomel BB (1998) New emerging zoonoses: a challenge and an opportunity for the veterinary profession. Comp Immunol Microbiol Infect Dis 21:1–14

Cleaveland S, Laurenson MK, Taylor LH (2001) Diseases of humans and their domestic mammals: pathogen characteristics, host range and the risk of emergence. Philos Trans R Soc Lond B Biol Sci 356:991–999

Cleaveland S, Hess GR, Dobson AP, Laurenson MK, McCallum HI, Roberts MG, Woodroffe R (2002) The role of pathogens in biological conservation. In: Hudson PJ, Rizzoli A,

Grenfell BT, Heesterbeek H, Dobson A (eds) The ecology of wildlife diseases. Oxford University Press, Oxford, pp 139–150
Daszak P, Cunningham AA (2002) Emerging infectious diseases: a key role for conservation medicine. In: Aguirre AA, Ostfeld RS, Tabor GM, House C, Pearl MC (eds) Conservation medicine: ecological health in practice. Oxford University Press, Oxford, pp 40–61
Daszak P, Cunningham AA, Hyatt AD (2000) Emerging infectious diseases of wildlife—threats to biodiversity and human health. Science 287:1756
Delpeyroux F, Guillot S, Szendroi A, Balanant J, Caro V, Cuervo N, Chevaliez S, Dahourou G, Crainic R (2000) What are the repercussions of the oral polio vaccine on the world of enteroviruses? Bull Soc Pathol Exot 93:193–197
Diamond J (2002) Evolution, consequences and future of plant and animal domestication. Nature 418:700–707
Dobson A, Foufopoulos J (2001) Emerging infectious pathogens of wildlife. Philos Trans R Soc Lond B Biol Sci 356:1001–1012
Dobson AP, May RM (1986) Disease and conservation. In: Soule ME (ed) Conservation biology: the science of scarcity and diversity. Sinauer Associates, Sunderland MA, pp 345–365
Domingo E, Holland JJ (1994) Mutation rates and rapid evolution of RNA viruses. In: Morse SS (ed) The evolutionary biology of viruses. Raven, New York, pp 151–184
Drake JW (1993) Rates of spontaneous mutation among RNA viruses. Proc Natl Acad Sci U S A 90:4171–4175
Funk SM, Fiorella CV, Cleaveland S, Gompper ME (2001) The importance of disease in carnivore conservation. In: Symposia of the Zoological Society of London. Cambridge University Press, Cambridge, pp 443–466
Gao F, Bailes E, Robertson DL, Chen YL, Rodenburg CM, Michael SF, Cummins LB, Arthur LO, Peeters M, Shaw GM, Sharp PM, Hahn BH (1999) Origin of HIV-1 in the chimpanzee *Pan troglodytes troglodytes*. Nature 397:436–441
Garnier T, Eiglmeier K, Camus JC, Medina N, Mansoor H, Pryor M, Duthoy S, Grondin S, Lacroix C, Monsempe C, Simon S, Harris B, Atkin R, Doggett J, Mayes R, Keating L, Wheeler PR, Parkhill J, Barrell BG, Cole ST, Gordon SV, Hewinson RG (2003) The complete genome sequence of *Mycobacterium bovis*. Proc Natl Acad Sci U S A 100:7877–7882
Goodhart CB (1988) Did virus transfer from harp seals to common seals? Nature 336:21
Grachev MA, Kumarev VP, Mamaev LV, Zorin VL, Baranova LV, Denikina NN, Belikov SI, Petrov EA, Kolesnik VS, Kolesnik RS, Dorofeev VM, Beim AM, Kudelin VN, Nagieva FG, Sidorov VN (1989) Distemper virus in Baikal seals. Nature 338:209
Hahn BH, Shaw GM, De Cock KM, Sharp PM (2000) AIDS—AIDS as a zoonosis: Scientific and public health implications. Science 287:607–614
Hart CA, Bennett M, Begon ME (1999) Zoonoses. J Epidemiol Community Health 53:514–515
Haydon DT, Cleaveland S, Taylor LH, Laurenson MK (2002) Identifying reservoirs of infection: a conceptual and practical challenge. Emerg Infect Dis 8:1468–1473
He JF, Peng GW, Min J, Yu DW, Liang WJ, Zhang SY, Xu RH, Zheng HY, Wu XW, Xu J, Wang ZH, Fang L, Zhang X, Li H, Yan XG, Lu JH, Hu ZH, Huang JC, Wan ZY, Hou

JL, Lin JY, Song HD, Wang SY, Zhou XJ, Zhang GW, Gu BW, Zheng HJ, Zhang XL, He M, Zheng K, Wang BF, Fu G, Wang XN, Chen SJ, Chen Z, Hao P, Tang H, Ren SX, Zhong Y, Guo ZM, Liu Q, Miao YG, Kong XY, He WZ, Li YX, Wu CI, Zhao GP, Chiu RWK, Chim SSC, Tong YK, Chan PKS, Tam JS, Lo YMD (2004) Molecular evolution of the SARS coronavirus during the course of the SARS epidemic in China. Science. 303:1666–1669

Holmes EC (2003) Error thresholds and the constraints to RNA virus evolution. Trends Microbiol 11:543–546

Holmes EC, Burch SS (2000) The causes and consequences of genetic variation in dengue virus. Trends Microbiol 8:74–77

Holmes EC, Rambaut A (2004) Viral evolution and the emergence of SARS coronavirus. Philos Trans R Soc Lond B Biol Sci 359:1059–1065

Hueffer K, Parker JSL, Weichert WS, Geisel RE, Sgro JY, Parrish CR (2003) The natural host range shift and subsequent evolution of canine parvovirus resulted from virus-specific binding to the canine transferrin receptor. J Virol 77:1718–1726

Hufnagel L, Brockmann D, Geisel T (2004) Forecast and control of epidemics in a globalized world. Proc Natl Acad Sci U S A 101:15124–15129

Institute of Medicine (1992) Emerging infections: microbial threats to health in the United States. National Academy, Washington, DC

Jackson AP, Charleston MA (2004) A cophylogenetic perspective of RNA-virus evolution. Mol Biol Evol 21:45–57

John RKS, King A, de Jong D, Bodie-Collins M, Squires SG, Tam TWS (2005) Border screening for SARS. Emerg Infect Dis 11:6–10

Keeling MJ, Grenfell BT (1997) Disease extinction and community size: Modeling the persistence of measles. Science 275:65–67

Keller LF, Waller DM (2002) Inbreeding effects in wild populations. Trends Ecol Evol 17:230–241

Kennedy S, Kuiken T, Jepson PD, Deaville R, Forsyth M, Barrett T, van de Bildt MWG, Osterhaus A, Eybatov T, Duck C, Kydyrmanov A, Mitrofanov I, Wilson S (2000) Mass die-off of Caspian seals caused by canine distemper virus. Emerg Infect Dis 6:637–639

Krebs JR, Anderson RM, Clutton-Brock T, Donnelly CA, Frost S, Morrison WI, Woodroffe R, Young D (1998) Policy: biomedicine—badgers and bovine TB: conflicts between conservation and health. Science 279:817–818

Kuiken T, Fouchier R, Rimmelzwaan G, Osterhaus A (2003) Emerging viral infections in a rapidly changing world. Curr Opin Biotechnol 14:641–646

Li KS, Guan Y, Wang J, Smith GJD, Xu KM, Duan L, Rahardjo AP, Puthavathana P, Buranathai C, Nguyen TD, Estoepangestie ATS, Chaisingh A, Auewarakul P, Long HT, Hanh NTH, Webby RJ, Poon LLM, Chen H, Shortridge KF, Yuen KY, Webster RG, Peiris JSM (2004) Genesis of a highly pathogenic and potentially pandemic H5N1 influenza virus in eastern Asia. Nature 430:209–213

Mackenzie JS, Chua KB, Daniels PW, Eaton BT, Field HE, Hall RA, Halpin K, Johansen CA, Kirkland PD, Lam SK, McMinn P, Nisbet DJ, Paru R, Pyke AT, Ritchie SA, Siba P, Smith DW, Smith GA, van den Hurk AF, Wang LF, Williams DT (2001) Emerging viral diseases of Southeast Asia and the western Pacific. Emerg Infect Dis 7:497–504

Mahy BWJ (1993) Seal plague virus. In: Morse EE (ed) Emerging viruses. Oxford University Press, Oxford, pp 184–193
Mamaev LV, Visser IKG, Belikov SI, Denikina NN, Harder T, Goatley L, Rima B, Edginton B, Osterhaus A, Barrett T (1996) Canine distemper virus in Lake Baikal seals (*Phoca sibirica*). Vet Rec 138:437–439
Mayr E (1942) Systematics and the Origin of Species. Columbia University Press, New York
McFadden G (2005) Poxvirus tropism. Nature Rev Microbiol 3:201–213
McMichael AJ (2004) Environmental and social influences on emerging infectious diseases: past, present and future. Philos Trans R Soc Lond B Biol Sci 359:1049–1058
McNamara TS (2002) Pathology and early recognition of zoonotic disease outbreaks. In: Burroughs T, Knobler S, Lederberg J (eds) The emergence of zoonotic diseases: understanding the impact on animal and human health. National Academy Press, Washington, DC, pp 64–66
McNeill WH (1976) Plagues and people. Amchor/Doubleday, Garden City,
Morimoto K, Hooper DC, Carbaugh H, Fu ZF, Koprowski H, Dietzschold B (1998) Rabies virus quasispecies: implications for pathogenesis. Proc Natl Acad Sci U S A 95:3152–3156
Morse SS (1995) Factors in the emergence of infectious diseases. Emerg Infect.Dis 1:7–15
Murphy FA (1998) Emerging zoonoses. Emerg Infect Dis 4:429–435
Murray DL, Kapke CA, Evermann JF, Fuller TK (1999) Infectious disease and the conservation of free-ranging large carnivores. Anim Conserv 2:241–254
Osburn BI (1996) Emerging diseases with a worldwide impact and the consequences for veterinary curricula. Vet Quart 18:S124–S126
Ostfeld RS, Keesing F (2000) Biodiversity and disease risk: the case of lyme disease. Conserv Biol 14:722–728
Palmer SR, Soulsby EJL, Simpson DIH (1998) Zoonoses: biology clinical practice, and public health control. Oxford University Press, New York
Parrish CR (1994) The emergence and evolution of canine parvovirus – an example of recent host-range mutation. Semin Virol 5:121–132
Rosenblatt DL, Heske EJ, Nelson SL, Barber DH, Miller MA, MacAllister B (1999) Forest fragments in east-central Illinois: Islands or habitat patches for mammals? Am Midl Nat 141:115–123
Schluter D (2000) The ecology of adaptive radiation. Oxford University Press, Oxford
Schrag SJ, Wiener P (1995) Emerging infectious disease – what are the relative roles of ecology and evolution. Trends Ecol Evol 10:319–324
Shears P (2000a) Communicable disease surveillance with limited resources: the scope to link human and veterinary programmes. Acta Trop 76:3–7
Shears P (2000b) Emerging and reemerging infections in Africa: the need for improved laboratory services and disease surveillance. Microbes Infect 2:489–495
Shu LP, Sharp GB, Lin YP, Claas ECJ, Krauss SL, Shortridge KF, Webster RG (1996) Genetic reassortment in pandemic and interpandemic influenza viruses—a study of 122 viruses infecting humans. Eur J Epidem 12:63–70
Simpson GG (1953) The major features of evolution. Columbia University Press, New York
Song HD, Tu CC, Zhang GW, Wang SY, Zheng K, Lei LC, Chen QX, Gao YW, Zhou HQ, Xiang H, Zheng HJ, Chern SWW, Cheng F, Pan CM, Xuan H, Chen SJ, Luo HM,

Zhou DH, Liu YF, He JF, Qin PZ, Li LH, Ren YQ, Liang WJ, Yu YD, Anderson L, Wang M, Xu RH, Wu XW, Zheng HY, Chen JD, Liang GD, Gao Y, Liao M, Fang L, Jiang LY, Li H, Chen F, Di B, He LJ, Lin JY, Tong SX, Kong XG, Du L, Hao P, Tang H, Bernini A, Yu XJ, Spiga O, Guo ZM, Pan HY, He WZ, Manuguerra JC, Fontanet A, Danchin A, Niccolai N, Li YX, Wu CI, Zhao GP (2005) Cross-host evolution of severe acute respiratory syndrome coronavirus in palm civet and human. Proc Natl Acad Sci U S A 102:2430–2435

Sprent JFA (1969) Helminth 'zoonoses'. Helminthol Abstracts 38:333–351

Taylor LH, Latham SM, Woolhouse MEJ (2001) Risk factors for human disease emergence. Philos Trans R Soc Lond.BBiol Sci 356:983–989

Tei S, Kitajima N, Takahashi K, Mishiro S (2003) Zoonotic transmission of hepatitis E virus from deer to human beings. Lancet 362:371–373

Telford SRI (2002) Deer tick-transmitted zoonoses in Eastern United States. In: Aguirre AA, Ostfeld RS, Tabor GM, House C, Pearl MC (eds) Conservation medicine: ecological health in practice. Oxford University Press, Oxford, pp 310–324

Thompson JN (1994) The evolutionary process. Chicago University Press? Chicago

Trevitt CR, Singh PN (2003) Variant Creutzfeldt-Jakob disease: pathology, epidemiology, and public health implications. Am J Clin Nutrit 78:651S–656S

Tyzzer EE (1938) *Cytoecetes microti*, n.g., n.sp., a parasite developing in granulocytes and infective for small rodents. Parasitolofy 30:242–257

Waterfield NR, Wren BW, French-Constant RH (2004) Invertebrates as a source of emerging human pathogens. Nature Rev Microbiol 2:833–841

World Health Organization (1959) Zoonoses: second report of the joint WHO/FAO expert committee

Wilson ME (2003) The traveller and emerging infections: sentinel, courier, transmitter. J Appl Microbiol 94:1S–11S

Wolfe ND, Escalante AA, Karesh WB, Kilbourn A, Spielman A, Lal AA (1998) Wild primate populations in emerging infectious disease research: the missing link? Emerg Infect Dis 4:149–158

Wolfe ND, Switzer WM, Carr JK, Bhullar VB, Shanmugam V, Tamoufe U, Prosser AT, Torimiro JN, Wright A, Mpoudi-Ngole E, McCutchan FE, Birx DL, Folks TM, Burke DS, Heneine W (2004) Naturally acquired simian retrovirus infections in central African hunters. Lancet 363:932–937

Woolhouse MEJ (2002) Population biology of emerging and re-emerging pathogens. Trends Microbiol 10:S3–S7

Woolhouse MEJ, Dye C (2001) Population biology of emerging and re-emerging pathogens—preface. Philos Trans R Soc Lond B Biol Sci 356:981–982

Woolhouse MEJ, Taylor LH, Haydon DT (2001) Population biology of multihost pathogens. Science 292:1109–1112

Woolhouse MEJ, Haydon DT, Anita R (2005) Emerging pathogens: the epidemiology and evolution of species jumps. Trends Ecol Evol 20(5):238–244

Worobey M, Holmes EC (1999) Evolutionary aspects of recombination in RNA viruses. J Gen Virol 80:2535–2543

CTMI (2007) 315:113–131
© Springer-Verlag Berlin Heidelberg 2007

Infection and Disease in Reservoir and Spillover Hosts: Determinants of Pathogen Emergence

P. W. Daniels (✉) · K. Halpin[1] · A. Hyatt[1] · D. Middleton[1]

[1]CSIRO Livestock Industries, Australian Animal Health Laboratory, Geelong, Australia,
Peter.Daniels@csiro.au

Abstract Infection and disease in reservoir and spillover hosts determine patterns of infectious agent availability and opportunities for infection, which then govern the process of transmission between susceptible species. In this chapter, using the zoonotic agents Hendra virus and Nipah virus as examples, the pathogenesis of infection in various species including the wildlife reservoirs and domestic spillover hosts is reviewed with an emphasis on the aspects of pathogenesis which contribute to the dissemination of infection. Through these discussions, the emergence of these zoonotic agents is explored.

1
Introduction

> The outbreak of fatal Nipah viral encephalitis of people in Malaysia and Singapore had as a necessary condition a "jump" of Nipah virus from a wildlife reservoir in fruit bats to establish infection in pigs. The jump from the wildlife reservoir was not sufficient to result in the outbreak, which also had as necessary preconditions a) the intensive pig farming industry, infection in which resulted in propagation and amplification of the virus among susceptible pigs, b) movement of infected pigs from farm to farm, to result in the infection of large naïve pig populations during the course of the outbreak, and c) close contact of large numbers of humans in the pig industry with infected pigs.
>
> P. Daniels (2000)

Emergence of a zoonosis such as Nipah virus disease will be a multifactorial process with progression through a number of possibilities, each of which constitute stop-go points, or "gates" (also referred to as transitions; see the chapter by Childs et al., this volume). Crucial to each step in the process is the need for transmission of the infectious agent. Some of the factors that facilitate transmission will be external, including anthropogenic factors. Essential for transmission, however, is the need for the agent to be biologically available for the infection of another host. This is typically via a method of excretion whereby the agent is externalised. The recipient host must also be susceptible to infection by a biologically plausible route.

Thus the pathogenic features of the agent in individual animals of the various host species determine patterns of availability and opportunities for infection that govern the process of transmission.

Below, the pathogenesis of henipavirus infection in various species will be reviewed and emergence of these zoonotic agents explored. It is important to remember that viruses do not have strategies for emergence, and that anthropomorphic descriptions are not appropriate. Rather, ongoing biological interactions between agent and host will undergo evolution and, with alignment of certain factors or events, this may allow progression of the opportunity for disease emergence.

2
The Agents and the Associated Diseases

The genus *Henipavirus* (see the chapter by Field et al., this volume) currently comprises two viruses, Hendra virus (HeV) and Nipah virus (NiV), non-segmented negative-strand RNA viruses. Taxonomically the genus is in the Paramyxovirinae subfamily of the family *Paramyxoviridae*, order *Mononegavirale*. Both viruses have caused zoonotic diseases affecting a broad range of hosts.

Hendra virus has been associated with deaths of horses and humans in Australia. It was first isolated in 1994 (Murray et al. 1995) from cases of acute respiratory disease in horses. In three of the seven recorded outbreaks, there was associated disease in people that came in contact with the equine cases and virus was isolated from one of these people (Selvey et al. 1995; O'Sullivan et al. 1997; Dunn, 2004; Taylor et al. 2005). Most recently, in 2004, two horses died of suspected Hendra virus respiratory disease and one human was shown to have antibodies to Hendra virus following a short flu-like illness (Dunn 2004; Taylor et al. 2005). To date, Hendra virus has not been identified outside Australia.

Nipah virus was isolated in Malaysia in 1999 and was the aetiological agent responsible for the deaths of 105 people to that time, most of whom suffered from a viral encephalitis (Chua et al. 2000). People became infected after close contact with Nipah virus-infected pigs, which showed a range of clinical manifestations including a respiratory disease (Mohd Nor et al. 2000; Daniels et al. 2004). The current distribution of Nipah virus extends beyond Southeast Asia into South Asia and Indochina (see the chapter by Field et al., this volume). In 2001 a strain of Nipah virus was identified as the agent responsible for a mysterious human illness in Bangladesh (Hsu et al. 2004). This new form of Nipah virus disease then re-emerged in early 2003, 2004 and again in 2005, with patients presenting with respiratory and neurological signs (Anonymous 2004, 2005; Hsu et al. 2004).

3
Pathogenesis in the Wildlife Reservoir

Maintenance of an infectious agent in a reservoir species requires transmission among animals of that species. The pathway may be vertical, without external excretion of the agent, or horizontal, usually but not necessarily with externalisation of the agent by excretion. Nonetheless, the agent must be presented to the new host animal for infection to occur. The modes of transmission of the henipaviruses in their reservoir hosts are not known. In this section, the various possibilities will be explored to indicate how these viruses might be transmitted.

Serological studies of pteropid bat populations in Australia and Malaysia (Field et al. 2000; Yob et al. 2001), Cambodia (Olson et al. 2002; Reynes et al. 2005), Thailand (Wacharapluesadee et al. 2005) and Indonesia (Sendow et al. 2006) have all shown seroprevalences of antibodies to henipaviruses varying from 10% to 50% of animals sampled. Hence it is clear that these paramyxoviruses are transmitted within pteropid bat populations, the identified reservoir hosts for these viruses.

Attempts to isolate or detect henipaviruses from several hundred wild caught bats have not yielded many isolates but have suggested biologically plausible theories of routes of transmission. Hendra virus was initially isolated from the uterine fluid of a *Pteropus poliocephalus* female that had miscarried, and from foetal tissues (Halpin et al. 2000). A third isolate was obtained from the lung of a foetus collected from an injured *P. alecto* (Halpin et al. 2000). Additionally, Hendra virus was isolated from the kidneys of three *P. scapulatus*, and from two *P. poliocephalus* (one from blood, and one from the lung of a neonate) (K. Halpin, unpublished results). Hence Hendra virus has demonstrated only a narrow tissue tropism in flying foxes, having been isolated from renal tissues in adults, in addition to foetal tissues and uterine fluids.

Using a different sampling strategy, Nipah virus has been detected in the excretions of free-living pteropid bats. It was isolated from pools of voided urine of *P. hypomelanus* collected by Chua et al. (2002) from under roosting bat colonies, and from urine of *P. lylei* in Cambodia, using similar techniques (Reynes et al. 2005). Wacharapluesadee et al. (2005) confirmed detection of Nipah virus genome in the urine of *P. lylei* by PCR, and also reported its detection in a saliva sample. Chua reported isolation of Nipah virus on a piece of fruit that had been chewed by a pteropid bat (Chua et al. 2002). Thus opportunities for horizontal transmission of this infection have been clearly identified for Nipah virus.

Where henipavirus pathogenesis studies have been conducted, possible routes of excretion or externalisation of viruses have been consistent with field observations. Horizontal transmission has yet to be demonstrated within an experimental setting.

In early experimental infections with parenteral inoculation of *P. poliocephalus* with Hendra virus, virus was recovered from organs at 10 days postinfection (pi) but not 21 days pi, and not from specimens collected from orifices where virus excretion may have occurred. In one bat, virus was recovered from the foetus, confirming that pathogenesis in pregnant animals results in infection crossing the placenta, supporting the possibility of vertical transmission. Immunohistochemistry indicated viral-induced lesions were associated mainly with vascular and lymphoid tissues in infected bats, although the kidney was the site of some such vascular lesions (Williamson et al. 2000).

In experimental infections of *P. poliocephalus* with Nipah virus, experimental protocols were developed that allowed sampling of individual animals every other day. Virus was isolated from urine, on three occasions from one of six animals sampled. In a subsequent experiment, virus was isolated at post mortem from the kidney of a male bat and the uterus of a female bat (Middleton et al. 2007). Subsequent Hendra virus infections of *P. alecto* have resulted in similar findings, with urine yielding virus (K. Halpin, personal communication). However, in all of these isolations, the amount of virus recovered was at the limit of detection, and could not be titrated. This raises questions about the opportunities for spillover of virus from bats to other susceptible species. There is little information regarding the minimum infectious dose for species which have been infected in previous outbreaks. Preliminary experiments indicate that pigs require quite large amounts of virus, delivered via a natural route of infection, before they become infected (K. Halpin, personal communication). This information put together challenges the presumed transmission of Nipah virus from pteropus bats directly to pigs.

Overall, the experimental infections of pteropid bats with Hendra and Nipah virus tend to corroborate the findings from wild caught animals. Experimentally, both viruses have been detected in urine, and Nipah virus has been isolated from urine specimens collected in the field. While Hendra virus has not been isolated from urine of bats in the field, it has been isolated from kidneys, supporting the suggestion that the urinary tract may play a role in the excretion of this virus. The possible source of the Nipah virus isolates from chewed fruit spat out by flying foxes remains undetermined.

In addition, the patterns of isolations of Hendra virus from infected bats and the lesions in experimentally infected animals have led to a hypothesis that pregnancy facilitates Hendra virus pathogenesis in the reservoir host. Opportunities for horizontal transmission of virus derive particularly from fluids externalised during abortions, miscarriages or normal births (Halpin et al. 2000; Williamson et al. 2000). Such fluids would contaminate pasture under roosts, as would urine and faeces, and also allow exposure of potential spillover domestic animal hosts that have been a necessary link in the chain of causation resulting in human infection. Opportunities for such transmission might occur only occasionally, as do cases of spillover of Hendra virus disease. A seasonal incidence of Hendra virus disease cases has been observed during the pteropid bat birthing season in Australia (Murray et al. 1995; Hooper et al. 1996; Field et al. 2000; Taylor et al. 2005). However, it should be borne in mind that during the birthing season, female pteropid bats form maternity colonies and have limited interactions with adult males. If virus transmission was limited to the birthing season, we might expect to see seasonal seroprevalence fluctuations. Data is currently being collected in a longitudinal study to test this hypothesis (R. Plowright, personal communication).

If the route of excretion made the virus more readily accessible in the environment (for example, in urine or faeces) one might expect an increased chance of exposure by susceptible hosts and hence a greater number of cases. Factors moderating transmission would be firstly temporal: there may be a narrow window of virus excretion, secondly virological: there will be a threshold level of virus excreted that constitutes an infectious dose, which may be influenced by season, species and/or individual animal variation, and thirdly physical: the excreted virus must survive in the environment until it is encountered by the new host.

The studies with Nipah virus more clearly suggest opportunities for transmission: excretion of infectious virus certainly in the urine and possibly in saliva. Observations of the Nipah virus outbreak in Malaysia were consistent with a single spillover from reservoir host to domestic animal host, which would have occurred some time before the first human cases (Bunning et al. 2000). Pig farms where Nipah virus disease emerged were among orchards visited by flying foxes. Fruit trees were immediately adjacent to pens on farms, and pens were open to the environment. There was no impediment to bat urine contamination of pens, to half-eaten fruit contaminated with saliva falling into or around the pens, or even to whole bat carcasses falling into pens by misadventure. Any of these three scenarios would give a biologically plausible source of exposure of pigs, or some other spillover host, to infection from the pteropid bat reservoir consistent with the known aspects of pathogenesis in those species.

4 Pathogenesis in the Spillover Host

4.1 Pathogenesis and Transmission of Infection

Spillover from the reservoir requires more than the availability of virus from the natural host. It also requires that the natural host be brought into biological proximity with a second species, and that this candidate spillover host be susceptible to infection (see the chapter by Childs et al., this volume). This process can be considered at the level of the whole organism, and at the level of molecular pathogenesis. The infectious agent must be delivered to the spillover host via a route which results in infection. Once the agent is in contact with the new host, it must be accessible to the host at the tissue level for internalisation, which results in viral replication. The issues of cellular receptors for henipaviruses and associated aspects of cell biology have been reviewed (Eaton et al. 2006) and the key findings are summarised in Sect. 6, "Molecular Pathogenesis").

Transmission of henipaviruses to spillover domestic animal hosts has clearly occurred, with cases of natural Hendra virus disease reported in horses (Murray et al. 1995; Hooper et al. 1996; Field et al. 2000; Taylor et al. 2005). The susceptibility of cats to infection with Hendra virus has been demonstrated experimentally (Westbury et al. 1996). Natural cases of Nipah virus disease were confirmed in pigs, dogs, cats and horses (Chua et al. 2000). To support the propagation of an epidemic there must then be onward transmission among spillover hosts (the basic reproductive number, R_0, exceeding unity; see the chapter by Real and Biek, this volume), implying a pathogenesis among these animals that results in cycles of excretion and then infection of naïve animals.

Hooper et al. (2001) described various tissue tropisms of the henipaviruses and commented on the role such tropisms may play in transmission of infection and hence disease emergence. These viruses consistently show a tropism for the vasculature, with arterial endothelial tissues being susceptible in many virus species host-parasite systems. Immunohistochemical studies showed that the vascular endothelium and tunica media are frequently the site of virus localisation. Both viruses are also neurotropic in some species, especially man where neural involvement is noted in degenerated lesions adjacent to affected blood vessels. Particular localisation has been noted within pulmonary vasculature and, to a lesser degree, the renal glomerulus. Subsequent inflammatory lesions, which may ultimately lead to necrotising alveolitis, induce loss of integrity of the alveolar wall, reflected in intra-alveolar haemorrhage and oedema, and offer a clear route of viral excretion via compromised distal airways. In addition, direct involvement of pulmonary epithelial tissues has also been identified, notably associated with Nipah virus infection of various species. Glomerulonephritis and excretion of virus in urine are also possible sequelae to infection of the glomerular endothelium. Vascular lesions of the placenta, infection of the foetus and viral contamination of uterine fluids have also been described (Mungall et al. 2006). The story of disease emergence pays particular attention to these aspects of pathogenesis.

4.2 Pathogenesis and Transmission of Hendra Virus Infection in Horses

The potential for lateral spread of Hendra virus within the major spillover host, the horse, is unclear. While horse-to-horse spread is not well supported epidemiologically, perhaps it was more a question of opportunities associated with husbandry. In the largest outbreak where 13 horses died, human mechanical intervention may have facilitated transmission of the virus from horse to horse. Evidence from the first two Hendra virus outbreaks indicated that in each outbreak transmission occurred from only one index case (Baldock et al. 1996). The subsequent outbreaks have only involved single horse cases. To understand the biological mechanisms via which the virus can potentially be transmitted from an infected host to another, it is important to appreciate some specific characteristics of the disease and the pathogenesis in the virus-host system.

In many of the field cases, the airways of infected horses were filled with thick foam occasionally tinged with blood. This clinical characteristic has to date not been observed in experimentally infected horses; the difference has been attributed to routes of infection and/or length of disease (Hooper et al. 2001). The lesions commonly observed in both field and experimentally infected horses included dilated pulmonary lymphatics, severe pulmonary oedema and congestion. Upon histological and electron microscope examination, endothelial syncytial cells were observed in many organs including the capillaries and arterioles of the lung (Hyatt and Selleck 1996; Hyatt et al. 2001; Hooper et al. 2001) (Fig. 1A, B). The consequences of endothelial tropism have been reported as a high probability of proteinaceous oedema, pulmonary oedema, meningitis, atrophic glomeruli and placental infections (Hooper et al. 2001).

Ward et al. (1996) and McCormack (1999) report there is little to no evidence for the transmission of HeV between infected horses. Hooper et al. (2001) interpret these findings as there being little evidence for contagiousness of the virus. However, there have been seven outbreaks with two fatal and two non-fatal human infections. This high rate of morbidity implies that HeV is readily transmitted. In the Malaysian Nipah virus outbreak, infection of cats, dogs, horses and humans was attributed to exposure to infected pigs in each case (Mohd Nor et al. 2000), making the pathogenesis of the Nipah virus in pigs particularly interesting and important.

4.3 Pathogenesis and Transmission of Nipah Virus Infection in Pigs

Necropsies of field cases in pigs gave essential information regarding the pathogenesis of the infection that had serious implications for transmission among pigs and to humans (Shahiruddin et al. 1999). Clinically a respiratory disease in pigs had been recognised, with a barking cough being reported (Mohd Nor et al. 2000). However, infection in pigs was frequently asymptomatic (Daniels et al. 2004).

In both field and experimentally NiV-infected pigs, lesions were observed in the lungs. Specifically lesions were observed in the trachea, bronchi, bronchioles and alveolar cells (Fig. 1C, D). The lesions consisted of hyperplasia of columnar epithelium, peribronchiolar and peribronchial infiltration of lymphocytes and the presence of epithelial syncytial cells containing ribonucleoproteins (RNPs) and viruses budding (egressing) into the airways. Immunohistochemistry confirmed the presence of NiV antigen within the airways, and experimentally virus has been isolated from nasal and pharyngeal swabs, suggesting a major route of excretion of infectious virus (Middleton et al. 2002; Weingartl et al. 2005).

Video footage of the behaviour of clinically infected pigs on an infected farm (J. Aziz, unpublished data) showed how such a source of infection would lead to transmission among pigs. Weaned and growing pigs were penned together in large numbers, and coughing pigs were in close proximity to pen mates. Clinically affected pigs also presented at the food troughs at feeding time and were observed coughing onto the feed while at the trough. Hence there were

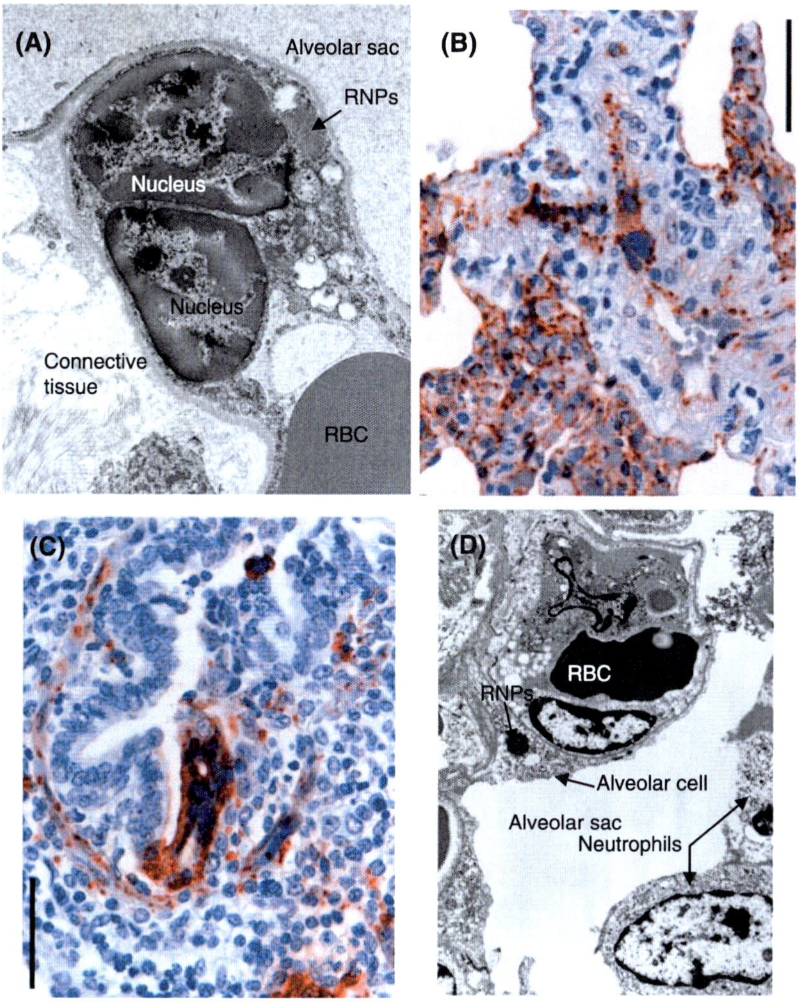

Fig. 1A–D Light and electron micrographs illustrating the different tropisms of HeV and NiV. **A** Transmission electron micrograph of the lung of an HeV-infected horse. *RBC* red blood cell; *RNPs* cytoplasmic aggregations of ribonucleoproteins. Note the infected cell is endothelial. **B** IPX-stained section of an HeV-infected horse lung. Note positive reaction in syncytial endothelial (and other) cell cytoplasms. **C** IPX-stained section of an NiV-infected pig lung. Note positive reaction in syncytial bronchiolar epithelial cell. **D** Transmission electron micrograph of the lung of an NiV-infected pig. *RBC* red blood cell; RNPs cytoplasmic aggregations of ribonucleoproteins. Note, the infected cell is epithelial. *Scale bars* in (**B**) and (**C**) represent 50 μm

numerous opportunities for the transfer of infected droplets and sputum from pig to pig. Susceptibility of pigs by the oral route of infection has been confirmed experimentally, and virus was frequently recovered from the oropharynx (tonsillar swabs) during the course of these experimental infections (Middleton et al. 2002). There have not yet been specific studies to determine if infected aerosols were produced during the course of the pathogenesis of the respiratory phase of the disease, but some level of aerosol spread would also have contributed to the high rate of contagion among pigs on farms.

Human exposure to this source of infection also seems probable. A case–control study on infected farms showed certain tasks were more likely to be associated with infection (Parashar et al. 2000). Importantly, just being resident on the farm was not a high risk factor (odds ratio [OR] 0.87), nor were tasks that involved walking around the pens under normal circumstances, such as hosing out pens (OR 1.48). Rather, feeding of pigs was identified as a high-risk occupation (OR 3.86). Observations of pigs at feeding time showed why this could be so. Under the intensive production systems of the farms, pens were quite close, separated by aisles wide enough to permit passage of barrows of feed. Operators would move the barrows down the aisles, manually distributing feed by bucket into troughs inside the pens. During this process, pigs in pens would stand with feet on the pen walls calling in expectation of the food, a posture that resulted in the heads of coughing pigs being at head height with the workers. Lorry drivers were also identified as being at increased risk. These workers were frequently required to manually lift apparently healthy pigs from pens onto trucks, a procedure that would likely have exposed them to respiratory excretions. This task would be analogous to that of pig chaser, the occupation in the Singapore abattoirs that had the highest risk of infection in that outbreak (Chew et al. 2000).

Other tasks that involved close handling of pigs, and particularly sick or dead pigs, also associated with elevated risks (assisting in breeding pigs, OR 3.86; assisting in the birth of pigs, OR 3.37; medicating pigs, OR 3.10; handling dead pigs, OR 3.89) (Parashar et al. 2000). Importantly, the study of natural and experimental Nipah virus disease in pigs showed not only a respiratory infection but also a generalised, systemic infection. Experimental Nipah virus disease in pigs induced a generalised vasculitis (Middleton et al. 2002). Clinically the range of signs reported on affected farms included fever and neurological disease in weaners and porkers, neurological disease and sudden death in adult breeding stock, abortions in sows, and a high incidence of neurological disease and mortality in suckling piglets. Exudation of blood-tinged respiratory exudates was a frequent finding in dead animals (Mohd Nor et al. 2000; Daniels et al. 2004). Histologically, meningitis was observed in disease on farms, as were syncytia in lymphoid organs (Hooper et al. 2001).

Although no specimens have been studied from reproductive disease in pigs, abortions and still births were widely reported by veterinary field investigations. Farm owners and workers would assist with difficult farrowings without the protection of gloves. They would also dismember carcasses without personal protection of the skin or mucous membranes. Hence unprotected contact with potentially infectious body fluids was an unintended feature of disease management on farms. Parashar et al. (2000) reported such occupations to be at high risk of infection on infected farms (OR 3.49). Hence the pathogenesis of Nipah virus disease in pigs created opportunities for spread among pigs and for pig-to-human transmission of infection.

4.4 Pathogenesis and Transmission of Infection in Other Spillover Hosts

For Hendra virus, the primary spillover host has been the horse, and for Nipah virus in Malaysia it was the pig. When considering other species, with both these viruses, cats have been shown to be susceptible. In the Nipah virus outbreak, dogs were affected. Cats have been infected experimentally with both Hendra and Nipah viruses by the oral route (Westbury et al. 1996; Middleton et al. 2002), and lateral transmission of Hendra virus between cats held experimentally in close contact has occurred. Virus was routinely isolated from the lung, kidney and urine of cases and less frequently from the trachea (Westbury et al. 1996). This pattern of virus isolation from infected cats was confirmed by Williamson et al. (1998), who also demonstrated the lateral transmission of Hendra virus from cats to a horse, with the urine being the probable source of infection. A field case of Nipah virus disease in a cat showed renal and pulmonary pathology (Hooper et al. 2001). In the experimental infection of cats with Nipah virus, the virus was isolated from specimens of urine and from tonsillar swabs collected during the clinical phase, and from lung, kidney and urine collected at necropsy in a fulminating case, as well as from other internal tissues. Immunohistochemistry demonstrated viral antigen in the epithelium of the trachea, bronchi, bronchioles and alveoli as well as extensively in renal cell types, and elsewhere in the body (Middleton et al. 2002). A recent finding of vertical transmission in a cat experimentally infected with Nipah virus parallels the similar observations with Hendra virus in pteropid bats (Mungall et al. 2007).

Hence the pathogenesis of henipavirus infections in cats shows clear potential for transmission of infections, both intraspecies and to other potential hosts, especially by contaminated urine but possibly also by oral and respiratory excretions.

Dogs were reported to suffer a high rate of disease with mortality on Nipah virus-affected pig farms (Asiah et al. 2001), and necropsy of a clinical case showing a distemper-like syndrome confirmed their involvement (P.W. Daniels and M. Bunning unpublished data). Histopathology revealed a systemic disease with both respiratory tract and urinary system involvement, as reported in other species, with demonstration of Nipah virus antigen by immunohistochemistry in the kidney (Hooper et al. 2001). However, a serological study of surviving dogs along a transect of known infected and uninfected areas showed no evidence of transmission among dogs, with seropositive animals being found only in the vicinity of previously infected farms (Asiah et al. 2001). It was believed that dogs on infected farms became infected through close contact with diseased pigs (Mohd Nor et al. 2000), with the eating of placental tissues following normal farrowings and abortions being an hypothesised route of transmission (J. Aziz, personal communication).

Horses are also susceptible to Nipah virus infection, with two seropositive animals being detected and archival nervous tissue from a third showing Nipah virus antigens by immunohistochemistry. All were from a stable in close proximity to infected pig farms and are believed to have been infected by transmission from pigs. Serological studies gave no evidence of transmission among horses (Mahendran et al. 1999).

4.5 Pathogenesis in the Human Host

Only four people have been diagnosed with Hendra virus infection and all had evidence of close contact to potentially infectious body fluids of equine cases (Taylor et al. 2005). There has been no evidence of person-to-person transmission of Hendra virus, either clinically or in serological studies.

There has been a similar lack of evidence of human-to-human transmission of Nipah virus infection in Malaysia. There were 265 human cases recorded, and 92% of these had definable exposure to infected pigs (Parashar et al. 2000). A large cohort serological study of human contacts of cases, such as healthcare workers, showed no infections (Mounts et al. 2001). Interestingly, Nipah virus has been readily detected in the saliva or throat swabs and urine of human patients (Chua et al. 2001), an aspect of the pathogenesis of the human disease that leaves open the opportunity for lateral spread.

Outbreaks of Nipah viral encephalitis in villagers in Bangladesh have resulted in different observations. Human cases have been diagnosed in circumstances where contact with infected pigs seems unlikely. In the first two series of outbreaks, in 2001 and 2003, no obvious domestic animal source of infection was identified (Hsu et al. 2004). Infection was attributed as possibly due to inadvertent direct contact with the pteropid bat, *P. giganteus*, which were present at outbreak sites and shown to have antibodies capable of neutralising Nipah virus.

The possibility of person-to-person spread was also seriously considered. There were clusters of cases in family households, with dates of symptom onset occurring over a range of time consistent with the presumed incubation period (Hsu et al. 2004). In a subsequent outbreak in 2004, 27 of 36 cases died, and the epidemiological evidence was taken to indicate person-to-person transmission. Cases were clustered in families, and prior to onset of illness, 92% of cases had contact with another case. Large droplet transmission was hypothesised. Acute respiratory disease syndrome was observed in a number of these cases (Anonymous 2004). Serological studies of various animal species showed only fruit bats to be seropositive.

Outbreaks of human Nipah virus disease continue to occur in Bangladesh, with a seasonal distribution in the period January to April, possibly indicating a factor in the pathogenesis of the infection in the reservoir host. In the most recent outbreak, a case–control study identified the drinking of raw date palm juice as a risk factor (OR 7.9), a product known to be exposed to contamination by pteropid bats (Anonymous 2005).

These new observations of Nipah virus infections of people in Bangladesh, of presumed transmission directly from the reservoir host to people and subsequent person-to-person spread, are of concern. The outbreaks are a regular occurrence, and recent human cases have emphasised the respiratory aspect of the disease. It would be of major concern if the pathogenesis in this human population resulted in greater transmission of the virus.

5
Aspects of Pathogenesis Contributing to Dissemination of Infection

Clinical disease has not been observed with henipaviral infections in pteropid bats, although the continuing detection of antibodies in these species indicates that virus is continually being transmitted in these populations. Individual animals are highly mobile, being able to fly long distances. They roost together in large numbers, including colonies of mixed species (Hall and Richards 2000). The absence of disease suggests such animals may be able to undertake normal activities of travel and mingling while infected, a convergence of factors that conceivably leads to opportunities for widespread dissemination in the reservoir host.

The distribution of the family *Pteropodidae* worldwide encompasses South-East Asia, the Pacific islands, India, Madagascar, and much of Africa, with representatives of the genus *Pteropus* found over the entire range except Africa (Mickleburgh et al. 1992). The ranges of neighbouring species overlap (Corbet and Hill 1992; Flannery 1995), making feasible the inter-species transmission of infectious agents and the possibility of related viruses in other pteropid species across the entire range. However, serology indicates that Australian species of pteropid bats are not naturally infected with Nipah virus, nor are Malaysian species infected with Hendra virus (H.E. Field and P.W. Daniels, unpublished observations). Although Australian species of pteropid bat (*P. poliocephalus*) can be experimentally infected with Nipah virus (Middleton et al. 2007), there has been no evidence of this happening in the wild. This observation raises the possibility that bat-virus ecosystems exist and that the pathogenesis of these viruses may be preferentially adapted to separate virus–bat species combinations. Nipah virus has been tentatively identified in *P. lylei* bats in Thailand, extending the number of bat species possibly susceptible to Nipah virus (Wacharapluesadee et al. 2005). The natural history of henipaviruses has yet to be fully described (see chapter by Field et al., this volume).

An aspect of the Nipah virus disease outbreak in Malaysia that caused international concern was the spread of disease to Singapore, where 22 abattoir workers were infected, with one fatality (Paton et al. 1999; Chan et al. 2002). Contact with live pigs was identified as the highest risk factor, although no unusual illnesses were reported among pigs processed during the presumed period of exposure (Chew et al. 2000). As reviewed above and elsewhere (Daniels et al. 2004), asymptomatic infections are a feature of porcine Nipah viral disease. Animals transported to abattoirs and accepted for processing at abattoirs are normally clinically healthy, and it was the asymptomatic presentation of the pathogenesis of the infection in pigs that allowed the infection to spread to different states in Malaysia and then internationally to Singapore.

6
Molecular Pathogenesis

Broad host distribution is a characteristic in which the henipaviruses differ from other paramyxoviruses (Eaton et al. 2006). This characteristic can be explained at the molecular level. Recently the cellular receptor which Hendra virus and Nipah virus use to gain entry to vertebrate cells was discovered (Bonaparte et al. 2005; Negrete et al. 2005). It is not known if this is the only cellular receptor for these viruses, but its wide distribution among vertebrate species provides an explanation behind the wide host distribution of henipaviruses. At the cellular level, ephrin B2 is located preferentially in arterial endothelial cells and the surrounding tunica media, but is not found in venous endothelial cells (Negrete et al. 2005). This explains the findings that systemic infections caused by henipaviruses display a tropism for arterial rather than venous endothelial cells (Negrete et al. 2005). Ephrin B2 is also found in neurons, which fits with the observations of encephalitis in human patients (Frisen et al. 1999).

Additional evidence to explain the systemic nature of henipavirus infections has come from studying the F protein of these viruses. This protein, which is found on the surface of the virus, requires proteolytic cleavage to generate a biologically active form. Studies have revealed that it is cleaved by cathepsin L, an endosomal protease (Pager and Dutch 2005). This protease is widely distributed, and it may prove to be crucial in the systemic spread of virus and the transmission of infectious virus within and between species.

7 Summary

An emerging disease event is defined by the context of its first recognition. Emergency management of such events may be very successful in the absence of a full understanding of how the event has occurred, including whether the dominant disease presentation does in fact reflect the primary spillover of agent to a novel host. Elimination of Nipah virus from the Malaysian pig population and control of the associated human epidemic was carried out by culling infected pig herds, removing the contaminated livestock environment, and by controlling the interface between pigs and people. However, greater understanding of the complexities of disease emergence is necessary to reduce even further the risk of re-emergence (see the chapter by Childs, this volume).

The pathogenesis of NiV disease in pigs explains how infectious virus can be transmitted to other pigs and humans. The primary route of NiV excretion is via the airways. Given high-density husbandry practices and a management system which resulted in the rapid transportation of pigs throughout Malaysia, it is not surprising there was efficient NiV transmission between cohorts, humans and to different herds.

Field data have identified bats from the genus *Pteropus* as the reservoir hosts for henipaviruses. In NiV experimentally infected bats, there is infrequent isolation of virus from these animals and Nipah virus has not been recovered at titres which would readily infect pigs via a natural route of infection. Conversely, cats are highly susceptible to Nipah virus, albeit via a parenteral route of inoculation, and high titres of virus have been isolated from these infected cats (Mungall et al. 2006). Experimental trials have been inconclusive about minimal infectious doses of HeV infection in horses, but in the outbreaks to date there does not appear to have been any horse-to-horse transmission. In HeV experimentally infected bats, only very low titres of virus have been recovered, again suggesting that a very sensitive spillover host would be needed for HeV transmission from bats, unless circumstances favour multiple simultaneous bat exposures. The question arises of whether pigs and horses are the secondary hosts of henipaviruses or whether intermediate hosts are involved. It may be hypothesised that the primary spillover host should be susceptible to low titres of henipavirus, should excrete high titres of virus and have an overlapping ecology with bats and the main indicator species, pigs in the case of NiV and horses in the case of HeV. Available data indicate that domestic cats could satisfy such criteria.

Further investigations to consolidate the current data may further clarify the relative importance of various domestic animals in the primary spillover event of henipavirus disease emergence, and generate more specific hypotheses as to how such events might occur.

References

Anonymous (2004) Nipah encephalitis outbreak over wide area of western Bangladesh (2004) Health Sci Bull 2:7–11

Anonymous (2005) Nipah virus outbreak from date palm juice. Health Sci Bull 3:1–5

Asiah NM, Mills JN, Ong BL, Ksiazek TG (2001) Epidemiological investigations of Nipah virus infection in peridomestic animals in peninsular Malaysia and future plans. In: Report of the regional seminar on nipah virus infection OIE representation for Asia and the Pacific. Tokyo, pp 47–50

Baldock FC, Douglas IC, Halpin K, Field HE, Young PL, Black PF (1996) Epidemiological investigations into the 1994 Equine Morbillivirus outbreaks in Queensland Australia. Singapore Vet J 2057–2061

Bonaparte MI, Dimitrov AS, Bossart KN, Crameri GS, Mungall BA, Bishop KA, Choudhry V, Dimitrov DS, Wang LF, Eaton BT, Broder CC (2005) Ephrin B2 ligand is a functional receptor for Hendra virus and Nipah Virus. Proc Natl Acad Sci U S A 102:10652–10657

Bunning M, Jamaluddin A, Cheang HT, Kitsutani P, Muhendren R, Olson J, Karim N, Field H, Johara Sharihuddin Choo Pow Yoon R, Daniels PW, Ksiazek T, Nordin MN (2000) Epidemiological trace-back studies of the Nipah virus outbreak in pig farms in the Ipoh district of Malaysia, 1997–1999. In: Cargill C, Mcorist S (eds) Proceedings, 16th International Pig Veterinary Society Congress, IPVS, Ocean Grove, p 551

Chan KP, Rollin PE, Ksiazek TG, Leo YS, Goh KT, Paton NI, Sng EH, Ling AE (2002) A survey of Nipah virus infection among various risk groups in Singapore. Epidemiol Infect 128:93–98

Chew MH, Arguin PM, Shay DK, Goh KT, Rollin PE, Shieh WJ, Zaki SR, Rota PA, Ling AE, Ksiazek TG, Chew SK, Anderson LJ (2000) Risk factors for Nipah virus infection among abattoir workers in Singapore. J Infect Dis 181:1760–1763

Chua KB, Bellini WJ, Rota PA, Harcourt BH, Tamin A, Lam SK, Ksiazek TG, Rollin PE, Zaki SR, Shieh WJ, Goldsmith CS, Gubler D, Roehrig JT, Eaton BT, Gould

AR, Olson J, Field H, Daniels P, Ling AE, Peters CJ, Anderson LJ, Mahy BJ (2000) Nipah virus: a recently emergent deadly Paramyxovirus. Science 288:1432–1435
Chua KB, Lam Sai Kit K, Goh KJ, Hooi PS, Ksiazek TG, Kamarulzaman A, Olson J, Tan CT (2001) The presence of Nipah virus in respiratory secretions and urine of patients during an outbreak of Nipah virus encephalitis in Malaysia. J Infect 42:40–43
Chua KB, Koh CL, Hooi PS, Wee KF, Khong JH, Chua BH, Chan YP, Lim ME, Lam SK (2002) Isolation of Nipah virus from Malaysian Island flying-foxes. Microbes Infect 4:145–151
Corbet GB, Hill JE (eds) (1992) The mammals of the Indomalay region: a systematic review. Oxford University Press, Oxford
Daniels P (2000) The Nipah virus outbreak in Malaysia: overview of the outbreak investigation and the issues that remain. In: Cargill C, Mcorist S (eds) Proceedings, 16th International Pig Veterinary Society Congress. IPVS, Ocean Grove, pp 553–554
Daniels PW, Shahirudin S, Aziz J, Ong BL (2004) Nipah virus disease. In: Coetzer JAW, Tustin RC (eds) Infectious diseases of livestock, 2nd edn. Oxford University Press, Oxford, pp 692–697
Dunn K (2004) A letter to the Queensland Racing Industry from the Chief Veterinary Officer Department of Primary Industries and Fisheries Ref 04/1(8929) Queensland Government Department of Primary Industries and Fisheries, Brisbane
Eaton BT, Broder CC, Middleton D, Wang LF (2006) Hendra and Nipah viruses: different and dangerous. Nat Rev Microbiol 4:23–35
Field HE, Yob JM, Azmin MR, Morrissy C, van der Heide B, White J, Daniels PW, Ksiazek TG (2000) Surveillance of wildlife for the source of Nipah virus – methodologies and outcomes. In: Proceedings of the 9th International Symposium on Veterinary Epidemiology and Economics (ISVEE), Colorado
Flannery TF (ed) (1995) Mammals of the South-West Pacific and Moluccan Islands. Australian Museum/Reed Books, Sydney
Frisen J, Holmberg J, Barbacid M (1999) Ephrins and their Eph receptors: multitalented directors of embryonic development. EMBO J 18:5159–5165
Hall L, Richards G (eds) (2000) Flying foxes: fruit and blossom bats of Australia. University of New South Wales Press, Sydney
Halpin K, Young PL, Field HE, Mackenzie JS (2000) Isolation of Hendra virus from pteropid bats: a natural reservoir of Hendra virus. J Gen Virol 81:1927–1932
Hooper PT, Gould AR, Russell GM, Kattenbelt JA, Mitchell G (1996) The retrospective diagnosis of a second outbreak of Equine Morbillivirus infection. Aust Vet J 74:244–245
Hooper PT, Zaki S, Daniels PW, Middleton D (2001) Comparative pathology of the diseases caused by Nipah and Hendra viruses. Microbes Infect 3:315–322
Hsu VP, Hossain MJ, Parashar UD, Ali MM, Ksiazek TG, Kuzmin I, Niezgoda M, Rupprecht C, Bresee J, Breiman RF (2004) Nipah virus encephalitis reemergence Bangladesh. Emerg Infect Dis 10:2082–2087
Hyatt AD, Selleck PW (1996) Ultrastructure of Equine Morbillivirus. Virus Res 43:1–15
Hyatt AD, Zaki SR, Goldsmith CS, Wise T, Hengstberger SG (2001) Ultrastructure of Hendra virus and Nipah virus within cultures cells and host animals. Microbes Infect 3:297–306

Mahendran R, Naseem M, Shahirudin S, Aziz AJ (1999) A preliminary study on the sero-prevalence of Nipah virus infection in horses in Malaysia. In: Seminar on Nipah virus: latest updates in the diagnosis, treatment and control in humans and animals. Veterinary Association of Malaysia, Melaka

McCormack JG (1999) Hendra virus: a new member of the paramyxoviridiae causing infections in humans, horses and fruit bats. Ippocrate 1:3

Mickleburgh SP, Hutson AM, Racey PA (eds) (1992) Old world fruit bats: an action plan for their conservation. IUCN, Gland, Switzerland

Middleton DJ, Westbury HA, Morrissy CJ, van der Heide BM, Russell GM, Braun MA, Hyatt AD (2002) Experimental Nipah virus infection in pigs and cats. J Comp Pathol 126:124–136

Middleton DJ, Morrissy CJ, van der Heide BM, Russell GM, Braun MA, Westbury HA, Halpin K, Daniels PW (2007) Experimental Nipah virus infection in pteropid bats (*Pteropus poliocephalus*) J Comp Pathology 136:266–272

Mohd Nor MN, Gan CH, Ong BL (2000) Nipah virus infection of pigs in peninsular Malaysia. Rev Sci Tech 19:160–165

Mounts AW, Kaur H, Parashar UD, Ksiazek TG, Cannon D, Arokiasamy JT, Anderson LJ, Lye MS, Nipah Virus Nosocomial Study Group (2001) A cohort study of health care workers to assess nosocomial transmissibility of Nipah virus Malaysia (1999) J Infect Dis 183:810–813

Mungall B, Middleton D, Crameri G, Halpin K, Bingham J, Eaton B, Broder C (2007) Vertical transmission and foetal replication of Nipah virus in an experimentally infected cat. Journal of infectious diseases (in press)

Mungall BA, Middleton D, Crameri G, Bingham J, Halpin K, Russell G, Green D, McEachern J, Pritchard LI, Eaton BT, Wang LF, Bossart KN, Broder CC (2006) Feline model of acute Nipah virus infection and protection with a soluble glycoprotein-based subunit vaccine. J Virol 80:12293–12302

Murray K, Selleck P, Hooper P, Hyatt A, Gould A, Gleeson L, Westbury H, Hiley L, Selvey L, Rodwell B, Ketterer P (1995) A morbillivirus that caused fatal disease in horses and humans. Science 268:94–97

Negrete OA, Levroney EL, Aguilar HC, Bertolotti-Ciarlet A, Nazarian R, Tajyar S, Lee B (2005) EphrinB2 is the entry receptor for Nipah virus, an emergent deadly paramyxovirus. Nature 436:401–405

Olson J, Rupprecht C, Rollin PE, An US, Niezgoda M, Clemins T, Walston J, Ksiazek TG (2002) Antibodies to Nipah-Like virus in bats (*Pteropus lylei*), Cambodia. Emerg Infect Dis 8:987–988

O'Sullivan JD, Allworth AM, Paterson DL, Snow TM, Boots R, Gleeson LJ, Gould AR, Hyatt AD, Bradfield J (1997) Fatal encephalitis due to novel paramyxovirus transmitted from horses. Lancet 349:93–95

Pager CT, Dutch RE (2005) Cathepsin L is involved in proteolytic processing of the Hendra virus fusion protein. J Virol 79:12714–12720

Parashar UD, Sunn LM, Ong F, Mounts AW, Arif MT, Ksiazek TG, Kamaluddin MA, Mustafa AN, Kaur H, Ding LM, Othman G, Radzi HM, Kitsutani PT, Stockton PC, Arokiasamy J, Gary HE Jr and Anderson LJ for the Nipah Encephalitis Outbreak

Investigation Team (2000) Case-control study of risk factors for human infection with a new zoonotic paramyxovirus Nipah virus, during a 1998–1999 outbreak of severe encephalitis in Malaysia. J Infect Dis 181:1755–1759

Paton NI, Leo YS, Zaki SR, Wong MC, Lee KE, Ling AE, Chew SK, Ang B, Rollin PE, Ksiazek TG, Auchus AP, Umapathi T, Sng I, Lee CC, Lim E, Kurup A, Lam MS, Wong SY (1999) Outbreak of Nipah virus infection among abattoir workers in Singapore. Lancet 354:1253–1256

Reynes JM, Counor D, Ong S, Faure C, Seng V, Molia S, Walston J, Georges-Courbot MC, Deubel V, Sarthou JL (2005) Nipah virus in Lyle's flying foxes Cambodia. Emerg Infect Dis 11:1042–1047

Selvey LA, Wells RM, McCormack JG, Ansford AJ, Murray PK, Rogers RJ, Lavercombe PS, Selleck PW, Sheridan J (1995) Infection of humans and horses by a newly described morbillivirus. Med J Aust 162:642–645

Sendow I, Field HE, Curran J, Darminto Morrissy C, Meehan G, Buick T, Daniels P (2006) Henipavirus in Pteropus bats Indonesia. Emerg Infect Dis 12:711–712

Shahirudin S, Zamri-Saad M, Shamshad SS, Mahani H, Norazian B, Daniels P, Aziz AJ (1999) A novel clinicopathological disorder associated with a zoonotic paramyxovirus infection in pigs. in: Proceedings of the Seminar on Nipah virus National Congress Animal Health and Production. Alor Gajah, Malaysia, 3–4 September, 1999

Taylor C, Smith I, Harrower B, Hanna J, Brookes D, Smith G (2005) Diagnosis of Hendra virus infection in a veterinarian following occupational exposure (abstract). Third Australian Virology Group Meeting, Cowes Phillip Island, Victoria. Abstract and Delegate Information, p 61

Wacharapluesadee S, Lumlertdacha B, Boongird K, Wanghongsa S, Chanhome L, Rollin P, Stockton P, Rupprecht CE, Ksiazek TG, Hemachudha T (2005) Bat Nipah virus Thailand. Emerg Infect Dis 11:1949–1951

Ward MP, Black PF, Childs AJ, Baldock FC, Webster WR, Rodwell BJ, Brouwer SL (1996) Negative findings from serological studies of Equine Morbillivirus in the Queensland horse population. Aust Vet J 74:241–243

Weingartl H, Czub S, Copps J, Berhane Y, Middleton DJ, Marszal P, Gren J, Smith G, Ganske S, Manning L, Czub M (2005) Invasion of the central nervous system in a porcine host by Nipah virus. J Virol 79:7528–7534

Westbury HA, Hooper PT, Brouwer SL, Selleck PW (1996) Susceptibility of cats to Equine Morbillivirus. Aust Vet J 74:132–134

Williamson MM, Hooper PT, Selleck PW, Gleeson LJ, Daniels PW, Westbury HA, Murray PK (1998) Transmission studies of Hendra virus (equine morbillivirus) in fruit bats, horses and cats. Aust Vet J 76:813–818

Williamson MM, Hooper PT, Selleck PW, Westbury HA, Slocombe RF (2000) Experimental Hendra virus infection in pregnant guinea-pigs and fruit bats (*Pteropus poliocephalus*). J Comp Pathol 122:201–207

Yob JM, Field H, Rashdi AM, Morrissy C, van der Heide B, Rota P, bin Adzhar A, White J, Daniels P, Jamaluddin A, Ksiazek T (2001) Nipah virus infection in bats (order *Chiroptera*) in peninsular Malaysia. Emerg Infect Dis 7:439–441

CTMI (2007) 315:133–159
© Springer-Verlag Berlin Heidelberg 2007

Henipaviruses: Emerging Paramyxoviruses Associated with Fruit Bats

H. E. Field[1] (✉) · J. S. Mackenzie[2] · P. Daszak[3]

[1]Department of Primary Industries and Fisheries, Brisbane, Australia
hume.field@dpi.qld.gov.au
[2]Australian Biosecurity Cooperative Research Centre, Curtin University of Technology, Perth, Australia
[3]Consortium for Conservation Medicine, 460 West 34th Street, New York, NY 10001, USA

Abstract Two related, novel, zoonotic paramyxoviruses have been described recently. Hendra virus was first reported in horses and thence humans in Australia in 1994; Nipah virus was first reported in pigs and thence humans in Malaysia in 1998. Human cases of Nipah virus infection, apparently unassociated with infection in livestock, have been reported in Bangladesh since 2001. Species of fruit bats (genus *Pteropus*) have been identified as natural hosts of both agents. Anthropogenic changes (habitat loss, hunting) that have impacted the population dynamics of *Pteropus* species across much of their range are hypothesised to have facilitated emergence. Current strategies for the management of henipaviruses are directed at minimising contact with the natural hosts, monitoring identified intermediate hosts, improving biosecurity on farms, and better disease recognition and diagnosis. Investigation of the emergence and ecology of henipaviruses warrants a broad, cross-disciplinary ecosystem health approach that recognises the critical linkages between human activity, ecological change, and livestock and human health.

1 Introduction

The apparent temporally clustered emergence of Hendra virus and Nipah virus in Australia and Malaysia, respectively, and the identification of species of fruit bats (*Pteropus* spp., commonly known as flying foxes) as likely reservoir hosts, poses a number of important questions on the ecology of henipaviruses. What factors precipitated their emergence? Why did they emerge at this time? What are the spillover mechanisms? What is their geographic occurrence? What are the potential impacts on humans and domestic species? The more recent description of Nipah virus-attributed disease in humans in Bangladesh reinforces the need for a comprehensive understanding of the ecology and, more broadly, epidemiology of these agents. This chapter describes the emergence of Hendra and Nipah viruses and the search for their natural hosts, discusses the impacts of emergence, and suggests factors putatively associated with emergence.

2 Emergence

2.1 Hendra Virus

Hendra virus was first described in 1994 in Australia when it caused an outbreak of severe acute respiratory disease with high mortality in thoroughbred horses in a training stable in the city of Brisbane (Murray et al. 1995b). A member of the family *Paramyxoviridae*, Hendra virus was initially called equine morbillivirus, but was later re-named Hendra virus, after the Brisbane suburb where the outbreak occurred.

To date there have been five known foci of Hendra virus infection in horses: Brisbane 1994, Mackay 1994, Cairns 1999, Cairns 2004, and Townsville 2004

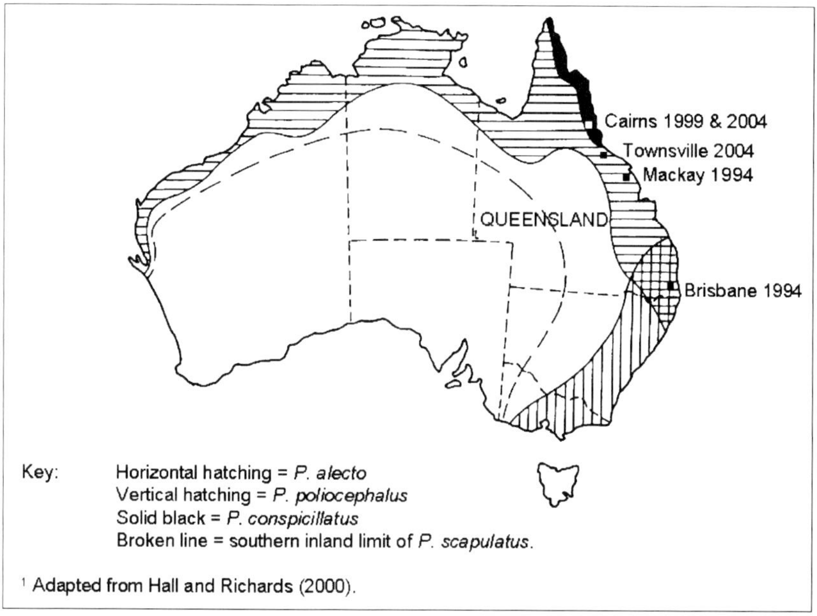

Fig. 1 Hendra virus spillover events and distribution of flying foxes (*Pteropus* spp.) on mainland Australia. (Adapted from Hall and Richards 2000)

(Fig. 1). The putative index case in the Brisbane outbreak was a heavily pregnant mare at pasture. She was observed to be ill on September 7, 1994 and moved to the Hendra training facility for intensive care, but died 2 days later. Over the following 14 days, 12 of 23 thoroughbreds in the facility and a neighbouring stable became ill and died acutely or were euthanised terminally (Fig. 2) (Murray et al. 1995a). Clinical signs included fever, facial swelling, severe respiratory distress, ataxia, and terminally, copious frothy (sometimes blood-tinged) nasal discharge. There were four non-fatal cases, two of which retained mild neurological signs. A further three horses in the stable were subsequently found to have seroconverted without apparent clinical signs. All seven were subsequently euthanised (Baldock et al. 1996; Douglas et al. 1997).

The trainer and a stable hand, both directly involved in nursing the index case, became ill with a severe influenza-like illness within a week of contacting the index case. The trainer was hospitalised and subsequently died after respiratory and renal failure. Infection with Hendra virus was demonstrated in both cases (Selvey et al. 1995).

The Mackay (1994) spillover chronologically preceded the Brisbane outbreak by 5 weeks, but was only retrospectively identified in October 1995 after the Hendra virus-attributed death of a thoroughbred stud-owner suffering a

September 1994								
	7	9	13	14	15	16	17	19-26
Horses								
Cannon Hill (Paddock)	2 horses moved							
Hendra (Stables)		Mare died				2 horses moved		10 horses dead 4 recovered
Hendra (Neighboring property)		1 horse moved						1 horse dead 1 recovered
Kenilworth (150 km distant)								1 horse dead 1 recovered
Samford (Paddock)								1 recovered
			New South Wales					
Humans								
Stablehand				Becomes ill				Slow recovery
Trainer					Becomes ill		Hospitalized	Died

Fig. 2 Chronology of equine and human cases of disease attributed to Hendra virus infection in the Brisbane outbreak. (From Murray et al. 1995a)

relapsing encephalitis. Two horses were infected on the Mackay property, both fatally (Hooper et al. 1996; Rogers et al. 1996). The first horse, a 10-year-old heavily pregnant thoroughbred mare died on August 1, 1994 after exhibiting severe respiratory distress, ataxia, and marked swelling of the cheeks and supra-orbital fossa over a 24-h period. The second horse, a two-year-old colt in an adjoining paddock was reported to have licked the muzzle of the dead mare. The colt died 11 days later, again after a 24-h clinical course, during which he exhibited aimless pacing, muscle trembling and haemorrhagic nasal discharge (Allworth et al. 1995).

Serological studies were an integral part of the outbreak investigations of the Brisbane and Mackay incidents. No evidence of Hendra virus infection was found in 800 domestic animals surveyed on the case properties or on in-contact properties. They included 387 horses, 287 cattle, goats and pigs, 23 dogs, 64 cats, and 39 poultry (Baldock et al. 1996; Rogers et al. 1996). Particular effort was directed towards surveying the broader horse population in the state of Queensland, with a further 2,024 horses from 166 properties sampled in a structured survey (Ward et al. 1996). With the exception of the seven horses that survived infection in the Hendra outbreak, none of the surveyed domestic animals showed serological evidence of exposure to Hendra virus. The negative surveillance findings (based on a highly sensitive serum neutralisation test) provided a high level of confidence that Hendra virus was not being sustained by in-contact domestic animal transmission, was not established in the Queensland horse population, and that the outbreak was unlikely to have originated from domestic species. Because of the temporal clustering of the Mackay and Brisbane incidents, efforts were made to identify possible links between the two properties. These investigations, undertaken in late 1995, and focused primarily on horse movements, personnel movements, and management practices, found no evidence to directly link the two outbreaks.

In January 1999, a third spillover was identified in a non-pregnant mare near Cairns in north Queensland. The horse deteriorated despite symptomatic treatment and was euthanised. A companion horse was unaffected on clinical and serological examination (Field et al. 2000).

In late 2004, again in north Queensland, in two spatially and temporally clustered events in Cairns and Townsville, a further two horses (both geldings) were fatally infected and a human case non-fatally infected. The human case, a veterinarian who undertook a necropsy on the first (Cairns) horse, and two assisting handlers, reported influenza-like symptoms 8–10 days after the necropsy. All three were negative for antibodies to Hendra virus by immunofluorescent antibody test (IFT) and enzyme-linked immunosorbent assay (ELISA) at that time. A follow-up sample taken from the veterinarian was positive by IFT and ELISA, and neutralising antibodies to Hendra virus were detected by serum neutralisation test. The two handlers remained seronegative. The veterinarian made an uneventful recovery. The horse was retrospectively diagnosed as a presumptive case on clinical grounds—no samples were available for laboratory confirmation (Field et al. 2007).

2.2 Nipah Virus

2.2.1 Malaysia

A major outbreak of disease in pigs and humans occurred in peninsular Malaysia between September 1998 and April 1999, resulting in the death of 105 of 265 human cases, and the culling of over 1 million pigs (Chua et al. 1999; Nor et al. 2000). Initially attributed to Japanese encephalitis virus, the primary disease aetiology was subsequently shown to be another previously undescribed virus of the family *Paramyxoviridae*. Preliminary characterisation of an isolate at the Centers for Disease Control and Prevention in Fort Collins, Colorado, and Atlanta, Georgia, USA, showed that the new virus, subsequently named Nipah virus, had ultrastructural, antigenic, serologic and molecular similarities to Hendra virus (CDC 1999).

The epidemic primarily impacted on pig and human populations. Infection in pigs was highly contagious, and clinical disease was characterised by acute fever with respiratory and/or neurological involvement. Incubation was estimated to be 7–14 days. Crude case mortality was low (<5%), and notably, a large proportion of infected pigs was asymptomatic. The clinical course appeared to vary with age. Sows primarily presented with neurological disease, but sometimes died suddenly without evident signs. In weaners and porkers, a respiratory syndrome predominated, frequently accompanied by a harsh non-productive (loud barking) cough. The predominant clinical syndrome in humans was encephalitic rather than respiratory, with clinical signs including fever, headache, myalgia, drowsiness, and disorientation sometimes proceeding to coma within 48 h (Chua et al. 1999; Goh et al. 2000). The majority of human cases had a history of direct contact with live pigs. Most were adult male Chinese pig farmers (Chua et al. 1999; Parashar et al. 2000).

Evidence of Nipah virus infection was also been found in dogs, cats and horses (Chua et al. 1999; Nor et al. 2000). The initially high prevalence of infection in dogs in the endemic area during and immediately following the removal of pigs suggested that dogs readily acquired infection from infected pigs. The much lower antibody prevalence and restriction of infection to within 5 km of the endemic area suggests that Nipah virus did not spread horizontally within dog populations, and that dogs were effectively a dead-end host (Field et al. 2001).

2.2.2 Bangladesh

Five outbreaks of Nipah virus-associated disease in humans were described in Bangladesh between April 2001 and February 2005 (Anonymous 2003, 2004a, 2004b, 2005b; Hsu et al. 2004). To 11 February 2005, a total of 122 cases were recognised by the Bangladesh Directorate of Health Services, at least 78 (64%) of which were fatal. A number of the characteristics of the Bangladesh outbreaks are similar to the outbreak in Malaysia; delayed recognition, a primary presentation with fever and central nervous system signs, and a high case fatality. However, in marked contrast to the Malaysian outbreak, infection in humans was not associated with disease in pigs (indeed pigs are uncommon in Bangladesh), and there was evidence of horizontal human transmission (discussed Sect. 4).

The first reported outbreak (13 cases, nine fatal) was in Meherpur in April–May 2001. The index case, a 33-year-old farmer, developed symptoms on April 20, and died 6 days later. Four other persons in the same household became cases 10–18 days after the index case. A further four of the cases were relatives of the index case. The second reported outbreak of 12 cases (eight fatal) occurred in Naogaon in January 2003. The index case was a 12-year-old boy. Cases occurred in eight households. A cluster of cases occurred in one household after the head of the household became ill and later died. Two weeks later, his wife and his three eldest daughters became ill; his wife and one daughter died. In both the Meherpur and Naogaon outbreaks, handling or exposure to patient secretions was a risk factor for illness (Hsu et al. 2004). The third reported outbreak occurred simultaneously in Goalanda (Rajbari district) and seven other districts between January and February 2004. A total of 29 cases were reported, of whom 22 died. There was a predominance of young boys in the Goalanda cluster, suggesting that a specific activity (such as climbing trees or ingesting fruits contaminated with the secretions of infected bats) may have led to exposure (Anonymous 2004a). The fourth reported outbreak occurred in the Faridpur district in April 2004. Of 36 identified cases, 27 were fatal. This outbreak differed from previous outbreaks in two important ways. Firstly, at least six patients developed an acute respiratory distress syndrome, in contrast to the previously observed predominant fever/neurological presentation; and secondly, the epidemiological evidence clearly indicated that person-to-person spread (possibly through large droplet transmission) was the primary mode of transmission. A fifth reported outbreak (12 cases, 11 fatal) was reported in January 2005, in the northern district of Tangail. Cases predominantly exhibited fever and neurological symptoms. Drinking raw date palm juice was the only surveyed exposure significantly associated with illness. Bats reportedly frequently drink from the open pots into which dripping juice is collected overnight, and bat excrement is reportedly common on or in the pots (Anonymous 2005b).

The pattern of the Bangladesh outbreaks suggests a sporadic, geographically scattered introduction of infection to humans. Nucleotide sequence data also supports a different epidemiology in Bangladesh. Overall, the nucleotide sequences of the genomes of the Nipah viruses isolated in Bangladesh in 2004 and in Malaysia in 1999 share 92% identity. While the size and distribution of the open reading frames and the sequences of key regulatory elements are conserved, the amount of genetic diversity present in sequences obtained in Malaysia and Bangladesh varies (Harcourt et al. 2005). Those obtained from human cases in Malaysia suggest a single source of human infection from the porcine amplifying host (AbuBakar et al. 2004; Chan et al. 2001; Chua et al. 2000); those from Bangladesh cases formed a cluster clearly distinct from the Malaysian sequences, but differed from each other by approximately 0.8%, suggesting possible multiple introductions of virus into humans. As yet, sequence data are unavailable from virus isolates obtained from putative person-to-person transmission chains to suggest genetic changes potentially associated with adaptation to the novel human host. Sequence changes in SARS coronavirus isolated from palm civets and from humans suggest active selection of novel genotypes, including genotypes potentially adapted to a novel human reservoir host (Liu et al. 2005; Song et al. 2005). Such genetic adaptation is a significant transition point in the evolution of specific human pathogens by which agents can emerge as pandemic threats, such as HIV and influenza A subtypes (see the chapter by Childs et al., this volume; Childs 2004).

3 Reservoir Studies

3.1 Hendra Virus

The emergence of Hendra virus caused consternation for both animal and public health authorities in Australia. Zoonotic infections of horses were previously unknown, yet it quickly became evident that the infection and consequent death of the trainer was attributable to his close contact with the index horse case. When the aetiological agent was established as a novel virus of the family *Paramyxoviridae*, the search for its origin began. The phylogenetic analysis suggested that the virus had not resulted from single or multiple point mutations from a closely related virus, and that emergence from a natural host was the most probable explanation of its origin (Murray et al. 1995b). Serosurveillance of wild-caught wildlife, initially in the Brisbane index case paddock, and later the Mackay index case property, was undertaken to evaluate this hypothesis. Negative findings prompted broadening of the sample base to include sick and injured wildlife in temporary captivity. Apart from increasing the number of species and locations sampled, this approach offered the advantages of a convenient and cost-effective sample, and access to that subset of the greater wildlife population in sub-optimal health. Access to the latter was of particular interest, because if infection in a natural host was associated with disease or debility, then infected animals would be over-represented in this group and thus more readily detected. Species that were common to the two locations and able to move readily between the two locations were given the highest surveillance priority (Young et al. 1996).

Flying foxes (genus *Pteropus*, order *Chiroptera*) were the only mammalian species meeting these criteria. The knowledge that viruses of the family *Paramyxoviridae* had previously been isolated from bats elsewhere (Henderson et al. 1995; Pavri et al. 1971) reinforced this focus. Of 27 flying foxes screened using this approach, 40% had antibodies neutralising Hendra virus (Field 2005)—a major breakthrough in the search for the origin of Hendra virus. Subsequently, virus was isolated from two species (Halpin et al. 2000). Investigations of the role of flying foxes in the ecology of the virus continued, and subsequent studies showed evidence of previous exposure to Hendra virus in all four mainland Australian flying fox species across their range. Species (*Pteropus alecto*) and increasing age were risk factors for infection in flying foxes. Retrospective studies identified evidence of infection in flying foxes well before the first known spillover to horses (Field 2005). These features suggest a major role for flying foxes in the ecology of Hendra virus, and are consistent a mature host-agent relationship. Subsequent studies identified neutralising antibodies to Hendra virus in multiple flying fox species in Papua New Guinea to the immediate north of Australia (Mackenzie et al. 2001). Current research priorities include modelling virus maintenance in flying fox populations (Plowright et al. 2005) and defining flying fox population dynamics by genetic and satellite telemetry studies (Daszak et al. 2006).

3.2
Nipah Virus

Investigation of the origin of Nipah virus was an integral part of the Malaysian outbreak response, and as the outbreak in pigs and humans came under control, the focus of part of the investigating team shifted to identifying the source of the infection in pigs. Molecular and serologic evidence indicating that Nipah and Hendra viruses were closely related made Malaysian bat species a logical surveillance priority. Malaysia has a great diversity of bat species: at least 13 described species of fruit bat (sub-order *Megachiroptera*), including two flying fox species, and 60 described species of insectivorous bats (sub-order *Microchiroptera*) in peninsular Malaysia alone (Medway 1978). Wildlife rescue networks are less extensive in Malaysia than in Australia, thus an opportunistic sampling methodology was not a realistic option in the Nipah virus investigations, as it was in Australia with Hendra virus, and wild-caught bats were the primary survey target. Over a 5-week period, bat populations at multiple locations across peninsular Malaysia were sampled, with sampling locations including but not limited to the outbreak areas. Neutralising antibodies to Nipah virus were found in bats from five species, but predominantly in *Pteropus hypomelanus* and *Pteropus vampyrus* (Johara et al. 2001). Subsequently, Chua et al. (2002) recovered Nipah virus from the urine of *P. hypomelanus* and from partially eaten fruit which had been contaminated by bat saliva or, less likely, bat urine. Current research priorities include investigation of the population dynamics of *P. vampyrus* in Malaysia and across its range, and the dynamics of Nipah virus infection in *P. vampyrus* and *P. hypomelanus* (Daszak 2005).

Ubiquitous peridomestic species were also extensively surveyed in seeking the origin of Nipah virus in Malaysia. The uniformly negative serology results from surveyed peridomestic rodents, insectivores, and birds in Malaysia (Asiah et al., unpublished data) indicate that these animals did not play a role as secondary reservoirs for Nipah virus. However, dogs did readily acquire infection following close association with infected pigs, and while horizontal transmission was not evident in dog populations, infected dogs possibly played a role in farm-to-farm transmission.

A serologic survey of domestic and wild animals undertaken after the 2001 Meherpur and 2003 Naogaon outbreaks in Bangladesh identified evidence of infection only in the flying fox *Pteropus giganteus*. Other (unidentified) bats showed no evidence of infection (Anonymous 2004a; Hsu et al. 2004). Concurrent serologic surveillance of *P. giganteus* in India in 2003 found that 54% had neutralising antibodies to Nipah virus (Epstein et al., unpublished observations), suggesting that Nipah virus or a cross-neutralising virus was widespread across the range of *P. giganteus*. Further, identification of neutralising antibodies to Nipah virus in *P. vampyrus* in Indonesia (Sendow et al. 2005) and Cambodia (Olson et al. 2002), and the isolation of Nipah virus from flying foxes in Cambodia (Reynes et al. 2005) strongly supported the hypothesis of Halpin et al. (2000) that henipaviruses likely exist across the entire global distribution of pteropid bats. A comprehensive investigation of the ecology of NiV in *P. giganteus* is needed to underpin risk management strategies in Bangladesh. Obvious research priorities include population dynamics of *P. giganteus*, Nipah virus infection dynamics in the species, potential modes of transmission to humans, and identification of factors precipitating emergence.

4 Modes of Spillover Transmissions

4.1 Hendra Virus

The mode of transmission of Hendra virus infection to horses in Australia has yet to be established. However, epidemiological investigations of natural infections in horses and flying foxes, and the outcomes of experimental infections in a range of species, provide useful information. Firstly, respiratory spread has not been demonstrated experimentally in any species, and the spatial pattern in naturally infected horses has not been consistent with respiratory spread. Secondly, Hendra virus has been isolated from the kidney and urine of horses and cats experimentally infected with Hendra virus, and cat-to-cat transmission and suspected cat-to-horse transmission have been attributed to exposure to infected urine (Westbury et al. 1996; Williamson et al. 1998). Thirdly, horses have been experimentally infected by the naso-oral route (Williamson et al. 1998).

Thus, hypotheses involving (1) the excretion of an infective dose of Hendra virus from a flying fox, (2) contamination of pasture, and (3) ingestion of the contaminated pasture by a susceptible horse are plausible. Young et al. (1997) proposed that transmission from flying foxes to horses was effected by contact with infected foetal tissues or fluids via the ingestion of recently contaminated pasture. This hypothesis was largely based on the August–September temporal overlay of the Brisbane and Mackay spillovers with the late gestation of *P. alecto* and *P. poliocephalus* in Queensland, the isolation of the virus from uterine fluid and foetal tissues of a naturally infected pregnant flying fox (Halpin et al. 2000), and on the absence of evidence of infection in flying fox carers regularly exposed to other potential routes of excretion such as urine, faeces and oro-nasal secretions.

Notwithstanding the latter, an alternative hypothesis is that the ingestion of pasture contaminated with infected flying fox urine is the mode of transmission to horses. Although Hendra virus has yet to be isolated from flying fox urine, the previously described isolation of Nipah virus from the urine of free-living *P. hypomelanus* (Chua et al. 2002) supports urine as a plausible route of excretion for Hendra virus. The urine hypothesis is also supported by the experimental studies described above that attribute cat-to-cat and probable cat-to-horse transmission to exposure to infected urine. Another plausible hypothesis is that the ingestion of spats (the fibrous remains of masticated fruit dropped by feeding flying foxes) is the mode of transmission to horses. The quantity of these spats under food trees bearing fibrous fruit (such as the *Ficus* species present in the Brisbane index case paddock) can be substantial, and they may represent an attractive source of saliva-laden virus to grazing horses. The viability of virus in spats is also likely to be prolonged due to slowed desiccation, heat and ultraviolet action. Hendra virus has been isolated from the oral cavity of experimentally infected horses (Williamson et al. 1998) and as previously noted, the closely related Nipah virus has been isolated from fruit partially eaten by bats (Chua et al. 2002), supporting saliva as a plausible route of excretion of Hendra virus from flying foxes.

It should be recognised that the mode of transmission between flying foxes and the mode of transmission from flying foxes to horses may differ. The infectious dose, the routes of infection, and the physiological risk factors for infection in both species are unknown or incompletely understood. Managing the risk of spillover to horses is further constrained by the lack of knowledge of the incidence of infection and the temporal pattern of infection (and thus excretion) in flying foxes. Regardless of the mode of transmission to horses, it is evident from natural infections and experimental studies that horse-to-horse transmission is not readily effected. The apparent exception is the first recognised outbreak in the Brisbane stables that involved 20 equine cases. However, the temporal pattern of infection in this outbreak suggests that the index case was the source of infection for all cases and that no secondary infection occurred (Baldock et al. 1996). Indeed, it is probable that horse-to-horse transmission in this instance was inadvertently facilitated by husbandry practices or other actions in the stable that resulted in the direct transmission of infected body fluids. Two plausible scenarios have been proposed: that a common syringe and needle was used to administer medication to the index case and to other (subsequently infected) horses; or that a cloth, bridle or other piece of equipment contaminated with infectious oral secretions from one horse was used on other (subsequently infected) horses. Likewise it is evident that horse-to-human transmission does not readily occur. Many people were potentially exposed to infection in the investigation of the Brisbane outbreak in particular, yet only the trainer and a strapper succumbed: both were closely involved in nursing the index case. It is also evident that bat-to-human transmission does not readily occur. After the identification of flying foxes as the origin of Hendra virus, a serologic survey of persons with high occupational or recreational exposure to flying foxes found none of 149 had neutralising antibodies to Hendra virus (Selvey et al. 1995). Whether the apparently low infectivity for horses and humans is a reflection of the innate infectivity of Hendra virus, the instability of the virus outside the host, or of ineffective contact is unclear, although experimental studies support the former.

4.2 Nipah Virus

The putative mode of transmission of Nipah virus in Malaysia is from flying foxes to pigs to humans. The epidemiological evidence indicates that the spillover from flying foxes to pigs occurred in northern peninsular Malaysia (Nor et al. 2000). Chua et al. (2002), having isolated Nipah virus from flying fox urine and from fruit partially eaten and discarded by flying foxes, hypothesised that transmission to pigs was effected by infected flying foxes feeding in trees overhanging pig pens. Epidemiological (Nor et al. 2000) and experimental (Middleton et al. 2002) findings indicate that pigs are highly susceptible to infection, and thus once infection is introduced to a farm, on-farm spread is rapid. The primary mode of on-farm transmission was believed to be via the oro-nasal route; the primary means of spread between farms and between regions was the movement of pigs. Secondary modes of transmission between farms within localised farming communities may have included roaming infected dogs and cats, and sharing of boar semen (although at present, virus has not been identified in porcine semen). Trucks transporting pigs may also have introduced the virus onto farms (Nor et al. 2000).

While the timing of the spillover (or spillovers) from flying foxes to pigs and the early epidemiology of infection in pig farms in northern peninsular Malaysia are unclear, retrospective investigations suggest that Nipah virus has been responsible for sporadic disease in pigs in peninsular Malaysia since late 1996. It is suggested that the disease was not recognised as a new syndrome because the clinical signs were not markedly different from those of several endemic pig diseases, and because morbidity and mortality were not remarkable (Bunning et al. 2000). Epidemiological modelling also supports an earlier spillover event (Daszak et al. 2006). They contend that a second introduction of infection was likely necessary for infection to become endemic in the 30,000-pig index-case farm and thus provide a sustained reservoir of Nipah virus from which to infect other farms.

Conclusive evidence of person-to-person transmission of Nipah virus was not found during investigations in Malaysia and Singapore. However, it should be noted that excretion of Nipah virus in urine and mucous obtained by throat swabs was readily demonstrable by isolation of virus from clinical samples obtained from acutely infected humans (Chua et al. 2001a). These data suggest the potential for person-to-person transmission of Nipah virus in southeastern Asia was epidemiologically plausible.

The evident horizontal human infection and the apparent absence of an intermediate domestic animal reservoir in the Bangladesh outbreaks are disturbing epidemiologic features not evident in Malaysia and Singapore. The earlier Bangladesh outbreaks suggested that close contact (handling or exposure to patient secretions) was necessary for person-to-person transmission, but the appearance of an acute respiratory syndrome in the 2004 Faridpur outbreak flagged the potential for much more efficient transmission. However, to date, person-to-person transmission appears to have been limited to a single generation, and no cases of transmission from patients to healthcare workers have been reported.

5
Putative Risk Factors for Emergence

A number of authors contend that a series of commonly occurring anthropogenic environmental changes drives disease emergence by pushing pathogens outside their normal population parameters(Krause 1992;

Lederberg et al. 1992; Morse 1995; Smolinski et al. 2003). They argue that the introduction of pathogens via global air travel and trade, the encroachment of human activities into wilderness regions, urbanisation, climatic changes and agricultural intensification are common drivers of emergence. For zoonotic diseases associated with wildlife reservoirs, anthropogenic factors that alter wildlife population structure, migration patterns and behaviour may also drive emergence of disease in human populations (Daszak et al. 2000, 2001). For example, human population encroachment into wildlife habitat may increase the risk of Lyme disease and other tick-borne encephalitides by driving the loss of less competent reservoir hosts and promoting a more efficient one, namely the white-footed mouse, *Peromyscus leucopus* (Ostfeld and Keesing 2000). Likewise, the introduction of a "new" infection into a human or domestic animal population may follow the incursion of humans (accompanied by their domestic animals) into previously remote natural habitats where unknown disease agents exist in harmony with wild reservoir hosts. Upon contact with new species, an agent may jump species barriers, thereby spilling over into humans or livestock. Unlike the natural host, the new host may have no natural immunity or evolved resistance. Additionally, high population densities and management practices may facilitate the rapid spread of pathogens throughout livestock populations. Infection may be transmitted to humans directly from the natural host or via an intermediate host.

The available evidence suggests that Hendra and Nipah viruses are phylogenetically distinct from other members of the *Paramyxoviridae* (Gould 1996; Murray et al. 1995b), well adapted to their natural flying fox hosts, and in whose populations they have long circulated (Field et al. 2001). The close phylogenetic relationship between Hendra and Nipah suggests a common progenitor. However, it also appears that flying fox populations in Australia and Malaysia have been separate for a length of time sufficient for the respective viruses to evolve further in geographic isolation. So what precipitated their emergence? Can environmental factors be identified that altered flying fox ecology and facilitated the movement of henipaviruses (and other bat-associated zoonoses) (Breed et al. 2005) beyond their natural ecological niche, precipitating their emergence? Disease emergence requires, in addition to the presence of an agent, an effective bridge from the natural host to a susceptible spillover host. Such bridges result from anthropogenic or natural changes to the agent, the host, or the environment. Available data on many fruit bat species suggests that populations in Australia and Asia are in decline and disruption throughout their range. In South-East Asia, anthropogenic activities (primarily habitat loss and hunting) have been identified as constituting the major threats. Deforestation, whether for agricultural land, commercial logging, or urban development, is widespread and results in loss or abandonment of roosting sites, and the loss of feeding habitats. Secondarily, habitat loss due to clearing is commonly exacerbated by tropical storms, the remnant forest being particularly prone to high wind damage (Mickleburg et al. 1992). Hunting, whether for consumption or crop protection, and at both a local and a commercial level, results in the abandonment of roost and feeding sites (Mickleburg et al. 1992). A scenario emerges of bat populations under stress, of altered foraging and behavioural patterns, of niche expansion, and of closer proximity to man. In eastern Australia, the increasing urban presence of flying foxes [thought to be due to more reliable and abundant food resources (Parry-Jones and Augee 2001)] and the associated changes in flying fox population dynamics, represents a similar emergence-promoting scenario for Hendra virus.

The emergence of Nipah virus disease clearly illustrates the two-step process described by Morse (1995). The establishment of pig farms within the range of the natural host supported the initial introduction into the pig population; the maintenance of high densities of pigs and the transport of pigs led to the establishment and rapid dissemination of infection within the pig population in peninsular Malaysia. Amplification of virus within pig populations then facilitated transmission to humans. A combination of factors likely increased the opportunity for effective contact between flying foxes and pigs, and thus the initial introduction of infection into the pig population. Plausible hypotheses include:

- The unsustainable hunting of *Pteropus* bat species has caused localised niche vacuums (sinks) with relative resource abundance, creating regional gradients along which neighbouring bat populations move, resulting in a net movement of virus into human-inhabited areas and so an increased probability of effective contact and spillover.
- Regional deforestation has changed the seasonal foraging movements of *Pteropus* bats and lead to an increased reliance on horticultural crops, resulting in a relative increased density of bats proximate to human and livestock populations. (Climatic changes, forest fires and associated haze events have similarly been hypothesised to influence flying fox movement patterns (Chua 2003).
- The marked increase in the number, density and distribution of the Malaysian pig population in the last 10 years has led to an increased probability of contact between flying foxes and pigs. This probability has been further increased by the practice of planting fruit orchards immediately adjacent to piggeries (Daszak et al. 2006).

Worldwide, there are approximately 60 species of bats in the genus *Pteropus* (family *Pteropodidae*, sub-order *Megachiroptera*). Their distribution extends from the west Indian Ocean islands of Mauritius, Madagascar and Comoro, along the sub-Himalayan region of Pakistan and India, through southeast Asia, the Philippines, Indonesia, New Guinea, southwest Pacific Islands as far east as the Cook Islands, and Australia (Fig. 3). Although other genera of *Pteropodidae* are present on mainland Africa (i.e. *Eidolon*, *Hypsignathus*, *Rousettus*, etc.) and in Asia (i.e. *Rousettus*), the genus *Pteropus* is restricted to Madagascar and surrounding islands in Africa; megachiropterans are absent from Europe and the Americas. Three of the four species of flying foxes found on mainland Australia are also found outside Australia. Black flying foxes (*P. alecto*), spectacled flying foxes (*P. conspicillatus*) and little red flying foxes (*P. scapulatus*) also occur in New Guinea, with the regional distribution of *P. alecto* extending to the Indonesian islands of Sulawesi, Lombok, Kangean and Baeween, and *P. conspicillatus* extending to the Indonesian island of Halmahera (Hall and Richards 2000; Mickleburg et al. 1992; Nowak 1994). Thus the distributions of two Australian species overlap with those of the island flying fox (*P. hypomelanus*) and the Malayan flying fox (*P. vampyrus*) in New Guinea and Indonesia. These species, at the northern extent of their range, overlap the Indian flying fox (*P. giganteus*), whose distribution extends eastward from Thailand and Burma across to India and Pakistan. Where distributions overlap, roosting camps are commonly shared. Such a scenario would facilitate the transmission of infectious agents between neighbouring species, leading to the plausible existence of related viruses in flying fox populations across their range, as previously hypothesised (Daszak et al. 2000; Halpin et al. 2000). Based

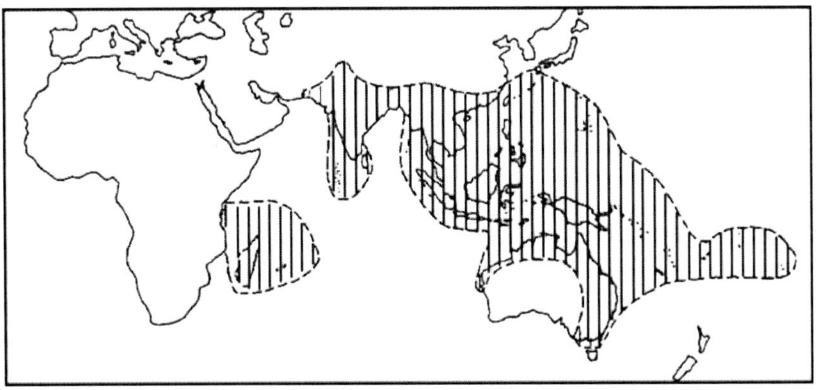

Fig. 3 World distribution of flying foxes (genus *Pteropus*). (From Hall and Richards 2000)

on maximum species diversity, flying foxes are believed to have originated from Sulawesi and eastern New Guinea (to the north of Australia) and to have radiated to their present distribution (Hall and Richards 2000; Mickleburg et al. 1992; Nowak 1994). Thus, by extension, if henipaviruses have co-evolved with flying foxes, it is likely that they exist across their entire geographic distribution. This hypothesis could readily be tested by screening long-isolated populations from the western extent of the global pteropid range.

6
Reservoir Management Strategies

Host management strategies have been discussed by the authors elsewhere (Mackenzie et al. 2003; Field et al. 2004). Effective disease management requires an understanding of the epidemiology of the disease (knowledge of its cause, maintenance and transmission, host range of the aetiologic agent, and the nature of the host–agent relationship), an ability to detect disease (surveillance and diagnostic capabilities) and political, public and industry support (see the chapter by Childs, this volume). Broadly, current strategies for the management of henipaviruses are directed at minimising direct or indirect contact with the natural host, monitoring intermediate hosts, improving biosecurity on farms, and better disease recognition and diagnosis.

The sporadic (and apparently rare) nature of Hendra virus spillover events from flying foxes to horses, the low infectivity for horses (and consequently the limited economic impact), and the apparent absence of direct transmission from flying foxes to people has resulted in more emphasis on management strategies for horses than for flying foxes. Quarantine of infected premises, movement controls on stock, and disinfection have proved effective strategies (Baldock et al. 1996). A Hendra virus vaccine is not currently available and development of one is not foreseen. Australian veterinarians have a high awareness of Hendra virus, and Hendra virus exclusion is routinely undertaken for horses exhibiting an acute respiratory syndrome. Veterinarians involved in these disease investigations wear appropriate protective equipment and use a limited necropsy approach, as horses have been the source of infection for all four human cases. Putative risk factors for infection in horses have been previously proposed – breed (thoroughbred), sex (female), age (>8 years), pregnancy status (late pregnancy), housing (paddocked), season (late gestation or the birthing season of local flying foxes), and the presence of favoured flying fox food trees in the index case paddock (Field et al. 2000). In the two 2004 cases, the putative association with age, paddock status, season and flying fox food source was maintained (Field et al. 2007). A considerable research focus on the ecology of Hendra virus has yet to define the route of virus excretion or any temporal pattern of infection in flying foxes. This information, and knowledge of the actual mode of flying fox-to-horse transmission would facilitate a risk management approach to spillover infection in horses.

In strong contrast to Hendra virus, the Nipah virus outbreak in peninsular Malaysia in 1999 had an enormous economic and social impact. Nipah virus was highly infectious for pigs, with all classes of pigs susceptible. The pattern of on-farm infection was consistent with respiratory transmission; between-farm spread was generally associated with the movement of pigs. The extensive post-outbreak surveillance program in Malaysia showed that farms that did not receive pigs generally remained uninfected even when neighbouring farms were infected. Human infections were predominantly attributed to contact with live pigs; none was attributed to contact with bats. Horizontal transmission was not a feature of infection in humans (although the potential for person-to-person transmission was noted previously). Recommended host management strategies primarily target pig-to-pig transmission, secondarily the flying fox-pig interface (that is, the natural host–spillover host interface). The central strategy is the implementation of sound farm management practices, such as monitoring herd health and early recognition of disease syndromes. The latter includes maintaining the high level of farmer and veterinarian awareness of the disease generated by the outbreak. A second key strategy is the strict application of farm-gate biosecurity (Daniels 2000), with clearly defined protocols for the introduction of new stock. These may include quarantine and/or exclusion testing. A Nipah virus vaccine is not currently available and development of one is unlikely in the near future. Overarching the above is a strategy of advanced planning for emergency management of disease outbreaks. This involves established surveillance, detection, and emergency response capabilities. The pre-existence of the latter in Malaysia enabled the implementation of effective quarantine, movement controls, and culling to bring the outbreak under control. The Malaysian pig population is now free of Nipah virus infection.

While strategies directed at the flying fox–pig interface are limited by our incomplete knowledge of the ecology of Nipah virus, several simple on-farm measures can be taken to reduce the likelihood of spillover events occurring. The removal of fruit orchards and other favoured flying fox food trees from the immediate vicinity of pig sheds greatly reduces the probability of flying fox–pig contact. Similarly, the wire screening of open-sided pig sheds is a simple and inexpensive strategy to prevent direct contact between flying foxes and pigs. Indirect contact (with flying fox urine, faeces or spats, or with partially eaten fruit) can be avoided by ensuring roof run-off does not enter pig pens (Chua 2003). The emergence (or detection) of apparently directly transmitted infection from the natural reservoir to humans and subsequent person-to-person transmission (as appears to be the case in Bangladesh) presents a new and formidable risk management challenge.

7 Phylogeny of Henipaviruses

Initial ultrastructural studies (Hyatt and Selleck 1996; Murray et al. 1995b) indicated that Hendra virus was a member of the family *Paramyxoviridae*, possibly genus *Paramyxovirus* or *Morbillivirus*. Comparative sequence analyses by PCR of a portion of the matrix protein supported this, with phylogenetic analysis indicating that the virus was distantly related to other known morbilliviruses (Murray et al. 1995b). Hence the name equine morbillivirus was tentatively ascribed to the virus. Subsequently the natural hosts of the virus were shown to be flying foxes (*Pteropus* spp.) rather than horses, and sequencing of the entire genome identified significant differences from morbilliviruses (including a larger genome size) that supported the creation of a new genus (Wang et al. 2000). The authors proposed *Henipavirus* as the new genus, with Hendra virus the type species and Nipah virus the second member. This was later accepted by the International Committee for the Taxonomy of Viruses.

Several other previously unknown members of the family *Paramyxoviridae* have been described in recent years. These include Phocine distemper virus and Cetacean morbillivirus (genus *Morbillivirus*), responsible for disease epidemics in marine mammals (Osterhaus et al. 1990; Taubenberger et al. 1996); Menangle virus (genus *Rubulavirus*), which caused severe reproductive disease in a commercial piggery in Australia in 1997 (Philbey et al. 1998); Salem virus (unclassified), possibly associated with a disease outbreak in horses in New Hampshire and Massachusetts, USA in 1992 (Renshaw et al. 2000); Tupaia paramyxovirus (unclassified), isolated from an apparently healthy tree shrew (*Tupaia belangeri*) in Thailand (Tidona et al. 1999); Tioman virus (genus *Rubulavirus*) and Pulau virus (unclassified) isolated from flying foxes in Malaysia during attempts to isolate Nipah virus (Chua et al. 2001b). Tioman and Menangle are phylogenetically closely related. Tupaia virus and Salem virus both share some sequence homology with Hendra and Nipah, yet have features that preclude their inclusion as henipaviruses or as morbilliviruses. While Palau virus has yet to be fully characterised, it too appears not to fit readily into either genus. Figure 4 presents a phylogenetic representation of the family *Paramyxoviridae*.

There are two reports of isolations of paramyxoviruses from bats prior to the description of Hendra virus in flying foxes in 1996; a sub-type of parainfluenza virus type 2 from *Rousettus leschenaulti* in India (Pavri et al. 1971) and

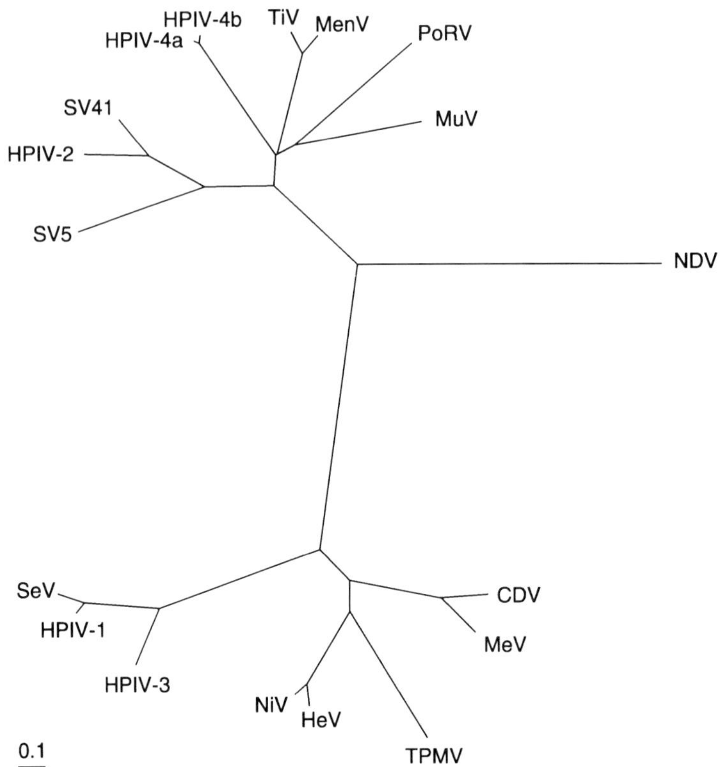

Fig. 4 A phylogenetic representation of the family *Paramyxoviridae*. A phylogenetic tree based on the deduced amino acid sequences of the matrix protein of members of the family *Paramyxoviridae*. Branch lengths represent relative evolutionary distances. *NDV* Newcastle disease; *CDV* canine distemper virus; *MeV* measles virus, *TPMV* Tupaia Paramyxovirus; *HeV* Hendra virus; *NiV* Nipah virus; *HPIV3* human parainfluenza virus 3; *HPIV1* human parainfluenza virus 1; *SeV* Sendai virus; *SV5* Simian virus 5; *HPIV2* human parainfluenza virus 2; *SV41* Simian virus 41; *HPIV4a* human parainfluenza virus 4a; *HPIV4b* human parainfluenza virus 4b; *TiV* Tioman virus; *MenV* Menangle virus; *PoRV* porcine rubulavirus; *MuV* mumps virus. (From Chua et al. 2002)

Mapuera virus from *Sturnira lilium* in Brazil (Henderson et al. 1995). Both of these viruses belong to the genus *Rubulavirus* (though unrelated to Menangle and Tioman viruses); the bat genera *Rousettus* (sub-order Megachiroptera) and *Sturnia* (sub-order Microchiroptera) are not closely related to flying foxes. A search for the ancestors of henipaviruses might best target bat species taxonomically closer to the genus *Pteropus*.

8
An Ecosystem Health Approach

Changes in biodiversity due to human activities were more rapid in the past 50 years than at any time in human history, and the drivers of change that cause biodiversity loss and lead to changes in ecosystem services are either steady, show no evidence of declining over time, or are increasing in intensity. Under the four plausible future scenarios developed by the Millennium Ecosystem Assessment Report (Anonymous 2005a), these rates of change in biodiversity are projected to continue or to accelerate.

There is increasing realisation of the interconnectedness of the ecosystem and human health, and the relationship between the environment, human and non-human hosts, and pathogens. Daszak et al. (2000) argue that most emerging diseases exist within a finely balanced host–agent continuum between wildlife, domestic animal and human populations. Taylor et al. (2001), in examining risk factors for disease emergence, conclude that emerging diseases are three times more likely to be associated with zoonotic pathogens than with non-zoonotic pathogens. An ecosystem health approach recognises the critical linkages between human activity, ecological change and health, and fosters a multidisciplinary approach that considers a range of influencing factors such as medical, environmental, economic and socio-political factors. The complexity of the emergence and epidemiology of the henipaviruses warrants such a broad, cross-disciplinary ecosystem health approach if the associated mechanisms are to be understood and future risks managed.

9
Conclusion

Henipaviruses appear to have only recently emerged. Their ability to dramatically impact human and animal health, and the associated societal and economic consequences, has been clearly illustrated. Horizontal transmission of henipaviruses in humans, absent in Australia and Malaysia, appears to be an alarming feature of Nipah virus outbreaks in Bangladesh. If transmission in humans becomes efficient, the potential exists for a worst-case emergence scenario. Further, if henipaviruses, and the necessary and sufficient precipitating emergence factors exist across the distribution of all pteropid species, the emergence of further novel agents can be expected unless factors associated with emergence are addressed.

Addendum

There were two further separate equine cases of Hendra virus infection in Australia in 2006: in the south-east of Queensland (June), and in the north of the adjacent state of New South Wales (October). These cases had a number of features in common with previous index cases, including apparent spatial and temporal clustering (Field et al, 2007). In addition, a cluster of human cases of suspect Nipah virus disease was reported in the Kushtia region of eastern Bangladesh and neighbouring West Bengal (India) in April, 2007. The disease presented as an acute neurological syndrome. (Promed 28, 30 April 2007).

Acknowledgements Hume Field is supported by the Department of Primary Industries and Fisheries, Queensland, Australia, the Australian Biosecurity Cooperative Research Centre for Emerging Infectious Disease, and by an NIH/NSF "Ecology of Infectious Diseases" award from the John E. Fogarty International Center (RO1-TW05869). John Mackenzie is supported by Western Australian Government Office of Science and Innovation, Curtin University, and the Australian Biosecurity Cooperative Research Centre for Emerging Infectious Disease. Peter Daszak is supported by the Consortium for Conservation Medicine and by NIH/NSF "Ecology of Infectious Diseases" award from the John E Fogarty International Center (RO1-TW05869).

References

Abu Bakar S, Chang LY, Ali AR, Sharifah SH, Yusoff K, Zamrod Z (2004) Isolation and Molecular Identification of Nipah Virus from Pigs. Emerg Infect Dis 10:2228–2230
Allworth A, O'Sullivan J, Selvey L, Sheridan J (1995) Equine morbillivirus in Queensland. Comm Disease Intell 19:575
Anonymous (2003) Outbreaks of encephalitis due to Nipah/Hendra-like viruses, western Bangladesh. Health Sci Bull 1:1–6
Anonymous (2004a) Nipah encephalitis outbreak over wide area of western Bangladesh (2004) Health Sci Bull 2:7–11
Anonymous (2004b) Person-to-person transmission of Nipah virus during outbreak in Faridpur District (2004) Health Sci Bull 2:5–9
Anonymous (2005a) Millennium Ecosystem Assessment Ecosystems and Human Well-Being report (May 2005). http://www.millenniumassessment.org/en/index.aspx. Accessed 19 July 2005
Anonymous (2005b) Nipah virus outbreak from date palm juice. Health Sci Bull 3:1–5
Baldock FC, Douglas IC, Halpin K, Field H, Young PL, Black PF (1996) Epidemiological investigations into the 1994 equine morbillivirus outbreaks in Queensland Australia. Sing Vet J 20:57–61
Breed AC, Field HE, Plowright RK (2005) Volant viruses: a concern to bats, humans and other animals. Microbiol Aust 26:59–62

Bunning M, Jamaluddin A, Cheang H et al. (2000) Epidemiological trace-back studies of the Nipah virus outbreak in pig farms in the Ipoh district of Malaysia, 1997–1999. In: Cargill C, McOrist S (eds) Proceedings of the 16th International Pig Veterinary Society Congress. Ocean Grove, p 551

Centers for Disease Control (1999) Outbreak of Hendra-like virus–Malaysia and Singapore. Morb Mort Wkly Rep 48:265–269

Chan YP, Chua KB, Koh CL, Lim ME, Lam SK (2001) Complete nucleotide sequences of Nipah virus isolates from Malaysia. J Gen Virol 82:2151–2155

Childs JE (2004) Zoonotic viruses of wildlife: hither from yon. Arch Virol Suppl 18:1–11

Chua KB (2003) Nipah virus outbreak in Malaysia. J Clin Virol 26:265–275

Chua KB, Goh KJ, Wong KT, Kamarulzaman A, Tan PS, Ksiazek TG, Zaki SR, Paul G, Lam SK, Tan CT (1999) Fatal encephalitis due to Nipah virus among pig-farmers in Malaysia. Lancet 354:1256–1259

Chua KB, Bellini WJ, Rota PA, Harcourt BH, Tamin A, Lam SK, Ksiazek TG, Rollin PE, Zaki SR, Shieh W, Goldsmith CS, Gubler DJ, Roehrig JT, Eaton B, Gould AR, Olson J, Field H, Daniels P, Ling AE, Peters CJ, Anderson LJ, Mahy BW (2000) Nipah virus: A recently emergent deadly paramyxovirus. Science 288:1432–1435

Chua KB, Lam SK, Goh KJ, Hooi PS, Ksiazek TG, Kamarulzaman A, Olson J, Tan CT (2001a) The presence of Nipah virus in respiratory secretions and urine of patients during an outbreak of Nipah virus encephalitis in Malaysia. J Infect 42:40–43

Chua KB, Wang LF, Lam SK, Crameri G, Yu M, Wise T, Boyle D, Hyatt AD, Eaton BT (2001b) Tioman virus, a novel paramyxovirus isolated from fruit bats in Malaysia. Virology 283:215–219

Chua KB, Koh CL, Hooi PS, Wee KF, Khong JH, Chua BH, Chan YP, Lim ME, Lam SK (2002) Isolation of Nipah virus from Malaysian Island flying foxes. Microbes Infect 4:145–151

Daniels PW (2000)The Nipah virus outbreak in Malaysia: Overview of the outbreak investigation and the issues that remain. In: Cargill C, McOrist S (eds) Proceedings of the 16th International Pig Veterinary Society Congress. Ocean Grove, p 551

Daszak P (2005)The Henipavirus Ecology Collaborative Research Group (HERG) project overview. http://www.henipavirus.net/project_overview/project_proposal.htm. Accessed 19 July 2005

Daszak P, Cunningham AA, Hyatt AD (2000) Emerging Infectious Diseases of Wildlife-threats to biodiversity and human health. Science 287:443–448

Daszak P, Cunningham AA, Hyatt AD (2001) Anthropogenic environmental change and the emergence of infectious diseases in wildlife. Acta Tropica 78:103–116

Daszak P, Plowright R, Epstein JH, Pulliam J, Abdul Rahman S, Field HE, Jamaluddin A, Sharifah SH, Smith CS, Olival KJ, Luby S, Halpin K, Hyatt AD, Cunningham AA, the Henipavirus Ecology Research Group (HERG) (2006) The emergence of Nipah and Hendra virus: pathogen dynamics across a wildlife-livestock-human continuum. In: Collinge SK, Ray C (eds) Disease ecology: community structure and pathogen dynamics. Oxford University Press, Oxford pp 186–201

Douglas IC, Baldock FC, Black P (1997) Outbreak investigation of an emerging disease (equine morbillivirus). Epidemiol Santé Animale: Procedings of 8th ISVEE conference, Paris, pp 04.08.1–04.08.08.3

Field HE (2005) The ecology of Hendra virus and Australian bat lyssavirus. PhD thesis, The University of Queensland, Brisbane, Australia. http://espace.library.uq.edu.au/view.php?pid=UQ:13859

Field HE, Barratt PC, Hughes RJ, Shield J, Sullivan ND (2000) A fatal case of Hendra virus infection in a horse in north Queensland: clinical and epidemiological features. Aust Vet J 78:279–280

Field HE, Breed AC, Shield J, Hedlefs RM, Pittard K, Pott B, Summers PM (2007) Epidemiological perspectives on Hendra virus infection in horses and flying foxes. Aust Vet J (in press).

Field H, Young P, Yob JM, Mills J, Hall L, Mackenzie J (2001) The natural history of Hendra and Nipah viruses. Microbes Infect 3:315–322

Field H, Mackenzie J, Daszak P (2004) Novel viral encephalitides associated with bats (*Chiroptera*) – host management strategies. Arch Virol Suppl 18:113–121

Goh KJ, Tan TC, Chew NK, Tan PSK, Karmarulzaman A, Sarji SA, Wong KT, Abdullah BJJ, Chua KB, Lam SK (2000) Clinical features of Nipah virus encephalitis among pig farmers in Malaysia. N Engl J Med 342:1229–1235

Gould AR (1996) Comparison of the deduced matrix and fusion protein sequences of equine morbillivirus with cognate genes of the Paramyxoviridae. Virus Res 43:17–31

Hall LS, Richards G (2000) Flying foxes: Fruit and Blossom Bats of Australia. University of New South Wales Press, Sydney

Halpin K, Young PL, Field HE, Mackenzie JS (2000) Isolation of Hendra virus from pteropid bats: a natural reservoir of Hendra virus. J Gen Virol 81:1927–1932

Harcourt BH, Lowe L, Tamin A, Lui X, Bankamp B, Bowden N, Rollin PE, Corner JA, Ksaizek TG, Hossain MJ, Gurley ES, Breiman RF, Bellini WJ, Rota PA (2005) Genetic characterisation of Nipah virus Bangladesh (2004) Emerg Infect Dis 11:1594–1597

Henderson GW, Laird C, Dermott E, Rima BK (1995) Characterization of Mapeura virus: structure, proteins and nucleotide sequence of the gene encoding nucleocapsid protein. J Gen Virol 76:2509–2518

Hooper PT, Gould AR, Russell GM, Kattenbelt JA, Mitchell G (1996) The retrospective diagnosis of a second outbreak of equine morbillivirus infection. Aust Vet J 74:244–245

Hsu VP, Hossain MJ, Parashar UD, Ali MM, Ksiazek TG, Kuzmin I, Niezgoda M, Rupprecht C, Bresee J, Breiman RF (2004) Nipah Virus Encephalitis Reemergence Bangladesh. Emerg Infect Dis 10:2082–2087

Hyatt AD, Selleck PW (1996) Ultrastructure of equine morbillivirus. Virus Res 43:1–15

Krause R (1992) The origins of plagues: old and new. Science 257:1073–1078

Lederberg J, Shope RE, Oaks SC (1992) Emerging infections: microbial threats to health in the United Sates. National Academy Press, Washington, DC

Liu J, Lim SL, Ruan Y, Ling AE, Ng LPF, Drosten C, Liu ET, Stanton LW, Hibberd ML (2005) SARS transmission pattern in Singapore reassessed by viral sequence variation analysis. PLoS Med 2:162–168

Mackenzie JS, Chua KB, Daniels PW, Eaton BT, Field HE, Hall RA, Halpin K, Johansen CA, Kirkland PD, Lam SK, McMinn P, Nisbet DJ, Paru R, Pyke AT, Ritchie SA, Siba P, Smith DW, Smith GA, van den Hurk AF, Williams DT (2001) Emerging viral diseases of southeast Asia and the western Pacific. Emerg Infect Dis 7:497–504

Mackenzie JS, Field HE, Guyatt KJ (2003) Managing emerging diseases borne by fruit bats (flying foxes) with particular reference to Henipaviruses and Australian bat lyssavirus. J Appl MicroBiol Suppl 94:59S–69S

Medway L (1978) The wild mammals of Malaya (Peninsular Malaysia) and Singapore. Oxford University Press, Kuala Lumpur

Mickleburg SP, Hutson AM, Racey PA (1992) Old World Fruit Bats: an action plan for their conservation. Internation Union for the Conservation of Nature and Natural Resources, Gland, Switzerland

Middleton DJ, Westbury HA, Morrissy CJ, Van der Heide BM, Russell GM, Braun MA, Hyatt AD (2002) Experimental Nipah virus infection in pigs and cats. J Comp Pathol 126:124–136

Morse SS (1995) Factors in the emergence of infectious diseases. Emerg Infect Dis 1:7–15

Murray K, Rogers R, Selvey L, Selleck P, Hyatt A, Gould A, Gleeson L, Hooper P, Westbury H (1995a) A novel morbillivirus pneumonia of horses and its transmission to humans. Emerg Infect Dis 1:31–33

Murray K, Selleck P, Hooper P, Hyatt A, Gould A, Gleeson L, Westbury H, Hiley L, Selvey L, Rodwell B, Ketterer P (1995b) A morbillivirus that caused fatal disease in horses and humans. Science 268:94–97

Nor MN, Gan CH, Ong BL (2000) Nipah virus infection of pigs in peninsular Malaysia. Rev Sci Off Int Epiz 19:160–165

Nowak RM (1994) Walker's Bats of the World. The Johns Hopkins University Press, Baltimore

Olson JG, Rupprecht C, Rollin PE, An US, Niezgoda M, Clemins T, Walston J, Ksiazek TG (2002) Antibodies to Nipah-like virus in bats (Pteropus lylei), Cambodia. Emerg Infect Dis 8:987–988

Osterhaus AD, Groen J, Spijkers HE, Broeders HW, UyteHaag FG, de Vries P, Teppema JS, Visser IK, van de Bildt MW, Vedder EJ (1990) Mass mortality in seals caused by a newly discovered morbillivirus. Vet Microbiol 23:343–350

Ostfeld RS, Keesing F (2000) Biodiversity and disease risk: The case of Lyme disease. Conserv Biol 14:722–728

Parashar UD, Sunn LM, Ong F, Mounts AW, Arif MT, Ksiazek TG, Kamaluddin MA, Mustafa AN, Kaur H, Ding LM, Othman G, Radzi HM, Kitsutani PT, Stockton PC, Arokiasamy J, Gary HE Jr, Anderson LJ (2000) Case-control study of risk factors for human infection with a novel zoonotic paramyxovirus Nipah virus, during a 1998–1999 outbreak of severe encephalitis in Malaysia. J Infect Dis 181:1755–1759

Parry-Jones KA, Augee ML (2001) Factors affecting the occupation of a colony site in Sydney New South Wales by the Grey-headed Flying-fox Pteropus poliocephalus (Pteropodidae). Aust Ecol 26:47–55

Pavri KM, Singh KR, Hollinger FB (1971) Isolation of a new parainfluenza virus from a frugivorous bat Rousettus leschenaulti, collected at Poona India. Am J Trop Med Hyg 20:125–130

Philbey AW, Kirkland PD, Ross AD, Davis RJ, Gleeson AB, Love RJ, Daniels PW, Gould AR, Hyatt AD (1998) An apparently new virus (family Paramyxoviridae) infectious for pigs, humans and fruit bats. Emerg Infect Dis 4:269–271

Plowright RK, Foley P, Field HE, Foley J (2005) Disease ecology of Hendra virus: epidemiological modelling to test theories for emergence. Abstracts Wildlife Disease Association International Conference (Wildlife Health in a shrinking world: ecology, management and conservation), June 26–July 1, 2005, Cairns, Australia

Renshaw RW, Glaser AL, van Campen H, Weiland F, Dubovi EJ (2000) Identification and phylogenetic comparison of Salem virus, a novel paramyxovirus of horses. Virology 270:417–429

Reynes JM, Counor D, Ong S, Faure C, Seng V, Molia S, Walston J, Georges-Courbot MC, Deubel V, Sarthou JL (2005) Nipah virus in Lyle's flying foxes Cambodia. Emerg Infect Dis 11:1042–1047

Rogers RJ, Douglas IC, Baldock FC, Glanville RJ, Seppanen KT, Gleeson LJ, Selleck PN, Dunn KJ (1996) Investigation of a second focus of equine morbillivirus infection in coastal Queensland. Aust Vet J 74:243–244

Selvey LA, Wells RM, McCormack JG, Ansford AJ, Murray K, Rogers RJ, Lavercombe PS, Selleck P, Sheridan JW (1995) Infection of humans and horses by a newly described morbillivirus. Med J Aust 162:642–645

Sendow I, Field HE, Curran J, Darminto Morrissey C, Buick T, Daniels P (2006) Henipavirus in *Pteropus vampyrus* bats Indonesia. Emerg Infect Dis 12:711–712

Smolinski M, Hamburg M, Lederberg J (2003) Microbial threats to health: Emergence Detection and Response. The National Academies Press, Washington, DC

Song HD, Tu CC, Zhang GW, Wang SY, Zheng K, Lei LC, Chen QX, Gao YW, Zhou HQ, Xiang H, Zheng HJ, Chern SW, Cheng F, Pan CM, Xuan H, Chen SJ, Luo HM, Zhou DH, Liu YF, He JF, Qin PZ, Li LH, Ren YQ, Liang WJ, Yu YD, Anderson L, Wang M, Xu RH, Wu XW, Zheng HY, Chen JD, Liang G, Gao Y, Liao M, Fang L, Jiang LY, Li H, Chen F, Di B, He LJ, Lin JY, Tong S, Kong X, Du L, Hao P, Tang H, Bernini A, Yu XJ, Spiga O, Guo ZM, Pan HY, He WZ, Manuguerra JC, Fontanet A, Danchin A, Niccolai N, Li YX, Wu CI, Zhao GP (2005) Cross-host evolution of severe acute respiratory syndrome coronavirus in palm civet and human. Proc Natl Acad Sci U S A 102:2430–2435

Taubenberger JK, Tsai M, Krafft AE, Lichy JH, Reid AH, Schulman FY, Lipscomb TP (1996) Two morbilliviruses implicated in bottlenose dolphin epizootics. Emerg Infect Dis 2:213–216

Taylor LH, Latham SM, Woolhouse ME (2001) Risk factors for human disease emergence. Philos Trans R Soc Lond B Biol Sci 356:983–989

Tidona CA, Kurz HW, Gelderblom HR, Darai G (1999) Isolation and molecular characterisation of a novel cytopathogenic paramyxovirus from tree shrews. Virology 258:425–434

Wang LF, Yu M, Hansson E, Pritchard LI, Shiell B, Michalski WP, Eaton BT (2000) The exceptionally large genome of Hendra virus: support for creation of a new genus within the family Paramyxoviridae. J Virol 74:9972–9979

Ward MP, Black PF, Childs AJ, Baldock RC, Webster WR, Rodwell BJ, Brouwer SL (1996) Negative findings from serological studies of equine morbillivirus in the Queensland horse population. Aust Vet J 74:241–243
Westbury HA, Hooper PT, Brouwer SL, Selleck PW (1996) Susceptibility of cats to equine morbillivirus. Aust Vet J 74:132–134
Williamson MM, Hooper PT, Selleck PW, Gleeson LJ, Daniels PW, Westbury HA, Murray PK (1998) Transmission studies of Hendra virus (equine morbillivirus) in fruit bats, horses and cats. Aust Vet J 76:813–818
Yob JM, Field HE, Rashdi AM, Morrissy C, van der Heide B, Rota P, bin Adzhar A, White J, Daniels P, Jamaluddin A, Ksiazek T (2001) Nipah virus infection in bats (order Chiroptera) in peninsular Malaysia. Emerg Infect Dis 7:439–441
Young PL, Halpin K, Selleck PW, Field H, Gravel JL, Kelly MA, Mackenzie JS (1996) Serologic evidence for the presence in Pteropus bats of a paramyxovirus related to equine morbillivirus. Emerg Infect Dis 2:239–240
Young PL, Halpin K, Field HE, Mackenzie JS (1997) Finding the wildlife reservoir of equine morbillivirus. In: Asche V (ed) Recent advances in microbiology. Australian Society of Microbiology Inc., Melbourne

CTMI (2007) 315:161–193
© Springer-Verlag Berlin Heidelberg 2007

Emergence of Lyssaviruses in the Old World: The Case of Africa

L. H. Nel[1] (✉) · C. E. Rupprecht[2]

[1]Department of Microbiology, Faculty of Natural and Agricultural Sciences, University of Pretoria, 0001 Pretoria, South Africa
louis.nel@up.ac.za
[2]Centers for Disease Control and Prevention, Division of Viral and Rickettsial Diseases, Poxvirus and Rabies Branch, Rabies Program, Atlanta, GA, USA

Abstract Rabies has a long history of occurrence throughout Africa, spanning hundreds of years. At least four distinct *Lyssavirus* species persist throughout the continent, among carnivores, bats and other mammals. Rabies virus is the most cosmopolitan member, with primary reservoirs within dogs and mongoose, but other wildlife vectors are important in viral maintenance, such as jackals. Besides a prominent toll on humans and domestic animals, the disease has an underappreciated role in conservation biology, especially for such highly endangered fauna as African wild dogs and Ethiopian wolves. Both Duvenhage and Lagos bat viruses are adapted to bats, but their epidemiology, together with Mokola virus, is poorly understood. Significantly, less than ideal cross-reactivity with modern biologicals used for veterinary and public health interventions is a major cause for concern among these emerging viral agents.

1 Rabies in Africa

1.1 North Africa Versus Sub-Saharan Africa

Rabies is an acute encephalitis caused by global RNA viruses in the genus *Lyssavirus*, family *Rhabdoviridae*. The diversity of extant viral species in Africa has suggested that this continent may have served as a motherland for primary lyssavirus emergence and diversification. Unfortunately, the history of rabies in Africa, prior to the twentieth century, is fragmentary, poorly recorded and ambiguous. With regard to this disease and in line with the history of the colonisation of Africa, it is nevertheless clear that this continent can be regarded in two halves: North Africa and sub-Saharan Africa. Rabies must have been present in North Africa for hundreds of years, but became epizootic in the countries of sub-Saharan Africa only well into the twentieth century, following the introduction of a cosmopolitan canine variant of rabies virus. In fact, this cosmopolitan lineage is thought to have originated from the Palearctic region that includes not only Europe, but also the Middle East and North Africa (Badrane and Tordo 2001). Thus, in northern Africa, rabies occurred principally as an urban disease – particularly of dogs in the northeastern regions of the continent, where it was also associated with rabies cycles in the Middle East. Today this is still the case and the virus persists over the Palearctic geographical domain in a suite of hosts that includes various canid species, most importantly dogs and foxes. Although not necessarily exclusively overland, a gradual southward migration of this particular, dog-associated cosmopolitan rabies virus is evident, to eventually cover the entire continent. Apart from its dominance in Africa, this lineage has spread fairly rapidly worldwide (Nadin-Davis and Bingham 2004) (Fig. 1).

Overland spread of rabies southward through Africa from the Palearctic region was probably slowed due to the large Saharan Desert forming an arid and sparsely populated buffer region to the south. Here, rabies has occurred only as scattered foci over a huge geographical area, occasionally involving camels as victim species, but one such focus (rabies virus of cosmopolitan origin) involves the Ethiopian wolf (*Canis simensis*), a rare carnivore seriously threatened with extinction by this very disease (Sillero-Zubiri et al. 1996). At the present time this specific lineage is maintained throughout Africa and was identified in all countries where molecular sequence data for virus isolates could be generated. It is clear that this lineage of rabies virus is dominant and cycles mainly in dogs, but it has also adapted to wildlife, with involvement of other carnivores, such as black-backed jackals (*Canis mesomelas*) and bat-eared foxes (*Otocyon megalotis*) in southern Africa. The more developed countries of

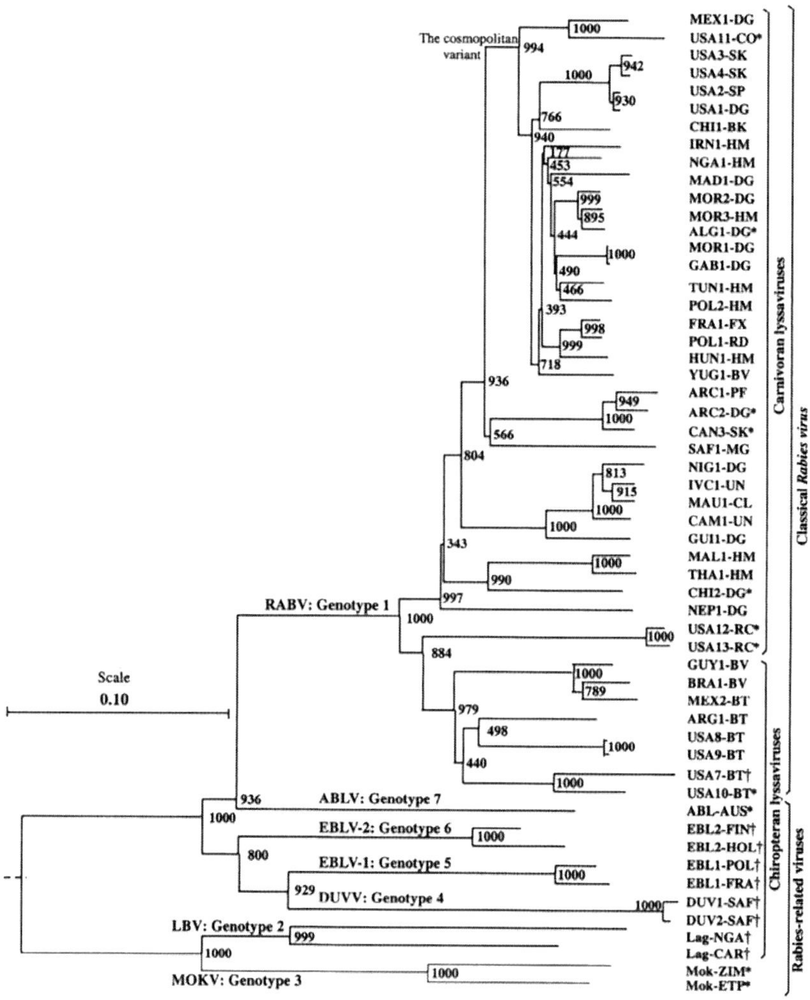

Fig. 1 Lyssavirus-rooted phylogenetic tree based on the ectodomain of the glycoprotein encoding nucleotide sequences. *RABV* Rabies virus; *EBLV-1* European bat lyssavirus 1; *EBLV-2* European bat lyssavirus 2; *ABLV* Australian bat lyssavirus; *DUVV* Duvenhage virus; *LBV* Lagos bat virus; *MOKV* Mokola virus. (From Badrane and Tordo 2001)

southernmost Africa may apply more ideal disease surveillance programs than elsewhere, probably accounting in part for high numbers of cases in a vast variety of wildlife reported locally. But here, and elsewhere in sub-Saharan Africa, epizootics of dog rabies and rabies in livestock also tend to be spread over large areas, co-incident with the wider distribution of humans and other vertebrates as compared to the Saharan region and North Africa.

However, clearly the African continent had not been rabies-free prior to the dissemination of the currently dominant cosmopolitan virus. At least two distinct rabies virus lineages can be distinguished from the cosmopolitan lineage. The first of these is a dog virus circulating in West Africa (Badrane and Tordo 2001; Kissi et al. 1995). Due to poor surveillance throughout the countries of this western part – the horn of Africa – the current status of this particular lineage is not clear. The second African lineage of rabies virus is, however, known to be well established and actively circulating, primarily as a virus adapted to various mongoose species, but also infecting a wide range of other wildlife species (Von Teichman et al. 1995; Nel et al. 2005) (Fig. 2).

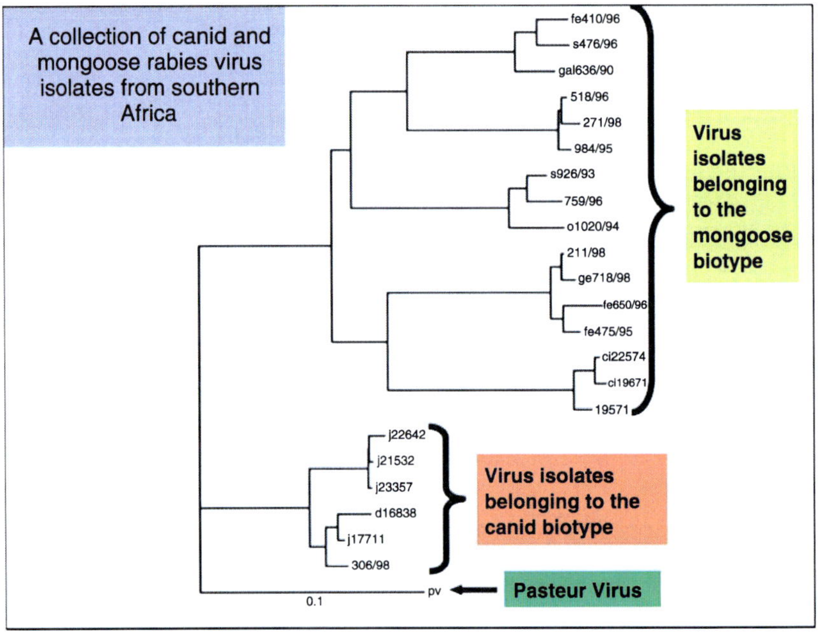

Fig. 2 Phylogenetic evidence demonstrates (*1*) the distance between mongoose and canine viruses of southern Africa, (*2*) large sequence variation within the mongoose lineage and (*3*) the close link between canine viruses of Africa and the cosmopolitan viruses of European origin (e.g. Pasteur virus)

1.2
West and Central Africa

It is possible that rabies is an ancient disease of West Africa. In part, the distinct lineage of dog rabies mentioned above provides virological evidence, but there are other indicators that rabies may have evolved and disseminated over this region for a considerable period of time, and certainly well before the introduction of cosmopolitan rabies more than a century ago. The generally densely vegetated tropical and subtropical western horn region of Africa yielded a number of different Rhabdoviruses and among the lyssaviruses, the rabies-related Mokola and Lagos Bat viruses were also first reported from here (Calisher et al. 1989). As far as rabies per se is concerned, there are from this general region undisputed historical accounts of rabies in dogs, the virus apparently less virulent than conventional street rabies virus (Blancou 1988). Recognition of this form of rabies came from Senegal and Niger (Bouffard 1912; Remlinger and Curasson 1924) but appeared to have extended over many countries in the horn of Africa (Snyman 1940; Thierry 1959) and eastward into Sudan and Ethiopia, where this form of the disease and a variant of rabies virus was isolated on several occasions since 1950 (Fekadu and Baer 1980; Fekadu 1972, 1982). It was suggested that dogs may be chronically infected, presenting what appeared to be a carrier state and transmitting rabies virus to humans over a number of years (Fekadu et al. 1983). The historical nonfatal form of rabies (known as oulou fato – mad dog disease) has not been described recently and the current status of this situation is not known.

Today, however, the cosmopolitan variant of rabies (of dogs) appears to be dominant and widespread throughout this region (Kissi et al. 1995), including in Ethiopia and Sudan (Johnson et al. 2004), but like elsewhere in Africa, governments in this region are generally under-resourced and surveillance is poor. It is worth re-iterating that it is in Ethiopia that the Ethiopian wolf is under threat of extinction by rabies (Sillero-Zubiri et al. 1996; Randall et al. 2004) from the relatively recently globally disseminated cosmopolitan dog rabies virus of European/Middle East/North African origin. It appears likely from molecular phylogenetic data that this virus was also introduced from Ethiopia into Sudan by both dog and wildlife vectors (Johnson et al. 2004). It is debatable whether introduction of this virus from the north, north-east may be related to increased interaction between Mediterranean Africa and the Sudan (north-east/central Africa) during the 300 years from 1500 to 1800. This was the period of Islamic spread into Africa and it is thought that Muslim merchants are unlikely to have taken dogs on these journeys (King et al. 1994). It may be more likely that, during this same period, (European) rabies may have been introduced into West Africa by slave-trading Europeans, this trans-Atlantic trade peaking during the 18th century, with similar repercussions in the New World as concerns the introduction of canine rabies.

1.3
East Africa and Malawi

In this general area, like elsewhere in Africa, endemic rabies of canines was recognised early in the twentieth century and onwards, following the availability of dependable diagnostic biologicals and technique. The first diagnosis of rabies thus came in 1912 in Kenya and was later recognised on the border of Kenya with Tanzania, in the 1930s. Nevertheless, it seems that rabies could have been known in this part of the world prior to the times of European colonisation and the advent of diseases associated with Europe (Kariuki and Ngulo 1985). In the middle to late 1950s, a serious rabies epidemic appeared to spread from the southern borders of Tanzania (with Malawi and Zambia) to the north-east and later the north-west, eventually covering the entire country (Magembe 1985). Here, and in bordering Kenya, dog rabies has persisted as a significant problem ever since (Siongok and Karama 1985). The disease has had a particularly devastating impact on wildlife. *Lycaon pictus*, the endangered African wild dog that once populated the Serengeti plains of Tanzania and Kenya, was severely depleted from these regions by rabies, possibly mediated by domestic dogs of nomadic cattle herders (Burrows 1994; East and Hofer 1996). Other wildlife regularly affected include common carnivores such as the black-backed jackal (Alexander et al. 1994) and bat-eared foxes (Creel et al. 1997). The persistent infection of the spotted hyena (*Crocuta crocuta*) has been described (East et al. 2001), but this suggestion has not been corroborated. In Malawi, rabies has been diagnosed since the 1920s and it has since been endemic in dogs, but also occurring in several other wild canids (Msiska 1988). In Uganda, the disease is no doubt as widespread and serious as in the neighbouring countries of East Africa. The unfortunate lack of data and any control effort locally may be ascribed to political unrest and the lack of veterinary and public health infrastructure (Rollinson 1956; Illango 1992).

1.4
Southern Africa

In southern Africa rabies affects many different species, but virus cycles are sustained primarily in carnivore hosts, principally the domestic dog (Bingham 2005). Apart from an indigenous virus biotype associated with mongooses, phylogenetic data have become available for canid rabies in southern Africa in recent years and it is clear that the associated viruses are strongly linked to the global cosmopolitan lineage of rabies virus variants (Sabeta et al. 2003; Nadin-Davies and Bingham 2004; Johnson et. al. 2004) (Fig. 2). This canid rabies virus lineage is capable of maintaining prolonged and independent cycles of disease throughout canid host populations of sub-Saharan Africa, particularly domestic dogs, and interactions with jackals and bat-eared foxes. It is also this lineage (emerging here only after the World War II) that causes the majority of human cases and human exposures requiring prophylaxis, although the mongoose virus biotype has been implicated in vaccine failures on at least two separate occasions (R. Swanepoel, personal communication).

There is some evidence of the presence of rabies in Zambia prior to the main influx of Europeans and the advent of definitive diagnostic methods. For example, an order of a tribal chief, Chief Lewanika, to destroy all dogs in the western part of the country was an attempt to control a serious dog rabies outbreak in the region at the time, around 1901 (Edmonds 1922; Snyman 1940; Shone 1962). The disease is still endemic throughout Zambia, involving mostly dogs, but also infection of cattle and wildlife (Zyambo et al. 1985). In a single case, a positive direct fluorescence assay was performed on the brain of a dead bat. It is likely that this may have represented a nonrabies lyssavirus infection, but the case was never further investigated (Ahmadu and Zulu 1998).

Angolans had been engaged in an ongoing civil war over many decades and it has only recently become reasonably peaceful but dreadfully disrupted and poor. A prolonged outbreak of Marburg virus (2004–2005) in northern Angola bears testimony to the limitations of proper public health infrastructure and resources. Consequently, very little is known about lyssaviruses here other than its occurrence in dogs, where it was also first confirmed in 1929. Outbreaks of dog rabies in the 1920s have been linked to cycles overlapping to the south, where Angola borders with Namibia. The republic of Namibia is a large country situated on the south Atlantic (west) coast of Africa, bordering South Africa in the south, Botswana in the east, and Angola in the north. Namibia is sparsely populated, with the Namib desert stretching along the western coast and the Kalahari Desert along the south-eastern border with Botswana. As early as 1887, a disease outbreak among dogs, cattle and other livestock was presumed to be rabies, given the disease characteristics (Schneider 1985; Hübschle 1988). It is particularly in the northern parts of Namibia and the Caprivi (bordering on Angola and Zambia) where sporadic reports of rabies (unconfirmed) involving dogs, cattle and humans occurred throughout the latter half of the 1920s. These reports of rabies coincided with sporadic reports from the southern parts of neighbouring Angola (Hübschle 1988; Swanepoel 2005). However, by the end of World War II, and towards the late 1940s, a major rabies epizootic ensued in Angola and Zambia (then northern Rhodesia) and spread southward into Namibia, Botswana and further (Courtin et al. 2000). It is probable that dogs were solely responsible initially, but jackals soon became an important species and remains so at the present time (Meredith 1982). It was evident that rabies occurrence in the black-backed jackal appeared soon after the introduction of dog rabies in Namibia (von Maltitz 1950). Thereafter, with the involvement of jackals, the disease spread southwards, past the Etosha National Park to reach the capital Windhoek, by 1951 (Alexander 1952). This canine epizootic continued to spread through Namibia and Botswana and into the northern provinces of the Republic of South Africa during the early 1950s, where it continued to involve jackals and in addition bat-eared foxes (King et al. 1993). By the mid-1970s, there was sporadic but endemic rabies throughout most of Namibia, generally with dog and human rabies in the more populous north, jackal and cattle rabies in the central ranching areas and sporadic canid or mongoose rabies in the arid sheep farming areas of the south (Hübschle 1988; Swanepoel 2005).

Typically, carnivore species serve as rabies reservoir hosts, and many taxa appear as incidental or dead-end infections from spillover activity. Domestic and wild herbivores are frequently reported as indicators of rabid domestic animal or wildlife activity. However, when epidemiological and ecological facets are ripe for rabies emergence and spread, unusual events may be appreciated. The elegant *Tragelaphus strepsiceros*, common name kudu (Fig. 3), is a wild ruminant with spiraling twisted horns, occurring over vast areas of southern Africa, where they prefer areas of broken rocky terrain with easy access to water (Hübschle 1988). The unusual occurrence of rabies in these antelopes was first observed in 1975 near Windhoek. Subsequently, an outbreak of rabies in kudu seems to have originated in 1977, in this general geographical area, but was also confirmed much further to north, in two kudu in the Etosha National Park. The number of confirmed cases rose steadily and peaked in 1980 after the disease had spread widely to the north, west and south of Namibia (Barnard and Hassel 1981; Hübschle 1988). This outbreak only subsided in 1985 and the latter part of the focus (1983–1984) coincided with the first cases of lions contracting the disease in the Etosha National Park. It is thought that the lions became infected from hunting rabid kudu, as all four reports of rabid lions were from an area of high kudu population density in eastern Etosha (Berry 1993). Eventually the disease caused an estimated loss of 30,000–50,000 antelope, or 20% of the population (Swanepoel 1993). However, during 2002 there was another substantial rabies outbreak in kudu, where an estimated 2,500 animals on more than 81 farms in Namibia died (OIE 2005). This recent outbreak continued into 2003. It is thought that both the density and social behavior of kudu, including their herd structuring, diet and eating habits, contributed to the rapid and effective spread of rabies. Kudu form small herds (four to six animals) that eat and move together, and have close contact with each other through activities such

Fig. 3 The elegant *Tragelaphus strepsiceros*, common name kudu, is a antelope that is widespread throughout southern Africa, but has only been associated with rabies outbreaks in Namibia. Here the largest of a few such known epidemics wiped out as many as 50,000 animals

as mutual grooming (Barnard and Hassel 1981). For example, mouth lesions from the browsing of thorn-bush (Acacia species such as *Acacia hereroensis*, or berg thorn) may have been a contributing factor due to the presence of rabies virus in saliva (Barnard and Hassel 1981). Contact between different social groups can occur at watering places and, in addition, farm fencing does not control the movement of kudu as these antelopes can easily clear a 2-m fence (Hübschle 1988). These rabies outbreaks may be interpreted as an example of non-bite transmission, with horizontal spread between kudus (Barnard et al. 1982), and posed a threat to human health via the game breeding and hunting industries in Namibia. The kudu is a popular game animal, venison is prized and rabid animals would thus constitute a risk to both hunters and butchers. Indeed, more than 60% of the game farming trade in Namibia revolved around kudu during the peak of the major outbreak of the late 1970s (Barnard and Hassel 1981).

The recognition of rabies in Botswana, a southern African country landlocked clockwise between Zambia (north), Zimbabwe, South Africa and Namibia, followed a similar pattern to that already described for most of the countries to the north. Thus rabies was first diagnosed within the first decades of the 1900s, followed by a steady dissemination and then a rapidly accelerated increase in the 1950s and a subsequent persistent endemic cycle in dogs, with infective spillover to herbivores and associated wildlife (Swanepoel 1993). A molecular epidemiological characterisation of a panel of virus isolates obtained between 1988 and 1992 confirmed the presence of two virus biotypes, i.e. viruses associated with canid species, predominating in the north and interspersed with a wildlife biotype (associated with mongooses) toward the south (Johnson et al. 2004). The sporadic occurrence of the mongoose biotype had been noticed earlier, demonstrated by monoclonal antibody typing, and will be further discussed in following sections (Maganu and Staugard 1985). The country of Zimbabwe (formerly Rhodesia) first reported rabies in 1902 (Edmonds 1922), coinciding with the outbreak in neighbouring Zambia and the dog control measures implemented there by the tribal ruler, Chief Lewanika. The disease spread through Zimbabwe quickly, but strict dog management measures brought the disease under control by 1913 (Sinclair 1914). Dog control measures on the border with South Africa prohibited the spread of this epizootic into South Africa, but it entered Mozambique (south-east), from where it was reported by 1908 (Valadao 1968). As mentioned above, canine rabies eventually invaded all of southern Africa after the World War II, spreading southward from the north-western countries of Angola and Zambia, through Namibia, Botswana, Zimbabwe, South Africa and Mozambique. Although the very first confirmed outbreak of rabies in sub-Saharan Africa came from the south-eastern coast of South Africa (Hutcheon 1894), this represented an isolated event not related to the manifestation of dog rabies in the subcontinent as evidenced by the gradual overland invasion from the north. This particular outbreak was caused by the importation of an infected dog from England during British occupation and settlement in the coastal area of Port Elizabeth (Britton 1894). In the immediate geographical vicinity, dogs were predominantly responsible and although some cases were also observed in cats and livestock, there was no evidence of wildlife involvement. This outbreak was extinguished by 1994 through stringent dog control. In the following period, leading up to the major introduction of cosmopolitan dog rabies into southern Africa, mongoose rabies was already present (Snyman 1940) and two fatal cases were confirmed in South Africa in 1928 (Hertzenberg 1928). The mongoose virus lineage is often referred to as the viverrid virus or biotype, because of older classification systems in which the mongooses were classified together with civets and genets under the family *Viverridae* (Skinner and Smithers 1990). However, the viruses belonging to this biotype are well adapted to infection of mongooses, most importantly the yellow mongoose (*Cynictis penicillata*), but also the slender mongoose (*Galerella sanguinea*) and many others (Swanepoel et al. 1993). These small carnivores (1–5 kg; 23–75 cm) are now classified in their own family (*Herpestidae*), based on structural differences with the members of the family *Viverridae*. In our understanding of rabies epidemiology in the region, it is significant that, as early as 1940, differences in host pathogenicity and epidemiology led Snyman (1940) to coin the designations mongoose type and European type. Indeed, many decades later, the very considerable antigenic and genetic variation amongst the mongoose viruses has been described and contrasted to that observed for European virus strains and the canine virus population in southern Africa (Wunner et al. 1988; Tuffereau et al. 1989; King et al. 1993; Nel et al. 1993; Von Teichman et al. 1995; Sabeta et al. 2003). Phylogenetic analysis (Fig. 4) was also found to reflect the geographical origin of the virus isolates, with the tree topology corresponding to five distinct and separate geographical regions of South Africa. These findings provided further evidence of the diverse origins and separate evolutionary paths of canine and mongoose rabies, lending strong support to the historical view that mongoose rabies, unlike the canine viruses (which belong to the cosmopolitan lineage), is indigenous to southern Africa and has been present in the region for an unknown number of centuries (Snyman 1940).

In the 1950s and in the subsequent decades, dog rabies became a more important public and veterinary health concern, spreading throughout the region and appearing in new wildlife principal species. The black backed jackal became an important species in Namibia (Sect 1.4), and also subsequently in northern South Africa, Botswana and Zimbabwe. Another species, *Canis adustus*, the side-striped jackal presented rabies cases from the 1950s together with the influx of dog rabies in Zimbabwe and was involved in several large epidemics since the 1960s (Sect. 1.4). Nevertheless, rabies in dogs remains the most important zoonotic threat of this region and throughout Africa. As a case in point, dog rabies is today endemic in one of South Africa's nine provinces, i.e. the province of KwaZulu-Natal on the eastern seaboard (Indian Ocean) of the the country. Being far south-east, the first documented case of canine rabies here was in 1961, as part of the steady southward spread of the virus, in this case from southern Mozambique (Mansvelt 1962). An intense epidemic followed, which was brought under control by 1968. However, another cycle was introduced from southern Mozambique during 1976 (unpublished records of the Onderstepoort Veterinary Institute) and was found to be enzootic ever since. The cases that followed provided further confirmation of the close relationship between the number of dog cases and human cases, which by then had become well understood throughout the continent (Fig. 5).

Most recently, the steadily emerging rabies problem of the Kwazulu-Natal province of South Africa appears to have been exacerbated by an explosion of ownerless feral dogs (Bateman 2005). Sub-Saharan Africa is tightly gripped in a devastating HIV/AIDS pandemic and in South Africa, the highest rates

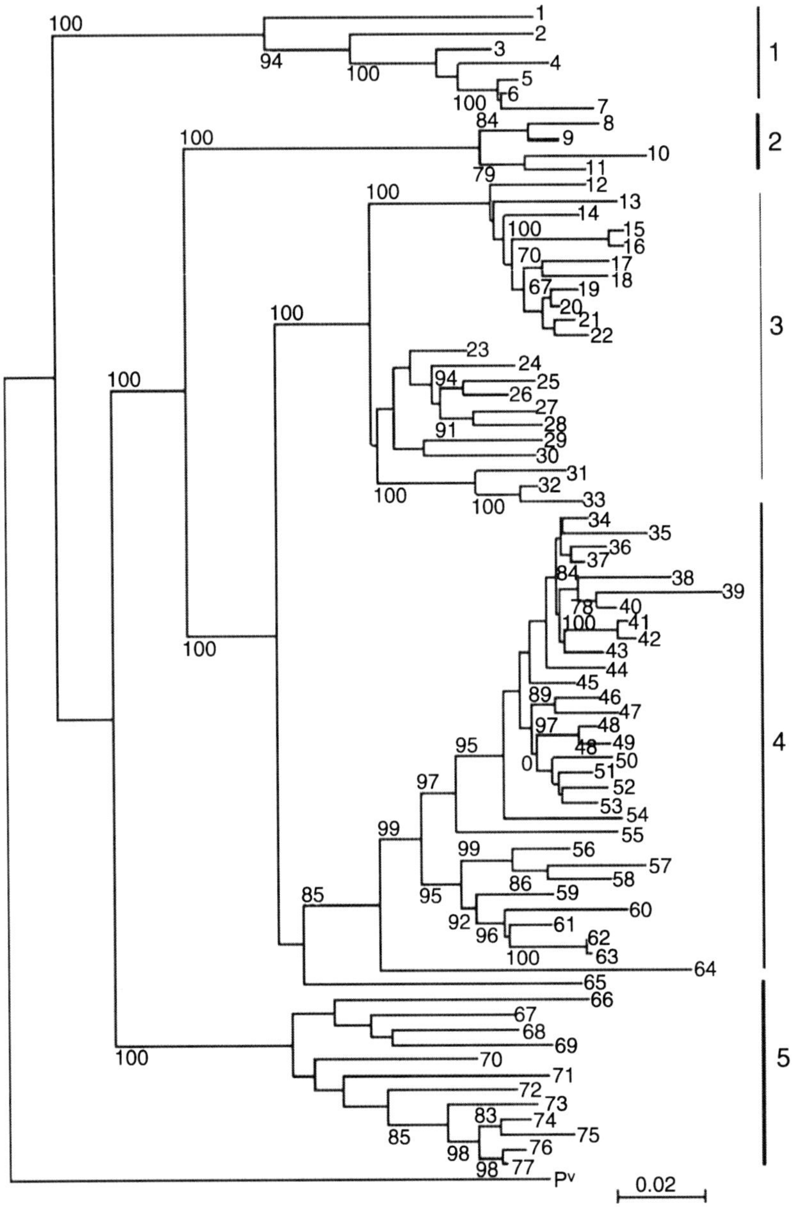

Fig. 4 An unrooted phylogenetic tree illustrating the genetic variability within the southern African mongoose rabies virus biotype. The sequence of Pasteur virus (*pv*) was included as an outgroup representative of the cosmopolitan lineage. The different virus clusters (groups 1–5), correspond to distinct geographical locations. (From Nel et al. 2003)

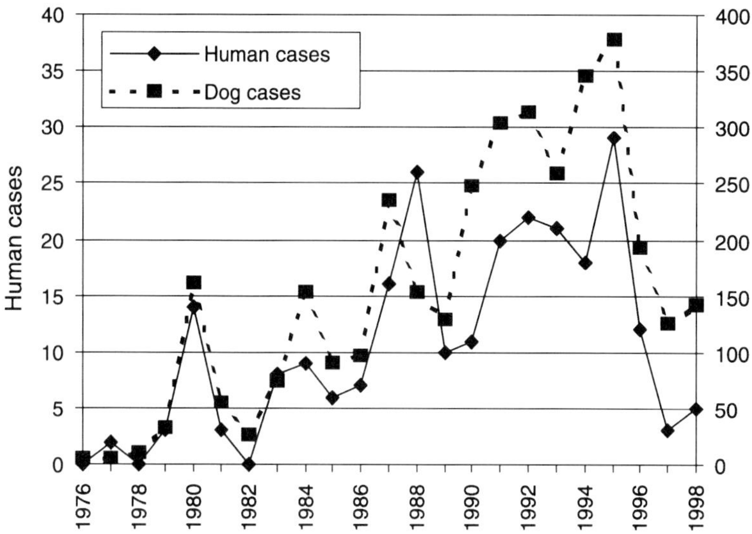

Fig. 5 The close relationship between the number of dog cases and human cases, as illustrated in the Kwazulu/Natal province of South Africa, 1976–1998. (From Bishop 1999)

of HIV/AIDS infections and deaths occur in the province of Kwazulu-Natal. It is the mortalities associated with the HIV/AIDS epidemic, together with a poverty-struck community, that is believed to be behind the dramatic emergence of packs of feral dogs that roam rural areas of the province. Indeed, feral dogs were directly responsible for the deaths of at least seven people in 2004, five of whom died through the transmission of rabies (the other two cases were the fatal maulings of an infant and an elderly man) (Bateman 2005). The population of ownerless feral dogs is not only much larger than previously thought (more than 800 dogs were destroyed in one small area of the province over a 12-month period), but seems to be expanding rapidly (Le Roux, Allerton veterinary laboratories, Pietermaritzburg South Africa, personal communication). Rural Zulu homes in Kwazulu-Natal traditionally consist of a *kraal*, a crudely fenced-off area where a family would reside in one or more mud huts. Where AIDS led to the demise of such family units, only the family-associated dogs may persist: up to eight dogs have been found in such a *kraal* that was abandoned by its human inhabitants (Bateman 2005). The radiating ownerless feral dog population is therefore growing in large numbers, in part because of a pre-existing poverty-induced lack of population control (leashing, spaying, etc.). These roving packs of feral dogs have now become much feared throughout the countryside, attacking livestock, domestic animals and children. The hunting, scavenging and survival success of the feral packs seems to be evident in the good physical condition of pack members. Attempts to control this emerging and complex problem include a revitalised primary dog census, extensive rabies vaccination and pro-active destruction of ownerless dogs (Bateman 2005).

1.5 Conclusions

There is little doubt that rabies virus evolved in the first instance, like the other lyssaviruses (with the possible exception of Mokola virus), as a virus of bats (Badrane and Tordo 2001). A transmission to terrestrial animals may have occurred as little as 1,000 years ago, but it is likely that such transmission events occurred more than once through time (presence of rabies in ancient Mesopotamia 3,000 years ago or more) and on different continents (accounting for the raccoon virus variant in North America). Historical records confirm that rabies in most of Africa – and in particular in sub-Saharan Africa – is a disease of modern times. It has become clear that the cosmopolitan variant of canine rabies virus was introduced from European territories into Africa during the years of colonisation, which peaked around 1900. Since then, this rabies virus lineage became well established principally in dogs, but also with wildlife species over vastly extended areas. The more distant variant of dog rabies observed in North Africa could have been a separate introduction some centuries ago, allowing for isolated evolution independent of European viruses. Little is known about the status of this viral cycle at present and it seems evident that the cosmopolitan lineage has become more successful in its widespread establishment in a multitude of host species. Following its widespread dissemination across the continent, dog rabies only emerged as a serious disease of epidemic proportions since the 1950s and later. In the last few decades, this variant of rabies virus has made an ever-increasing impact as an endemic zoonotic agent throughout Africa. Colonial introduction, translocation, and emergence among feral dogs have been the historical factors and epidemiological drivers of rabies over most of the continent.

In terms of the involvement of wildlife, jackals (both *Canis adustus* and *Canis mesomelas*) and bat-eared foxes (*Otocyon megalotis*) have emerged as important incidental species for this cosmopolitan virus in southern Africa. In particular, *Canis mesomelas* and *Otocyon megalotis* appear to have emerged as maintenance hosts for this variant of rabies, although the mechanisms and ecological factors leading to the maintenance of rabies is not fully understood (Bingham et al. 1999; Bingham 2005). It is nevertheless evident that jackal rabies predominantly occur in commercial farming areas where jackal populations reach high densities, implying that such areas and activities are ecologically favourable for these species and consequently for the maintenance of rabies (Bingham et al. 1999).

In recent years, both African wild dogs (*Lycaon pictus*) and Ethiopian wolves (*Canis simensis*) have fallen victim to the engulfing rabies epidemic and these species are now, more than ever, seriously endangered (Hofmeyer et al. 2004; Sillero-Zubiri et al. 1996; Randall et al 2004). Neither of these carnivores appears to sustain rabies independently of dogs. For another of the wildlife carnivores affected by cosmopolitan rabies, the spotted hyena (*Crocuta crocuta*), a rather controversial mechanism has been proposed (East et al. 2001), but since the initial study, no further evidence could be produced in support.

The wild ruminant, kudu (*Tragelaphus strepsiceros*), has sustained several rabies epidemics in the last three decades. The virus could be identified as cosmopolitan and was found to be related to a jackal variant present in this region. The largest of the kudu epidemics peaked in 1980, with seemingly very effective kudu-to-kudu transmission, but subsided by 1985 (Hübschle 1988). This epidemic caused an estimated loss of 30,000–50,000 antelope, or 20% of the population (Swanepoel 2005). During 2002 and continuing into 2003, there was another substantial outbreak in kudu, where an estimated 2,500 animals over large areas of Namibia succumbed to rabies (OIE 2005). It is evident that human-induced ecological changes, with the inclusion of commercial farming activities, have contributed significantly to the establishment of rabies in various wildlife species, with the correspondingly increased and varied threat of zoonosis. This trend is set to continue.

A striking example of the impact of a HIV/AIDS on the population and demographics of not only humans, but also on other species, comes from Kwazulu-Natal in South Africa. It is here that, through lack of human ownership, stray domestic dogs form highly successful hunting packs and emerge not only as a direct threat to human and animal health, but significantly contribute to an ever-radiating rabies problem of the region. The number of these roving feral packs seems to be on the increase and is thought to be fuelled by the dramatic mortalities associated with the AIDS epidemic of this region, as owners succumb (Bateman 2005).

Although the now well-established mongoose rabies virus lineage of southern Africa is today less of a human disease threat than the cosmopolitan dog rabies variant, it was responsible for approximately half of the human rabies cases encountered before 1950 (Swanepoel 2005). Sporadic cases due to the mongoose virus still occur and, notably, this virus variant has been associated with unexplained failure of post-exposure prophylaxis in humans on more than one occasion during the past two decades (R. Swanepoel, unpublished observations). The origin of this apparently unique variant of rabies virus in southern Africa is not clear and there seems to be at least two possible explanations:

1. A separate introduction from bats to small herpestid carnivores of southern Africa, sometime after the original establishment of cosmopolitan dog rabies and North American raccoon rabies (phylogenetically, the mongoose variant seems to be closer to the cosmopolitan variant than to the raccoon variant). Presumably, this would mean that a bat variant of rabies virus, with ancestral links to the European progenitor, would have to have been present in southern Africa a few centuries ago. If this is the case, the bat virus itself must have become extinct in southern Africa, since there is no evidence of rabies in bats in southern Africa, or anywhere on the African continent. Assuming that there are no extant true rabies viruses in African or other Old World bats (although there may still be a question, given scant surveillance attempts over the last century), it has to be considered unlikely that a well-established bat rabies virus will have become extinct in a stable reservoir(s) during the recent past. This scenario leads to a perhaps more likely second possibility of origin.
2. Introduction of terrestrial mongoose rabies into southern Africa at some time before the dissemination of the cosmopolitan variant. However, given the efficiency with which dog rabies has manifested in dogs and a huge variety of wildlife during the past 50 years, the mongoose virus is unlikely to ever have been a dog virus, given the specificity and adaptation of this virus for species of the *Herpestidae* and its tendency to cause dead-end infections in other hosts, including canids. If this is the case, it would constitute a very different scenario from the only other form of mongoose rabies known, i.e., rabies in mongooses in the Caribbean. These mongooses were imported into the Caribbean from India in the 1870s and 1880s and genetic analysis indicate that these mongooses acquired cosmopolitan dog rabies from endemically infected Caribbean dogs, resulting in a first major mongoose rabies outbreak in 1950 (Smith and Seidel 1993). Globally, the epizootiology of rabies in mongoose is poorly understood, outside of southern Africa and the secondary foci in the Caribbean.

2 The Nonrabies Lyssaviruses of Africa

2.1 The Three African Nonrabies Lyssavirus Genotypes

The three nonrabies lyssavirus species (genotypes) present in Africa – Lagos Bat virus (LBV; GT2), Mokola virus (GT3) and Duvenhage virus (GT4) – have not been encountered outside of Africa. As eluded to in the previous sections, rabies virus has only been associated with infections of terrestrial mammals on the African continent, although rabies virus infection of bats is well known in the Americas (Belotto et al. 2005). Mokola virus has also not been isolated from bats but indeed from various terrestrial species (Nel et al. 2000). Of the remaining African lyssaviruses, both LBV and Duvenhage viruses are thought to be bat viruses, although LBV infection of terrestrial animals has been known to occur (King and Crick 1988). Duvenhage and Mokola viruses, but not LBV as of yet, have also been responsible for rare zoonotic events (Swanepoel et al. 1993; Nel et al. 2000). Due to limited attention given to these viruses, little is known about their epidemiology. All three are present in southern Africa, where they have been isolated in the countries of South Africa and Zimbabwe. LBV and Mokola virus have been isolated from countries in tropical West Africa and from Ethiopia (Figs. 6, 7, and 8). It is of interest to note that the geographical domains (tropical West Africa and southern Africa) of the nonrabies lyssaviruses of Africa and the indigenous rabies virus variants of Africa overlap (i.e. Africa dog virus, mongoose rabies and the three nonrabies lyssaviruses).

LBV (Fig. 6) was first isolated from a brain pool of *Eidolon helvum* fruit bats in 1956 at Lagos island in Nigeria (Boulger and Porterfield 1958), but it was not until 1970 that it was identified in its present form (Shope et al. 1970). In this case, a total of six of these large fruit bats were killed while resting, with collection via a shotgun. Although virus must have been encephalitic in at least one of the bats, allowing isolation from the brain pool, there was no reported evidence of clinical neurological disease at the time. These bats are the largest of the African Megachiroptera and they have an interesting migratory pattern between the latitudes 10°N and 34°S, with a breeding range in the African forest tropical belt 9°N to 18°S, but strangely also occurring in arid locations lacking fruit plants (King et al. 1994). In 1974, a second isolate followed from Bozo, Central African Republic, where it was found in another species of frugivorous bat, *Micropterus pusillus* (Sureau et al. 1980). Sporadically, a few more isolations of LBV were made, seemingly as a matter of course in those areas where the virus was considered with competent technique. In June 1980, the first LBV isolation from southern Africa came from the town of Pinetown in the Kwazulu/Natal province of the Republic of South Africa from an *Epomophorus wahlbergi* fruit bat which displayed abnormal behaviour (Meredith and Standing 1981). At the time, this region experienced a heightened awareness of rabies and the potential implication of bats, due to a raging dog rabies epidemic in this region as well as reports from the USA on the role of bats in rabies from North America. A total of 282 bats from Pinetown and parts of Durban in Kwazulu/Natal were submitted for testing and 13 of these (4.6%) were found positive for lyssavirus antigens. During this period, virus from three brains were cultured and all

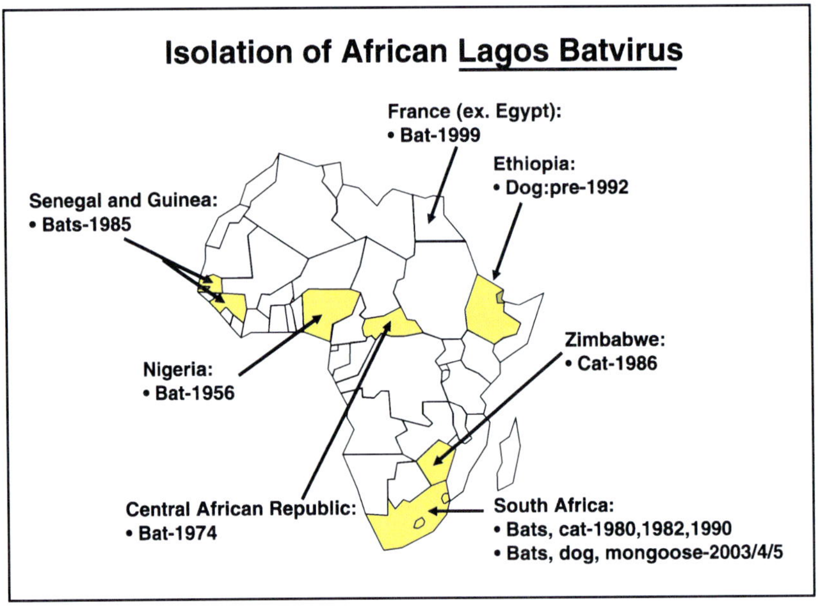

Fig. 6 The sites, hosts and time of isolation of LBV (GT 2), as described in detail in the relevant sections of the text

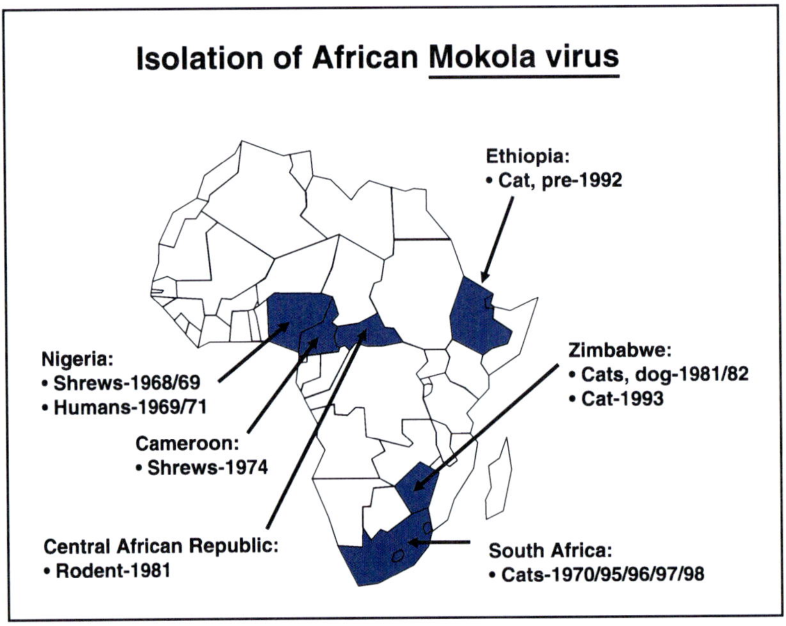

Fig. 7 The sites, hosts and time of isolation of Mokola virus (GT 3), as described in detail in the relevant sections of the text

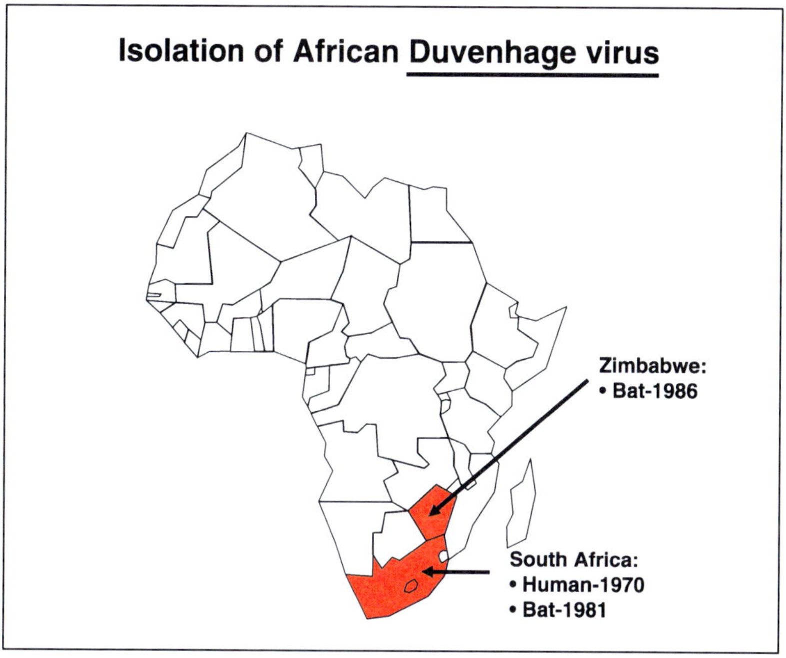

Fig. 8 The sites, hosts and time of isolation of Duvenhage virus (GT 4), as described in detail in the relevant sections of the text

three were identified as Lagos bat by serological typing (Van der Merwe 1982; Crick et al. 1982; King and Crick 1988).

In 1982, an isolation of LBV was made for the first time from a terrestrial animal. The host in this case was a domestic cat from Stanger, Kwazulu/Natal in South Africa, i.e. the same region associated with the *Epomorphorus* outbreak described above (King and Crick 1988). After this isolation, the virus was not encountered again in South Africa until 1990, when a further isolation was made from *Epomophorus wahlbergii* in Durban, Kwazulu/Natal (G. Bishop and R. Swanepoel, unpublished data). Meanwhile, from the Pasteur Institute in Dakar came two reports of LBV isolations in Senegal in 1985, one made from *Nycteris cambiensis* in Guinea and the other from *Eidolon helvum* in Dakar. A second isolation from a domestic cat was reported from Zimbabwe in 1986 (Foggin 1988) and in a rabid dog from Ethiopia, following a program for the isolation and characterisation of street rabies virus isolates in this country (Mebatsion et al. 1992). From France, an isolate of LBV was reported in 1999. This unusual case was in an Egyptian tomb bat that had been imported from the African continent, via Belgium, as an exotic pet (Aubert 1999). Following a passive surveillance program involving bat interest groups in the KZN province of South Africa, five more isolates of LBV have been obtained from *Epomophorus wahlbergii* since 2003 (L. Nel, unpublished data). Moreover for the first time, LBV was recently isolated from terrestrial wildlife, i.e. a water mongoose (*Atilax paludonsis*) and retrospectively from a rabies vaccine failure case (domestic dog), both from the Kwazulu/Natal province, South Africa (L. Nel, Unpublished, 2005).

The isolation of Mokola virus has also been reported rather infrequently from only a small number of African countries, where appropriate surveillance and laboratory investigations have been initiated (Fig. 7). Mokola virus is the only lyssavirus species never isolated from bats, but it has been found in a surprisingly diverse host range, considering the small number of virus isolates available. The first isolation of the virus was made in 1968, from organ pools of shrews in Nigeria (Shope et al. 1970; Kemp et al. 1972). Thereafter, the virus was isolated from two naturally occurring human cases, one of which was fatal (Nigeria, 1969–1971; Familusi and Moore 1972; Familusi et al. 1972), again from shrews (Cameroon, 1974; Le Gonidec et al. 1978), from a rodent (Central African Republic, 1983; Saluzzo et al. 1984), from domestic cats and a dog (Zimbabwe, 1981–1982; Foggin 1983) and again in 1989 from a cat in Ethiopia (Mebatsion et al. 1992) and in 1993 from a cat in Zimbabwe (Bingham et al. 2001). The first isolation of Mokola virus in South Africa was made from a domestic cat in 1970, although its identity as Mokola virus was only affirmed in the 1980s, prompted by the Mokola virus isolations in Zimbabwe (Schneider et al. 1985). The next isolates were made only many years later, from 1995 to 1998 (Von Teichman et al. 1998; Nel et al. 2000). It is likely that improved rabies surveillance programs and lyssavirus diagnostics have led to these recent isolations in South Africa, all made from domestic cats with suspicious rabies-like clinical signs. Most of these animals were in fact vaccinated against rabies, a long-time standard practice for pet owners and required by law since 2000 in this province of South Africa. Of all the current lyssavirus genotypes, Mokola virus is genetically one of the most distant from rabies virus, as demonstrated with serological studies (King and Crick 1988) and through analyses of specific genomic nucleotide sequences (Bourhy et al. 1993). Poor cross-reactivity between sera specific for Mokola and rabies viruses is observed experimentally (Shope 1975). These observations are supported by the failure of rabies virus-specific vaccines to protect against Mokola virus infection, as observed experimentally with laboratory mice (Wiktor 1985; Bahloul et al. 1998).

Our poor understanding of the epidemiology of this lyssavirus is corroborated by the reports of the presence of serum antibodies to Mokola in Nigerian dogs, albeit with a low incidence (Aghomo et al. 1990) and in a variety of other species, including domestic herbivores and the bat *Eidolon helvum* (Kemp et al. 1972). From Zimbabwe, serum antibodies against Mokola virus were also reported for the gerbil, *Tatera leucogaster* (Foggin 1988). A report of a positive rabies fluorescent antibody test on the brains of two greater cane rats (*Thryonomys swinderianus*) from northern South Africa (Swart 1989) has been noted as suggestive of a possible Mokola virus infection, although virus isolation and characterisation has not been attempted (Swanepoel 2005). The fact that Mokola virus has never been isolated from bats, but only from terrestrial animals, is indeed unusual in the context of the close association of all the other lyssaviruses with bats and when the evidence of their potential co-evolution is considered (Badrane and Tordo 2001). A phylogenetic analysis of the full-length glycoprotein sequences of Mokola virus and comparison with the glycoproteins of a wide range of rabies virus isolates indicates a comparable degree of variation within the two genotypes (L. Nel, unpublished data). The significance of the finding lies in the fact that the genetic variation among only four isolates of Mokola (three from southern Africa and one from Ethiopia) more or less equals the variation found among the most diverse classical rabies virus isolates from various host species throughout the world (Fig. 9). Clearly, the obscure epidemiology of this African lyssavirus is not understood and its true reservoir(s) remains to be determined (Von Teichman et al. 1998; Nel et al. 2000, 2003).

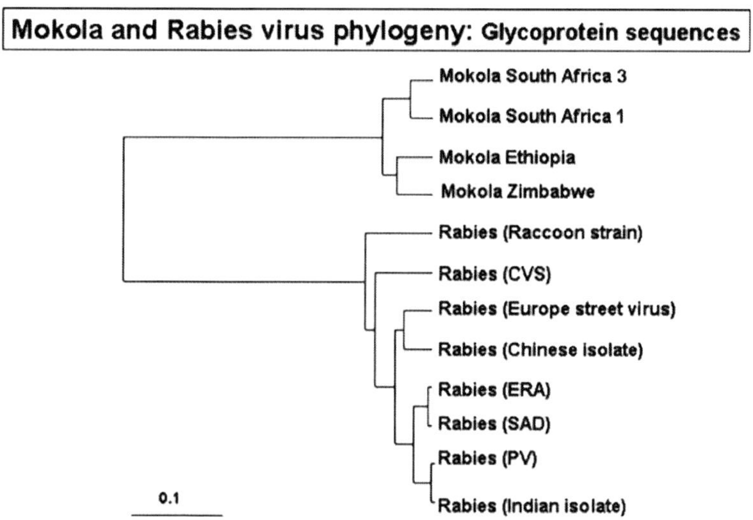

Fig. 9 A phylogenetic reconstruction indicating the relationship among Mokola virus isolates from South Africa, Ethiopia and Zimbabwe and with the most genetically diverse rabies virus isolates from all over the world. The phylogeny is based on full-length sequences of the glycoprotein-encoding genes (L. Nel, unpublished data)

Duvenhage virus (GT 4) was only encountered historically on three occasions (Fig. 8), the first of which was in 1970 in a fatal human case from South Africa, but from the northern parts of the country as opposed to the province of Kwazulu/Natal, associated with LBV and Mokola cases (Meredith et al. 1971). From this case history, it is evident that the deceased person, an adult male (Mr. Duvenhage) was bitten on the mouth by a small bat which may have been *Miniopterus schreibersii*, a common small insectivorous bat of the region. The biting incident took place during the night, while Mr. Duvenhage was asleep. After about 4–5 weeks, rabies symptoms developed and death followed about 2 days after the diagnosis of clinical rabies. More than a decade later, this virus was again reported from northern South Africa, in 1981, and from an insectivorous bat that may have been *Miniopterus schreibersii* (Van der Merwe 1982). During a survey about 400 km further north from South Africa, in Zimbabwe, Foggin (1988) succeeded in isolating Duvenhage virus from a fairly common insectivorous bat, the slit faced bat, *Nycteris thebaica*. These were the only three cases of Duvenhage virus ever reported and the only clinical picture was that of Mr. Duvenhage, the original 1970 case which presented as classical human rabies, inclusive of hydrophobia and progressive neurological disease until a recent report (Paweska et al. 2006).

Over the years, there were a number of unqualified reports of bat lyssaviruses from southern Africa. In a survey of bats carried out in the late 1950s and early 1960s in South Africa, one bat (*Nycteris thebaica*) collected from north-eastern South Africa was reported to be rabid on the basis of the mouse inoculation test and histopathology, but no isolate was kept (Onderstepoort Veterinary Institute, unpublished data; Swanepoel, 1994). A more recent South African survey for rabies in bats, applying the RREID test, could not demonstrate any suggestion of rabies in 530 bats (Oelofsen and Smith 1993). In 1992, however, rabies was diagnosed in a bat brain from Messina in northern South Africa. The brain had been fixed in formalin and was tested by histology, but the isolate was never characterised (Onderstepoort Veterinary Institute, unpublished data). In 1996, rabies was detected from an unidentified Zambian bat that had been found dead (Ahmadu and Zulu 1998). Fluorescent antibody tests were positive for this animal and also for the brains of mice that died after inoculation with the bat brain suspension. No further characterisation was carried out.

2.2
Disease Potential of the African Nonrabies Lyssaviruses

Based on phylogeny, pathogenicity and immunogenicity, the Lyssavirus genus was suggested for division into two distinct Phylogroups (Badrane et al. 2001). Phylogroup 1 is the largest, comprising genotypes 1 (rabies virus), 4 (Duvenhage virus), 5 and 6 (European bat lyssaviruses 1 and 2) and the Australian bat lyssavirus (GT7). The status of rabies as one of the world's leading zoonotic diseases and the emergence of canine rabies throughout Africa during the latter half of the last century has been described in previous sections.

However, Phylogroup 2 is composed currently solely of the two African lyssaviruses: Lagos bat virus and Mokola virus (GT 2 and 3), although phylogenetic evidence suggests that one of the newly discovered Asian lyssaviruses, the West Caucasian bat virus, may also belong to this group (Botvinkin et al. 2003; Kuzmin et al. 2005). Experimental findings of pathogenicity in mice seem to indicate that the Phylogroup 1 and 2 viruses differ by virtue of the latter being pathogenic only by the intracerebral route, while the Phylogroup 1 viruses are also pathogenic when administered at peripheral locations (Badrane et al. 2001). As shown for Mokola virus (Fig. 10), the absence (Phylogroup 2) or presence (Phylogroup 1) of the virulence-associated Arginine 333 in the viral glycoproteins (Tuffereau et al. 1989) is undoubtedly important, but there may be other differences in the trimerisation and fusion properties of the glycoproteins of the Phylogroup 1 and 2 members that could also contribute to differences in their pathogenicity (Desmezieres et al. 2003). Nevertheless, all the lyssaviruses (including Lagos bat virus) have shown the capacity to cause human and

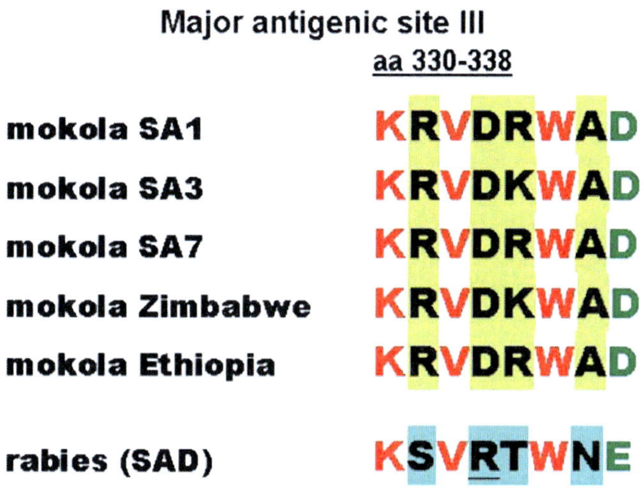

Fig. 10 Dissimilarities in the crucial glycoprotein antigenic domain III of Mokola virus isolates and rabies virus. Note that arginine (a basic amino acid residue), present at position 333 of rabies viruses (Phylogroup 1), is substituted by aspartic acid in all of the Mokola virus (Phylogroup 2) isolates investigated

animal disease and nonrabies and rabies lyssaviruses should, from a public and veterinary health perspective, be treated equally (Rupprecht et al. 2002). Considering that virulence may vary based in part upon dose, route, isolate and host model, additional experimental studies of comparative pathogenesis are needed.

All experimental evidence to date indicates that rabies virus vaccines (in Phylogroup 1) may provide adequate protection against other members within Phylogroup 1, but not against the members of the Phylogroup 2. Whereas results of protection by rabies vaccination for members of Phylogroup 1 lyssaviruses seem to be variable and dependent on the vaccine strain and type, the consensus is that neutralising antibody is at a sufficient level to offer at least some, if not complete, cross-reactivity against all the current members of this group. The situation is different for the Phylogroup 2 viruses and it has been shown that rabies vaccines do not offer any real protection against viruses in this group, leading to the conclusion that protection is inversely proportional to the genetic distance between them and rabies virus (Fekadu et al. 1988; Bahloul et al. 1998; Badrane et al. 2001). Failure of rabies virus vaccination to protect against Mokola virus has also been demonstrated by the most recent cases of the disease in vaccinated cats in South Africa (Von Teichman et al. 1998), as well as with a dog in Zimbabwe (Foggin 1983). The same applies to LBV, with a rabies vaccine failure in a dog established to be due to a fatal Lagos bat virus infection (unpublished data from South Africa, 2005).

Although LBV (GT2) may also infect terrestrial species, Mokola virus is also the only member of this group to have been implicated as zoonotic to date (Familusi et al. 1972a, 1972b) and it follows that this is the only Phylogroup 2 member to have been considered in the development of new lyssavirus vaccines. The first potential Mokola virus vaccine, a recombinant baculovirus expressing the Mokola virus glycoprotein, was reported in 1993 (Tordo et al. 1993). It was demonstrated that the baculovirus expressed glycoprotein was less strongly glycosylated in *Spodoptera frugiperda* (SF) cells than the native viral glycoprotein in BHK-21 cells, but that it was antigenically and immunologically similar. Used as a vaccine in laboratory mice, the baculovirus Mokola glycoprotein-expressing SF cells elicited a protective immunological response against Mokola virus. This vaccine failed to protect against heterologous rabies virus challenge. In a comparison of DNA vaccines for Mokola virus, it was shown that such vaccines were able to protect against Mokola, but could not fully protect using a single immunisation only (Nel et al. 2003). It was also shown that these vaccines did not offer any protection against rabies virus, nor was any cross-protective immunity achieved in a combined prime-boost approach, using a Mokola DNA vaccine and the poxvirus recombinant rabies vaccine, VRG. In addition, the co-expression of Mokola virus glycoprotein and nucleoprotein did not indicate any synergy resulting in a protective advantage over DNA vaccines that express the glycoprotein only (Nel et al. 2003). An experimental DNA vaccine that expresses a Mokola and rabies virus chimeric glycoprotein gene demonstrated effective cross-protection against both viruses, the success of the strategy lying in the observation that lyssavirus glycoproteins can be divided into two parts, separated by a flexible hinge. Each of these parts contains one of the two most important antigenic domains in the elicitation of a protective immune response, i.e. the antigenic site II (NH2 half) and the antigenic site III (COOH part) (Anilionis 1981; Jallet et al. 1999).

2.3 Conclusions

Clearly, we have a poor understanding of the African lyssaviruses. For Mokola virus, the reservoir is not even known, although it likely has a wide host range, proven zoonotic potential, is not protected against by rabies virus vaccines and no prophylaxis is available for the treatment of exposures, contrary to the case with traditional rabies viruses. Its extended history on the continent is further suggested by the degree of variation observed in sequence analysis of only a few isolates within this genotype (Fig. 9). Similarly, although LBV appears to be rare and has not been identified anywhere in the previous 13 years, a small-scale passive surveillance effort in Kwazulu/Natal, South Africa, enabled us to make six new isolations of LBV in a relatively short period of time, four from bats and two of these from terrestrial animals, one being domestic and the other a wildlife species. This finding re-emphasises our lack of understanding of the true spatio-temporal occurrence of lyssaviruses throughout Africa, due to poor surveillance of nonrabies viruses (and in fact, rabies virus) throughout the continent. Even if more active surveillance programs were to be considered, very few laboratories in Africa would be capable of detecting nonrabies lyssaviruses and be able to differentiate between rabies and nonrabies viruses with any degree of certainty. The employment of a small passive surveillance plan demonstrated to us that LBV can still be readily identified and isolated from bat species in South Africa despite having not been reported from any species in this region for more than a decade.

Cumulatively, all available evidence indicates that LBV is highly likely to be persistently maintained in Megachiroptera populations in South Africa and probably elsewhere in Africa from where LBV has been reported in the past. For example, a recent survey from Ethiopia also yielded isolates of both Mokola virus and LBV (Mebatsion et al. 1992), seemingly indicating that these viruses may be readily and invariably encountered when and where a surveillance effort has been made in sub-Saharan Africa. Providing informed advice on the control and prevention of any disease when basic epidemiological data is as scarce as for the African lyssaviruses is not possible. However, it is clear that a very real possibility of lyssavirus infections and consequently zoonoses has to be recognised. Thus it has to be strongly recommended that appropriate care should be taken when interacting with any of these species, particularly by bat handlers and interest groups. However, other risk groups would be pet owners, veterinarians, laboratory personnel (particularly those associated with rabies diagnoses), animal control officers, and others in regular contact with animals or tissue samples, which may support exposure to these lyssaviruses. Even though the value of rabies vaccination is doubtful, it may also be considered in lieu of the possibility of some cross-reactivity (Hanlon et al. 2005) and the lack of any alternative.

Within the Mononegavirales, the *Rhabdoviridae* is a family whose members cumulatively infect more than 200 species, which include a broad variety of plants, insects, fish, birds and mammals. Although, at the present time, some of the genera, such as the *Lyssavirus, Vesiculovirus* and *Ephemerovirus* genera, are only known to infect warm-blooded animals, there are strong arguments for the origin of the Rhabdovirus family from an insect virus progenitor (Badrane and Tordo 2001), arguing for its subsequent dissemination among the wide diversity of species that include not only insects but plants and animals. Among the lyssaviruses, for instance, Mokola virus has been found in insectivorous shrews and the virus is known to be able to replicate in insects and insect cells, while all the other lyssaviruses can either be transmitted by insectivorous bats (GT 1, 2, 7) or are exclusive to insectivorous bats (GT 4, 5, 6). This last category also includes the putative Asian genotypes Aravan, West-Caucasian bat virus, Khujand and Irkut (Kuzmin et al. 2005). Three other rhabdoviruses were once considered as putative lyssaviruses, and have been encountered in insects only, i.e. Rochambeau (Digoutte 1975), Obodhiang and kotonkan (Bauer and Murphy 1975). It is therefore tempting to speculate that bat lyssaviruses emerged from an insect rhabdovirus, possibly around 7,000–12,000 years ago (Badrane and Tordo 2001). One of these (GT1) then spilled over to terrestrial animals on different occasions, leading to the global dissemination of classical dog rabies, a disease which emerged only in recent decades over much of the world, including Africa, where it has established wildlife reservoirs connected with dog rabies cycles, and in so doing significantly increasing the frequency of animal and human rabies.

If the above scenario resembles the true chain of events, several questions regarding the appropriate circumstances and conditions of the spillover events that led to new virus populations in new reservoirs are relevant. Indeed, information on the ecological/population dynamics and virus adaptation/evolution before and during such an event would greatly facilitate an understanding of the likelihood of lyssaviruses other than rabies following similar dramatic emergence and global dissemination events. For lyssaviruses other than rabies, such spillover into domestic animals and terrestrial wildlife can be demonstrated in modern times, but probably occurred over many centuries, apparently without successfully establishing terrestrial reservoirs. It is true, however, that zoonotic disease emergence in particular is strongly linked to changes in ecosystems which, in our modern-day world, can only be regarded as particularly dramatic. The profound ecological changes associated with urbanisation, commercial agriculture and general overpopulation leads to species disruption, which includes exploitation by species that may be favoured by the new ecological setting. In this way, rabies has only recently become established in Africa and other lyssaviruses can be regarded as candidates for similar events, given the appropriate conditions for introduction and persistence/maintenance in new host metapopulations.

Although the lack of appropriate pre- and post-exposure prophylactic biologicals against newly emerging lyssaviruses will have serious public health implications, the low apparent current occurrence of these viruses and their confinement to the sub-Saharan African subcontinent will continue to inhibit any progress in this regard. The development and production of any given vaccine is ultimately a function of commercial viability. This reality affects not only the African lyssavirus diseases, but also other viral diseases that appear to be sporadic, such as those caused by the filoviruses, Ebola and Marburg. However, the establishment of lyssavirus vaccines with expanded range can be regarded as a worthwhile objective. Combined or cross-reactive vaccines would be of obvious specific benefit to laboratory diagnosticians worldwide and to high-risk groups in those areas where nonrabies lyssaviruses are endemic. There has been progress in rabies-specific biologicals: for example, the advent of reverse genetics technology and the possibility of direct genetic manipulation of rabies virus does not only improve our understanding of the molecular mechanisms of lyssavirus pathobiology, but also holds promise for the development of novel and cross-protective lyssavirus vaccines (Schnell et al. 2005). Also relevant to the treatment of lyssavirus infections would be the innovations around the use of monoclonal antibodies instead of rabies immune globulin in the post-exposure prophylaxis of rabies exposures (Hanlon et al. 2005). Not only does there exist a global crisis around the supply of human rabies immunoglobulin, but effective monoclonal antibody combinations may also be found not only for rabies, but they may also be expanded to cover other, more distant lyssaviruses. Like the development of new vaccines for rabies virus and the rare lyssaviruses of Africa, the development of monoclonal antibody panels for rabies as well as other lyssaviruses, remains an issue of commercial viability. From both a scientific and a public health point of view, it could well be argued that vigilance and a sustained improvement in the surveillance for lyssaviruses in Africa should at this time be the foremost priority in qualifying the extent of threat and potential threat from rabies-causing viruses. Our poor understanding of the African lyssaviruses is due to our lack of knowledge of their epidemiology and pathogenicity, a direct result of poor surveillance with a corresponding low detection and isolation rate.

References

Aubert FA (1999) Rabies in individual countries: France. Rabies Bull Eur 23:6
Aghomo HO, Tomori O, Oduye OO, Rupprecht CE (1990) Detection of Mokola virus neutralizing antibodies in Nigerian dogs. Res Vet Sci 48:264
Alexander KA, Kat PW, Wayne RK, Fuller TK (1994) Serologic survey of selected canine pathogens among free-ranging jackals in Kenya. J Wild Dis 30:486–491
Alexander RA (1952) Rabies in South Africa: a review of the present position. J South Afr Vet Med Assoc 23:135–139
Anilionis A, Wunner WH, Curtis PJ (1981) Structure of the glycoprotein gene in rabies virus. Nature 294:275–278
Aspden K, van Dijk A, Bingham J, Cox D, Passmore JA, Williamson AL (2002) Immunogenicity of a recombinant lumpy skin disease virus (neetling vaccine strain) expressing the rabies virus glycoprotein in cattle. Vaccine 20:2693–2701
Aspden K, Passmore J, Tiedt F, Williamson AL (2003) Evaluation of lumpy skin disease virus, a capripoxvirus, as a replication-deficient vaccine vector. J Gen Virol 84:1985–1996
Badrane H, Tordo N (2001) Host switching in Lyssavirus history from the Chiroptera to the Carnivora orders. J Virol 17:8096–104
Badrane H, Bahloul C, Perrin P, Tordo N (2001) Evidence of two Lyssavirus phylogroups with distinct pathogenicity and immunogenicity. J Virol 75:3268–3276
Bahloul C, Jacob Y, Tordo N, Perrin P (1998) DNA-based immunization for exploring the enlargement of immunological cross-reactivity against the lyssaviruses. Vaccine 16:417–425
Barnard BJH, Hassel RH (1981) Rabies in kudus (*Tragelaphus strepsiceros*) in South West Africa/Namibia. J South Afr Vet Assoc 52:309–314
Barnard BJH, Hassel RH, Geyer HJ and de Koker WC (1982) Non-bite transmission of rabies in kudu (*Tragelaphus strepsiceros*). Onderstepoort J Vet Res 49:191–219
Bateman C (2005) AIDS fuels ownerless feral dog populations. SAMJ 95:78–79
Bauer SP, Murphy FA (1975) Relationship of two arthropod-borne rhabdoviruses (kotonkan and Ododhiang) to the rabies serogroup. Infect Immun 12:1157–1172
Belotto A, Leanes LF, Schneider MC, Tamayo H, Correa E (2005) Overview of rabies in the Americas. Virus Res 111:5–12
Bingham J (2005) Canine rabies ecology in southern Africa. Emerg Infect Dis 11:1337–1342

Bingham J, Foggin CM, Wandeler AI, Hill FWG (1999) The epidemiology of rabies in Zimbabwe. 1. Rabies in dogs (*Canis familiaris*). Onderstepoort J Vet Res 66:1–10
Bingham J, Javangwe S, Sabeta CT, Wandeler AI, Nel LH (2001) Report of isolations of unusual lyssaviruses (rabies and Mokola virus) Identified retrospectively from Zimbabwe. J South Afr Vet Assoc 72:92–94
Blancou J (1988) Epizootiology of rabies: Eurasia and Africa. In: Campbell JB, Charlton KM (eds) Rabies (eds) Kluwer Academic, Boston
Botvinkin AD, Poleschuk EM, Kuzmin IV, Borisova TI, Gazaryan SV, Yager P, Rupprecht CE (2003) Novel lyssaviruses isolated from bats in Russia. Emerg Infect Dis 9:1623–1625
Bouffard G (1912) Sur l'existence de la rage dans le Haut-Senégal et le Niger. Ann Inst Pasteur 26:727–731
Boulger LR, Porterfield JS (1958) Isolation of a virus from Nigerian fruit bats. Trans R Soc Trop Med Hyg 52:421–424
Bourhy H, Kissi B, Tordo N (1993) Molecular diversity of the *Lyssavirus* genus. Virology 194:70–81
Britton TA (1894) Report upon the outbreak of rabies at Port Elizabeth during the year 1893. Cape of Good Hope Department of Agriculture
Burrows R (1994) Rabies in African wild dogs of Tanzania. J Wild Dis 30:297–302
Calisher CH, Karabatsos N, Zeller H, Digoutte JP, Tesh RB, Travassos da Rosa AP, St George TD (1989) Antigenic relationships among rhabdoviruses from vertebrates and hematophagous arthropods. InterVirol 30:241–257
Courtin F, Carpenter TE, Paskin RD, Chomel BB (2000) Temporal patterns of domestic and wildlife rabies in central Namibia stock-ranching area, 1986–1996. Prev Vet Med 43:13–28
Creel S, Creel NM, Munson L, Sanderlin D, Appel MJ (1997) Serosurvey for selected viral diseases and demography of African wild dogs in Tanzania. J Wild Dis 33:823–832
Crick J, Tigmor GH, Moreno K (1982) A new isolate of Lagos bat virus from the Republic of South Africa. Trans R Soc Trop Med Hyg 76:211–213
Desmezieres E, Maillard AP, Gaudin Y, Tordo N, Perrin P (2003) Differential stability and fusion activity of Lyssavirus glycoprotein trimers. Virus Res 91:181–187
Digoutte JP (1975) Rapport Annuelo de l'Institut Pasteur de la Guyane Francaise, 31–32. Institute Pasteur French Guiana
East ML, Hofer H (1996) Wild dogs in the Serengeti. Science 271:275–276
East ML, Hofer H, Cox JH, Wulle U, Wiik H, Pitra C (2001) Regular exposure to rabies virus and lack of symptomatic disease in Serengeti spotted hyenas. Proc Natl Acad Sci U S A 98:15026–15031
Edmonds CR (1922) Diseases of animals in South Africa. Ballière Tindall & Cox, London, pp 195–212
Familusi JB, Moore DL (1972) Isolation of a rabies related virus from the CSF of a child with "aseptic meningitis". Afr J Med Sci 3:93–96
Familusi JB, Osunkoya BO, Moore DL, Kemp GE, Fabiyi A (1972) A fatal human infection with Mokola virus. Am J Trop Med Hyg 21:959–963
Fekadu M (1972) Atypical rabies in dogs in Ethiopia. Ethiop Med J 10:79–86

Fekadu M, Baer GM (1980) Recovery from clinical rabies of two dogs inoculated with a rabies virus strain from Ethiopia. Am J Vet Res 41:1632–1634

Fekadu M, Shaddock JH, Baer GM (1982) Excretion of rabies virus in the saliva of dogs. J Infect Dis 145:715–719

Fekadu M, Shaddock JH, Chandler FW, Baer GM (1983) Rabies virus in the tonsils of a carrier dog. Arch Virol 78:37–47

Fekadu M, Shaddock JH, Sanderlin DW, Smith JS (1998) Efficacy of rabies vaccines against Duvenhage virus isolated from European house bats (*Eptesicus serotinus*), classic rabies and rabies-related viruses. Vaccine 6:533–539

Foggin CM (1983) Mokola virus infection in cats and a dog in Zimbabwe. Vet Rec 113:115

Foggin CM (1988) Rabies and rabies-related viruses in Zimbabwe: Historical, virological and ecological aspects. PhD thesis, University of Zimbabwe

Foley HD, McGettigan JP, Siler CA, Dietzschold B, Schnell MJ (2000) A recombinant rabies virus expressing vesicular stomatitis virus glycoprotein fails to protect against rabies virus infection. Proc Natl Acad Sci U S A 97:14680–14685

Fooks AR (2004) The challenge of new and emerging lyssaviruses. Exp Rev Vaccines 3:89–92

Hanlon CA, DeMattos CA, DeMattos CC, Niezgoda M, Hooper DC, Koprowski H, Notkins A, and Rupprecht CE (2001) Experimental utility of rabies virus-neutralizing human monoclonal antibodies in post-exposure prophylaxis. Vaccine 19:3834–3842

Hanlon CA, Kuzmin IV, Blanton JD, Weldon WC, Manangan JS, Rupprecht CE (2005) Efficacy of rabies biologics against new lyssaviruses from Eurasia. Virus Res 111:44–54

Hersenberg L (1928) Two cases of hydrophobia. J Med Assoc South Afr 2:659–661

Hofmeyr M, Hofmeyr D, Nel L, Bingham J (2004) A second outbreak of rabies in African wild dogs (*Lycaon pictus*) in Madikwe Game Reserve South Africa, demonstrating the efficacy of vaccination against natural rabies challenge. Anim Conser 7:193–198

Hübschle OJB (1988) Rabies in the kudu antelope (*Tragelaphus strepsiceros*). Rev Infect Dis Suppl 4:S629–S633

Hutcheon D (1894) Report of the Colonial Veterinary Surgeon and Assistant Veterinary Surgeons for the Year (1893) Department of Agriculture Cape of Good Hope, pp 7–10

Illango J (1992) National report on rabies in Uganda. Proceedings of a Joint CVRI, WHO, FAO, OIE Seminaar on Rabies in Southern Africa Lusaka Zambia, 2–5 June (1992) World Health Organization, Geneva

Jallet C, Jacob Y, Bahloul C, Drings A, Desmezieres E, Tordo N, Perrin P (1999) Chimeric lyssavirus glycoproteins with increased immunological potential. J Virol 73:225–233

Johnson N, McElhinney LM, Smith J, Lowings P, Fooks AR (2002) Phylogenetic comparison of the genus Lyssavirus using distal coding sequences of the glycoprotein and nucleoprotein genes. Arch Virol 147:2111–2123

Johnson N, Letshwenyo M, Baipoledi EK, Thobokwe G, Fooks AR (2004) Molecular epidemiology of rabies in Botswana: a comparison between antibody typing and nucleotide sequence phylogeny. Vet Microbiol 101:31–38

Kariuki DP, Ngulo WK (1985) Epidemiology of animal rabies in Kenya (1900–1983). In: Kuwert E, Mérieux C, Koprowski H, Bögel K (eds) Rabies in the tropics. Springer-Verlag, Berlin Heidelberg New York

Kemp GE, Causey OR, Moore DL, Odeola A, Biyi A (1972) Mokola virus. Further studies on IbAn 27377, a novel agent of zoonosis in Nigeria. Am J Trop Med Hyg 21:356–359

King A, Crick J (1988) Rabies-related viruses. In: Rabies (eds) Campbell JB, Charlton KM. Kluwer Academic Publishers, Boston, pp 177–200

King AA, Meredith CD, Thomson GR (1993) Canid and viverrid rabies viruses in South Africa. Onderstepoort J Vet Res 60:295–299

King AA, Meredith CD, Thomson GR (1994) The biology of southern African lyssavirus variants. In: Rupprecht CE, Dietzschold B, Koprowski H (eds) Lyssaviruses. Springer, Berlin Heidelberg New York

Kissi B, Tordo N, Bourhy H(1995) Genetic polymorphism in the rabies virus nucleoprotein gene. Virology 209:526–537

Kuzmin IV, Hughes GJ, Botvinkin AD, Orciari LA, Rupprecht CE (2005) Phylogenetic relationships of Irkut and West Caucasian bat viruses within the lyssavirus genus and suggested quantitative criteria based on the N gene sequence for lyssaviruses genotype definition. Virus Res 111:28–43

Le Gonidec G, Rickenbach A, Robin Y, Heme G (1978) Isolement d'une souche de virus Mokola au Cameroun. Ann Microbiol 129A:245–249

Maganu ET, Staugard F (1985) Epidemiology of rabies in Botswana. In: Kuwert E, Mérieux C, Koprowski H, Bögel K (eds) Rabies in the tropics. Springer, Berlin Heidelberg New York

Magembe SR (1985) Epidemiology of rabies in the United Republic of Tanzania. In: In: Kuwert E, Mérieux C, Koprowski H, Bögel K (eds) Rabies in the tropics. Springer, Berlin Heidelberg New York

Mansvelt PR (1962) Rabies in South Africa. Field control of the disease. J South Afr Vet Med Assoc 33:313–319

Mebatsion T, Cox JH, Frost JW (1992) Isolation and characterization of 115 street rabies virus isolates from Ethiopia by using monoclonal antibodies: Identification of 2 isolates as Mokola and Lagos Bat viruses. J Inter Virol 166:972–977

Meredith CD (1982) Wildlife rabies: past and present in South Africa. South Afr J Sci 78:411–415

Meredith CD, Standing E (1981) Lagos Bat virus in South Africa. Lancet 1:832–833

Meredith CD, Rossouw AP, Koch HP (1971) An unusual case of human rabies though to be of chiropteran origin. South Afr Med J 45:767–769

Msiska JGM (1988) The epidemiology and control of rabies and brucellosis in Malawi. Proceedings of the International Conference on Epidemiology Control and Prevention of Rabies and Brucellosis in Eastern and Southern African Countries, Gabarone, Botswana, 23–25 November 1988

Nadin-Davis SA, Bingham J (2004) Europe as a source of rabies for the rest of the world. In: King AA, Fooks AR, Aubert M, Wanderler AI (eds) Historical perspective of rabies in Europe and the Mediterranean basin. OIE Publications, pp 259–280

Nel L, Jacob J, Jaftha J, von Teichman B, Bingham J (2000) New cases of Mokola virus infection in South Africa: A genotypic comparison of southern african virus isolates. Virus Genes 20:103–106

Nel LH, Niezgoda M, Hanlon CA, Morril PA, Yager PA, Rupprecht CE (2003) A comparison of DNA vaccines for the rabies-related virus Mokola. Vaccine 21:2598–2606

Nel LH, Sabeta CT, Von Teichman B, Jaftha JB, Rupprecht CE, Bingham J (2005) Mongoose rabies in southern Africa: a re-evaluation based on molecular epidemiology. Virus Res 109:165–173

Office Internationale des Epizooties: HANDISTATUS II Namibia/rabies multiannual animal disease status. (www.oie.int/hs2). Cited 26 February 2007

Paweska JT, Blumberg LH, Liebenberg C, Hewlett RH, Grobbelaar AA, Leman PA, Croft JE, Nel LH, Nutt L, Swanepoel R. Fatal human infection with rabies-related Duvenhage virus, South Africa. Emerg Infect Dis. 2006; 12:1965–1967

Randall DA, Williams SD, Kuzmin IV, Rupprecht CE, Tallents LA, Tefera Z, Argaw K, Shiferaw F, Knobel DL, Sillero-Zubiri C, Laurenson MK (2004) Rabies in endangered Ethiopian wolves. Emerg Infect Dis 10:214–217

Remlinger P, Curasson M (1924) Identité de l'oulou fato (maladie du chien fou de l'ouest africain) et de la rage. Bull Acad Natl Méd 92:1112–1117

Rollinson DHL (1956) Problems of rabies control in Africa. Bul Epiz Dis Afr 4:7–16

Rupprecht CE, Blass L, Smith K, Orciari LA, Niezgoda M, Whitfield SG, Gibbons RV, Guerra M, and Hanlon CA (2001) Human infection due to recombinant vaccinia-rabies glycoprotein virus. N Engl J Med 345:582–586

Rupprecht CE, Hanlon CA, Hemachudha T (2002) Rabies re-examined. Lancet Infect Dis 2:327–343

Sabeta CT, Bingham J, Nel LH (2003) Molecular epidemiology of canid rabies in Zimbabwe and South Africa. Virus Res 91:203–211

Saluzzo JF, Rollin PE, Daugard C, Digiutte JP, Georges AJ, and Sureau P (1984) Premier isolement du virus Mokola a partir d'un rongeur (Lophuromys sikapusi). Ann Inst Pasteur Virol 135E:57–66

Schneider LG (1985) Oral immunization of wildlife against rabies. Ann Inst Pasteur Virol 136:469–473

Schneider LG, Barnard BJH, Schneider HP (1985) Application of monoclonal antibodies for epidemiological investigations and oral vaccination studies. I African viruses In: Kuwert E, Mérieux C, Koprowski H, Bögel K (eds) Rabies in the tropics. Springer, Berlin Heidelberg New York, pp 47–59

Schnell MJ, Tan GS, Dietzschold B (2005) The application of reverse genetics technology in the study of rabies virus (RV) pathogenesis and for the development of novel RV vaccines. J Neuro Virol 11:76–81

Shone DK (1962) Rabies in Southern Rhodesia 1900–(1961) J South Afr Vet Med Assoc 33:567–580

Shope R (1975) Rabies virus antigenic relationships. In: Baer GM (ed) The natural history of rabies, 1st edn. Academic, New York, pp 141–152

Shope RE, Murphy FA, Harrison AK, Causey OR, Kemp GE, Simpson DIH, Moore DL (1970) Two African viruses serologically and morphologically related to rabies virus. J Virol 6:690–692

Sillero-Zubiri C, King AA, Macdonald DW (1996) Rabies and mortality in Ethiopian wolves (*Canis simensis*). J Wild Dis 32:80–86
Sinclair JM (1914) Report of the Director of Agriculture for the Year 1913, Southern Rhodesia
Siongok TKA, Karama M (1985) Epidemiology of human rabies in Kenya. In: Kuwert E, Mérieux C, Koprowski H, Bögel K (eds) Rabies in thc tropics. Springer, Berlin Heidelberg New York
Snyman PS (1940) The study and control of vectors of rabies in South Africa. Onderstepoort J Vet Sci Anim Husband 66:296–307
Sureau P, Tignor GH, Smith AL (1980) Antigenic characterization of the Bangui strain (ANCB-672D) of Lagos bat virus. Ann Virol 131:25–32
Swart H (1989) Hondsdolheid in Suid Afrika. South Afr Vet Med 2:163–166
Swanepoel R, Barnard BJH, Meredith CD, Bishop GC, Bruchner GK, Foggin CM, Hubschle OJB (1993) Rabies in southern Africa. Onderstepoort J Vet Res 60:323–346
Swanepoel R (2005) Rabies. In: Coetzer JAW, Tustin RC (eds) Infectious diseases of livestock with special reference to southern Africa. Oxford University Press Southern Africa, Cape Town
Thierry G (1959) Particularitiés de la rage dans l'ouest africaine. Bul Epiz Dis Afr 7:265
Tordo N, Bourhy H, Sather S, Ollo R (1993) Structure and expression in baculovirus of the Mokola virus glycoprotein: An efficient recombinant vaccine. Virology 194:59–69
Tuffereau C, LeBlois H, Bënëjean J, Coulon P, Lafay F, Flamand A (1989) Arginine or lysine in position 333 of ERA, CVS glycoprotein is necessary for rabies virulence in adult mice. Virology 172:206–212
Valadao F (1968) The most important aspects of the rabies problem in Mozambique. Veterin Moçambique 2:13–20
Van der Merwe M (1982) Bats as vectors of rabies. South Afr J Sci 78:421–422
Von Maltitz L (1950) Rabies in the northern districts of South West Africa. J South Afr Vet Med Assoc 21:4–12
Von Teichman BF, Thomson GR, Meredith CD, Nel LH (1995) Molecular epidemiology of rabies virus in South Africa: evidence for two distinct virus groups. J Gen Virol 76:73–82
Von Teichman B, de Koker WC, Bosch SJE, Meredith CD, Bingham J (1998) Mokola virus infection: description of recent South African cases and a review of virus epidemiology. J South Afr Vet Med Assoc 69:169–171
Wiktor TJ (1985) Is a special vaccine required against rabies-related viruses and variants of rabies? In: Vodopija I (ed) Improvements in rabies post-exposure treatment. Zagreb Institute of Public Health, Zagreb, pp 9–14
Zyambo GCN, Sinyangwe PG, Bussein NA (1985) Rabies in Zambia. In: Kuwert E, Mérieux C, Koprowski H, Bögel K (eds) Rabies in the tropics. Springer, Berlin Heidelberg New York

CTMI (2007) 315:195–215
© Springer-Verlag Berlin Heidelberg 2007

Tuberculosis: A Reemerging Disease at the Interface of Domestic Animals and Wildlife

M. V. Palmer (✉)

Bacterial Diseases of Livestock Research Unit, National Animal Disease Center, Agricultural Research Service, USDA, 2300 Dayton Avenue, Ames, IA 50010, USA
mpalmer@nadc.ars.usda.gov

Abstract Complex interactions involving humans, domestic animals, and wildlife create environments favorable to the emergence of new diseases. Today, reservoirs of *Mycobacterium bovis*, the causative agent of tuberculosis in animals and a serious zoonosis, exist in wildlife. The presence of these wildlife reservoirs is the direct result of spillover from domestic livestock in combination with anthropogenic factors such as translocation of wildlife, supplemental feeding of wildlife and wildlife populations reaching densities beyond normal habitat carrying capacities. As many countries attempt to eradicate *M. bovis* from domestic livestock, efforts are impeded by spillback from wildlife reservoirs. It will not be possible to eradicate *M. bovis* from livestock until transmission between wildlife and domestic animals is halted. Such an endeavor will require a collaborative effort between agricultural, wildlife, environmental and political interests.

1 Introduction

The emergence of newly recognized diseases in wildlife is often the result of complex and sometimes unintended interactions between wildlife, domestic animals, and humans. Wild animals are susceptible to infection with many of the same disease agents that afflict domestic animals, and transmission between domestic animals and wildlife can occur in both directions. Transmission of *Mycobacterium bovis* from domestic animal populations to wildlife (spillover) and subsequent transmission from wildlife back to domestic animals (spillback) is a theme common in most parts of the world currently attempting eradication of *M. bovis* infection among animal populations. In most cases, both spillover and spillback have been facilitated by anthropogenic factors such as human and domestic animal encroachment on traditional wildlife habitat, translocation of animals, or supplemental feeding of wildlife.

Critical to control of tuberculosis is the understanding of maintenance hosts and spillover hosts. Among spillover hosts, disease does not persist without an external source of reinfection. This external source of infection may be any other population of susceptible hosts, wild or domestic. However, in most cases *M. bovis* was originally introduced by spillover from domestic cattle to a susceptible wild population. Spillover hosts may be dead-end hosts and play no role in disease transmission or may be amplifying hosts that can increase transmission to other wildlife hosts or back to livestock. Disease in spillover hosts will gradually disappear as disease is eliminated in the species acting as the source of infection. In contrast, among maintenance hosts, disease persists without any external source of reinfection. Maintenance hosts may be domestic or wild, but are critical in disease epidemiology and control because without intervention, disease will persist among a population of maintenance hosts (see the chapters by Childs et al. and Childs, this volume). The most efficient disease control efforts are generally aimed at maintenance hosts. There is general acceptance that among wildlife species the European badger (*Meles meles*) in the United Kingdom, the brushtail possum (*Trichosurus vulpecula*) in New Zealand, and the white-tailed deer (*Odocoileus virginianus*) in the United States represent true maintenance hosts.

In the early part of the twentieth century, there were large numbers of tuberculous cattle in industrialized nations in North America, Europe and Australia.

Often an association was made between the number of *M. bovis* -infected humans and the prevalence of tuberculosis in the local cattle population. Infected cattle were generally considered the source of human infection with *M. bovis*, transmission being through direct inhalation or ingestion of unpasteurized dairy products (Grange and Yates 1994; Wigle et al. 1972). Abattoir workers have been infected during the slaughter and processing of cattle (Robinson et al. 1988; Cousins and Dawson 1999). More recently, exposure to tuberculous elk (*Cervus elaphus*) resulted in human infection (Fanning and Edwards 1991). With mandatory pasteurization of milk, tuberculin skin testing of cattle, and slaughter of infected cattle, the incidence of human tuberculosis due to *M. bovis* has declined dramatically in developed countries. However, it is estimated that worldwide approximately 50 million cattle remain infected with *M. bovis*, with a cost to the agricultural community of US $3–4 billion per annum (Steele 1995). In underdeveloped countries, such as many of those in Africa, tuberculosis in cattle is still widespread, as is *M. bovis* infection in humans. Even in developed countries where bovine tuberculosis eradication efforts have been in place for decades, successful eradication of disease from livestock is hampered by several factors, not least of which is the presence of wildlife reservoirs of *M. bovis*. Generally, countries with a documented wildlife reservoir of *M. bovis* have not been successful in eradication of *M. bovis* infection from domestic livestock. Several factors are critical in the development of a wildlife reservoir of disease: disease prevalence, clinical course of the disease, and host ecology. The following three examples illustrate the complex interaction of wildlife, domestic animal, and human factors in the creation and maintenance of wildlife reservoirs of tuberculosis.

2 United Kingdom

2.1 History of *Mycobacterium bovis* Infection in European Badgers

In the 1970s, tuberculosis had been removed from large areas of Great Britain, and eradication was predicted. In 1981, the Wildlife and Countryside Act provided protection to badger populations and resulted in a large increase in the number of badgers. Over the past 10 years, Great Britain has experienced a rising incidence of tuberculosis in cattle, especially in the southwest of England, South Wales, and also in the Republic of Ireland. *Mycobacterium bovis* is endemic among badgers, and although most of the evidence is indirect, it is hypothesized that badgers are a source of infection for cattle and responsible for the increase in tuberculosis among domestic cattle herds. *Mycobacterium bovis* was first isolated from badgers in Switzerland in 1957 (Bouvier 1963). It is postulated that these badgers were infected by contact with tuberculous roe deer (*Capreolus capreolus*). In 1971, the first tuberculous badger was identified in England (Muirhead et al. 1974) and in 1975 an infected badger was reported in Ireland (Noonan et al. 1975). It is believed that badgers in England became infected with *M. bovis* during the late nineteenth and early twentieth centuries when a large percentage of British cattle were infected with *M. bovis* and infection spilled over from cattle to badgers.

2.2
Badger Ecology

The badger's natural habitat is such that it lives on or near pastures used by cattle where it digs for earthworms and dung beetles. Badgers live in groups of up to 35 animals that defend a communal territory that may include several setts, described as complex, long-lasting networks of tunnels and channels (Tuyttens et al. 2000). Setts provide ideal conditions for the spread of respiratory diseases. Badger social groups may remain stable for years with a low rate of dispersal (Tuyttens et al 2000). Such stability decreases the likelihood of disease transmission between groups, an idea supported by the observation that in undisturbed badger populations disease prevalence is highly localized in clusters (Cheeseman et al. 1988). In extreme cases, badger density can be as high as 25.3 adults per square kilometer; however, there appears to be no correlation between badger density and the prevalence of *M. bovis* infection among badgers (Cheeseman et al. 1989; Rogers et al. 1998).

2.3
Pathology and Transmission

Lesions in tuberculous badgers may be found in the lungs and associated lymph nodes, pharyngeal lymph nodes, mesenteric lymph nodes and kidneys (Gallagher et al. 1976; Gavier-Widen et al. 2001). However, several characteristics distinguish tuberculous lesions in badgers from those typically seen in cattle and may have important implications in disease pathogenesis and transmission. While caseous necrosis, mineralization and peripheral fibrosis are often associated with tuberculous lesions in cattle they are the exception in badgers. Langhans type giant cells, commonly seen in bovine lesions, are rare in badgers, while acid fast bacilli are often numerous. Renal lesions are more common in badgers than in cattle. These lesions can be extensive, involving several regions of the nephron, and acid fast bacilli can be numerous (Gallagher et al. 1976). Experimental studies demonstrate that badgers can transmit *M. bovis* to cattle (Little et al. 1982); however, the exact route of transmission is unknown. Infected badgers shed large numbers of *M. bovis* in saliva, urine, feces, and exudates from draining lesions (Gavier-Widen et al. 2001). It is suggested that cattle may become infected by inhalation of bacilli from grass contaminated with infected badger urine, feces, or exudates from superficial draining lesions (Hutchings and Harris 1997). Urine is believed to be of greatest risk due to the high numbers of *M. bovis* bacilli present. Badgers urinate either at localized areas used for urination and defecation known as latrines or on pastures where badger paths cross linear features such as hedgerows or ditches known as crossing points (Scantlebury et al. 2004). Both latrines and crossing points are generally accessible to cattle. Moreover, infected badgers can live 3–4 years following the first documented episode of shedding of *M. bovis* (Little et al. 1982), making badgers an ideal maintenance host of *M. bovis*. Experimentally, calves have been infected from contact with experimentally infected, as well as naturally infected badgers (Little et al. 1982), and epidemiological studies have shown that areas with the greatest density of badgers have the highest incidence of tuberculosis among cattle (Muirhead et al. 1974; Cheeseman et al. 1989; Krebs et al. 1998). Badger-to-badger transmission is most likely respiratory and to a lesser extent cutaneous through bite wounds (Cheeseman et al 1989).

2.4
Zoonotic Potential

Recently, the first documented cases of spillover of bovine tuberculosis from animals to humans were reported since the resurgence of the disease in the United Kingdom (Smith et al. 2004). Two siblings residing on a farm were diagnosed with tuberculosis due to *M. bovis*. Cattle on the farm also had been diagnosed with *M. bovis*. The cattle isolate was indistinguishable from the isolates from the two siblings when examined by restriction fragment length polymorphism (RFLP) analysis, spacer oligonucleotide typing (spoligotyping), and variable number tandem repeat (VNTR) analysis, suggesting transmission between cattle and humans. Moreover, the farm supported a large badger population where *M. bovis* infection had been previously diagnosed. It is suggested, although not proven, that cattle became infected through contact with badgers and that humans became infected through contact with cattle.

2.5 Disease Control

Efforts to remove badgers from cattle farming areas have resulted in a decline in bovine tuberculosis (Little et al. 1982). Following the first suggested links between badgers and bovine tuberculosis, farmers were licensed to cull badgers; from 1975 to 1981 hydrogen cyanide gas was used to kill badgers (Donnelly et al. 2003). Gassing with hydrogen cyanide was later replaced with a strategy to identify and remove clusters of infected badgers. From 1986 to 1998, culling occurred only on land where tuberculin test-positive cattle were present (Donnelly et al. 2003). The effectiveness of large-scale culling as opposed to selective culling remained unknown until recently. In 1998, a large experiment was implemented to compare the effects of three different control strategies: no culling of badgers, localized selective culling of badgers in response to identified cases of tuberculosis in cattle, and proactive culling aimed at reducing badger densities to low levels across entire trial areas. Five years into the study, it was determined that reactive culling of badgers resulted in increased levels of tuberculosis in cattle within the trial areas (Donnelly et al. 2003). In response to these findings, reactive culling was discontinued as part of the study, while proactive culling and no culling continue as experimental treatments within the study. The reason for an increased level of tuberculosis in cattle in reactive culling treatment areas is unknown. However, it is known that badger social structures are complex and selective removal of some but not all badgers may result in increased badger movement with badgers using latrines far distant from their original sett, resulting in enlarged social groups with overlapping boundaries (Tuyttens et al. 2000). Such social restructuring among populations with *M. bovis*-infected badgers may result in increased disease transmission among badgers and between badgers and cattle. Increased social restructuring and badger movement has been correlated with increased incidence of *M. bovis* infection among badger populations (Rogers et al. 1998).

Complete removal of any wildlife reservoir of infection is extremely difficult and in the long term (see the chapters by Childs and by Stallknecht, this volume), most believe that the best prospect for control of bovine tuberculosis in Great Britain is a vaccine for cattle, combined with improved diagnostic tests to distinguish vaccinated from infected cattle (Krebs et al. 1998). However, some also contend that a vaccine for badgers should be kept as an option (Anonymous 1997).

Cattle husbandry practices aimed at separating cattle and badgers have also been proposed as a means of tuberculosis control, including keeping cattle away from badger setts, urination trails, and latrines and keeping badgers away from cattle feed troughs and buildings. Studies to design elevated feed troughs that would exclude badgers concluded that the maximum height to which badgers would climb into a trough was beyond that which would be useable for younger cattle (Garnett et al. 2003). Public attitudes favor no culling of badgers and surveys show that the public generally ranks conservation and animal welfare concerns over those of disease control.

2.6
Other Species as Potential Wildlife Reservoirs of *M. bovis*

In 2004, the results were released of a study to examine numerous species of wildlife in the UK for tuberculosis. Over 4,700 animal carcasses were examined and tissue samples processed for isolation of *M. bovis*. Infection was confirmed in foxes (*Vulpes vulpes*), stoats (*Mustela erminea*), polecats (*Mustela putorius*), common shrews (*Sorex araneus*), yellow-necked mice, squirrels (*Sciurus carolinensis*), roe deer, red deer (*Cervus elaphus*), fallow deer (*Dama dama*), and muntjac deer (*Muntiacus reevesi*). Sample size varied widely, but the highest prevalences were seen in foxes (3.2% of 756), stoats (3.9% of 78), polecats (4.2% of 24), common shrews (2.4% of 41), roe deer (1.0% of 885), red deer (1.0% of 196), fallow deer (4.4% of 504) and muntjac deer (5.2% of 58). A qualitative risk assessment based on prevalence, likelihood of excretion, likelihood of contact with cattle and animal biomass identified fallow deer and red deer as the highest risk, with scores of 0.75 and 0.5, respectively (a score of 1.0 being the highest risk). However, with a regional tuberculosis prevalence as high as 20.5% in badgers, they remain a primary concern for tuberculosis control in the UK. However, other species, particularly deer, may also pose significant risk, especially in regions where deer density is high (http://www.defra.gov.uk/science/project_data/DocumentLibrary/SE3010/SE3010_1628_FRP.doc).

3
New Zealand

3.1
History of Brushtail Possums and *M. bovis* Infection in New Zealand

Prior to the arrival of the first humans to New Zealand, the only mammals present were two species of bats (O'Neil and Pharo 1995). Cattle were introduced approximately 200 years ago and large areas of forest were cleared in the early nineteenth century to accommodate pastoral farming. About the same time, several species of deer were introduced for recreational hunting purposes. By the middle of the twentieth century, deer numbers had climbed to such levels that deer were considered by many as nuisance pests. Deer farming began in the 1970s as wild deer were captured to establish breeding herds (O'Neill and Pharo 1995). Brushtail possums were first taken to New Zealand from Australia in the mid-nineteenth century to establish a fur trade. Between 1837 and 1922, over 30 groups of possums were imported, maintained in captivity for breeding, and released in over 160 different sites around New Zealand (O'Neill and Pharo 1995). The lack of natural predators combined with abundant food sources resulted in a rapid rise in possum numbers. Currently, possums occupy over 90% of New Zealand land area with an estimated 60–70 million possums nationwide. Possum density estimates range from 1.5 to 25 per hectare, where in some areas the possum density is 20 times greater than that seen in Australia (O'Neill and Pharo 1995).

Mycobacterium bovis was likely introduced to New Zealand with the importation of cattle in the nineteenth century. By the early twentieth century, tuberculosis was recognized as a serious animal and human health problem. Tuberculosis was first diagnosed in farmed deer in 1978 and subsequently spread by movement of untested farmed deer and capture of infected wild deer. The first reported case of tuberculosis in a wild possum in New Zealand was in 1967 (Ekdahl et al. 1970). However, the susceptibility of brushtail possums to infection with *M. bovis* had been determined much earlier (Bolliger and Bolliger 1948). Epidemiological evidence has linked possum tuberculosis and tuberculosis in cattle (Collins et al. 1988). It is likely that possums in New Zealand acquired *M. bovis* from other animals, likely cattle, as *M. bovis* infection has never been seen in Australian possums, the original source of New Zealand's possums.

3.2
Pathology and Transmission

Tuberculous possums often develop disseminated disease, with lymph nodes and lungs being the most common sites of infection. Additionally, one study reported that at least 45% of affected possums had a discharging sinus from a superficial lymph node lesion (Cooke et al. 1995). Lesions can also be seen in the liver, spleen, kidneys, adrenal glands, and bone marrow, suggesting generalized hematogenous spread of bacilli. In one study, lesions were present in one or more of these sites in 86% of 73 tuberculous possums, suggesting that hematogenous dissemination of disease is common in possums (Jackson et al. 1995a). In contrast to lesions in cattle, fibrosis, mineralization and Langhans type giant cells are uncommon, while acid fast bacilli are numerous. The character of the lesions suggests an ineffective host immune response to infection, unable to sequester infection, thereby allowing rapid hematogenous dissemination. In spite of disseminated disease, normal growth of the possum is not significantly affected until late stages of the disease (Jackson et al. 1995a). Infections among terminally ill possums are, however, characterized by widespread lesions involving numerous organ systems, resulting in a profound effect on behavior and survivability. The disseminated nature of the disease and limited effect on possum growth combined with pulmonary lesions and draining superficial lesions, all of which contain large numbers of *M. bovis*, make possums an ideal maintenance host capable of efficient transmission to other susceptible hosts.

Transmission among possums occurs between mother and offspring as well as direct horizontal transmission among adults. Respiratory secretions are thought to be most important in possum to possum transmission; however, some transmission to offspring through milk also occurs (Jackson et al. 1995b). Infected possums shed *M. bovis* primarily in respiratory secretions and exudate from draining lesions (Jackson et al. 1995b). Shared use of dens would seem a logical point at which transmission of *M. bovis* would occur and indeed in studies using captive possums; den-sharing provided the greatest risk of transmission between possums (Corner et al. 2003). Den-sharing has not been commonly observed in free-living possums; however, sequential den use by different possums has been observed (Paterson et al. 1995) and *M. bovis* has been shown to survive inside possum dens for 7–28 days depending on environmental conditions (Jackson et al. 1995c). The dynamics of possum-to-possum transmission of *M. bovis* appear to be complex and involve individual possum social status. Evidence of this is found in studies demonstrating that naturally infected possums tend to be possums that are central and prominent in the local social hierarchy. Furthermore, experimental infection of such socially dominant possums results in higher levels of disease transmission than experimental infection of possums ranked lower in the societal structure (Corner et al. 2003).

Healthy possums generally avoid contact with cattle (Paterson et al. 1995). Terminally ill possums exhibit abnormal behavior such as increased daytime activity, stumbling, rolling and falling, which attracts attention of inquisitive cattle. Studies using sedated possums to simulate terminally ill possums demonstrated that both deer and cattle exhibit profound interest in abnormally behaving possums. Cattle were seen to be attracted from as far as 50 m to investigate sedated possums (Paterson and Morris 1995). Deer and cattle were shown to spend significant amounts of time within a distance compatible with aerosol transmission (approximately 1.5 m) and to even sniff, touch, lick, roll, lift, chew and kick the possum, creating opportunity for direct transmission (Paterson and Morris 1995; Sauter and Morris 1995). In studies where cattle have been excluded from areas used for denning by tuberculous possums, decreased transmission of *M. bovis* from possums to cattle has been demonstrated. In contrast, where cattle are allowed to graze areas used for denning by tuberculous possums, transmission to cattle continues unabated (Paterson et al. 1995).

3.3 Disease Control

No widespread eradication of a vertebrate host has ever been successful in New Zealand. The attitudes toward possums in New Zealand differ from those of other wildlife reservoirs of tuberculosis in other countries. In New Zealand, possums are viewed as non-native, invasive pests, and widespread removal of possums is desirable for many reasons apart from tuberculosis control. Possums have had a disastrous impact on New Zealand's native flora and fauna. Every night, an estimated 70 million possums consume approximately 21,000 tons of green shoots, leaves, and berries. Possums are omnivorous and also consume bird's eggs, chicks, and insects. While browsing in the forest canopy on fruits and flowers, possums are in direct competition with native nectar feeding birds. While on the ground, possums compete with native kiwi for dens and have been seen eating kiwi eggs. Theoretically, widespread removal of possums from New Zealand's ecosystem would be more socially palatable than removal of native wildlife reservoirs of tuberculosis in other countries. Early control measures included a bounty system on possums, which was minimally effective. Bounties did not allow for prioritization of control efforts and although many possums were removed they were generally not removed from the right places. Possums were generally taken for bounty from easily accessible locations leaving many critical areas unchanged. Recently, aerial distribution of possum baits containing 1080 poison (sodium monofluoracetate) has achieved 90% death rates in some areas (Caley et al. 1999). An effective poison, 1080 causes possums to die of cardiac or respiratory failure. Other poisons that have been used to control possums include brodifacoum, pindone, cyanide, and cholecalciferol. In areas where 1080 baits have been used to decrease possum numbers, tuberculin reactor rates in cattle herds and numbers of tuberculous possums have decreased, only to return to elevated levels in 8–10 years as possum numbers recover through breeding and immigration from surrounding areas (Barlow 1991; Tweddle and Livingstone 1994). Long term (>10 years) maintenance of possum populations below 40% of precontrol densities over widespread areas may be required to affect significant change in cattle tuberculin reactor rates and eradicate tuberculosis from possum populations (Caley et al. 1999).

Although widespread removal of possums through poisoning may decrease the prevalence of tuberculosis in cattle, complete removal of possums from New Zealand may be impractical. It has been suggested that the most promising option for long-term control of tuberculosis in possums is the development of a vaccine combined with a strategy for biological control of possums. *Mycobacterium bovis* BCG vaccine has been administered to possums by subcutaneous, intranasal, and intraduodenal routes (Aldwell et al. 1995a, 1995b; Buddle et al. 1997; Corner et al. 2001). All routes provide some protection against aerosol challenge with virulent *M. bovis*, evidenced by reduced disease severity, reduced loss of body weight, fewer lung lesions, and decreased bacterial colonization.

Other species such as red deer, feral pigs (*Sus scrofa*), feral cats (*Felis catus*), ferrets (*Mustela furo*), and stoats, goats (*Capra hircus*), rabbits (*Oryctolagus cuniculus*), hares (*Lepus europaeus*), and hedgehogs (*Erinaceus europaeus*) have been found infected with *M. bovis* (Jackson 2002; Tweddle and Livingstone 1994). The role most of these species play in the epidemiology of bovine tuberculosis in New Zealand is not clear; however, of these species, red deer may be another maintenance host of tuberculosis in New Zealand.

4 United States

4.1 History of *M. bovis* Infection in White-Tailed Deer in Michigan, USA

Prior to 1994, there had been only isolated case reports of tuberculosis in white-tailed deer in the United States (Levine 1934; Ferris et al. 1961; Belli 1962; Friend et al. 1963). All reports involved one to two animals and were seen in captive deer, hunter-killed deer, or deer dying of accidental causes. In almost all cases, it was postulated that *M. bovis* had spilled over from tuberculous livestock in the area; however, no follow-up surveys were conducted and no strain comparisons were made to confirm such a hypothesis. In 1975, a free-ranging white-tailed deer in northern Michigan was diagnosed with tuberculosis due to *M. bovis*. Michigan had been declared free of *M. bovis* in livestock in 1975 and was granted TB-free status by the United States Department of Agriculture in 1979. The tuberculous white-tailed deer was thought to be an anomaly and no follow-up surveys of free-ranging deer were conducted. In 1994, a free-ranging, hunter-killed white-tailed deer was identified with tuberculosis due to *M. bovis*. This deer was located just 13 km from the site where the tuberculous deer had been identified in 1975. Subsequent surveys conducted by the Michigan Department of Natural Resources and Michigan State University Animal Health Diagnostic Laboratory identified a focus of *M. bovis* infection in free-ranging white-tailed deer in northeast Michigan (Schmitt et al. 1997). This represented the first known reservoir of *M. bovis* in free-living wildlife in the United States and the first known epizootic of tuberculosis in white-tailed deer in the world. Several factors are thought to have contributed to the establishment and persistence of *M. bovis* in this wildlife reservoir. It is postulated that

M. bovis was transmitted from cattle to deer at some time during the early to mid 1900s when a large number of Michigan cattle were infected with *M. bovis* (Frye 1995). Statistical models estimate that spillover from cattle to deer occurred around 1955 (McCarty and Miller 1998). During this same period, Michigan's deer population was steadily increasing beyond normal habitat carrying capacity. In 1930, there were an estimated 592,000 deer in Michigan. By 1998, the number of deer had grown to over 1.7 million statewide, with focal concentrations of 19–23 deer per square kilometer. The regions of highest deer density were later found to be the center of the current tuberculosis outbreak (Schmitt et al. 1997; O'Brien et al. 2002; Miller et al. 2003). Transmission and maintenance of *M. bovis* among deer is thought to have been facilitated by the common practice in Michigan of long-term winter feeding of large volumes of sugar beets, carrots, corn, apples, pumpkins, and pelleted feed to deer by private citizens to prevent migration and decrease winter mortality in order to keep deer numbers high for hunting purposes (Schmitt et al. 1997). The resulting increased population, combined with prolonged crowding of deer around feeding sites provided increased opportunity for deer-to-deer contact and enhanced transmission of tuberculosis. Supplemental feeding has been documented as a contributing factor to *M. bovis* infection in deer (Miller et al. 2003). Specific risk factors associated with increasing risk of tuberculosis were location of a feeding site near hardwood forest, number of deer fed per year, presence of other nearby feeding sites, and the quantity of grain, fruits, or vegetables fed. DNA fingerprinting through RFLP analysis of *M. bovis* isolates from Michigan white-tailed deer showed that the majority of deer were infected with a common strain of *M. bovis*, suggesting a single source of infection (Whipple et al. 1997). By 2003, over 123,249 deer had been tested by gross necropsy, bacteriologic culture, and histopathology since the identification of the first case in 1994. Of these, 481 cases of confirmed *M. bovis* infection had been identified in 12 counties in northern Michigan.

4.2 Transmission

The presence of *M. bovis* in wildlife is not only detrimental to the health of this wildlife population, but also represents a serious threat to domestic livestock. Thirty-two *M. bovis* -infected cattle herds have been identified in Michigan since the identification of tuberculosis in white-tailed deer. Restriction fragment length polymorphism analysis of *M. bovis* isolates from deer and cattle show that they are identical, suggesting cattle became infected through contact with free-ranging white-tailed deer (Whipple et al. 1999). Surveys of carnivores and omnivores in Michigan have confirmed *M. bovis* infection in coyotes (*Canis latrans*), bobcats

(*Felis rufus*), foxes (*Vulpes vulpes*), black bears (*Ursus americanus*), opossums (*Didelphis virginiana*), raccoons (*Procyon lotor*), and domestic cats (Bruning-Fann et al. 1998, 2001; Kaneene et al. 2002). Restriction fragment length polymorphism analysis suggests that deer and other wildlife are infected with a common strain of *M. bovis* and likely became infected through scavenging of dead deer carcasses; however, infection with limited lesion development in these scavenger species suggests that they are true spillover hosts and not important in the maintenance of the epizootic in deer or transmission to other susceptible hosts.

White-tailed deer experimentally infected with *M. bovis* shed bacilli in saliva and nasal secretions and less frequently in urine and feces (Palmer et al. 1999, 2001). Research has also shown that experimentally infected deer can transmit *M. bovis* to other deer or cattle through indirect contact such as sharing of feed (Palmer et al. 2001, 2004a, b). Furthermore, white-tailed deer experimentally inoculated by the aerosol route do not develop a pattern of lesions similar to that seen in naturally infected deer in Michigan, suggesting that aerosol transmission may not be the primary means of *M. bovis* transmission among Michigan deer (Palmer et al. 2003). Saliva and nasal secretions containing *M. bovis* contaminate feed that can act as a source of infection for other animals. *Mycobacterium bovis* is relatively resistant to environmental factors and under appropriate conditions (cool and protected from sunlight), *M. bovis* may persist in the environment for weeks or months, increasing the likelihood of transmission to other animals (Duffield and Young 1985; Jackson et al. 1995c; Tanner and Michel 1999; Palmer and Whipple 2006). Transmission from doe to fawn, although possible, is probably not important in the maintenance of the disease. Research has shown that fawns can be experimentally infected through consumption of milk containing *M. bovis* (Palmer et al. 2002); however, mammary gland lesions in naturally infected deer have been reported only rarely (O'Brien et al. 2001).

Epidemiologic modeling suggests a two-stage model of transmission. Stage 1 involves transmission within matriarchal groups, allowing disease to persist in the population at a low level (O'Brien et al. 2002). The social structure of white-tailed deer is characterized by family groups consisting of a matriarchal doe and several generations of her daughters and their fawns. Fawns from the previous year leave the dam when she nears parturition. Yearling does often rejoin their dam and her fawns in the fall. Stage 2 involves both supplemental feeding, with resultant increased deer density, and male fawns that disperse to join male groups that travel together at all times except during breeding season (O'Brien et al. 2002). Higher disease prevalence has been observed in adult male deer (Schmitt et al. 2002). Shifting membership by many males in these groups results in males temporarily belonging to several different groups and increased contact with numerous susceptible animals.

4.3 Pathology

Tuberculous white-tailed deer most commonly develop lesions in retropharyngeal lymph nodes, and in lung and pulmonary lymph nodes (Schmitt et al. 1997, Palmer et al. 2000, Fitzgerald et al. 2000). Similar to other species of Cervidae, lesions may grossly resemble abscesses, making differential diagnosis important. Unlike red deer, elk, and fallow deer, draining fistulae from superficial lymph node lesions have not been reported in white-tailed deer (Robinson et al. 1989; Lugton et al. 1998; Beatson 1985; Whiting and Tessaro 1994). Such lesions may be important in disease transmission among these other species of deer.

Microscopically, lesions consist of foci of caseous necrosis with or without mineralization, surrounded by infiltrates of epithelioid macrophages, lymphocytes, and Langhans type multinucleated giant cells. Lesions are often surrounded by variable amounts of fibrous connective tissue and low numbers of acid fast bacilli may be present within the caseum, macrophages, or multinucleated giant cells. Microscopically, lesions in white-tailed deer are similar to those seen in cattle; although subjectively, lesions in cattle may be surrounded by greater amounts of fibrous connective tissue.

4.4 Zoonotic Potential

Although *M. bovis* is a recognized zoonotic agent, no change in incidence of *M. bovis* infections in Michigan's human population has been detected since the epizootic was recognized (Wilkins et al. 2003), and only one case of *M. bovis* in humans has been directly attributed to contact with infected wildlife. Nevertheless, there are potential risks as hunters are exposed to *M. bovis* during the field dressing of deer or the consumption of undercooked venison products. Michigan's Departments of Community Health, Natural Resources and Agriculture have worked cooperatively to educate hunters, farmers, and other Michigan residents on the identification of tuberculosis in deer, personal protective measures hunters can take while field dressing deer, and the importance of thorough cooking of venison prior to consumption (Wilkins et al. 2003).

4.5 Disease Control

In Michigan, wildlife and domestic animal health authorities have adopted control measures that (1) reduce deer density and population through increased hunting, (2) restrict or eliminate supplemental feeding of deer, and (3) monitor both wildlife and domestic livestock through hunter-killed deer surveys, selected carnivore and omnivore removal and surveillance, and whole-herd cattle testing. These control measures appear to have succeeded in preventing increasing prevalence and geographic spread of tuberculosis in white-tailed deer in Michigan. Supplemental feeding of deer has been banned since 1998 in counties where tuberculous deer have been identified. Enforcement of such a ban has been problematic and universal compliance has not been achieved. Deer numbers have been reduced by 50% in the endemic areas through increased hunting pressure and unlimited harvest of female deer. However, progress toward eradication will likely require further action and more time. Epidemiological modeling suggests that further decreases in deer density and a strictly enforced ban on supplemental feeding will be required to achieve TB-free status.

5 Italy and Spain

Recently, *M. bovis* has been identified in wild boars in Italy. Restriction fragment length polymorphism analysis and spoligotyping have shown that many of the strains isolated from boars are identical to isolates obtained from cattle in the same region (Serraino et al. 1999). The exact means of interspecies transmission is unknown; however, it is speculated that boars are contaminating pastures and thus transmitting the disease to cattle. Similarly, *M. bovis* has been identified in wildlife in Spain, including red deer, fallow deer, wild boar, Iberian lynx (*Lynx pardina*), and hare. Again, transmission between cattle and wildlife is implicated due to similar spoligotyping patterns between livestock and wildlife species (Aranaz et al. 2004).

6 Conclusions

The complex interactions of domestic animals, wildlife, and humans that create emerging disease situations dictate that approaches to disease control will not be simple. Any single approach directed at only one area is not likely to succeed. The test and slaughter policy of tuberculosis, which has been relatively effective in control of tuberculosis in domestic livestock, is recognized as insufficient in areas where wildlife reservoirs exist. Measures to prevent disease transmission are more efficient than efforts required to eliminate an established disease from wildlife or domestic animals. Human involvement in risk reduction strategies such as education and promotion of biosecurity practices that limit interactions between livestock and wildlife will be required. Serious risk analysis should be conducted prior to introduction or re-introduction of wildlife to new geographic areas. In areas where tuberculosis is endemic in wildlife, certain agricultural practices such as allowing wildlife access to livestock feed may no longer be tolerable if disease is to be eradicated.

The elimination of tuberculosis from free-ranging wildlife is a difficult goal. It will require the cooperation of agricultural and wildlife agencies, legislative bodies, private landowners, and citizens. Because of limited resources, multiple agencies must work collaboratively; assessing blame to one group or organization will be counterproductive. The idea of organizations from different backgrounds working together to address diseases transmitted between domestic animals and wildlife is gaining momentum. This movement is evidenced by a number of recent symposia on diseases at the interface of domestic animals and wildlife, and the creation of wildlife disease committees in traditionally agriculturally based producer groups. Furthermore, resolutions from groups such as the Wildlife Disease Association and the Society for Tropical Veterinary Medicine have come forward that call for funding organizations to encourage projects that foster integration of livestock production and natural resource management, address wildlife, livestock, and rangeland health in environmental impact statements, and use science-based advice when contemplating projects involving wildlife and livestock (Anonymous 2002).

References

Aldwell FE, Keen DL, Stent VC, Thomson A, Yates GF, de Lisle GW, Buddle BM (1995a) Route of BCG administration in possums affects protection against bovine tuberculosis. N Z Vet J 43:356–359

Aldwell FE, Pfeiffer A, de Lisle GW, Jowett G, Heslop J, Keen D, Thomson A, Buddle BM (1995b) Effectiveness of BCG vaccination in protecting possums against bovine tuberculosis. Res Vet Sci 58:90–95

Anonymous (1997) TB in cattle and badgers: review group advocates an investigative approach. Vet Rec 141:636

Anonymous (2002) Resolution by the Wildlife Disease Association and the Society for Tropical Veterinary Medicine calling for international donor community recognition of animal health sciences as critical for the design and management of sustainable wildlife and/or livestock based programs. Ann N Y Acad Sci 969: 364–365

Aranaz A, de Juan L, Montero N, Sanchez C, Galka M, Delso C, Alvarez J, Romero B, Bezos J, Vela AI, Briones V, Mateos A, Dominguez L (2004) Bovine tuberculosis (*Mycobacterium bovis*) in wildlife in Spain. J Clin Microbiol 42:2602–2608

Barlow NE (1991) A spatially aggregated disease/host model for bovine TB in New Zealand possum populations. J Appl Ecol 28:777–793
Beatson NS (1985) Tuberculosis in red deer in New Zealand. In: Fennessy PF, Drew KR (eds) Biology of deer production. R Soc N Z 22:147–150
Belli LB (1962) Bovine tuberculosis in a white-tailed deer. Can Vet J 3:356–358
Bollinger A, Bolliger W (1948) Experimental transmission of tuberculosis to *Trichosurus vulpecula*. Aust J Sci 10:182–183
Bouvier G (1963) Possible transmission of tuberculosis and brucellosis from game animals to man and to domestic animals. Bull Off Inter Epiz 59:433–436
Bruning-Fann CS Schmitt SM, Fitzgerald SD, Payeur JB, Whipple DL, Cooley TM, Carlson T, Friedrich P (1998) *Mycobacterium bovis* in coyotes from Michigan. J Wildl Dis 34:632–636
Bruning-Fann CS, Schmitt SM, Fitzgerald SD, Fjerke JS, Friedrich PD, Kaneene JB, Clarke KA, Butler KL, Payeur JB, Whipple DL, Cooley TM, Miller JM, Muzo DP (2001) Bovine tuberculosis in free-ranging carnivores from Michigan. J Wildl Dis 37:58–64
Buddle BM, Aldwell FE, Keen DL, Parlane NA, Yates G, de Lisle GW (1997) Intraduodenal vaccination of brushtail possums with bacilli Calmette-Guerin enhances immune responses and protection against *Mycobacterium bovis* infection. Int J Tub Lung Dis 1:377–383
Caley P, Hickling GJ, Cowan PE, Pfeiffer DU (1999) Effects of sustained control of brushtail possums on levels of *Mycobacterium bovis* infection in cattle and brushtail possum population from Hohotaka New Zealand. N Z Vet J 47:133–142
Cheeseman CL, Wilesmith JW, Stuart FA, Mallinson PJ (1988) Dynamics of tuberculosis in a naturally infected badger population. Mammal Rev 18:61–72
Cheeseman CL, Wilesmith JW, Stuart FA (1989) Tuberculosis: the disease and its epidemiology in the badger, a review. Epidem Infect 103:113–125
Collins DM, Gabric DM, de Lisle GW (1988) Typing of *Mycobacterium bovis* isolates from cattle and other animals in the same locality. N Z Vet J 36:45–46
Cooke MM, Jackson R, Coleman JD, Alley MR (1995) Naturally occurring tuberculosis caused by *Mycobacterium bovis* in brushtail possums (*Trichosurus vulpecula*): II. Pathology. N Z Vet J 43:315–321
Corner LAL, Buddle B.M, Pfeiffer DU, Morris RS (2001) Aerosol vaccination of the brushtail possum (*Trichosurus vulpecula*) with bacilli Calmette-Guerin: the duration of protection. Vet Microbiol 81:181–191
Corner LAL, Pfeiffer DU, Buddle B.M, and Morris RS (2003) Social-network analysis of *Mycobacterium bovis* transmission among captive brushtail possums (*Trichosurus vulpecula*) Prev Vet Med 59:147–167
Cousins DV, Dawson DJ (1999) Tuberculosis due to *Mycobacterium bovis* in the Australian population: cases recorded during 1970–1994. Int J Tub Lung Dis 3:715–721
Donnelly CA, Woodroffe R, Cox ER, Bourne J, Gettinby G, Le Fevre AM, McInerney JP, Morrison WI (2003) Impact of localized badger culling on tuberculosis incidence in British cattle. Nature 426:834–837
Duffield BJ, Young DA (1985) Survival of *Mycobacterium bovis* in defined environmental conditions. Vet Microbiol 10:193–197

Ekdahl MO, Smith BL, Money DFL (1970) Tuberculosis in some wild and feral animals in New Zealand. N Z Vet J 18:44–45

Fanning A, Edwards S (1991) *Mycobacterium bovis* infection in human beings in contact with elk (*Cervus elaphus*) in Alberta Canada. Lancet 338:1253–1255

Ferris DH, Beamer PD, Alberts JO, Trainer D (1961) Tuberculosis in transported deer. J Am Vet Med Assoc 138:326–328

Fitzgerald SD, Kaneene JB, Butler KL, Clarke KR, Fierke JS, Schmitt SM, Bruning-Fann CS, Mitchell RR, Berry DE, Payeur JB (2000) Comparison of postmortem techniques for the detection of *Mycobacterium bovis* in white-tailed deer (*Odocoileus virginianus*). J Vet Diag Invest 12:322–327

Friend M, Kroll ET, Gruft H (1963) Tuberculosis in a wild white-tailed deer. N Y Fish Game J 10:118–123

Frye GH (1995) Bovine tuberculosis eradication: the program in the United States. In: Thoen CO, Steele JH (eds) *Mycobacterium bovis* infection in animals and humans. Iowa State University Press, Ames IA, pp 119–129

Gallagher J, Muirhead RH, Burn KJ (1976) Tuberculosis in wild badgers (*Meles meles*) in Gloucestershire: Pathology. Vet Rec 98:9–14

Garnett BT, Roper TJ, Delahay RJ (2003) Use of cattle troughs by badgers (*Meles meles*): a potential route for the transmission of bovine tuberculosis (*Mycobacterium bovis*) to cattle. Appl Anim Beh Sci 80:1–8

Gavier-Widen D, Chambers MA, Palmer N, Newell DG, Hewinson RG (2001) Pathology of natural *Mycobacterium bovis* infection in European badgers (*Meles meles*) and its relationship with bacterial excretion. Vet Rec 148:299–304

Grange JM, Yates MD (1994) Zoonotic aspects of *Mycobacterium bovis* infection. Vet Microbiol 40:137–151

Hutchings MR, Harris S (1997) Effects of farm management practices on cattle grazing. Behaviour and the potential for transmission of bovine tuberculosis from badgers to cattle. Vet J 153:149–162

Jackson R (2002) The role of wildlife in *Mycobacterium bovis* infection of livestock in New Zealand. N Z Vet J 50:49–52

Jackson R, Cooke MM, Coleman JD, Morris RS (1995a) Naturally occurring tuberculosis caused by *Mycobacterium bovis* in brushtail possums (*Trichosurus vulpecula*): I. An epidemiologic analysis of lesion distribution. N Z Vet J 43:306–315

Jackson R, Cooke MM, Coleman JD, Morris RS, de Lisle GW, Yates GF (1995b) Naturally occurring tuberculosis caused by *Mycobacterium bovis* in brushtail possums (*Trichosurus vulpecula*): III. Routes of infection and excretion. N Z Vet J 43:322–327

Jackson R, de Lisle GW, Morris RS (1995c) A study of the environmental survival of *Mycobacterium bovis* on a farm in New Zealand. N Z Vet J 43:346–352

Kaneene JB, Bruning-Fann CS, Dunn J, Mullaney TP, Berry D, Massey JP, Thoen CO, Halstead S, Schwartz K (2002) Epidemiologic investigation of *Mycobacterium bovis* in a population of cats. Am J Vet Res 63:1507–1511

Krebs JR, Anderson RM, Clutton-Brock T, Donnelly CA, Frost S, Morrison WI, Woodroffe R, Young D (1998) Badgers and bovine TB: conflicts between conservation and health. Science 279:817–818

Levine PP (1934) A report on tuberculosis in wild deer (*Odocoileus virginianus*) Cornell Vet 24:264–266
Little TWA, Naylor PF, Wilesmith JW (1982) Laboratory study of *Mycobacterium bovis* infection in badgers and calves. Vet Rec 111:550–557
Lugton IW, Wilson PR, Morris RS, Nugent G (1998) Epidemiology and pathogenesis of *Mycobacterium bovis* infection of red deer (*Cervus elaphus*) in New Zealand. N Z Vet J 46:147–156
McCarty CW, Miller MW (1998) A versatile model of disease transmission applied to forecasting bovine tuberculosis dynamics in white-tailed deer populations. J Wildl Dis 34:722–730
Miller R, Kaneene JB, Fitzgerald SD, Schmitt SM (2003) Evaluation of the influence of supplemental feeding of white-tailed deer (*Odocoileus virginianus*) on the prevalence of bovine tuberculosis in the Michigan wild deer population. J Wildl Dis 39:84–95
Muirhead RH, Gallagher J, Burn KJ (1974) Tuberculosis in wild badgers in Gloucestershire: epidemiology. Vet Rec 95:552–555
Noonan NL, Sheane WD, Harper LR, Ryan PJ (1975) Wildlife as a possible reservoir of bovine tuberculosis. Irish Vet J 29:1
O'Brien DJ, Fitzgerald SD, Lyon TJ, Butler JL, Fierke JS, Clarke KR, Schmitt SM, Cooley TM, Berry DE (2001) Tuberculous lesions in free-ranging white-tailed deer in Michigan. J Wildl Dis 37:608–613
O'Brien DJ, Schmitt SM, Fierke JS, Hogle SA, Winterstein SR, Cooley TM, Moritz WE, Diegel KL, Fitzgerald SD, Berry DE, Kaneene JB (2002) Epidemiology of *Mycobacterium bovis* in free-ranging white-tailed deer Michigan USA, 1995–2000. Prev Vet Med 54:47–63
O'Neill BD, Pharo HJ (1995) The control of bovine tuberculosis in New Zealand. N Z Vet J 43:249–255
Palmer MV, Whipple DL, Olsen SC (1999) Development of a model of natural infection with *Mycobacterium bovis* in white-tailed deer. J Wildl Dis 35:450–457
Palmer MV, Whipple DL, Payeur JB, Alt DP, Esch KJ, Bruning-Fann CS, Kaneene JB (2000) Naturally occurring tuberculosis in white-tailed deer. J Am Vet Med Assoc. 216:1921–1924
Palmer MV, Whipple DL, Waters WR (2001) Experimental deer to deer transmission of *Mycobacterium bovis*. Am J Vet Res 62:692–696
Palmer MV, Waters WR, Whipple DL (2002) Milk containing *Mycobacterium bovis* as a source of infection for white-tailed deer fawns (*Odocoileus virginianus*). Tuberculosis 82:161–165
Palmer MV, Waters WR, Whipple DL (2003) Aerosol exposure of white-tailed deer (*Odocoileus virginianus*) to *Mycobacterium bovis*. J Wildl Dis 39:817–823
Palmer MV, Waters WR, Whipple DL (2004a) Investigation of the transmission of *Mycobacterium bovis* from deer to cattle through indirect contact. Am J Vet Res 65:1483–1489
Palmer MV, Waters WR, Whipple DL (2004b) Shared feed as a means of deer-to-deer transmission of *Mycobacterium bovis*. J Wildl Dis 40:87–91

Palmer MV, Whipple DL (2006) Survival of *Mycobacterium bovis* on feedstuff commonly used as supplemental feed for white-tailed deer (*Odocoileus virginianus*). J Wild Dis 42:853–858

Paterson BM, Morris RS (1995) Interactions between beef cattle and simulated tuberculous possums on pasture. N Z Vet J 43:289–293

Paterson BM, Morris RS, Weston J, Cowan PE (1995) Foraging and denning patterns of brushtail possums, and their possible relationship to contact with cattle and the transmission of bovine tuberculosis. N Z Vet J 43:281–288

Robinson P, Morris D, Antic R (1988) *Mycobacterium bovis* as an occupational hazard in abattoir workers. Aust N Z J Med 18:701–703

Robinson RC, Phillips PH, Stevens G, Storm PA (1989) An outbreak of *Mycobacterium bovis* infection in fallow deer (*Dama dama*). Aust Vet J 66:195–197

Rogers LM, Delahay R, Cheeseman CL, Langton S, Smith GC, Clifton-Hadley RS (1998) Movement of badgers (*Meles meles*) in a high density population: individual, population and disease effects. Proc Ro Soc Lond B 265:1269–1276

Sauter CM, Morris RS (1995) Behavioural studies on the potential for direct transmission of tuberculosis from feral ferrets (*Mustela furo*) and possums (*Trichosurus vulpecula*) to farmed livestock. N Z Vet J 43:294–300

Scantlebury M, Hutchings MR, Allcroft DJ, Harris S (2004) Risk of disease from wildlife reservoirs: badgers, cattle and bovine tuberculosis. J Dairy Sci 87:330–339

Schmitt SM, Fitzgerald SD, Cooley TM, Bruning-Fann CS, Sullivan L, Berry D, Carlson T, Minnis RB, Payeur JB, Sikarskie J (1997) Bovine tuberculosis in free-ranging white-tailed deer from Michigan. J Wildl Dis 33:749–758

Schmitt SM, O'Brien DJ, Bruning-Fann CS, Fitzgerald SD (2002) Bovine tuberculosis in Michigan wildlife and livestock. Ann N Y Acad Sci 969:262–268

Serraino A, Marchetti G, Sanguinetti V, Rossi MC, Zanoni RG, Catozzi L, Bandera A, Dini W, Mignone W, Franzetti F, Gori A (1999) Monitoring of transmission of tuberculosis between wild boars and cattle: genotypical analysis of strains by molecular epidemiology techniques. J Clin Microbiol 37:2766–2771

Smith RMM, Drobniewski F, Gibson A, Montague JDE, Logan MN, Hunt D, Hewinson G, Salmon RL, O'Neill B (2004) *Mycobacterium bovis* infection United Kingdom. Emerg Infect Dis 10:539–541

Steele JH (1995) Regional and country status reports. In: Thoen CO, Steele JH (eds) *Mycobacterium bovis* infection in animals and humans. Iowa State University Press, Ames, IA, pp 169–171

Tanner M, Michel AL (1999) Investigation of the viability of *Mycobacterium bovis* under different environmental conditions in the Kruger National Park. Onderst J Vet Res 66:185–190

Tuyttens FAM, Delahay RJ, MacDonald DW, Cheeseman CL, Long B, Donnelly CA (2000) Spatial perturbation caused by a badger (*Meles meles*) culling operation: implications for the function of territoriality and the control of bovine tuberculosis (*Mycobacterium bovis*). J Anim Ecol 69:815–828

Tweddle NE, Livingstone P (1994) Bovine tuberculosis control and eradication programs in Australia and New Zealand. Vet Microbiol 40:23–29

Whipple DL, Palmer MV (2000) Survival of *Mycobacterium bovis* on feeds used for baiting white-tailed deer (*Odocoileus virginianus*) in Michigan. In: Proceedings of the = 49 th Annual Wildlife Disease Assocociation, p 21
Whipple DL, Meyer RM, Berry DF, Jarnigan JL, Payeur JB (1997) Molecular epidemiology of tuberculosis in wild white-tailed deer in Michigan and elephants. In: Proc U S Anim Health Assoc 543–546
Whipple DL, Jarnagin JL, Payeur JB (1999) DNA fingerprinting of *Mycobacterium bovis* isolates from animals in northeast Michigan. In: Proceedings for the IX International Symposium and World Association of Veternary Laboratory Diagnosis, p 83
Whiting TL, Tessaro SV (1994) An abattoir study of tuberculosis in a herd of farmed elk. Can Vet J 35:497–501
Wigle W, Ashley MJ, Killough EM, Cosens M (1972) Bovine tuberculosis in humans in Ontario. Am Rev Resp Dis 106:528–534
Wilkins MJ, Bartlett PC, Frawley B, O'Brien DJ, Miller CE, Boulton ML (2003) *Mycobacterium bovis* (bovine TB) exposure as a recreational risk for hunters: results of a Michigan Hunter Survey, 2001. Int J Lung Dis 7:1001–1009

CTMI (2007) 315:217–252
© Springer-Verlag Berlin Heidelberg 2007

Emergence and Persistence of Hantaviruses

S. L. Klein[1] (✉) C. H. Calisher[2]

[1]W. Harry Feinstone Department of Molecular Microbiology and Immunology, Johns Hopkins Bloomberg School of Public Health, Baltimore, MD 21205, USA
saklein@jhsph.edu
[2]Department of Microbiology, Immunology, and Pathology, Colorado State University, Fort Collins, CO 80523, USA

Abstract Hantaviral diseases have been recognized for hundreds of years but, until 1976, they had not been associated with an infectious agent. When Lee and colleagues isolated what is now known as Hantaan virus, the techniques they introduced allowed further investigations into the etiology of the classical hantavirus disease, hemorrhagic fever with renal syndrome (HFRS), now known to be caused by any of multiple hantaviruses. The discovery of hantavirus pulmonary syndrome (HPS) in the New World, and that it also can be caused by any of multiple hantaviruses (family *Bunyaviridae*, genus *Hantavirus*), has opened an entire field of epidemiologic, virologic, molecular, behavioral, and ecologic studies of these viruses. There appears to be a single hantavirus-single rodent host association, such that understanding the idiosyncrasies of each rodent host species and the ecologic variables that affect them are recognized as critical if we are to reduce human risk for infection. This chapter summarizes what is known about hantaviruses with regard to history of these viruses, their taxonomy, recognized geographical distribution, ecologic factors impacting their maintenance and spread of hantaviruses, effect of rodent behavior on hantavirus transmission, influence of host factors on susceptibility to and transmission of hantaviruses, and transmission of hantaviruses from rodents to humans. In addition, we summarize all these complexities and provide suggestions for future research directions.

1 Introduction and History of Hantaviruses

More than 1,000 years ago, Chinese physicians described a disease now known as hemorrhagic fever with renal syndrome (HFRS). HFRS is characterized by an incubation period of 2–3 weeks (range, 4–42 days) and a triad of fever, hemorrhagic manifestations, and renal impairment (Lee 1991). Poorly substantiated records from previous wars suggested that this or a similar disease had been recognized. Clinically compatible diseases associated with trench warfare were recorded during the US Civil War and World War I, Songo fever was described in the Japanese-Chinese war, and "feldnephritis" was described in German soldiers during 1941–1942 in Russia and Finland. HFRS, known as Korean hemorrhagic fever in Korea and as epidemic hemorrhagic fever in China, is now recognized to affect as many as 200,000 people each year in Asia and Europe.

Although many hypotheses as to the cause of this disease were suggested, it was not until 1976 that Lee and collaborators isolated an etiologic agent, which they named Hantaan virus, after a river in Korea (Lee et al. 1978). The first confirmed isolate was from lung tissue of a striped field mouse (*Apodemus agrarius*) (Lee et al. 1978). Subsequent to this discovery, other viruses antigenically related to Hantaan were characterized and classified in a newly established genus of viruses in the family *Bunyaviridae* (Hung et al. 1983; McCormick et al. 1982; Schmaljohn and Dalrymple 1983; Schmaljohn et al. 1985; White et al. 1982), the genus *Hantavirus* (Elliott et al. 2000).

With the single exception of Thottapalayam virus (for which the natural host appears to be an insectivore (Zeller et al. 1989), hantaviruses are maintained and transmitted by rodents and are found essentially throughout the world (Table 1). Once infected, reservoir rodents remain persistently infected, which may be caused by changes in the regulation of virus replication or in the ability of the virus to evade host immune responses (Meyer and Schmaljohn 2000a, 2000b). Although hantaviruses have existed for millions of years (Plyusnin and Morzunov 2001), recent interest in these viruses was stimulated by the 1993 outbreak of human hantavirus pulmonary syndrome (HPS) cases caused by Sin Nombre virus in the southwestern United States (Childs et al. 1994).

Table 1 Hantaviruses and their rodent reservoir hosts

Virus	Rodent host	Human disease	Location first detected	Reference
Order Rodentia, family Muridae, subfamily Murinae				
Hantaan (HTN)	*Apodemus agrarius*	Severe HFRS	Republic of Korea	Lee et al. 1978
Seoul (SEO)	*Rattus norvegicus*	Mild to moderate HFRS	Republic of Korea	Lee et al. 1982
Sangassou virus	*Hylomyscus simus*	None recognized	Guinea	Klempa et al. 2006
Soochong	*Apodemus peninsulae*	None recognized	Korea	Baek et al 2004
Dobrava (DOB)	*Apodemus flavicollis*	Severe HFRS	Slovenia	Avsic-Zupanc et al. 1992
Thai	*Bandicota indicus*	None recognized	Thailand	Elwell et al. 1985
Saarema	*Apodemus agrarius*	Mild HFRS	Finland	Plyusnin et al 2001
Amur	*Apodemus peninsulae*	HFRS	Russia	Yashina et al. 2001
Order Rodentia, family Muridae, subfamily Sigmodontinae				
Sin Nombre	*Peromyscus maniculatus*	HPS	New Mexico, USA	Nichol et al. 1993
New York	*Peromyscus leucopus*	HPS	New York, USA	Song et al. 1994
Black Creek Canal	*Sigmodon hispidus*	HPS	Florida, USA	Rollin et al. 1995
Bayou	*Oryzomys palustris*	HPS	Louisiana, USA	Morzunov et al. 1995
Muleshoe	*Sigmodon hispidus*	HPS	Texas, USA	Rawlings et al. 1996
Monongahela	*Peromyscus maniculatus*	HPS	Pennsylvania, USA	Song et al. 1996
Limestone Canyon	*Peromyscus boylii*	None recognized	Arizona, USA	Sanchez et al. 2001
Blue River	*Peromyscus leucopus*	None recognized	Indiana, USA	Morzunov et al. 1998
El Moro Canyon	*Reithrodontomys megalotis*	None recognized	New Mexico, USA	Hjelle et al. 1994
Rio Segundo	*Reithrodontomys mexicanus*	None recognized	Costa Rica	Hjelle et al. 1995a
Caño Delgadito	*Sigmodon alstoni*	None recognized	Venezuela	Fulhorst et al. 1997
Juquitiba	*Oligoryzomys nigripes*	HPS	Brazil	Johnson et al. 1999
Araraquara	*Bolomys lasiurus*	HPS	Brazil	Johnson et al. 1999
Castelo dos Sonhos	(unknown)	HPS	Brazil	Johnson et al. 1999

(*Continued*)

Table 1 Hantaviruses and their rodent reservoir hosts—cont'd.

Virus	Rodent host	Human disease	Location first detected	Reference
Rio Mamoré	*Oligoryzomys microtis*	HPS	Bolivia	Bharadwaj et al. 1997
Laguna Negra	*Calomys laucha*	HPS	Paraguay	Johnson et al. 1997
Andes	*Oligoryzomys longicaudatus*	HPS	Argentina	Lopez et al. 1997
Lechiguanas	*Oligoryzomys flavescens*	HPS	Argentina	Levis et al. 1998
Bermejo	*Oligoryzomys chacoensis*	HPS	Argentina	Levis et al. 1998
Orán	*Oligoryzomys longicaudatus*	HPS	Argentina	Levis et al. 1998
Maciel	*Bolomys obscurus*	None recognized	Argentina	Levis et al. 1998
Hu39694	(unknown)	HPS	Argentina	Levis et al. 1998
Pergamino	*Akodon azarae*	None recognized	Argentina	Levis et al. 1998
Choclo	*Oligoryzomys fulvescens*	HPS	Panama	Vincent et al. 2000
Calabazo	*Zygodontomys brevicauda*	None recognized	Panama	Vincent et al. 2000
Maporal	*Oecomys bicolor*	None recognized	Venezuela	Fulhorst et al. 2004
Order Rodentia, family Muridae, subfamily Arvicolinae				
Puumala	*Myodes glareolus*	HFRS	Sweden	Brummer-Korvenkontio et al. 1980
Prospect Hill	*Microtus pennsylvanicus*	None recognized	Maryland, USA	Lee et al. 1982
Bloodland Lake	*Microtus ochrogaster*	None recognized	Missouri, USA	Hjelle et al 1995b
Isla Vista	*Microtus californicus*	None recognized	California, USA	Song et al. 1995
Tula	*Microtus arvalis/ Microtus rossiaemeridionalis*	None recognized	Russia	Plyusnin et al. 1994
Khabarovsk	*Microtus fortis*	None recognized	Eastern Russia	Horling et al. 1996
Topografov	*Lemmus sibericus*	None recognized	Siberia	Plyusnin et al. 1996
Order Insectivora, family Soricidae				
Thottapalayam	*Suncus murinus*	None recognized	India, Southeast Asia	Carey et al. 1971

HFRS hemorrhagic fever with renal syndrome; *HPS* hantavirus pulmonary syndrome

Hantaviruses are single-stranded, negative-sense RNA viruses with a tripartite genome, the segments of which are designated: small (S), medium (M), and large (L), and which encode the viral nucleocapsid (N), envelope glycoproteins (G_C and G_N), and polymerase (L). The consensus nucleotide sequences of all three segments are AUCAUCAUCUG... at the 3' end and UAGUAGUA... at the 5' end, distinguishing them from other viruses in the family (Elliott et al. 2000). These viruses primarily infect pulmonary endothelial cells, monocytes, and macrophages and replicate in the cytoplasm of the cell (Mackow and Gavrilovskaya 2001; Nagai et al. 1985; Raftery et al. 2002; Temonen et al. 1993). Several host proteins, including integrins and other cell adhesion molecules (e.g., ICAM-1), have been implicated as receptors for hantavirus entry into host cells (Mackow and Gavrilovskaya 2001; Singh et al. 2001; Song et al. 1999). Hantaviruses are known to cause natural disease only in humans.

Our knowledge and understanding of hantavirus pathogenesis in rodents and humans has improved with increased interest in emerging infectious diseases. The primary goals of this chapter are to: (1) review our knowledge of the ecological and environmental factors that lead to increased spread of hantaviruses from rodents to humans; (2) establish that social behaviors among rodents contribute to increased transmission of hantaviruses; (3) examine host factors that contribute to increased susceptibility and transmission of hantaviruses; and (4) examine the immunological factors that may contribute to persistence of infections in rodent reservoir hosts. Persistent infection of rodents with hantaviruses, the release of virus into the environment, and social contact contribute to transmission of hantaviruses between rodents and from rodents to humans and, thus, are factors that require further investigation if we are to prevent human infections with these viruses.

2 Taxonomy and Geographical Distribution of Rodent Host–Hantavirus Systems

The flurry of activity following the outbreak of HPS in 1993 led to keener surveillance for similar illnesses throughout the Americas. Many newly recognized hantaviruses were detected in rodents of various species and, as of this writing, there are 47 recognized hantaviruses and these have considerable geographic diversity (Table 1) (Calisher et al. 2003).

Unlike other viruses of the family *Bunyaviridae*, hantaviruses are not transmitted by arthropods. Hantaviruses are horizontally transmitted between rodents and each hantavirus appears to have co-evolved with a primary rodent host species (Table 1) (Plyusnin and Morzunov 2001). For example, in the

Americas, Sin Nombre virus infects deer mice (*Peromyscus maniculatus*), New York virus infects white-footed mice (*Peromyscus leucopus*), Black Creek Canal virus infects hispid cotton rats (*Sigmodon hispidus*), Seoul virus infects Norway rats (*Rattus norvegicus*), and Andes virus is carried by rice rats (*Oligoryzomys longicaudatus*) (Table 1) (Plyusnin and Morzunov 2001; Schmaljohn and Hjelle 1997). Phylogenetically, hantaviruses are associated with three subfamilies of murid rodents: Old World hantaviruses are carried by Old World rats and mice (Order Rodentia, family Muridae, subfamily Murinae) and New World hantaviruses are maintained by arvicolid and sigmodontine rodents (Order Rodentia, family Muridae, subfamilies Arvicolinae and Sigmodontinae) (Table 1) (Plyusnin and Morzunov 2001). Genetically-related rodents carry hantaviruses with nearly identical gene and protein sequences and analyses of rodent host mitochondrial genes and viral gene sequences produce similar phylogenetic trees (Plyusnin and Morzunov 2001).

Rodents infected with species-specific hantaviruses remain persistently infected and shed virus in saliva, urine, and feces, but exhibit no overt signs of disease (Botten et al. 2002; Hutchinson et al. 2000; Klein et al. 2001; Lee et al. 1981). When spillover to nonhuman mammals occurs, the result usually is a nonproductive infection; when spillover to humans occurs, however, the result can be profound morbidity or mortality. Human disease caused by hantaviruses occurs when people come in contact with excrement or secretions from infectious rodents (Glass et al. 1993; Johnson 2001; Klein et al. 2000 2001). Andes virus is an exception in that there is evidence to suggest that it is transmitted by person-to-person contact (Padula et al. 1998). Among the approximately 200,000 human cases of hantavirus infection reported annually, case-fatality rates range from 0.1% to 10% worldwide (37% in the case of SNV in North America) (Schmaljohn and Hjelle 1997).

The emergence of hantaviruses in human populations is of concern because these viruses cause diseases for which there currently are no cures. Hantaviruses cause two clinical syndromes in humans: hemorrhagic fever with renal syndrome (HFRS) or its milder form, called nephropathia epidemica (NE), and hantavirus pulmonary syndrome (HPS). The severity of disease in humans depends on the strain of hantavirus involved: Hantaan, Dobrava, and Amur viruses are associated with severe HFRS; Puumala, Seoul and Saarema viruses cause NE; Sin Nombre, New York, Black Creek Canal, Bayou, Muleshoe, and Monongahela viruses are the agents of HPS in North America; and Andes, Lechiguanas, Bermejo, Oran, Juquitiba, Araraquara, Castelo dos Sonhos, Laguna Negra, Choclo, and Hu39694 viruses have been implicated in cases of HPS in Central and South Americas (Table 1). Because hantavirus–rodent host relationships are relatively host-specific, the geographic distribution of the rodent host limits the geographic distribution of the particular hantavirus and can be used to identify locations where hantaviruses are likely to emerge (Mills et al. 1997).

3
Ecological Factors Impact Maintenance and Spread of Hantaviruses

Current data suggest that emergence of hantavirus infection in humans is correlated with increased rodent population densities (Childs et al. 1995; Yates et al. 2002). Therefore, understanding the ecological and environmental factors that impact rodents, their resources, and their habitat may provide insights to how the environment affects persistence and spread of hantaviruses between rodents and from rodents to humans (Calisher et al. 2005a, 2005b). Environmental factors that cause fluctuations in rodent populations such as precipitation, temperature, habitat quality, food availability, and food source may affect the prevalence of hantaviruses (Biggs et al. 2000; Hjelle and Glass 2000). Environmental factors may directly affect host immune responses against infection or act indirectly through changes in population densities and subsequent exposure to and transmission of these viruses (Biggs et al. 2000; Hjelle and Glass 2000; Nelson et al. 2002).

The relationship among environment, hantavirus infection, and rodent populations is exemplified by the 1993 outbreak of HPS in the Four Corners region of the United States (i.e., New Mexico, Colorado, Utah, and Arizona). In May 1993, a cluster of fatalities in adult humans with acute respiratory failure was recognized in northwestern New Mexico and, shortly thereafter, in Colorado, Arizona, and Utah. Molecular epidemiologic and virologic investigations of the 1993 outbreak indicated that the etiologic agent was Sin Nombre virus and that deer mice were the primary rodent hosts (Nichol et al. 1993).

Analyses of weather conditions prior to the 1993 outbreak of Sin Nombre virus revealed that elevated precipitation associated with El Niño southern oscillation led to increased habitat and food resources conducive to increased rodent population densities (Hjelle and Glass 2000). Elevated precipitation can trigger a trophic cascade that affects availability of food and habitat resources that, in turn, affect rodent reservoir populations (Parmenter et al. 1999; Yates et al. 2002). Examination of the relationship between environmental conditions and hantavirus infection has increased our ability to predict high-risk environmental factors that influence outbreaks of hantavirus infection (Glass et al. 2002).

Annual transmission of hantaviruses generally occurs between rodents during times of the year when vegetation is abundant and reproductive activities are high (Abbott et al. 1999; Mills et al. 1999). Consequently, annual variation in hantavirus prevalence is observed among seasonally breeding rodents, but not among opportunistic rodents, such as Norway rats, which breed throughout the year (Klein et al. 2002a). Among deer mouse populations in New Mexico, Colorado, Utah, and Arizona, elevated population densities and reduced food availability are associated with increased seroprevalence of Sin Nombre virus and possibly increased spread of infection from rodents to humans (Biggs et al. 2000). Similar observations also are reported for bank voles (*Myodes glareolus*) infected with Puumala virus (Ahlm et al. 1997; Bernshtein et al. 1999).

In addition to food availability, food content may influence the prevalence of hantaviruses in rodent reservoir populations. Bioactive plant compounds, including 6-methoxy-2-benzoxazolinone (6-MBOA), can affect rodent reproduction. 6-MBOA occurs in certain rapidly growing green plants, principally grasses. When there has been a relatively warm, wet winter, and grasses emerge earlier in the year than usual, the presence of 6-MBOA facilitates earlier breeding, otherwise dependent on diurnal periodicity (Nelson and Shiber 1990; Nelson 1991). During warmer winter conditions, if rodents begin breeding earlier than usual, rodent population densities and transmission of hantaviruses may increase relative to population densities following cold or dry winters, as has been shown in southeastern Colorado (Calisher et al. 2005a).

Phytoestrogens, such as genistein from leguminous plants and coumestrol from alfalfa and clover sprouts, can also affect rodent reproduction and immune function. Phytoestrogens are structurally similar to the mammalian estrogen, estradiol, and can have incidental estrogenic effects by binding to estrogen receptors (Flynn et al. 2000; Fritz et al. 2002). Exposure to phytoestrogens can alter morphological and functional development of tissues that are sensitive to the effects of sex steroid hormones. Exposure to genistein during gestation and lactation demasculinizes the reproductive system and elevates immune function in rats, possibly via suppression of testosterone production (Klein et al. 2002c; Wisniewski et al. 2003). Ingestion of specific plant compounds may alter reproductive potential, perhaps serving to increase or decrease population size and hantavirus transmission. Alternatively, exposure to these plant compounds may alter immune responses to infection and influence persistence of hantaviruses in rodent reservoirs. Whether exposure to plant compounds, including phytoestrogens, affects responses to infections has not been investigated.

Among nontropical rodents, most challenging environmental conditions occur during the winter months when food availability and ambient temperatures are low. The stress of coping with energetically demanding conditions can result in increased viral persistence, possibly through glucocorticoid-mediated immunosuppression

Table 2 Evidence for behavioral transmission of hantaviruses among rodent reservoirs

Virus	Host	Route of transmission[a]	Reference
Andes	*Oligoryzomys longicaudatus*	Aerosol	Padula et al. 2004
Black Creak Canal	*Sigmodon hispidus*	Aerosol, wounding, social contact	Hutchinson et al. 2000
Hantaan	*Apodemus agrarius*	Aerosol, social contact	Lee et al. 1981
Puumala	*Myodes glareolus*	Aerosol, social contact	Bernshtein et al. 1999; Escutenaire et al. 2002; Yanagihara et al. 1985
Sin Nombre	*Peromyscus maniculatus*	Aerosol, social contact	Botten et al. 2002; Calisher et al. 2000
Seoul	*Rattus norvegicus*	Aerosol, wounding	Glass et al. 1988; Hinson et al. 2004; Kariwa et al. 1998; Klein et al. 2004b

[a] Each hantavirus listed is shed in saliva from infectious rodents

(Demas et al. 1997; Nelson et al. 2002; Sinclair and Lochmiller 2000; Webster et al. 2002). The extent to which high population densities and low food availability serve as environmental stressors affecting host immunity has not been examined in relation to hantavirus infection. Increased population densities also may serve as a stressor for rodents by increasing utilization of available resources (habitat and food), causing displacement of habitat and increasing social contact. Hantaviruses can be transmitted within rodent reservoir populations via wounding (Table 2; Fig. 1) (Escutenaire et al. 2002; Glass et al. 1988; Hinson et al. 2004).

4 Behavior Facilitates Transmission of Hantaviruses

Among directly transmitted pathogens, such as hantaviruses, social behavior can facilitate transmission from infected to susceptible individuals (Table 2) (Calisher et al. 2000; Glass et al. 1988). Intraspecific transmission of hantaviruses appears to occur through contact with saliva during aggressive encounters (Glass et al. 1988; Hinson et al. 2004). Although hantaviruses can be aerosolized, infection of laboratory rats with Seoul virus by subcutaneous or intramuscular injection is more effective than is inhalation at causing infection (Nuzum et al. 1988).

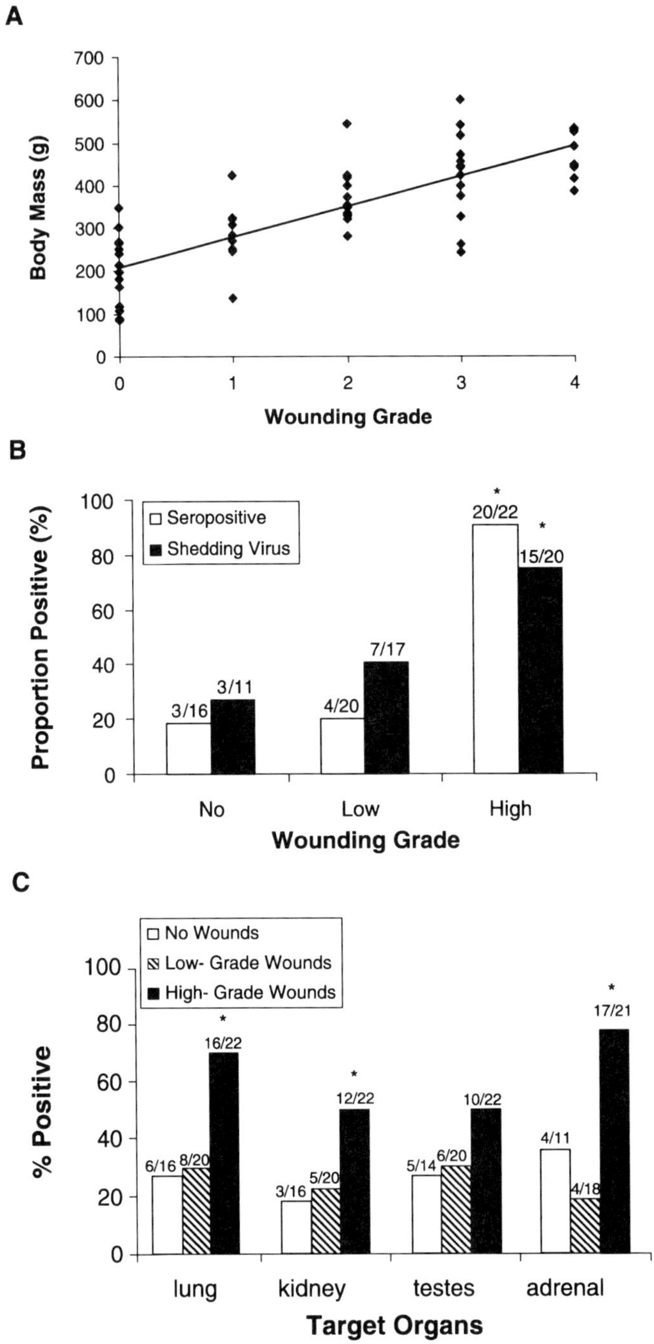
A
700
600
500
400
300
200
100
0
Body Mass (g)
0
1
2
3
4
Wounding Grade
B
100
80
60
40
20
0
Proportion Positive (%)
Seropositive
Shedding Virus
3/16
3/11
4/20
7/17
20/22
15/20
No
Low
High
Wounding Grade
C
No Wounds
Low- Grade Wounds
High- Grade Wounds
100
80
60
40
20
0
% Positive
6/16
8/20
16/22
3/16
5/20
12/22
5/14
6/20
10/22
4/11
4/18
17/21
lung
kidney
testes
adrenal
Target Organs

Thus, bite wounds may be a more productive route of transmission than is inhalation. In natural populations of Norway rats, adult males are more likely to have severe wounds than are either females or juvenile males (Glass et al. 1988; Hinson et al. 2004). Male rats with more severe wounds also are more likely to have anti-Seoul virus IgG, to shed virus in saliva, urine, and feces, and to have viral RNA in organs (Fig. 1) (Hinson et al. 2004). Whether engaging in aggressive behavior increases exposure to hantaviruses (i.e., host-mediated hypothesis) or whether infection increases the propensity to engage in aggression (i.e., parasite-mediated hypothesis) remains unclear.

In several host–pathogen systems, pathogens can manipulate the proximate mechanisms that mediate the expression of social behaviors, presumably to facilitate transmission (Klein 2003; Moore 2002). Laboratory studies of male Norway rats infected with Seoul virus reveal that during the onset of the persistent phase of infection (i.e., 30 days after inoculation with Seoul virus) males spend more time engaged in aggression during resident-intruder tests than either uninfected males or males tested during the acute phase of infection (i.e., 15 days after inoculation) (Klein et al. 2004b). Males that engage in aggression for a longer duration of time have more virus present in lung, kidney, and testis than do males that are less aggression (Klein et al. 2004b). Although certain individuals may be more likely to engage in behaviors that increase the probability of infection, these data suggest that infection with hantaviruses can alter behavior in the host.

Hantaviruses do not cross the blood–brain barrier of adult rodents and, therefore, virus is not usually present in the central nervous system (CNS) (Botten et al. 2000; Hinson et al. 2004; Kawamura et al. 1991); the few instances in which hantaviruses have been detected in rodent brains have been attributed to direct injection of virus, aberrant infection, or immature development of the blood–brain barrier, as in young animals. Unlike rabies viruses, which enter the CNS and infect brain regions involved in aggression, hantaviruses may cause changes in host aggressive behavior by replicating in peripheral target tissues, such as the testes, and altering the hormonal signals relayed to the CNS. Intermale aggression is mediated, in part, by circulating androgens (Nelson and Chiavegatto 2001). Consequently,

Fig. 1 **A** Correlation between body mass and wounding grade (*0* no wounds, *1* minor wounds on tail, *2* larger tail wounds and small body wounds, *3* larger wounds, *4* many extensive wounds on tail and body) in male Norway rats (*Rattus norvegicus*). **B** The proportion of wild-caught male rats with anti-Seoul virus IgG and shedding Seoul virus in saliva, urine, and feces by wounding grade (*no* wounding score of 0; *low* wounding score of 1–2; *high* wounding score of 3–4). **C** Prevalence of Seoul virus RNA (number of animals with detectable virus/total number of animals tested) in lung, kidney, testes, and adrenals from wild-caught rats with no wounds, low-grade wounds, or high-grade wounds. An *asterisk* indicates that the proportion of positive rats with high-grade and/or low-grade wounds was higher compared with the other group(s) of animals; $P < 0.05$. (Data adapted from [41])

wild-caught male rats with more severe wounds have higher testosterone concentrations than do males with no wounds or with low-grade wounds (Hinson et al. 2004). Additionally, viral RNA and viral protein are detected in the testes of wild-caught and laboratory-inoculated male Norway rats (Hinson et al. 2004; Klein et al. 2004b). Black Creek Canal virus and Sin Nombre virus RNA have been localized in the gonads of hispid cotton rats and deer mice, respectively (Hutchinson et al. 1998; Botten et al. 2000). Whether hantavirus infection of the gonads directly causes increased production of androgens requires further investigation.

From an evolutionary perspective, hantaviruses may exploit the proximate mechanisms that modulate social behaviors in rodents to increase the likelihood of transmission. Because social behaviors facilitate interactions between conspecifics, these behaviors can increase the transmission of hantaviruses from infected to susceptible individuals. During host–parasite co-evolution, host populations have evolved adaptations to evade infection or to reduce their susceptibility to infection, and pathogens have evolved counter-adaptations to overcome host defense mechanisms. In many cases, these counter-adaptations involve direct manipulation of host behavior to increase contact between infected and susceptible individuals (Moore 2002). This is not to say that all behavioral modifications following infection are parasite-mediated; host-mediated changes in behavior also may influence the probability of exposure to hantaviruses (Hinson et al. 2004). Thus, members of the same species may be differentially infected with hantaviruses because they vary in the expression of behaviors associated with sex or social status.

5
Host Factors Influence Susceptibility to and Transmission of Hantaviruses

5.1
Age-Dependent Pathology

In adult rodent reservoirs, most hantaviruses appear to cause chronic infection with no evidence of pathology or disease (Kariwa et al. 1996). In contrast, Seoul virus is reportedly pathogenic for young animals infected during the 1st week of life (Kariwa et al. 1996). When adult (50 days of age) and newborn (24 h after birth) Norway rats are inoculated with Seoul virus, virus replication occurs in a greater number of tissues and for a longer duration of time in newborn than in adult rats, leading to growth retardation and death of young rats (Kariwa et al. 1996; Yamanouchi et al. 1984). Infection of young rodents can also lead to increased viral persistence (Kariwa et al. 1996; Yanagihara et al. 1985); whether this is mediated by host immune responses or changes in virus replication requires further investigation.

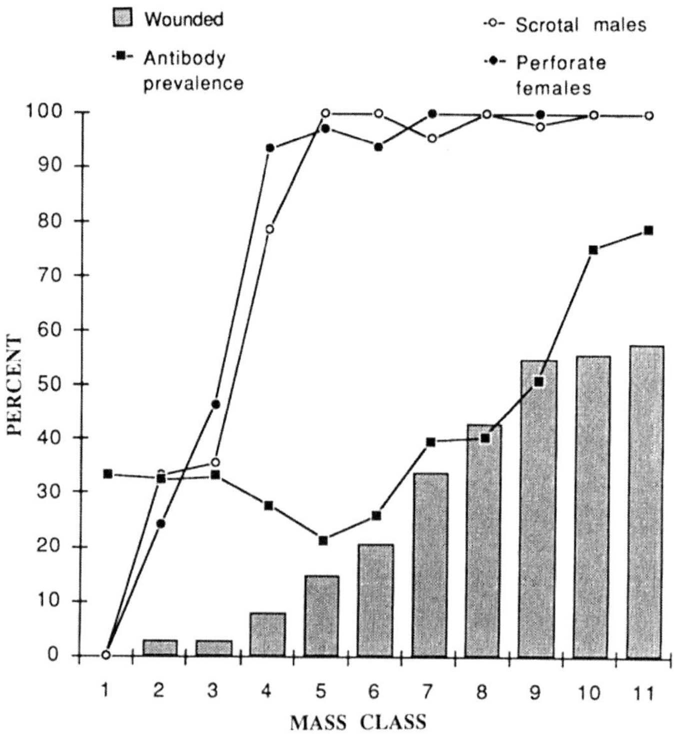

Fig. 2 Sexual maturation (i.e., puberty) corresponds with presence of severe wounds and antibodies against Seoul virus in Norway rats (*Rattus norvegicus*) trapped in Baltimore, Maryland. Mass classes are represented as 50-g intervals beginning at 0–49 g. Sexual maturation was defined as scrotal decent in males and vaginal opening in females. (Data adapted from [21])

In natural populations of rodents, including Norway rats, deer mice, meadow voles (*Microtus pennsylvanicus*), bank voles, and hispid cotton rats, body mass and, hence, physiological maturity, predicts the likelihood of being infected with their associated hantaviruses (Fig. 2) (Bernshtein et al. 1999; Calisher et al. 1999; Childs et al. 1988; Glass et al. 1998; Mills et al. 1998). The onset of sexual maturity in both male and female rats, for example, corresponds with an increased prevalence of antibody against Seoul virus (Fig. 2) (Childs et al. 1985, 1988). The sexual and aggressive behaviors associated with sexual maturity may increase the likelihood of exposure to hantaviruses (Fig. 2) (Childs et al. 1988). Alternatively, hormonal changes associated with puberty (e.g., increased production of sex steroids) may increase susceptibility to hantavirus infection by causing changes in immune responses against infection (Klein 2000). Although rodent reservoirs do not develop signs of illness, viral shedding and persistence are likely influenced by these age-dependent changes in response to infection.

5.2 Sex Differences in Response to Hantavirus Infections

The prevalence and intensity of many infectious diseases is higher in males than females (Klein 2000, 2004; Roberts et al. 2001). Reported human hantavirus infections in the Americas and Europe, as well as field observations of several rodent species, including Limestone Canyon virus in brush mice (*P. boylii*), Sin Nombre virus in deer mice, El Moro Canyon virus in western harvest mice (*Reithrodontomys megalotis*), Puumala virus in bank voles, Black Creek Canal virus in hispid cotton rats, and, to a lesser extent, Seoul virus in Norway rats, indicate that more males than females are infected with hantaviruses and that these differences in prevalence become apparent only after puberty (Bernshtein et al. 1999; Childs et al. 1994; Glass et al. 1998; Mills et al. 1997, 1998; Weigler et al. 1996; White et al. 1996; Williams et al. 1997). Although sex differences in hantavirus infection are reported for many species, few studies have examined the mechanisms that mediate sex differences in these, or other zoonotic, infections (Klein 2000; Klein et al. 2000, 2001, 2002b). Because sex differences in hantavirus infection become apparent only after puberty, sex steroid hormones, including testosterone in males and estradiol in females, are hypothesized to underlie the dimorphism in infection (Childs et al. 1988; Mills et al. 1997). Sex steroids can modulate sex differences in infection through effects on the immune system or on the expression of behaviors (e.g., aggression, group foraging) that increase the likelihood of being exposed to a pathogen (Klein 2000, 2004; Root et al. 1999; Zuk and McKean 1996).

Laboratory studies of Seoul virus infection of Norway rats reveal that, when given the same challenge, male and female rats are equally likely to become infected (Klein et al. 2000). After inoculation, however, male rats exhibit higher anti-Seoul virus IgG responses and elevated Th1 responses (i.e., IgG2a, IL-2, and IFNγ) compared with females (Fig. 3) (Klein et al. 2000, 2001). Although males have elevated immune responses, males shed Seoul virus longer and via more routes (i.e., a combination of saliva, urine, and feces) and have more viral RNA copies present in target organs, such as the lungs, than females (Klein et al. 2000, 2001, 2002b). Additionally, the expression of key transcriptional factors (e.g., eIF-2α, NF-κB, IRF-1, NF-IL-6, and STAT6) and genes that encode for proinflammatory (e.g., TNFαR, IL-1R, and

Fig. 3 Plasma anti-Seoul virus IgG (**A**), IgG2a (**B**), IgG1 (**C**) responses (mean ± SE) in male and female Long Evans rats. Blood samples were collected 0, 10, 15, 20, 30, and 40 days following inoculation with Seoul virus. Data are presented as IgG, IgG2a, or IgG1 units, in which the mean OD of each test sample was divided by the OD of the positive control sample run on the same microtiter plate. *Asterisk* indicates that males had higher responses than females; $P < 0.05$. (Data adapted from [64])

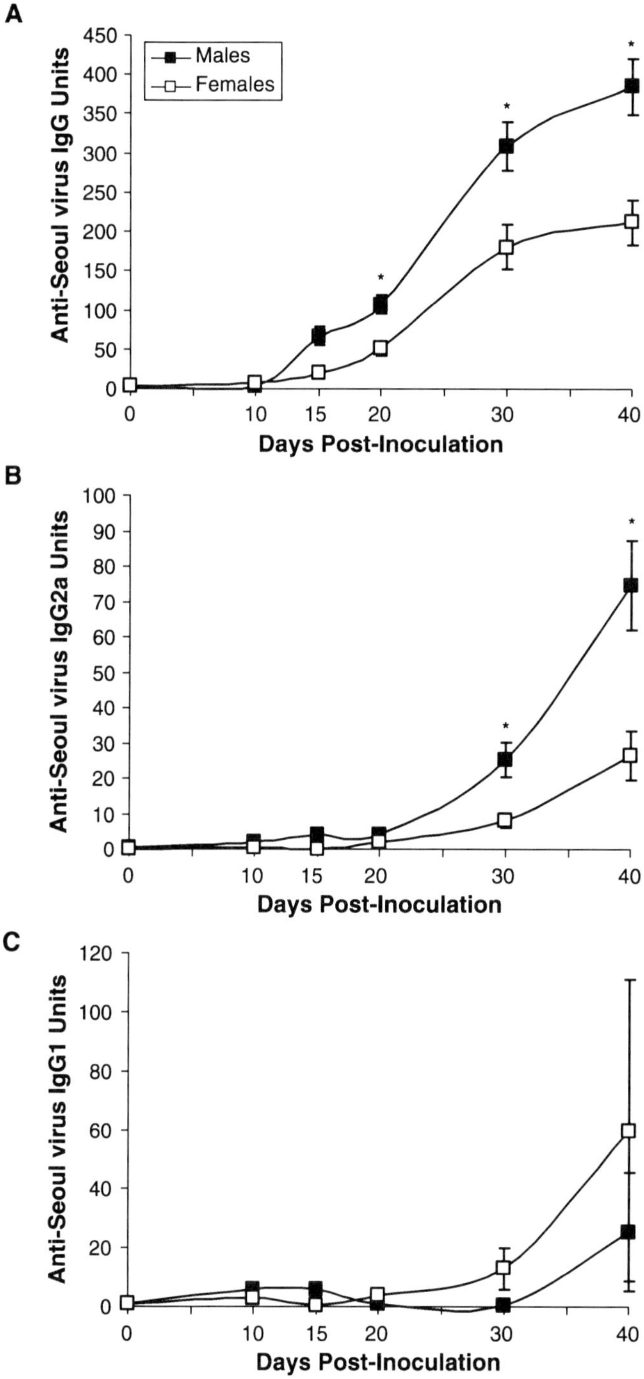
A
Males
Females
Anti-Seoul virus IgG Units
450
400
350
300
250
200
150
100
50
0
0
10
15
20
30
40
Days Post-Inoculation
B
Anti-Seoul virus IgG2a Units
100
90
80
70
60
50
40
30
20
10
0
0
10
15
20
30
40
Days Post-Inoculation
C
Anti-Seoul virus IgG1 Units
120
100
80
60
40
20
0
0
10
15
20
30
40
Days Post-Inoculation

IL-1RAcP), antiviral (e.g., IFNγR and Mx proteins), T cell (e.g., CD3 and TCR), and Ig superfamily (e.g., IgM, IgG, and MHC class I and II) proteins are higher in females than males (Klein et al. 2004a). Thus, females may be more efficient at transcribing genes that encode for immune responses against Seoul virus infection and that reduce virus replication and viral protein synthesis.

Mx proteins are induced by interferons and possess important antiviral properties. Human MxA and rodent Mx2, in particular, confer resistance against hantaviruses, including Seoul virus, Puumala virus, Hantaan virus, and Andes virus, in vitro (Jin et al. 2001; Khaiboullina et al. 2005; Temonen et al. 1995). The reduced expression of Mx genes in males and the increased expression of IFNγR and related transcriptional factors in females may contribute to increased virus shedding and virus replication in lung tissue observed among males compared with females (Klein et al. 2000, 2001, 2002b, 2004a). Conversely, males have higher expression of heat shock protein genes (e.g., hsp70) suggesting that cellular stress is elevated in males after infection (Klein et al. 2004a). Hantaviruses increase hsp70 levels in cells in vitro (Ye et al. 2001). Because heat shock proteins are molecular chaperones, sex differences in the production of heat shock proteins may affect how antigens are processed for recognition by the immune system and, in turn, viral persistence.

Manipulation of sex steroids in adulthood does not alter sex differences in antibody responses or virus shedding in rats following Seoul virus infection (Klein et al. 2000). Conversely, manipulation of sex steroid concentrations during perinatal development alters the sex difference in antibody responses against Seoul virus, but does not affect Seoul virus replication in target tissues, suggesting that differences in hantavirus infection may not be solely contingent on sex steroids (Klein et al. 2002b). The effects of sex steroids on innate responses to hantavirus infection, however, have not been reported and require additional investigation. Also, whether dimorphic production of stress-related hormones, including glucocorticoids, influences responses to hantavirus infection is currently under investigation.

5.3
Maternal Antibody Protects Offspring Against Hantaviruses

Rodents transmit hantaviruses horizontally; there is no evidence that hantaviruses are transmitted vertically, from mother to young. Females of all rodent species examined do, however, transfer antibodies against hantaviruses to their offspring (Bernshtein et al. 1999; Borucki et al. 2000; Childs et al. 1985; Dohmae et al. 1993; Dohmae and Nishimune 1995, 1998; Hutchinson et al. 2000; Zhang et al. 1988). Consequently, the age-dependent pathology caused by hantaviruses, such as Seoul virus, can be mitigated by transmission of antibodies against the virus from infected dams to their offspring. Cross-foster studies reveal that females transmit protective immunity against hantaviruses to offspring in utero and through breast milk. Specifically, offspring born to infected females and nursed by uninfected females as well as offspring born to uninfected females and nursed by infected females have high antibody concentrations, reduced virus replication in target tissues, and reduced mortality following challenge with Seoul virus on the day of birth (Dohmae et al. 1993; Dohmae and Nishimune 1995, 1998). Maternal immunity is transient and is only detectable in offspring for up to 8 weeks (Zhang et al. 1988). After this 8-week period, offspring are susceptible to Seoul virus infection; by 8 weeks of age, however, rats do not exhibit signs of disease-related pathology. Thus, maternal immunity persists until the young are resistant to the detrimental effects of Seoul virus infection.

Transfer of maternal immunity to offspring has also been reported in natural populations of rodents. Maternal antibodies appear to be transferred from infected female deer mice to offspring because antibody against Sin Nombre virus is prevalent among young deer mice and typically declines with age (Borucki et al. 2000; Mills et al. 1997). Although some weanling deer mice have detectable antibody against Sin Nombre virus, there is no detectable viral RNA in their organs, suggesting that these animals are not infected (Borucki et al. 2000). These data have been interpreted to illustrate that infected female mice transfer protective antibody to offspring but do not transmit virus vertically (Borucki et al. 2000; Mills et al. 1997). Similar observations also have been reported for female hispid cotton rats infected with Black Creek Canal virus and for wild-caught female Norway rats infected with Seoul virus (Childs et al. 1988; Glass et al. 1998). To date, there are no field studies examining whether offspring of infected females survive subsequent hantavirus infection better than do offspring of uninfected females. Because maternally derived antibody may increase the likelihood of offspring survival resistance to infection with hantaviruses later in life, future studies should consider whether these mechanisms contribute to persistence of hantaviruses, in the absence of pathology, among adult rodents.

5.4
Host Immune Responses to Hantaviruses

The role of host immunity against hantavirus infection is well established (Table 3). Natural killer (NK) cells and macrophages are more abundant in the respiratory tracts of patients infected with hantaviruses compared with those of control patients (Linderholm et al. 1993; Mori et al. 1999). Serum concentrations of proinflammatory cytokines, including TNF, IL-1, and IL-6, are elevated during the acute phase of hantavirus infection (Krakauer et al. 1995; Linderholm et al. 1996). In patients, immunohistochemical staining reveals that cytokine

Table 3 Immunological changes following exposure to hantaviruses in humans and rodent reservoirs

Immune Factor	Host	Tissue/cell	Hantavirus	Response to infection	Reference
MHC	Humans	PBMC	PUUV	Susceptibility alleles	Plyusnin et al. 1997
		Dendritic cells	HTNV	Upregulated	Raftery et al. 2002
Mx2	Mice	Fibroblasts	HTNV, SEOV	Prevents accumulation of virus	Jin et al. 2001
MxA	Humans	Monocytes	PUUV	Elevated, antiviral	Temonen et al. 1995
		Endothelial cells	ANDV	Elevated, antiviral	Khaiboullina et al. 2005
TNF	Humans	Endothelial cells, MΦ	SNV	Reduces virus	Khaiboullina et al. 2000
		Lung, spleen	SNV	Increased production	Mori et al. 1999
		Kidney	PUUV	Increased production	Temonen et al. 1996
	Monkeys	Blood	PUUV	Increased production	Klingstrom et al. 2002
IL-1β	Humans	Blood	PUUV	Reduced in NE patients	Makela et al. 2001
		Lung, spleen	PUUV	Increased production	Mori et al. 1999
IL-6	Humans	Blood, lung, spleen	PUUV, SNV	Increased production	Linderholm et al. 1996; Mori et al. 1999
	Monkeys	Blood	PUUV	Increased production	Klingstrom et al. 2002
IL-10	Humans	Blood	PUUV	Increased production	Linderholm et al. 1996
IL-2	Humans	Lung, spleen	SNV	Increased production	Mori et al. 1999
IFNα	Humans	Blood	HTNV	No change	Krakauer et al. 1994
			PUUV	Reduced virus growth	Temonen et al. 1995
IFNβ	Humans	Endothelial cells, MΦ	PUUV	Elevated, inhibits virus replication	Pensiero et al. 1992; Tamura et al. 1987
IFNγ	Humans	Blood, lung, spleen	HTNV, SNV	Increased production	Krakauer et al. 1994; Mori et al. 1999

	Rodents	Spleen	SEOV, HTNV	Increased production	Araki et al. 2003; Klein et al. 2001
Ccl5	Humans	Lung endothelial	SNV, HTNV	Increased in infected cells	Khaiboullina et al. 2004; Sundstrom et al. 2001
Cxcl10	Humans	Lung endothelial	SNV, HTNV	Increased in infected cells	Sundstrom et al. 2001;
IRF-3, IRF-7	Humans	Lung endothelial	SNV, HTNV	Increased nuclear translocation	Sundstrom et al. 2001
CD8+	Humans	Blood, lung,	DOBV, PUUV, HTNV	Elevated expression	Huang et al. 1994; Linderholm et al. 1993; Markotic et al. 1999; Zaki et al. 1995
	Rodents	Spleen	HTNV	Downregulation	Araki et al. 2003
CD4+CD25+	Humans	Blood	DOBV, PUUV	Elevated expression	Markotic et al. 1999
TGFβ	Humans	Kidney	PUUV	Elevated production	Temonen et al. 1996
β3 integrins	Humans	Endothelial cells	NYV, SNV, HTNV, SEOV, PUUV, PHV	Receptor for entry into cells	Mackow et al. 2001
ICAM-1, VCAM, PECAM	Humans	Kidney	PUUV	Elevated production	Temonen et al. 1996
IL-1RA	Humans	Blood	PUUV	Reduced susceptibility to NE	Makela et al. 2001
IgM, IgA, IgG, IgE	Mammals	Blood	All HFRS/ HPS hantaviruses	Neutralization, diagnostics	Khaiboullina and St Joer 2002; Vapalahti et al. 2001

(*Continued*)

Table 3 Immunological changes following exposure to hantaviruses in humans and rodent reservoirs—cont'd.

Immune Factor	Host	Tissue/cell	Hantavirus	Response to infection	Reference
E-selectin	Humans	Blood	PUUV	Elevated, sign of inflammation	Takala et al. 2000
CD11b	Humans	Blood	PUUV	Elevated, sign of inflammation	Takala et al. 2000
CD40, CD80, CD86	Humans	Dendritic cells	HTNV	Upregulated expression	Raferty et al. 2002
sIL-2R	Humans	Blood	PUUV	Elevated expression	Takala et al. 2000
Nitric oxide	Humans, monkeys	Blood	PUUV	Elevated production	Groeneveld et al. 1995; Klingstrom et al. 2002
Creatinine	Humans	Blood	PUUV	Elevated production	Groeneveld et al. 1995; Takala et al. 2000
C-reactive protein	Humans	Blood	PUUV	Elevated production	Takala et al. 2000
hsp70	Monkeys, rodents	Kidney, lung	HTNV, SEOV	Elevated expression	Klein 2004a; Ye et al. 2001
	Humans	Endothelial cells	SNV	Elevated expression	Khaiboullina et al. 2004

ANDV Andes virus; *DOBV* Dobrava virus; *HTNV* Hantaan virus; *NYV* New York virus; *PHV* Prospect Hill virus; *PUUV* Puumala virus; *SEOV* Seoul virus; *SNV* Sin Nombre virus; *HFRS* hemorrhagic fever with renal syndrome; *HPS* hantavirus pulmonary syndrome

producing cells are recruited to the sites of virus infection, and that cytokines, including IL-1α, IL-1β, IL-2, IL-4, IL-6, lymphotoxin, TNF, and IFNα may cause capillary leakage and pathology seen in HPS (Mori et al. 1999). Polymorphisms of cytokine genes in patients infected with Puumala virus are associated with severity of illness, suggesting a host genetic basis for viral pathogenesis (Makela et al. 2001). Concentrations of chemokines, including Ccl5 (i.e., RANTES) and Cxcl10 (i.e., IP-10;10-kDa interferon-inducible protein), are elevated in human endothelial cells within 3 days after exposure to hantaviruses in vitro (Sundstrom et al. 2001). Expression of anti-viral proteins, including IFNβ, IFNγ, and Mx proteins, is upregulated during the acute phase of infection (Jin et al. 2001; Pensiero et al. 1992; Tamura et al. 1987; Temonen et al. 1995). Little is known about the immune-related genes and proteins that constrain infection with Seoul virus (see Klein et al. 2004a), Sin Nombre virus, or other hantaviruses in their rodent reservoirs. There is also scarce information about how cytokine or other immune responses may influence persistence of hantaviruses in rodents. To date, most studies examining hantavirus-induced changes in cytokine and chemokine synthesis have been conducted using cells from patients exposed to hantaviruses or cell lines exogenously stimulated with hantaviruses in vitro (Table 3).

As with most viruses, antibody production is initiated at the onset of hantaviral infection and persists for the duration of infection in humans and in rodent reservoirs (Table 3). Although neutralizing antibody is important for reducing plasma viremia, antibody does not eliminate virus replication in tissue. T cell-mediated immunity is important for elimination of hantaviruses in tissue (Table 3) (Asada et al. 1987; Vapalahti et al. 2001). Both cytotoxic T cells and helper T cells (Th) are involved in the control of hantavirus replication (Asada et al. 1987; Vapalahti et al. 2001). Patients diagnosed with hantavirus infections have elevated Th1 responses (i.e., IFNγ and IgG3) early during infection and elevated Th2 responses (i.e., IL-6, IL-10, and IgG1) during later phases of infection, at least in response to Puumala and Hantaan viruses (Groen et al. 1994; Krakauer et al. 1995; Linderholm et al. 1996; Lundkvist et al. 1993). After Seoul virus infection in rats, splenic IFNγ and IL-4 production increase in both sexes. Males, however, have higher Th1 immune responses (i.e., IgG2a, IFNγ, and IL-2) than females; in contrast, Th2 immune responses (i.e., IgG1, IL-4, and IL-10) are similar between the sexes (Klein et al. 2001). In humans, vigorous Th1 responses are correlated with high virus load and increased risk for severe Puumala virus infection (Vapalahti et al. 2001). Elevated cell-mediated responses to Sin Nombre virus are hypothesized to cause capillary leakage in humans (Mori et al. 1999).

Despite the presence of effector immune responses, rodents can remain persistently infected with hantaviruses (Meyer and Schmaljohn 2000b). The mechanisms mediating hantavirus persistence are not completely understood, but may involve virus evasion of host immunity or suppression of host immunity by the virus (Araki et al. 2003; Meyer and Schmaljohn 2000b). Arenaviruses, including lymphocytic choriomeningitis virus, cause immunosuppression, specifically downregulation of CD8+ T cells, in their rodent hosts (Moskophidis et al. 1993). Newborn BALB/c mice infected with Hantaan virus, and which become persistently infected, exhibit downregulation of IFNγ-producing CD8+ T cells, suggesting that hantaviruses, like arenaviruses, may induce immunosuppression in rodent hosts (Araki et al. 2003). Whether these mechanisms are involved in persistence of hantavirus infection of rodent reservoir hosts has not been reported. Comparisons of immune responses between humans (for whom hantaviruses are pathogenic) and rodents (for whom hantaviruses are nonpathogenic) may provide insight into the reasons hantavirus infections can be fatal in humans, but only cause persistent, subclinical infections in rodents. Examination of the effects of Sin Nombre virus (i.e., a pathogenic hantavirus) and Prospect Hill virus (i.e., an apparently nonpathogenic hantavirus) on gene expression profiles in human endothelial cells reveals that Sin Nombre virus causes increased cellular transcription compared with Prospect Hill virus (Khaiboullina et al. 2004). With increased availability of reagents and genetic and protein sequences for rodent hosts (Blanco et al. 2004; Schountz et al. 2004), cross-host species analyses of gene expression profiles in response to hantaviruses should be conducted (Klein et al. 2004a).

6
Transmission of Hantaviruses from Rodents to Humans

With an HPS case-fatality rate of approximating 40% in humans infected with Sin Nombre virus and the possibility that this, or other hantaviruses, could be used as biowarfare weapons, it is important for us to understand the mechanisms by which hantaviruses persist in and are transmitted between rodents. Our current understanding of these mechanisms is rudimentary; we do, however, know that physiological and environmental factors interact and contribute to hantavirus prevalence in natural rodent populations. Many field workers (mammalogists, ranchers, electricians, plumbers, and others who work outdoors or in closely confined quarters) have been exposed to aerosolized virus in rodent feces, urine, and saliva, yet HPS is a rather rare occurrence and incidence rates vary not only from state to state but from location to location (Douglass et al. 2005). Thus, acquisition of a hantavirus infection may simply be a question of probability. Alternatively, epidemiologic investigations have illustrated that victims of hantavirus exposure and infection engaged in risky activities or behaviors, including working in spaces without adequate ventilation, working without adequate respiratory protection, failure to reduce aerosol content of the work area, or living or working in a space infested with rodents. Increased education about the risk factors associated with acquiring hantavirus infection may reduce spread of infection from rodents to humans. A complementary approach is to increase our understanding of the environmental and physiological factors that mediate persistence and transmission of zoonotic agents within rodent populations.

7 Conclusions and Future Directions

Hantaviruses cause persistent infections in rodent reservoirs, which directly impact transmission of these zoonotic agents between rodents and from rodents to humans. As noted throughout this chapter, environmental (e.g., precipitation, food availability, habitat, and ambient temperature), demographic (e.g., age and sex), and physiological (e.g., hormones and immune responses) factors likely interact to mediate responses to hantaviruses in natural rodent populations.

Changes in the environment may impact the prevalence of hantaviruses in natural populations of rodents by affecting population structure and size. Seasonal fluctuations in rodent populations and, hence, hantavirus prevalence, are dependent on weather conditions, plant productivity (e.g., total mass, type of plants, plant maturation, and general plant health), soil conditions, availability of alternative food sources, habitat destruction, disastrous meteorological events, fires, and other factors. As shown in Fig. 4, deer mouse populations fluctuate over time, as does prevalence of antibody to Sin Nombre virus (C.H. Calisher and B.J. Beaty, unpublished data). Specifically, in a study of deer mice at one site in Colorado from 1999 to 2004, population peaks occurred in summer 1999, fall 2003 and fall 2004; conversely, major peaks in antibody prevalence occurred in spring 1999, 2002, and 2004. Increases in deer mouse population densities have a 12- 18-month delayed dependency on precipitation totals (data not shown). These data illustrate the complexity of the trophic cascade leading from increased precipitation to increased virus prevalence. Although considerable progress has been made with regard to Sin Nombre virus infection of deer mice, how precipitation and subsequent changes in food and habitat affect host responses to other hantaviruses has not been reported. Future longitudinal studies (i.e., studies over time and with other hantavirus–host systems),

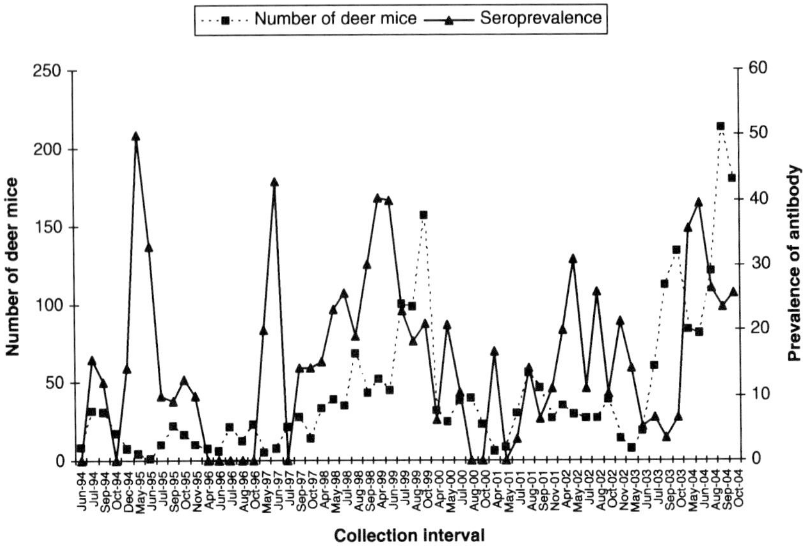

Fig. 4 Annual fluctuations in numbers of deer mice and proportion of deer mice (*Peromyscus maniculatus*) with antibodies against Sin Nombre virus at a montane shrubland site in southwestern Colorado from 1999 to 2004 [14]. (C.H. Calisher and B.J. Beaty, unpublished data)

must be conducted so that the impact of the trophic cascade on the prevalence of hantaviruses in many geographic areas can be recognized. Thus, the trophic cascade hypothesis illustrates that factors associated with biodiversity can impact host populations and, presumably, affect persistence and transmission of hantaviruses.

Although speculative, with the improvement of molecular biological tools for rodents other than rats and mice, future studies should evaluate the possibility that specific genes, groups of genes, or even allelic forms of genes, may impact responses to hantaviruses in natural rodent populations. Genetic polymorphisms typically are examined in relation with disease susceptibility (Wright et al. 1999). Whether genetic polymorphisms in host populations affect persistence or shedding of hantaviruses has not been adequately addressed. Studies of gene flow in rodent populations illustrate that genetic bottlenecks may result in changes in susceptibility of deer mice to Sin Nombre virus within certain populations (Root et al 2003). Additionally, utilization of microsatellite and mitochondrial DNA markers in populations of grey red-backed voles (*Clethrionomys rufocanus bedfordiae*) reveals that kinship does not predict hantavirus transmission (Iwasa et al. 2004). Male grey red-backed voles infected with this Puumala-related hantavirus, however, share an uncommon mitochondrial haplotype, which may explain the increased prevalence of hantaviruses in male rodents.

Host age as well as sex appear to play a pivotal role in predicting who within a rodent population is likely to be infected with hantaviruses. Generally, within rodent host populations, adult males are most likely to be infected with their species-specific hantavirus. Whether age- and sex-related differences in hantavirus infections represent differences in exposure to infection, susceptibility to infection, or both requires further examination. The role of behaviors, such as aggression, in the transmission and maintenance of hantaviruses in rodent communities requires additional investigation. Cross-species analyses may assist with establishing generalities, as well as distinctions, among rodent–hantavirus systems.

Age-related differences in the prevalence of hantavirus infections may be associated with the presence of maternal immunity and the age of an individual at the time of exposure to hantaviruses. As noted previously in this chapter, infected dams do not pass virus to offspring, which are protected by maternal antibody. As maternal antibody wanes, the offspring become susceptible to the hantavirus with which they are co-evolutionarily associated. In contrast, offspring from an uninfected dam may become infected as juveniles, possibly by an infectious (i.e., shedding virus) male entering the nest. Whether offspring that do not possess protective maternal antibodies, but that become infected prior to puberty, are more or less susceptible to infection or are more or less likely to be persistently infected has not been established. Finally, if neither dam nor sire is infected with their species-specific hantavirus and the pups become infected after weaning, then the offspring may either persistently shed virus or clear infection. These possible scenarios illustrate that the outcome of infection with hantaviruses is dependent on maternal infection status and the age of offspring at the time of infection. The notion that exposure to maternal antibody and/or virus early in life may influence responses to infection later in life has not been adequately considered in either field or laboratory studies.

Whether hantaviral persistence is mediated by changes in virus replication, the effects of hantaviruses on host immunity, the ability of hantaviruses to evade host immunity, or a combination of factors requires additional investigation. Future studies should continue to examine rodent host immune responses to hantaviruses to better understand the mechanisms that contribute to viral persistence. Whether hantaviruses, like arenaviruses, cause immunosuppression in rodent hosts has not been reported for natural populations of rodents. Comparisons of human and rodent immune responses to hantaviruses may provide insights into why hantaviruses often are fatal for humans but cause only persistent infection, in the absence of disease, in rodents.

References

Abbott KD, Ksiazek TG, Mills JN (1999) Long-term hantavirus persistence in rodent populations in central Arizona. Emerg Infect Dis 5:102–112

Ahlm C, Alexeyev OA, Elgh F, Aava B, Wadell G, Tarnvik A, Juto P, Palo T (1997) High prevalence of hantavirus antibodies in bank voles (*Clethrionomys glareolus*) captured in the vicinity of households afflicted with nephropathia epidemica. Am J Trop Med Hyg 56:674–678

Araki K, Yoshimatsu K, Lee BH, Kariwa H, Takashima I, Arikawa J (2003) Hantavirus-specific CD8(+)-T-cell responses in newborn mice persistently infected with Hantaan virus. J Virol 77:8408–8417

Asada H, Tamura M, Kondo K, Okuno Y, Takahashi Y, Dohi Y, Nagai T, Kurata T, Yamanishi K (1987) Role of T lymphocyte subsets in protection and recovery from Hantaan virus infection in mice. J Gen Virol 68:1961–1969

Avsic-Zupanc T, Xiao SY, Stojanovic R, Gligic A, van der Groen G, LeDuc JW (1992) Characterization of Dobrava virus: a Hantavirus from Slovenia Yugoslavia. J Med Virol 38:132–137

Baek LJ, Kariwa H, Lokugamage K, Yoshimatsu K, Arikawa J, Takashima I, Kang JI, Moon SS, Chung SY, Kim EJ, Kang HJ, Song KJ, Klein TA, Yanagihara R, Song JW (2006) Soochong virus: an antigenically and genetically distinct hantavirus isolated from Apodemus peninsulae in Korea. J Med Virol 78:290–297

Bernshtein AD, Apekina NS, Mikhailova TV, Myasnikov YA, Khlyap LA, Korotkov YS, Gavrilovskaya IN (1999) Dynamics of Puumala hantavirus infection in naturally infected bank voles (*Clethrionomys glareolus*). Arch Virol 144:2415–2428

Bharadwaj M, Botten J, Torrez-Martinez N, Hjelle B (1997) Rio Mamore virus: genetic characterization of a newly recognized hantavirus of the pygmy rice rat *Oligoryzomys microtis*, from Bolivia. Am J Trop Med Hyg 57:368–374

Biggs JR, Bennett KD, Mullen MA, Haarman TK, Salisbury M, Robinson RJ, Keller D, Torrez-Martinez N, Hjelle B (2000) Relationship of ecological variables to Sin Nombre virus antibody seroprevalence in populations of deer mice. J Mammal 81:676–682

Blanco JC, Pletneva L, Boukhvalova M, Richardson JY, Harris KA, Prince GA (2004) The cotton rat: an underutilized animal model for human infectious diseases can now be exploited using specific reagents to cytokines, chemokines, and interferons. J Interferon Cytokine Res 24:21–28

Borucki MK, Boone JD, Rowe JE, Bohlman MC, Kuhn EA, DeBaca R, St Jeor SC (2000) Role of maternal antibody in natural infection of *Peromyscus maniculatus* with Sin Nombre virus. J Virol 74:2426–2429

Botten J, Mirowsky K, Kusewitt D, Bharadwaj M, Yee J, Ricci R, Feddersen RM, Hjelle B (2000) Experimental infection model for Sin Nombre hantavirus in the deer mouse (*Peromyscus maniculatus*). Proc Natl Acad Sci U S A 97:10578–10583

Botten J, Mirowsky K, Ye C, Gottlieb K, Saavedra M, Ponce L, Hjelle B (2002) Shedding and intracage transmission of Sin Nombre hantavirus in the deer mouse (*Peromyscus maniculatus*) model. J Virol 76:7587–7594

Brummer-Korvenkontio M, Vaheri A, Hovi T, von Bonsdorff CH, Vuorimies J, Manni T, Penttinen K, Oker-Blom N, Lahdevirta J (1980) Nephropathia epidemica: detection

of antigen in bank voles and serologic diagnosis of human infection. J Infect Dis 141:131–134
Calisher CH, Sweeney W, Mills JN, Beaty BJ (1999) Natural history of Sin Nombre virus in western Colorado. Emerg Infect Dis 5:126–134
Calisher CH, Childs JE, Sweeney WP, Canestrop KM, Beaty BJ (2000) Dual captures of Colorado rodents: implications for transmission of hantaviruses. Emerg Infect Dis 6:363–369
Calisher CH, Mills JN, Root JJ, Beaty BJ (2003) Hantaviruses: etiologic agents of rare, but potentially life-threatening zoonotic diseases. J Am Vet Med Assoc 222:163–166
Calisher CH, Mills JN, Sweeney W, Root JJ, Reeder SA, Jentes ES, Beaty BJ (2005a) Population dynamics of a diverse rodent assemblage in mixed grass-shrub habitat, southeastern Colorado, 1995–2000. J Wildl Dis 41:12–28
Calisher CH, Root JJ, Mills JN, Rowe JE, Reeder SA, Jentes ES, Wagoner K, Beaty BJ (2005b) Epizootiology of Sin Nombre and El Moro Canyon hantaviruses, southeastern Colorado, 1995–2000. J Wildl Dis 41:1–11
Carey DE, Reuben R, Panicker KN, Shope RE, Myers RM (1971) Thottapalayam virus: a presumptive arbovirus isolated from a shrew in India. Ind J Med Res 59:1758–1760
Childs JE, Korch GW, Smith GA, Terry AD, LeDuc JW (1985) Geographical distribution and age related prevalence of antibody to Hantaan-like virus in rat populations of Baltimore Maryland USA. Am J Trop Med Hyg 34:385–387
Childs JE, Korch GW, Glass GE, Leduc JW, Shah KV (1987) Epizootiology of Hantavirus infections in Baltimore: isolation of a virus from Norway rats, and characteristics of infected rat populations. Am J Epidemiol 126:55–68
Childs JE, Glass GE, Korch GW, LeDuc JW (1988) The ecology and epizootiology of hantaviral infections in small mammal communities of Baltimore: a review and synthesis. Bull Soc Vector Biol 13:113–122
Childs JE, Ksiazek TG, Spiropoulou CF, Krebs JW, Morzunov S, Maupin GO, Gage KL, Rollin PE, Sarisky J, Enscore RE, Frey JK, Peters CJ, Nichol ST (1994) Serologic and genetic identification of *Peromyscus maniculatus* as the primary rodent reservoir for a new hantavirus in the southwestern United States. J Infect Dis 169:1271–1280
Childs JE, Krebs JW, Ksiazek TG, Maupin GO, Gage KL, Rollin PE, Zeitz PS, Sarisky J, Enscore RE, Butler JC, Peters CJ (1995) A household-based, case-control study of environmental factors associated with hantavirus pulmonary syndrome in the southwestern United States. Am J Trop Med Hyg 52:393–397
Demas GE, DeVries AC, Nelson RJ (1997) Effects of photoperiod and 2-deoxy-D-glucose-induced metabolic stress on immune function in female deer mice. Am J Physiol. 272:R1762–R1767
Dohmae K, Nishimune Y (1995) Protection against hantavirus infection by dam's immunity transferred vertically to neonates. Arch Virol 140:165–172
Dohmae K, Nishimune Y (1998) Maternal transfer of Hantavirus antibodies in rats. Lab Anim Sci 48:395–397
Dohmae K, Koshimizu U, Nishimune Y (1993) In utero and mammary transfer of hantavirus antibody from dams to infant rats. Lab Anim Sci 43:557–561
Douglass RJ, Calisher CH, Bradley KC (2005) State-by-state incidences of hantavirus pulmonary syndrome in the United States of America, 1993–2004. Vect Borne Zoon Dis 5:189–192

Elliott RM, Bouloy M, Calisher CH, Goldbach R, Moyer JT, Nichol ST, Pettersson R, Plyusnin A, Schmaljohn CS (2000) Bunyaviridae. In: Van REgenmortel MHV, Fauquet CM, Bishop DHL, Carstens EB, Estes MK, Lemon SM, Maniloff J, Mayo MA, McGeoch DJ, Pringle CR, Wickner RB (eds) Seventh Report of the International Committee on Taxonomy of Viruses. Academic, San Diego, pp 599–621

Elwell MR, Ward GS, Tingpalapong M, LeDuc JW (1985) Serologic evidence of Hantaan-like virus in rodents and man in Thailand. Southeast Asian J Trop Med Public Health 16:349–354

Escutenaire S, Chalon P, De Jaegere F, Karelle-Bui L, Mees G, Brochier B, Rozenfeld F, Pastoret PP (2002) Behavioral, physiologic, and habitat influences on the dynamics of Puumala virus infection in bank voles (*Clethrionomys glareolus*). Emerg Infect Dis 8:930–936

Flynn KM, Ferguson SA, Delclos KB, Newbold RR (2000) Effects of genistein exposure on sexually dimorphic behaviors in rats. Toxicol Sci 55:311–319

Fritz WA, Wang J, Eltoum IE, Lamartiniere CA (2002) Dietary genistein down-regulates androgen and estrogen receptor expression in the rat prostate. Mol Cell Endocrinol 186:89–99

Fulhorst CF, Monroe MC, Salas RA, Duno G, Utrera A, Ksiazek TG, Nichol ST, de Manzione NM, Tovar D, Tesh RB (1997) Isolation, characterization and geographic distribution of Cano Delgadito virus, a newly discovered South American hantavirus (family Bunyaviridae). Virus Res 51:159–171

Fulhorst CF, Cajimat MN, Utrera A, Milazzo ML, Duno GM (2004) Maporal virus, a hantavirus associated with the fulvous pygmy rice rat (*Oligoryzomys fulvescens*) in western Venezuela. Virus Res 104:139–144

Glass GE, Childs JE, Korch GW, LeDuc JW (1988) Association of intraspecific wounding with hantaviral infection in wild rats (*Rattus norvegicus*). Epidemiol Infect 101:459–472

Glass GE, Watson AJ, Leduc JW, Kelen GD, Quinn TC, Childs JE (1993) Infection with a ratborne hantavirus in US residents is consistently associated with hypertensive renal disease. J Infect Dis 167:614–620

Glass GE, Livingstone W, Mills JN, Hlady WG, Fine JB, Biggler W, Coke T, Frazier D, Atherley S, Rollin PE, Ksiazek TG, Peters CJ, Childs JE (1998) Black Creek Canal Virus infection in *Sigmodon hispidus* in southern Florida. Am J Trop Med Hyg 59:699–703

Glass GE, Yates TL, Fine JB, Shields TM, Kendall JB, Hope AG, Parmenter CA, Peters CJ, Ksiazek TG, Li CS, Patz JA, Mills JN (2002) Satellite imagery characterizes local animal reservoir populations of Sin Nombre virus in the southwestern United States. Proc Natl Acad Sci U S A 99:16817–16822

Groen J, Gerding M, Jordans JG, Clement JP, Osterhaus AD (1994) Class and subclass distribution of Hantavirus-specific serum antibodies at different times after the onset of nephropathia epidemica. J Med Virol 43:39–43

Groeneveld PH, Colson P, Kwappenberg KM, Clement J (1995) Increased production of nitric oxide in patients infected with the European variant of hantavirus. Scand J Infect Dis 27:453–456

Hinson ER, Shone SM, Zink MC, Glass GE, Klein SL (2004) Wounding: the primary mode of Seoul virus transmission among male Norway rats. Am J Trop Med Hyg 70:310–317

Hjelle B, Glass GE (2000) Outbreak of hantavirus infection in the Four Corners region of the United States in the wake of the 1997–1998 El Nino-southern oscillation. J Infect Dis 181:1569–1573
Hjelle B, Chavez-Giles F, Torrez-Martinez N, Yates T, Sarisky J, Webb J, Ascher M (1994) Genetic identification of a novel hantavirus of the harvest mouse *Reithrodontomys megalotis*. J Virol 68:6751–6754
Hjelle B, Anderson B, Torrez-Martinez N, Song W, Gannon WL, Yates TL (1995a) Prevalence and geographic genetic variation of hantaviruses of New World harvest mice (*Reithrodontomys*): identification of a divergent genotype from a Costa Rican *Reithrodontomys mexicanus*. Virology 207:452–459
Hjelle B, Jenison SA, Goade DE, Green WB, Feddersen RM, Scott AA (1995b) Hantaviruses: clinical, microbiologic, and epidemiologic aspects. Crit Rev Clin Lab Sci 32:469–508
Horling J, Chizhikov V, Lundkvist A, Jonsson M, Ivanov L, Dekonenko A, Niklasson B, Dzagurova T, Peters CJ, Tkachenko E, Nichol S (1996) Khabarovsk virus: a phylogenetically and serologically distinct hantavirus isolated from *Microtus fortis* trapped in far-east Russia. J Gen Virol 77:687–694
Huang C, Jin B, Wang M, Li E, Sun C (1994) Hemorrhagic fever with renal syndrome: relationship between pathogenesis and cellular immunity. J Infect Dis 169:868–870
Hung T, Xia SM, Zhao TX, Zhou JY, Song G, Liao GX, Ye WW, Chu YL, Hang CS (1983) Morphological evidence for identifying the viruses of hemorrhagic fever with renal syndrome as candidate members of the Bunyaviridae family. Arch Virol 78:137–144
Hutchinson KL, Rollin PE, Peters CJ (1998) Pathogenesis of a North American hantavirus Black Creek Canal virus, in experimentally infected *Sigmodon hispidus*. Am J Trop Med Hyg 59:58–65
Hutchinson KL, Rollin PE, Shieh WJ, Zaki S, Greer PW, Peters CJ (2000) Transmission of Black Creek Canal virus between cotton rats. J Med Virol 60:70–76
Iwasa MA, Kariwa H, Cui BZ, Lokugamage K, Lokugamage N, Hagiya T, Mizutani T, Takashima I (2004) Modes of hantavirus transmission in a population of Clethrionomys rufocanus bedfordiae inferred from mitochondrial and microsatellite DNA analyses. Arch Virol 149:929–941
Jin HK, Yoshimatsu K, Takada A, Ogino M, Asano A, Arikawa J, Watanabe T (2001) Mouse Mx2 protein inhibits hantavirus but not influenza virus replication. Arch Virol 146:41–49
Johnson AM, Bowen MD, Ksiazek TG, Williams RJ, Bryan RT, Mills JN, Peters CJ, Nichol ST (1997) Laguna Negra virus associated with HPS in western Paraguay and Bolivia. Virology 238:115–127
Johnson AM, de Souza LT, Ferreira IB, Pereira LE, Ksiazek TG, Rollin PE, Peters CJ, Nichol ST (1999) Genetic investigation of novel hantaviruses causing fatal HPS in Brazil. J Med Virol 59:527–535
Johnson KM (2001) Hantaviruses: history and overview. Curr Top Microbiol Immunol 256:1–14
Kariwa H, Kimura M, Yoshizumi S, Arikawa J, Yoshimatsu K, Takashima I, Hashimoto N (1996) Modes of Seoul virus infections: persistency in newborn rats and transiency in adult rats. Arch Virol 141:2327–2338

Kariwa H, Fujiki M, Yoshimatsu K, Arikawa J, Takashima I, Hashimoto N (1998) Urine-associated horizontal transmission of Seoul virus among rats. Arch Virol 143:365–374

Kawamura K, Zhang XK, Arikawa J, Takashima I, Dempo K, Hashimoto N (1991) Susceptibility of laboratory and wild rodents to *Rattus* or *Apodemus* -type hantaviruses. Acta Virol 35:54–63

Khaiboullina SF, St Jeor SC (2002) Hantavirus immunology. Viral Immunol 15:609–625

Khaiboullina SF, Netski DM, Krumpe P, St Jeor SC (2000) Effects of tumor necrosis factor alpha on Sin Nombre virus infection in vitro. J Virol 74:11966–11971

Khaiboullina SF, Rizvanov AA, Otteson E, Miyazato A, Maciejewski J, St Jeor S (2004) Regulation of cellular gene expression in endothelial cells by Sin Nombre and Prospect Hill viruses. Viral Immunol 17:234–251

Khaiboullina SF, Rizvanov AA, Deyde VM, St Jeor SC (2005) Andes virus stimulates interferon-inducible MxA protein expression in endothelial cells. J Med Virol 75:267–275

Klein SL (2000) The effects of hormones on sex differences in infection: from genes to behavior. Neurosci Biobehav Rev 24:627–638

Klein SL (2003) Parasite manipulation of the proximate mechanisms that mediate social behavior in vertebrates. Physiol Behav 79:441–449

Klein SL (2004) Hormonal and immunological mechanisms mediating sex differences in parasite infection. Parasite Immunol 26:247–264

Klein SL, Bird BH, Glass GE (2000) Sex differences in Seoul virus infection are not related to adult sex steroid concentrations in Norway rats. J Virol 74:8213–8217

Klein SL, Bird BH, Glass GE (2001) Sex differences in immune responses and viral shedding following Seoul virus infection in Norway rats. Am J Trop Med Hyg 65:57–63

Klein SL, Bird BH, Nelson RJ, Glass GE (2002a) Environmental and physiological factors associated with Seoul virus infection among urban populations of Norway rats. J Mammal 83:478–488

Klein SL, Marson AL, Scott AL, Ketner G, Glass GE (2002b) Neonatal sex steroids affect responses to Seoul virus infection in male but not female Norway rats. Brain Behav Immun 16:736–746

Klein SL, Wisniewski AB, Marson AL, Glass GE, Gearhart JP (2002c) Early exposure to genistein exerts long-lasting effects on the endocrine and immune systems in rats. Mol Med 8:742–749

Klein SL, Cernetich A, Hilmer S, Hoffman EP, Scott AL, Glass GE (2004a) Differential expression of immunoregulatory genes in male and female Norway rats following infection with Seoul virus. J Med Virol 74:180–190

Klein SL, Zink MC, Glass GE (2004b) Seoul virus infection increases aggressive behaviour in male Norway rats. Anim Behav 67:421–429

Klempa B, Fichet-Calvet E, Lecompte E, Auste B, Ainiskin V, Meisel H, Denys C, Koivogui L, ter Meulen J, Kruger DH (2006) Hantavirus in African Wood Mouse Guinea. Emerg Infect Dis 12:838–840

Klingstrom J, Plyusnin A, Vaheri A, Lundkvist A (2002) Wild-type Puumala hantavirus infection induces cytokines C-reactive protein, creatinine, and nitric oxide in cynomolgus macaques. J Virol 76:444–449

Krakauer T, LeDuc JW, Morrill JC, Anderson AO, Krakauer H (1994) Serum levels of alpha and gamma interferons in hemorrhagic fever with renal syndrome. Viral Immunol 7:97–101
Krakauer T, Leduc JW, Krakauer H (1995) Serum levels of tumor necrosis factor-alpha, interleukin-1, and interleukin-6 in hemorrhagic fever with renal syndrome. Viral Immunol 8:75–79
Lee HW, Lee PW, Johnson KM (1978) Isolation of the etiologic agent of Korean Hemorrhagic fever. J Infect Dis 137:298–308
Lee HW, Lee PW, Baek LJ, Song CK, Seong IW (1981) Intraspecific transmission of Hantaan virus, etiologic agent of Korean hemorrhagic fever, in the rodent *Apodemus agrarius*. Am J Trop Med Hyg 30:1106–1112
Lee HW, Baek LJ, Johnson KM (1982) Isolation of Hantaan virus, the etiologic agent of Korean hemorrhagic fever, from wild urban rats. J Infect Dis 146:638–644
Lee JS (1991) Clinical features of hemorrhagic fever with renal syndrome in Korea. Kidney Int Suppl 35: S88–S93
Lee PW, Amyx HL, Gajdusek DC, Yanagihara RT, Goldgaber D, Gibbs CJ Jr (1982) New hemorrhagic fever with renal syndrome-related virus in rodents in the United States. Lancet 2:1405
Levis S, Morzunov SP, Rowe JE, Enria D, Pini N, Calderon G, Sabattini M, St Jeor SC (1998) Genetic diversity and epidemiology of hantaviruses in Argentina. J Infect Dis 177:529–538
Linderholm M, Bjermer L, Juto P, Roos G, Sandstrom T, Settergren B, Tarnvik A (1993) Local host response in the lower respiratory tract in nephropathia epidemica. Scand J Infect Dis 25:639–646
Linderholm M, Ahlm C, Settergren B, Waage A, Tarnvik A (1996) Elevated plasma levels of tumor necrosis factor (TNF)-alpha, soluble TNF receptors, interleukin (IL)-6, and IL-10 in patients with hemorrhagic fever with renal syndrome. J Infect Dis 173:38–43
Lopez N, Padula P, Rossi C, Miguel S, Edelstein A, Ramirez E, Franze-Fernandez MT (1997) Genetic characterization and phylogeny of Andes virus and variants from Argentina and Chile. Virus Res 50:77–84
Lundkvist A, Bjorsten S, Niklasson B (1993) Immunoglobulin G subclass responses against the structural components of Puumala virus. J Clin Microbiol 31:368–732
Mackow ER, Gavrilovskaya IN (2001) Cellular receptors and hantavirus pathogenesis. Curr Top Microbiol Immunol 256:91–115
Makela S, Hurme M, Ala-Houhala I, Mustonen J, Koivisto AM, Partanen J, Vapalahti O, Vaheri A, Pasternack A (2001) Polymorphism of the cytokine genes in hospitalized patients with Puumala hantavirus infection. Nephrol Dial Transplant 16:1368–1673
Markotic A, Dasic G, Gagro A, Sabioncello A, Rabatic S, Kuzman I, Zgorelec R, Smoljan I, Beus I, Zupanc TA, Dekaris D (1999) Role of peripheral blood mononuclear cell (PBMC) phenotype changes in the pathogenesis of haemorrhagic fever with renal syndrome (HFRS). Clin Exp Immunol 115:329–334
McCormick JB, Sasso DR, Palmer EL, Kiley MP (1982) Morphological identification of the agent of Korean haemorrhagic fever (Hantaan virus) as a member of the Bunyaviridae. Lancet 1:765–768

Meyer BJ, Schmaljohn CS (2000a) Accumulation of terminally deleted RNAs may play a role in Seoul virus persistence. J Virol 74:1321–1331
Meyer BJ, Schmaljohn CS (2000b) Persistent hantavirus infections: characteristics and mechanisms. Trends Microbiol 8:61–67
Mills JN, Ksiazek TG, Ellis BA, Rollin PE, Nichol ST, Yates TL, Gannon WL, Levy CE, Engelthaler DM, Davis T, Tanda DT, Frampton JW, Nichols CR, Peters CJ, Childs JE (1997) Patterns of association with host and habitat: antibody reactive with Sin Nombre virus in small mammals in the major biotic communities of the southwestern United States. Am J Trop Med Hyg 56:273–284
Mills JN, Johnson JM, Ksiazek TG, Ellis BA, Rollin PE, Yates TL, Mann MO, Johnson MR, Campbell ML, Miyashiro J, Patrick M, Zyzak M, Lavender D, Novak MG, Schmidt K, Peters CJ, Childs JE (1998) A survey of hantavirus antibody in small-mammal populations in selected United States National Parks. Am J Trop Med Hyg 58:525–532
Mills JN, Ksiazek TG, Peters CJ, Childs JE (1999) Long-term studies of Hanta virus Reservoir populations in the southwestern United States: a synthesis. Emerg Infect Dis 5:135–142
Moore J (2002) Parasites and the behavior of animals. Oxford University Press, Oxford
Mori M, Rothman AL, Kurane I, Montoya JM, Nolte KB, Norman JE, Waite DC, Koster FT, Ennis FA (1999) High levels of cytokine-producing cells in the lung tissues of patients with fatal hantavirus pulmonary syndrome. J Infect Dis 179:295–302
Morzunov SP, Feldmann H, Spiropoulou CF, Semenova VA, Rollin PE, Ksiazek TG, Peters CJ, Nichol ST (1995) A newly recognized virus associated with a fatal case of hantavirus pulmonary syndrome in Louisiana. J Virol 69:1980–1983
Morzunov SP, Rowe JE, Ksiazek TG, Peters CJ, St Jeor SC, Nichol ST (1998) Genetic analysis of the diversity and origin of hantaviruses in *Peromyscus leucopus* mice in North America. J Virol 72:57–64
Moskophidis D, Lechner F, Pircher H, Zinkernagel RM (1993) Virus persistence in acutely infected immunocompetent mice by exhaustion of antiviral cytotoxic effector T cells. Nature 362:758–761
Nagai T, Tanishita O, Takahashi Y, Yamanouchi T, Domae K, Kondo K, Dantas JR, Jr, Takahashim M, Yamanishi K (1985) Isolation of haemorrhagic fever with renal syndrome virus from leukocytes of rats and virus replication in cultures of rat and human macrophages. J Gen Virol 66:1271–1278
Nelson RJ (1991) Maternal diet influences reproductive development in male prairie vole offspring. Physiol Behav 50:1063–1066
Nelson RJ, Shiber JR (1990) Photoperiod affects reproductive responsiveness to 6-methoxy-2-benzoxazolinone in house mice. Biol Reprod 43:586–591
Nelson RJ, Chiavegatto S (2001) Molecular basis of aggression. Trends Neurosci 24:713–719
Nelson RJ, Demas GE, Klein SL, Kriegsfeld LJ (2002) Seasonal patters of stress, immune function, and disease. Cambridge University Press, Cambridge
Nichol ST, Spiropoulou CF, Morzunov S, Rollin PE, Ksiazek TG, Feldmann H, Sanchez A, Childs J, Zaki S, Peters CJ (1993) Genetic identification of a hantavirus associated with an outbreak of acute respiratory illness. Science 262:914–917

Nuzum EO, Rossi CA, Stephenson EH, LeDuc JW (1988) Aerosol transmission of Hantaan and related viruses to laboratory rats. Am J Trop Med Hyg 38:636–640
Padula P, Figueroa R, Navarrete M, Pizarro E, Cadiz R, Bellomo C, Jofre C, Zaror L, Rodriguez E, Murua R (2004) Transmission study of Andes hantavirus infection in wild sigmodontine rodents. J Virol 78:11972–11979
Padula PJ, Edelstein A, Miguel SD, Lopez NM, Rossi CM, Rabinovich RD (1998) Epidemic outbreak of Hantavirus pulmonary syndrome in Argentina. Molecular evidence of person to person transmission of Andes virus. Medicina (B Aires) 58 Suppl 1:27–36
Parmenter RR, Yadav EP, Parmenter CA, Ettestad P, Gage KL (1999) Incidence of plague associated with increased winter-spring precipitation in New Mexico. Am J Trop Med Hyg 61:814–821
Pensiero MN, Sharefkin JB, Dieffenbach CW, Hay J (1992) Hantaan virus infection of human endothelial cells. J Virol 66:5929–5936
Plyusnin A, Morzunov SP (2001) Virus evolution and genetic diversity of hantaviruses and their rodent hosts. Curr Top Microbiol Immunol 256:47–75
Plyusnin A, Vapalahti O, Lankinen H, Lehvaslaiho H, Apekina N, Myasnikov Y, Kallio-Kokko H, Henttonen H, Lundkvist A, Brummer-Korvenkontio M, Gavrilovskaya I, Vaheri A (1994) Tula virus: a newly detected hantavirus carried by European common voles. J Virol 68:7833–7839
Plyusnin A, Vapalahti O, Lundkvist A, Henttonen H, Vaheri A (1996) Newly recognised hantavirus in Siberian lemmings. Lancet 347:1835
Plyusnin A, Horling J, Kanerva M, Mustonen J, Cheng Y, Partanen J, Vapalahti O, Kukkonen SK, Niemimaa J, Henttonen H, Niklasson B, Lundkvist A, Vaheri A (1997) Puumala hantavirus genome in patients with nephropathia epidemica: correlation of PCR positivity with HLA haplotype and link to viral sequences in local rodents. J Clin Microbiol 35:1090–1096
Plyusnin A, Kruger DH, Lundkvist A (2001) Hantavirus infections in Europe. Adv Virus Res 57:105–136
Raftery MJ, Kraus AA, Ulrich R, Kruger DH, Schonrich G (2002) Hantavirus infection of dendritic cells. J Virol 76:10724–10733
Rawlings JA, Torrez-Martinez N, Neill SU, Moore GM, Hicks BN, Pichuantes S, Nguyen A, Bharadwaj M, Hjelle B (1996) Cocirculation of multiple hantaviruses in Texas, with characterization of the small (S) genome of a previously undescribed virus of cotton rats (*Sigmodon hispidus*). Am J Trop Med Hyg 55:672–679
Roberts CW, Walker W, Alexander J (2001) Sex-associated hormones and immunity to protozoan parasites. Clin Microbiol Rev 14:476–488
Rollin PE, Ksiazek TG, Elliott LH, Ravkov EV, Martin ML, Morzunov S, Livingstone W, Monroe M, Glass G, Ruo S, Khan AS, Childs JE, Nichol S, Peters CJ (1995) Isolation of Black Creek Canal virus, a new hantavirus from *Sigmodon hispidus* in Florida. J Med Virol 46:35–39
Root JJ, Calisher CH, Beaty BJ (1999) Relationships of deer mouse movement, vegetative structure, and prevalence of infection with Sin Nombre virus. J Wildl Dis 35:311–318
Root JJ, Black WC, Calisher CH, Wilson WR, Mackie RS, Schountz T, Mills JN, Beaty BJ (2003) Analyses of gene flow among populations of deer mice (*Peromyscus maniculatus*)

at sites near hantavirus pulmonary syndrome case-patient residences. J Wildl Dis 39:287–298

Sanchez AJ, Abbott KD, Nichol ST (2001) Genetic identification and characterization of Limestone Canyon virus, a unique *Peromyscus* -borne hantavirus. Virology 286:345–353

Schmaljohn CS, Dalrymple JM (1983) Analysis of Hantaan virus RNA: evidence for a new genus of Bunyaviridae. Virology 131:482–491

Schmaljohn CS, Hjelle B (1997) Hantaviruses: a global disease problem. Emerg Infect Dis 3:95–104

Schmaljohn CS, Hasty SE, Dalrymple JM, Leduc JW, Lee HW, von Bonsdorff CH, Brummer-Korvenkontio M, Vaheri A, Tsai TF, Regnery HL Goldgaber D, Lee PW (1985) Antigenic and genetic properties of viruses linked to hemorrhagic fever with renal syndrome. Science 227:1041–1044

Schountz T, Green R, Davenport B, Buniger A, Richens T, Root JJ, Davidson F, Calisher CH, Beaty BJ (2004) Cloning and characterization of deer mouse (*Peromyscus maniculatus*) cytokine and chemokine cDNAs. BMC Immunol 5:1

Sinclair JA, Lochmiller RL (2000) The winter immunoenhancement hypothesis: associations among immunity, density, and survival in prairie vole (*Microtus ochrogaster*) populations. Can J Zool 78:254–264

Singh B, Rawlings N, Kaur A (2001) Expression of integrin alphavbeta3 in pig, dog and cattle. Histol Histopathol 16:1037–1046

Song JS, Min CH, Kang E, Yu SH (1999) Expression of ICAM-1 on the Hantaan virus-infected human umbilical vein endothelial cells. Korean J Intern Med 14:47–54

Song JW, Baek LJ, Gajdusek DC, Yanagihara R, Gavrilovskaya I, Luft BJ, Mackow ER, Hjelle B (1994) Isolation of pathogenic hantavirus from white-footed mouse (*Peromyscus leucopus*). Lancet 344:1637

Song JW, Baek LJ, Nagle JW, Schlitter D, Yanagihara R (1996) Genetic and phylogenetic analyses of hantaviral sequences amplified from archival tissues of deer mice (*Peromyscus maniculatus nubiterrae*) captured in the eastern United States. Arch Virol 141:959–967

Song W, Torrez-Martinez N, Irwin W, Harrison FJ, Davis R, Ascher M, Jay M, Hjelle B (1995) Isla Vista virus: a genetically novel hantavirus of the California vole *Microtus californicus*. J Gen Virol 76:3195–3199

Sundstrom JB, McMullan LK, Spiropoulou CF, Hooper WC, Ansari AA, Peters CJ, Rollin PE (2001) Hantavirus infection induces the expression of RANTES, IP-10 without causing increased permeability in human lung microvascular endothelial cells. J Virol 75:6070–6085

Takala A, Lahdevirta J, Jansson SE, Vapalahti O, Orpana A, Karonen SL, Repo H (2000) Systemic inflammation in hemorrhagic fever with renal syndrome correlates with hypotension and thrombocytopenia but not with renal injury. J Infect Dis 181:1964–1970

Tamura M, Asada H, Kondo K, Takahashi M, Yamanishi K (1987) Effects of human and murine interferons against hemorrhagic fever with renal syndrome (HFRS) virus (Hantaan virus). Antiviral Res 8:171–178

Temonen M, Vapalahti O, Holthofer H, Brummer-Korvenkontio M, Vaheri A, Lankinen H (1993) Susceptibility of human cells to Puumala virus infection. J Gen Virol 74:515–518
Temonen M, Lankinen H, Vapalahti O, Ronni T, Julkunen I, Vaheri A (1995) Effect of interferon-alpha and cell differentiation on Puumala virus infection in human monocyte/macrophages. Virology 206:8–15
Temonen M, Mustonen J, Helin H, Pasternack A, Vaheri A, Holthofer H (1996) Cytokines, adhesion molecules, and cellular infiltration in nephropathia epidemica kidneys: an immunohistochemical study. Clin Immunol Immunopathol 78:47–55
Vapalahti O, Lundkvist A, Vaheri A (2001) Human immune response, host genetics, and severity of disease. Curr Top Microbiol Immunol 256:153–169
Vincent MJ, Quiroz E, Gracia F, Sanchez AJ, Ksiazek TG, Kitsutani PT, Ruedas LA, Tinnin DS, Caceres L, Garcia A, Rollin PE, Mills JN, Peters CJ, Nichol ST (2000) Hantavirus pulmonary syndrome in Panama: identification of novel hantaviruses and their likely reservoirs. Virology 277:14–19
Webster JI, Tonelli L, Sternberg EM (2002) Neuroendocrine regulation of immunity. Annu Rev Immunol 20:125–163
Weigler BJ, Ksiazek TG, Vandenbergh JG, Levin M Sullivan WT (1996) Serological evidence for zoonotic hantaviruses in North Carolina rodents. J Wildl Dis 32:354–357
White DJ, Means RG, Birkhead GS, Bosler EM, Grady LJ, Chatterjee N, Woodall J, Hjelle B, Rollin PE, Ksiazek TG, Morse DL (1996) Human and rodent hantavirus infection in New York State: public health significance of an emerging infectious disease. Arch Intern Med 156:722–726
White JD, Shirey FG, French GR, Huggins JW, Brand OM, Lee HW (1982) Hantaan virus, aetiological agent of Korean haemorrhagic fever, has Bunyaviridae-like morphology. Lancet 1:768–771
Williams RJ, Bryan RT, Mills JN, Palma RE, Vera I, De Velasquez F, Baez E, Schmidt WE, Figueroa RE, Peters CJ, Zaki SR, Khan AS, Ksiazek TG (1997) An outbreak of hantavirus pulmonary syndrome in western Paraguay. Am J Trop Med Hyg 57:274–282
Wisniewski AB, Klein SL, Lakshmanan Y, Gearhart JP (2003) Exposure to genistein during gestation and lactation demasculinizes the reproductive system in rats. J Urol 169:1582–1586
Wright AF, Carothers AD, Pirastu M (1999) Population choice in mapping genes for complex diseases. Nat Genet 23:397–404
Yamanouchi T, Domae K, Tanishita O, Takahashi Y, Yamanishi K, Takahashi M, Kurata T (1984) Experimental infection in newborn mice and rats by hemorrhagic fever with renal syndrome (HFRS) virus. Microbiol Immunol 28:1345–1353
Yanagihara R, Amyx HL, Gajdusek DC (1985) Experimental infection with Puumala virus, the etiologic agent of nephropathia epidemica, in bank voles (*Clethrionomys glareolus*). J Virol 55:34–38
Yashina L, Mishin V, Zdanovskaya N, Schmaljohn C, Ivanov L (2001) A newly discovered variant of a hantavirus in *Apodemus peninsulae*, far Eastern Russia. Emerg Infect Dis 7:912–913

Yates TL, Mills JN, Parmenter CA, Ksiazek TG, Parmenter RR, Vande Castle JR, Calisher CH, Nichol ST, Abbott KD, Young JC, Morrison ML, Beaty BJ, Dunnum JL, Baker RJ, Salazar-Bravo J, Peters CJ (2002) The ecology and evolutionary history of an emergent disease: hantavirus pulmonary syndrome. Bioscience 52:989–998

Ye L, Liu Y, Yang S, Liao W, Wang C (2001) Increased expression of Hsp70 and co-localization with nuclear protein in cells infected with the Hantaan virus. Chin Med J (Engl) 114:535–539

Zaki SR, Greer PW, Coffield LM, Goldsmith CS, Nolte KB, Foucar K, Feddersen RM, Zumwalt RE, Miller GL, Khan AS, Rollin PE, KsiazekTG, Nichol ST, Mahy BWJ, Peters CJ (1995) Hantavirus pulmonary syndrome. Pathogenesis of an emerging infectious disease. Am J Pathol 146:552–579

Zeller HG, Karabatsos N, Calisher CH, Digoutte JP, Cropp CB, Murphy FA, Shope RE (1989) Electron microscopic and antigenic studies of uncharacterized viruses. II. Evidence suggesting the placement of viruses in the family Bunyaviridae. Arch Virol 108:211–227

Zhang XK, Takashima I, Hashimoto N (1988) Role of maternal antibody in protection from hemorrhagic fever with renal syndrome virus infection in rats. Arch Virol 103:253–265

Zuk M, McKean KA (1996) Sex differences in parasite infections: patterns and processes. Int J Parasitol 26:1009–1023

CTMI (2007) 315:253–288
© Springer-Verlag Berlin Heidelberg 2007

Arenaviruses

J. P. Gonzalez[1,3] (✉) · S. Emonet[1] · X. de Lamballerie[2,3] · R. Charrel[2,3]

[1]IRD U178, Conditions et Territoires d'Emergence des Maladies, Thaïlande, *frjpg@mahidol.ac.th*
[2]Unité des Virus Emergents (EA3292, IFR48), Université de la Méditerranée, Marseille, France
[3]Unité Pathologies Virales Emergentes, Univ Méditerranée-IRD

Abstract The *Arenaviridae* family contains 22 recognized virus species, each of them strongly associated with a rodent species (except Tacaribe virus which is associated with a species of bat), suggesting an ancient co-evolutionary process. Although the concept of co-evolution between rodents and arenaviruses is now largely accepted, little has been uncovered in terms of dating the phenomenon and the mechanisms of evolution, including speciation and pathogenicity. These questions are targeted in the present chapter. Old World arenaviruses are associated with the Eurasian rodents in the family Muridae. New World arenaviruses are associated with American rodents in the subfamily Sigmodontinae. The correlation between the rodent host phylogeny and the viruses suggests a long association and a co-evolutionary process. Furthermore, three distinct New World arenaviruses share a common ancestor, demonstrating a unique recombination event that probably occurred in that ancestor. This shows that recombination among arenaviruses of different lineages might occur in nature. Recombination and co-evolutionary adaptation appear as the main mechanisms of arenavirus evolution, generating a high degree of diversity. The diversity among rodent host reservoir and virus species and the potential to exchange genomic material provide a basis for the emergence of new viruses and the risk of these becoming pathogenic for humans.

1 Introduction

The *Arenaviridae* family consists of a unique *Arenavirus* genus that currently contains 22 recognized virus species (Salvato et al. 2005). Arenaviruses are enveloped single-stranded RNA viruses, with a genome consisting of two RNA segments, designated large (L) and small (S). The L genomic segment (~7.2 kb) encodes the viral RNA-dependent RNA polymerase and a zinc-binding protein. The S genomic segment (~3.5 kb) encodes the nucleocapsid protein and envelope glycoproteins in nonoverlapping open reading frames of opposite polarities. The genes on both S and L segments are separated by an intergenic noncoding region with the potential of forming one or more hairpin configurations. The 5′ and 3′ untranslated terminal sequences of each RNA segment possess a relatively conserved reverse complementary sequence spanning 19 nucleotides at each extremity. Nucleocapsid antigens are shared by most arenaviruses, and quantitative relationships show the basic split between viruses of Africa and viruses of the Western Hemisphere. Individual viruses are immunologically distinct by neutralization assays, which depend on the specificity of epitopes contained in the envelope glycoproteins (Salvato et al. 2005).

Virions are spherical to pleomorphic with a diameter of 50–300 nm (average diameter for spherical particles is 120 nm). They possess a dense lipid-containing envelope covered with 8- 10-nm-long club-shaped projections. Host cell ribosomes present in the viral particles, are responsible for the sandy appearance of the virus by electron microscopy, hence the name arenavirus (Latin: *arena*, sand). Buoyant density is 1.17–1.18 g/cm^3 in sucrose and 1.19–1.20 g/cm^3 in CsCl. Virus is rapidly inactivated at 56°C, at pH below 5.5 or above 8.5, or by exposure to UV and gamma irradiation (Table 1).

Lymphocytic choriomeningitis virus (LCMV) was first isolated in the 1930s (Armstrong and Lillie 1934) but it is only in the late 1960s that LCMV was found to be related to the already existing Tacaribe group, which then led to the creation of the *Arenaviridae* family (Murphy et al. 1969). The arenaviruses have been classified into two groups according to their antigenic properties: (1) the Tacaribe serocomplex (including viruses indigenous to rodents of the New World) and the prototype Tacaribe virus (TCRV) isolated from *Artibeus* bats in Trinidad (Downs et al. 1963), and (2) the Lassa-lymphocytic choriomeningitis (LCM) serocomplex (including the viruses indigenous to rodents of Africa and the ubiquitous lymphocytic choriomeningitis virus (LCMV), recognized as the Old World group) (Fig. 1).

Table 1 The *Arenaviridae* family

Virus		Acronym	Country of prototype virus isolate	Human significance[a]	Historical reference
1	Allpahuayo	ALLV	Peru	NE	Moncayo et al. 2001
2	Amapari	AMAV	Brazil	NE	Pinheiro et al. 1966
3	Bear canyon	BCNV	USA, California	NE	Peters et al. 1996
5	Cupixi	CPXV	Brazil	NE	Charrel et al. 2002
6	Flexal	FLEV	Brazil	LI	Pinheiro et al. 1977
7	Guanarito	GTOV	Venezuela	HF, LI	Salas et al. 1991
8	Ippy	IPPYV	Central African Republic	NE	Swanepoel et al. 1985
9	Junin	JUNV	Argentina	HF	Parodi et al.1958
10	Lassa	LASV	Nigeria	HF	Buckley et al. 1970
11	Latino	LATV	Bolivia	NE	Webb et al. 1973
12	Lymphocytic choriomeningitis	LCMV	Europe, USA	NS	Amstrong and Lilly 1934
13	Machupo	MACV	Bolivia	HF, LI	Johnson et al. 1965
14	Mobala	MOBV	Central African Republic	NE	Gonzalez et al. 1983
15	Mopeia	MOPV	Mozambique	NE	Wulff et al. 1977
16	Oliveros	OLVV	Argentina	NE	Mills et al. 1996
16[b]	Pampa		Argentina	NE	Lozano et al. 1997
17	Parana	PARV	Paraguay	NE	Webb et al. 1970
18	Pichinde	PICV	Colombia	LI	Trapido and Sanmartin 1971
19	Pirital	PIRV	Venezuela	NE	Fulhorst et al. 1997
20	Sabia	SABV	Brazil	HF, LI	Lisieur et al. 1994
21	Tacaribe	TCRV	Trinidad	LI	Downs et al. 1963
22	Tamiami	TAMV	USA Florida	NE	Calisher et al. 1970
23	Whitewater Arroyo	WWAV	USA, south West	NE	Fulhorst et al. 1996

Acronyms are attributed by the ICTV (Salvato et al. 2000). Countries are where the virus was first isolated and the associated reference is also the first report of the prototype virus. For arenaviruses known to be human pathogens virus, the primary clinical syndrome is indicated

[a] *BSL* biosafety level

[b] *HF* hemorrhagic fever; *NS* neurological syndrome; *LI* laboratory infection; *NE* No evidence of natural human infection

[c] Pampa virus should be considered as a genotype of Oliveros virus

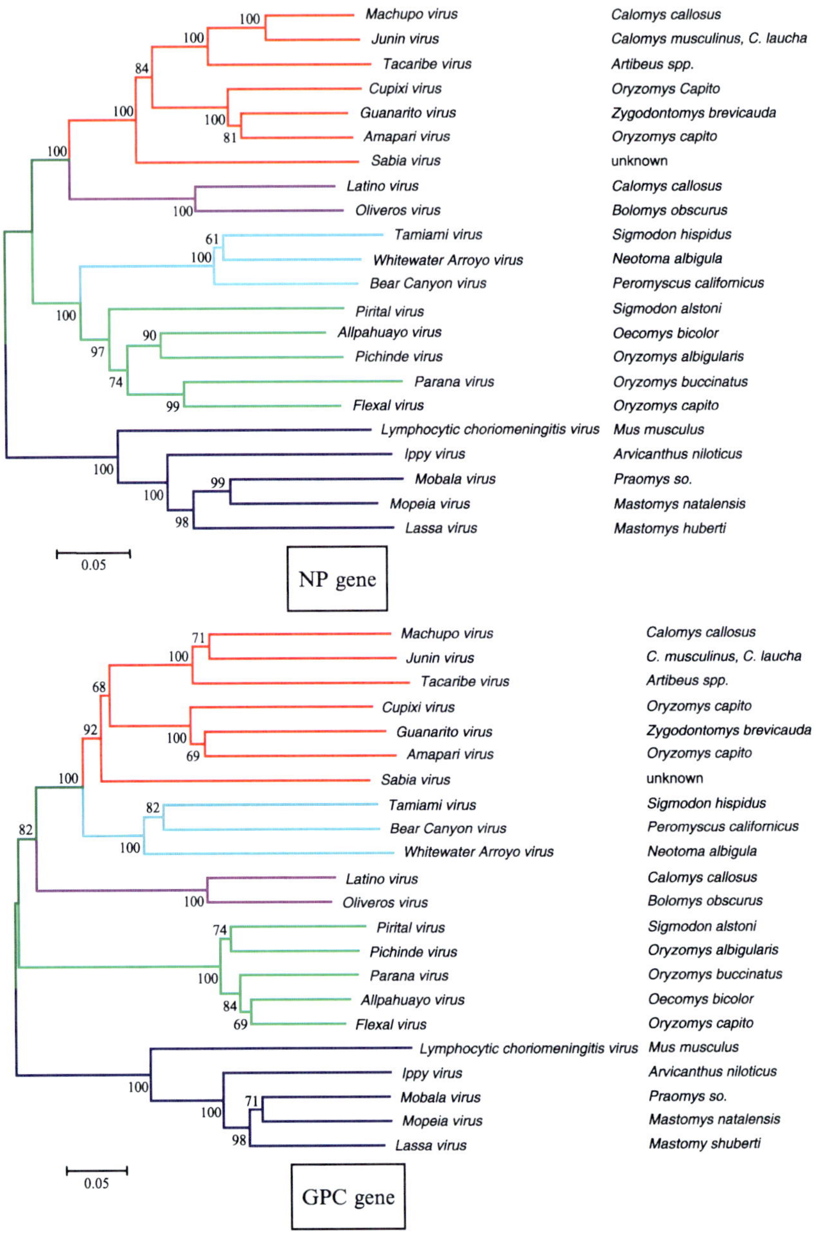

Fig. 1 *Arenavirus* phylogeny and rodent reservoir

Genetic studies of arenaviruses are congruent with comparative serological analyses. Both methods indicate that the 22 arenaviruses represent four phylogenetic lineages. The Old World (Lassa-LCM serocomplex) lineage comprises five viruses: LCMV, Lassa (LASV), Mopeia (MOPV), Mobala (MOBV) (Buckley et al. 1970; Wulff et al. 1977; Gonzalez et al. 1983), and Ippy (IPPYV) (Swanepoel et al. 1985) and is deeply rooted to the three New World (Tacaribe serocomplex) lineages, designated A, B, and C. Lineage A includes five South American viruses, Pirital (PIRV), Pichindé (PICV) (Fulhorst et al. 1997; Trapido and Sanmartin 1971), Flexal (FLEV), Paraná (PARV), and Allpahuayo (ALLV) (Pinheiro et al. 1977; Webb et al. 1970; Moncayo et al. 2001). Lineage B includes seven South American viruses including Sabiá (SABV), Junín (JUNV), Machupo (MACV), Guanarito (GTOV), Amapari (AMAV) (Lisieux et al. 1994; Parodi et al. 1958, Johnson et al. 1965; Salas et al. 1991; Pinheiro et al. 1966), Tacaribe (TCRV) (Downs et al. 1963), and Cupixi (CPXV) (Charrrel et al. 2002). Lineage C comprises three South American viruses: Oliveros (OLVV) (Mills et al. 1996), Latino (LATV) (Webb et al. 1975), and Pampa (PAMV), which is a genotype of OLVV and does not represent a taxonomic species (Salvato et al. 2005). Phylogenetic studies conducted with complete gene sequences recently demonstrated that discrepancies observed in the topology of phylograms reconstructed from nucleoprotein and envelope glycoprotein genes are attributed to the recombinant nature of the S RNA segment of the three North American viruses: Whitewater Arroyo (WWAV), Tamiami (TAMV), and Bear Canyon (BCNV) (Fulhorst et al. 1996; Calisher et al. 1970; Fulhorst et al. 2002) (Table 2).

LASV, JUNV, MACV, GTOV, and SABV are known to cause a severe hemorrhagic fever, in western Africa, Argentina, Bolivia, Venezuela, and Brazil, respectively (Peters et al. 1996), and were first recovered during investigations of human disease in 1969 (Buckley et al. 1970), 1958 (Parodi et al.1958), in 1963 (Johnson et al. 1965), 1989, (Salas et al. 1991), and 1990 (Coimbra et al. 1994), respectively. They are included in the Category A Pathogen List as defined by the CDC, and listed as Biosafety Level 4 (BSL-4) agents. The family prototype, LCMV, was first isolated in 1933 during serial monkey passage of human material obtained from a fatal infection in the first documented epidemic of St. Louis encephalitis. LCMV is an agent of acute central nervous system disease (Barton and Hyndman 2000) and is also responsible for congenital malformations (Barton et al. 1993). FLEV and TCRV viruses have caused febrile illnesses in laboratory workers. WWAV has been associated with three fatal cases of infection in California in 2000 (CDC 2000), but further cases have not been documented since.

LCMV, LASV, and related viruses from the Old World are associated with rodents from the family *Muridae*, subfamily *Murinae*. New World arenaviruses are associated with New World rodents in the family Muridae, subfamily

Table 2 Geographic and reservoir characteristics of arenaviruses

Acronym	Evolutionary lineage[a]	Distribution[b]	Biogeographic domain	Reservoir
Old World arenaviruses				
LASV	OW	Nigeria, Guinea, Liberia, Sierra Leone[c]	Paleartic	*Mastomys huberti*
MOBV	OW	Central African Republic	Paleartic	*Praomys* spp.
MOPV[d]	OW	Mozambique, Tanzania	Paleartic	*Mastomys natalensis*
IPPYV	OW	Central African Republic	Paleartic	*Arvicanthus niloticus.*
LCMV	OW	Eurasia, USA, Canada	Holoartic	*Mus musculus*
New World arenaviruses (North Central America)				
BCNV[e]	NW-rec-A/B	USA, California	Neartic	*Peromyscus californicus*
TAMV[e]	NW-rec-A/B	USA, Florida Everglades	Neartic	*Sigmodon hispidus*
WWAV[e]	NW-rec-A/B	Southwestern USA	Neartic	*Neotoma albigula*
New World arenaviruses (South America)				
Lineage A				
ALLV[e]	NW-A	Peru	Neotropic	*Oecomys bicolor*
FLEV	NW-A	Brazil	Neotropic	*Oryzomys capito*
PARV	NW-A	Paraguay	Neotropic	*Oryzomys buccinatus*
PICV	NW-A	Colombia	Neotropic	*Oryzomys albigularis*
PIRV	NW-A	Venezuela	Neotropic	*Sigmodon alstoni*
Lineage B				
AMAV	NW-B	Brazil	Neotropic	*Oryzomys capito*
CPXV	NW-B	Brazil, northeastern	Neotropic	*Oryzomys capito*
JUNV	NW-B	Argentina	Neotropic	*Calomys musculinus*
GTOV	NW-B	Venezuela	Neotropic	*Zygodontomys brevicauda*

(*Continued*)

Table 2 Phylogeny and rodent vector reservoir —cont'd.

Acronym	Evolutionary lineage[a]	Distribution[b]	Biogeographic domain	Reservoir
MACV	NW-B	Bolivia	Neotropic	*Calomys callosus*
SABV	NW-B	Brazil	Neotropic	unknown
TCRV	NW-B	Trinidad	Neotropic	*Artibeus spp.* (bat)
Lineage C				
LATV	NW-C	Bolivia	Neotropic	*Calomys callosus*
OLVV	NW-C	Argentina	Neotropic	*Bolomys obscurus*

[a] *OW* Old World; *NW* New World
[b] Listed countries are included on the basis of virus isolation only, no serology
[c] One case was probably generated between Ivory Coast and Burkina Faso; the place of origin remains unknown
[d] Morogoro virus, which is a genotype of Mopeia virus, has recently been isolated from Mastomys rodents in Tanzania and is under study (Gunther et al., unpublished data)
[e] Recombinant lineage as previously reported (Charrel et al. 2001)

Sigmodontinae Wilson and Reeder 2005). The correspondence between the phylogeny of the hosts and of the viruses suggests a long association and co-evolution (Gonzalez 1986a, 1986b; Bowen et al. 1998). TCRV isolated from bats is the only member of the family that is not known to be a chronic, inapparent infection of rodents (Fig. 2).

2 Arenaviruses and Their Natural Hosts

Arenavirus species and rodent species are strongly associated in a specific manner, suggestive of a possible co-evolutionary process. Although the concept of co-evolution between rodents and arenaviruses is now largely accepted within the scientific community, little information has been found in terms of dating the phenomenon and detailed leading mechanisms (Gonzalez et al. 1986b; Bowen et al. 1997, 1998; Charrel et al. 2001) (Table 3).

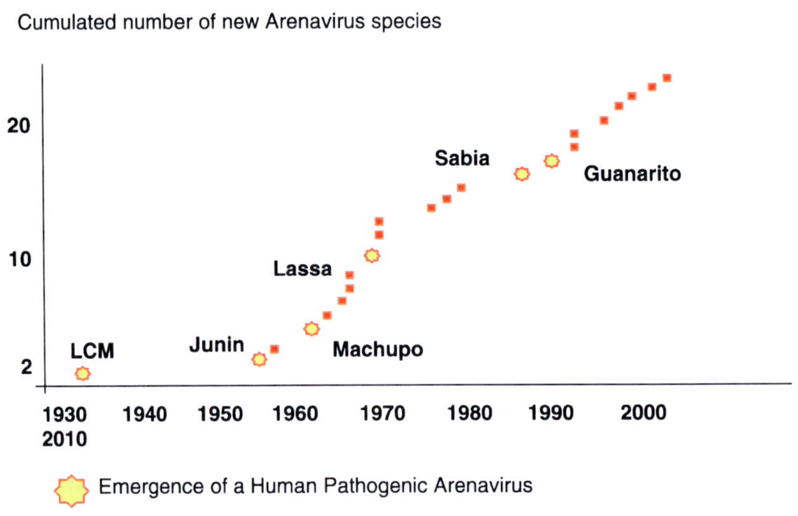

Fig. 2 The time scale of *Arenavirus* emergence. Each *red circle* represents the time of the first isolation of a new arenavirus; *stars* are those pathogenic for humans

Table 3 Arenaviruses and their natural reservoir hosts

Virus	Place of isolation	Primary host[a], Hp/ host reservoir, Hr	Secondary host[a]	Hr, main biotope
Old World Arenaviruses				
LCMV	Worldwide	*Mus musculus*	*Apodemus sylvaticus; Mus domesticus*	Domestic environment
LASV	Nigeria	*Mastomys huberti*	*Mastomys erythroleucus*	Savannah and forest galleries
IPPYV	Central African Republic (north)	*Arvicanthis niloticus*	*Lemniscomys striatus*	Sudanese dry savanna
MOBV	Central African Republic (south)	*Praomys jacksoni*	*Mastomys erythroleucus*	Sub-Sudanese wet savanna
MOPV	Mozambique, Tanzania	*Mastomys natalensis*	*Mastomys huberti*	Dry savannah
New World Arenaviruses (North Central America)				
BCNV	USA, California	*Peromyscus californicus*	*Neotoma fuscipes; Peromyscus boylii*	
TAMV	USA, Florida Everglades	*Sigmodon hispidus*		Marshes

(*Continued*)

Table 3 Arenaviruses and their natural reservoir hosts —cont'd.

Virus	Place of isolation	Primary host[a], Hp / host reservoir, Hr	Secondary host[a]	Hr, main biotope
WWAV	USA, southwest	*Neotoma albigula*	*Neotomys mexicana* *N. cinerea,* *N. micropus,* *N. fuscipes*	
New World Arenaviruses (South America)				
ALLV	Peru	*Oecomys bicolor*	*Oecomys paricola*	
AMAV	Brazil, northeastern Amapa	*Oryzomis capito*	*Neacomys guianae, Oryzomis gaeldi; Neacomys spinosus*	Amazonian tropical forest
CPXV	Brazil, northeastern Amapa	*Oryzomis megacephalus*	*Oryzomys capito*	Forest
FLEV	Brazil	*Oryzomys capito*	*Oryzomys* spp.	Tropical forest
JUNV	Argentina	*Calomys musculinus Calomys laucha*	*Calomys musculinus; Akodon azarae*	Extensive agricultural area (corn fields)
GTOV	Venezuela	*Zygodontomys brevicaudata*	*Sigmodon alstoni; Zygodon longicaudatus*	
LATV	Bolivia, Brazil	*Calomys callosus*		Low tropical savanna
MACV	Bolivia, eastern	*Calomys callosus*		Low tropical savanna
OLVV	Argentina	*Bolomys obscurus*		Pampa
PARV	Paraguay	*Oryzomys buccinatus*	*Bolomys obscurus*	
PICV	Colombia: Cali, Medellin, Popaya	*Oryzomys albigularis*	*Thomasomys fuscatus, Zygodontomys* spp.	Primary fog forest (elevation 1,500 m)
PIRV	Venezuela	*Sigmodon alstoni*	*Zygodontomys brevicaudata*	
SABV	Brazil, central	unknown	unknown	Secondary clearing forest
TCRV	Trinidad	*Artibeus lituratus palmarum*	*Artibeus jamaicensis trinitatus*	Tropical forest

[a]Primary hosts are those most commonly infected in nature by the virus, secondary hosts are those that have been accidentally infected or have been consistently found with reactive antibody to specific arenaviral antigens. The habitat refers to that of the primary host

2.1
Co-evolution Process

Specific rodents are the principal hosts of arenaviruses (Childs and Peters 1993; Bowen et al. 1997). Usually one rodent species, less often two closely related species, act as the principal host(s) (virus reservoir) of each arenavirus species, in which natural infection is usually a chronic mild or inapparent infection. The only exception is Tacaribe virus, which has only been associated with a chronic infection of bats. It is now widely recognized that the diversity of arenaviruses is the result of a long-term, shared evolutionary relationship (termed co-evolution or co-speciation) between viruses of the family *Arenaviridae* and rodents of the family *Muridae* (Johnson et al. 1965; Gonzalez 1986a; Bowen et al. 1996). The time scale of the co-evolutionary divergence of specific arenaviruses and their rodent hosts is still under discussion. From our observations and analyses, we strongly favor an ancient co-evolutionary process with several transfers, parallel and diffuse evolution. Our hypothesis is that an ancestral arenavirus type was chronically infecting a common rodent ancestor before New World sigmodontine and Old World murids diverged, approximately 35 million years before the present (Mybp). Each lineage (i.e., New World sigmodontine and Old World murid rodents) evolved independently with their own arenaviruses (co-evolution and co-speciation) resulting in a specific association between rodent species and arenavirus type, as we see today. In addition, when rodent and virus phylogenies are compared, major rodent subfamilies (*Sigmontinae* and *Murinae*) correspond with the major arenavirus clades (i.e., New World arenaviruses vs Old World arenaviruses). A similar association is also evident among South American arenavirus strains and among the South American neotropical *Sigmodontiae* due to the same evolutionary processes. However, discrepancies from the general hypothesis of co-evolution have also been observed, suggesting that spillover from one species or genus to another might occur, and that genomic segments might also be exchanged in some instances (Gonzalez et al. 1986b, 1996a, 1996b; Hugot et al. 2001). Thus the emergence of new virus types and pathogen transmission to humans appears likely to be associated with specific rodent species and their ecology and behavior (Figs. 3, 4).

2.2
A Brief Ancient History of Rodents

We use the most common theory on rodent radiation to support part of our hypothesis. From the Eurasian continent, cricetid rodents, ancestors of murid rodents, spread into the Americas, and then, from Asia, murid rodents spread to Europe and Africa. The term "murid" corresponds to the *Murinae* subfamily of the family *Muridae* (Wilson and Reeder 2005). The term "sigmodontine"

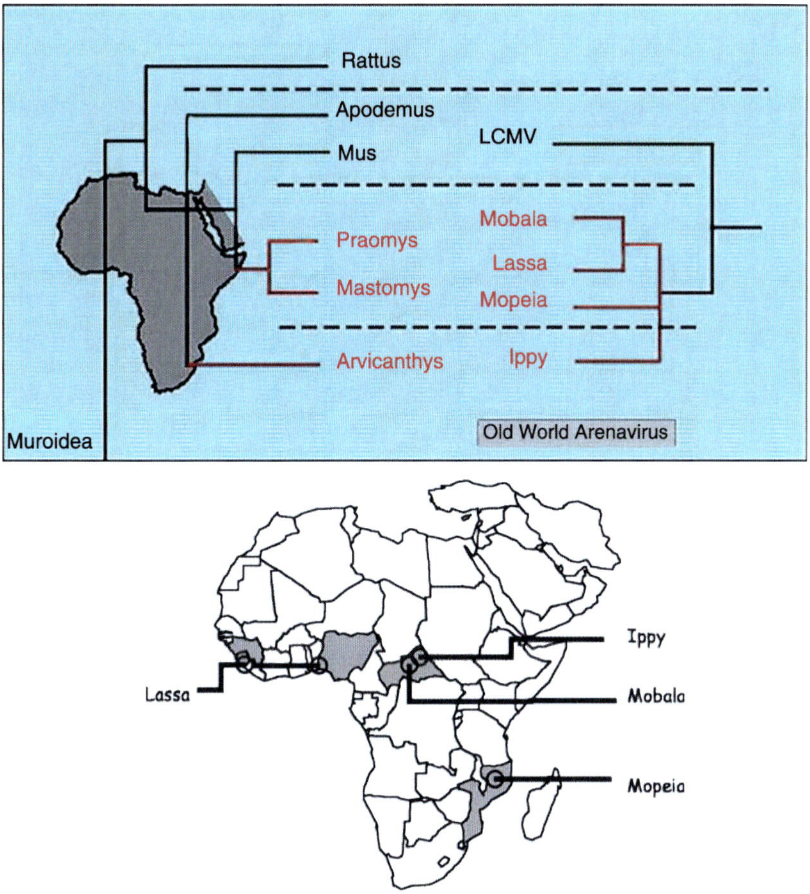

Fig. 3 Phylogeography of Old World arenaviruses and hosts. A specific association between virus and rodent host is exemplified by a diffuse co-evolution process of Old World arenaviruses and their murid rodent hosts

refers specifically to New World rodents of the subfamily *Sigmodontinae* of the family *Muridae* (previously classified as being in the family *Cricetidae*) and includes the New World rats and mice. As early as the Eocene, 65 Mybp, a rodent ancestor bearing *Muridae* characters, *Simimys* , was recognized within North America. During the Oligocene (37 Mybp), the *Muridae* distribution became holoartic. The New World *Sigmodontinae* colonized the Americas by waves of migration northward and southward. As a result, the sigmodontine fauna of South America derived from North America and today, the South

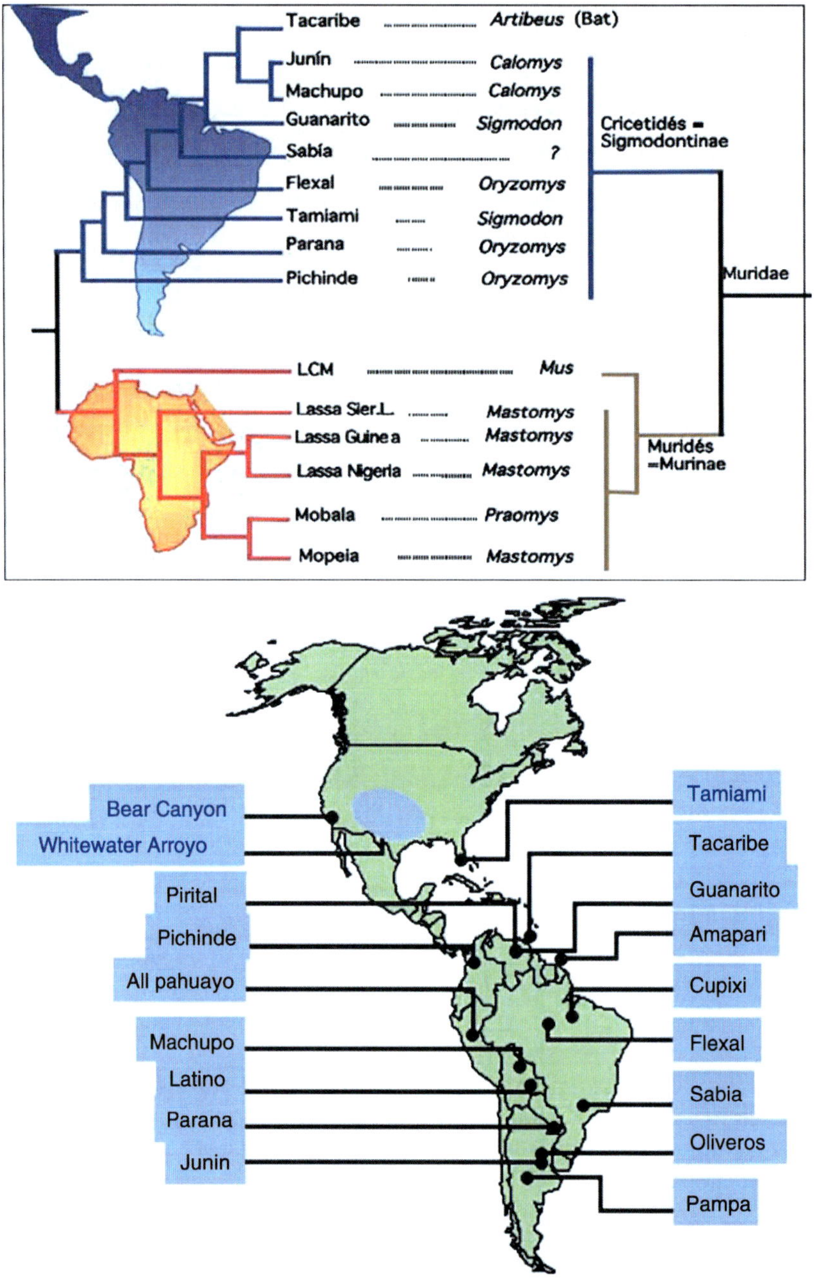

Fig. 4 Phylogeography of New World arenaviruses. The geographic distribution is shown for three indigenous arenaviruses from North America and 14 indigenous arenavirus types from of South America

American group can be distinguished from the less diversified sigmodontine rodents of genera such as *Neotoma* and *Peromyscus* of North America.

In Asia, murid rodents probably came from North America and were present during the Oligocene (35 Mybp). Arising from an original pool, successive waves of murids spread to Europe during the late Miocene (15 Mybp), but there was only a limited extension into Africa where they became underrepresented.

From Asia, murid rodents spread around the Mediterranean basin to Europe approximately 14 Mybp. During that period, the subfamily Murinae extended from Europe into North Africa and rapidly became the most widely distributed rodents in Africa.

From the Pleistocene era (2 Mybp), murid rodents were present in northern Africa. They then spread southward, although their species radiation was severely influenced by arid climate and geomorphology. During that time, speciation reached its highest point influenced by climate variation and physical isolation because of physical barriers such as the Rift Valley and the division of the African continent by the Sahara. More recently, humans have played an important role in the spread of rodents, particularly commensal species such as *Mus*. Some rodent genera from the Pleistocene are still present in East Africa, while others from North Africa have disappeared. However, it likely that murid ancestors were very closely related to the present extant genera (Gonzalez 1996a) (Fig. 5).

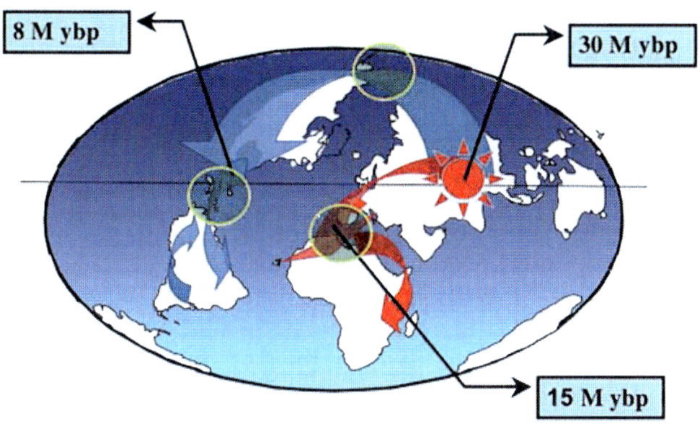

Fig. 5 The rodent migration and emergence and spread of arenaviruses and their rodent hosts. Rodent expansion shows the path of virus dispersion. The subfamilies *Murinae* and *Sigmodontinae* are indicated by *red* and *blue*, respectively, by the approximate time of expansion and speciation. After 34 Mybp (Oligocene), the Eurasian continent became colder and arid with a general shrinking of forest cover; 30 Mybp, rodents probably emerged from Central Eastern Asia and started their Asian radiation journey through Europe. Temperature changes and warmer periods (15 Mybp) would have helped separate the original rodent lineages of Asia and Europe and further their spread in Africa and the Americas. However, back-migrations occurred by way of the Bering Strait and other land bridges such as the Panamanian isthmus in the Americas

2.3
Rodent Migration Within the Americas and an Astonishing Diversity

Comparative phylogenetic analyses of N and GPC proteins showed that the three North American arenaviruses (Whitewater Arroyo, Tamiami, Bear Canyon) group together; however, depending upon the gene used for analysis, these viruses group within different lineages. They are more closely related to lineage A viruses in N protein-based analyses, whereas they are more closely related to lineage B viruses in GPC protein-based phylograms. This suggests that WWAV, TAMV, and BCNV share a common ancestor, which must have been a recombinant of lineages A and B (Fig. 6).

According to the history of rodent migrations within America, rodents migrated across the Panamanian isthmus, from North America to South America, where rodent diversification was able to expand explosively because of the absence of predators and highly favorable ecological conditions. It is postulated that recombination events among arenaviruses most likely occurred in South America and the resulting chimeric viruses were then introduced into North America during the back migration of certain rodent populations

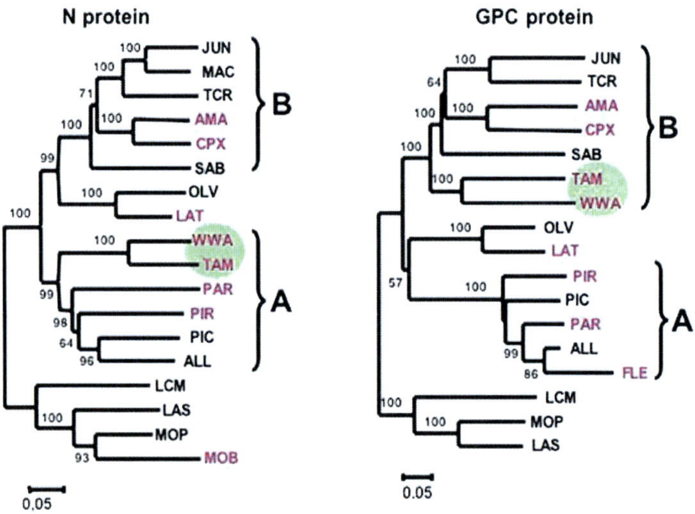

Fig. 6 Comparative New World *Arenavirus* phylogeny using N and GPC sequences demonstrating recombination processes in evolution. *Left*, the capsid protein (N gene) of TAMV and WWAV are inherited from an ancestor virus belonging to lineage A; *right*, the GPC protein of TAMV and WWAV are inherited from an ancestor that belonged to lineage B

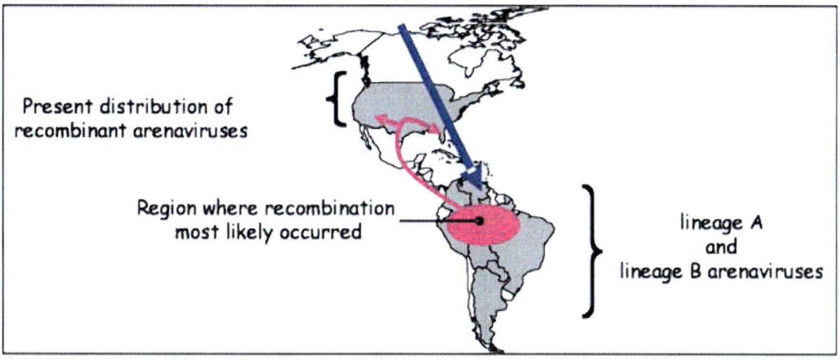

Fig. 7 Rodent migration and American arenavirus recombinations. Rodent diffusion in the Americas. *Blue arrow* shows the first migration of murids from North to South America, corresponding to the split between Old World and New World rodents, estimated at 35 Mybp. First migration from North to South America estimated at 15 Mybp. *Purple ovals* indicate rodent speciation in South America (an explosive radiation of species, 10 Mybp), and *arrows* indicate the back-migration (across the Panamanian land bridge between 10 and 8.6 Mybp) of extant rodent species into Central and North America harboring recombinant arenaviruses (Fig. 8)

(e.g., *Sigmodon* spp.) across the Panamanian land bridge. Although dating the period of recombination is difficult because of controversial data for time estimation of rodent migrations, the paleobiogeography of sigmodontines suggests that recombination could have occurred as far as back as10 Mybp (Engel et al. 1998) (Fig. 7, 8).

2.4 Mechanisms of Virus Evolution

There are three possible mechanisms driving the evolution of arenaviruses: (1) accumulation of point mutations; (2) intersegmental reassortment; and (3) intrasegmental recombination.

2.4.1 Accumulation of Mutations

In the *Arenaviridae* family, the accumulation of mutations appears to be the mechanism most often responsible for virus diversity observed between isolates within a given viral species. By analogy to other RNA viruses, it is believed that mutations are caused by the absence of proofreading activity of the viral RNA-dependent RNA polymerase during virus replication. With respect to the

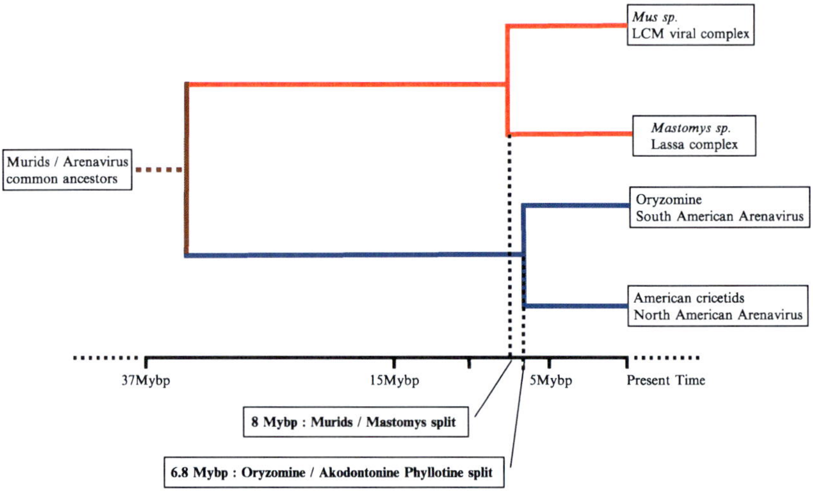

Fig. 8 Proposed time scale of *arenavirus* and rodent co-evolution/cospeciation

rate at which rate mutations are produced and accumulated in arenaviruses, experimental data generated in vitro with a partial region of the polymerase of LCMV suggest mutation frequencies ranging from 1.2 to 3.5×10^{-4} substitutions per nucleotide per genome replication (Grande-Perez et al. 2005). These mutations lead to the generation of virions exhibiting various fitness patterns, and only the best-adapted virions are presumably selected and maintained. The factors driving the selection are multiple and complex and change over time. The occurrence of mutations together with natural selection account for the creation of the genetic diversity observed within a virus species.

2.4.2 Intersegmental Recombination (Reassortment)

Reassortants of arenaviruses have been generated experimentally (Lukashevich et al. 1992; Rivière and Oldstone 1986), with the genome of the reassortant virus containing one genomic segment from each parent. This mechanism has not been described in nature so far for arenaviruses. Experimental generation of a reassortant arenaviruses consisting of the L RNA segment of Mopeia virus and the S RNA segment of Lassa virus has demonstrated that an exchange of genetic material is possible despite a genetic diversity of 28% at the amino acid level (Lukashevich et al. 1992). It is worth noting that these reassortant viruses were produced by co-cultivation on Vero cell monolayers, without the use of sophisticated equipment or complicated molecular techniques. During the current atmosphere of heightened bioterrorism surveillance, this data would suggest that the generation of chimeric viruses is not so complicated and may be attempted with very basic equipment.

Recently, fatal cases of acute hemorrhagic fever in Kenya and Somalia have been attributed to a reassortant bunyavirus—the family *Bunyaviridae* contain tri-segmented genomes—comprising genomic segments from Bunyamwera virus and from a novel bunyavirus; both were previously unknown as etiologic agents of hemorrhagic fever (Gerrard et al. 2004; Bowen et al. 2001). Bunyavirus reassortment under laboratory conditions had previously documented that exchange of M segments of LaCrosse (LACV) and Snowshoe Hare (SSH) viruses created chimeric viruses; those containing the M segment of LACV, irrespective of S and L segments of SSHV, showed an enhanced capability to disseminate and be transmitted by *Aedes trisereriatus* mosquitoes (Beaty et al. 1982; Beaty et al. 1981). These field and laboratory findings highlight the ability of viral reassortment in creating a chimeric new virus that exhibits increased pathogenicity for humans (as compared to the two parental strains) or specific, novel biologic properties (not displayed by the parental strains).

To identify virus reassortment, complete sequence characterization of viral genomes is a necessary prerequisite. Until recently, the lack of genetic data for the L segment of arenaviruses in all but a handful of virus species hampered the quest for identifying natural reassortment. Recently, however, large genomic programs dedicated to arenaviruses have provided significant sequence data sets containing the complete genomes of almost all arenaviruses. Subsequent sequence analyses and phylogenetic studies, however, were unable to detect any evidence of the natural occurrence of reassortment among arenavirus species, despite an exhaustive search using full-length genomes. Thus, although demonstrated experimentally, it is believed that reassortment may not play a major role in evolution of the *Arenaviridae.*

2.4.3
Intrasegmental Recombination

Intragenic recombination is one of the well-documented mechanisms of evolution of positive-strand, double-stranded and negative-strand RNA viruses (Lai 1992; Hahn et al. 1998; Worobey et al. 1999; Desselberg et al. 1986; Suzuki et al. 1998; Bergman et al. 1992; Orlich et al. 1994; Sibold et al. 1999). Intrasegmental recombination was recently demonstrated for the three North American arenaviruses (WWAV, TAMV, and BCNV) (Charrel et al. 2001, 2002, Archer and Rico-Hesse 2002), indicating common derivation from a recombinational event between ancestors in both lineage A and lineage B viruses. Analysis of complete genome sequences for all recognized members of the genus *Arenavirus* suggest that there are no other examples of intrasegmental recombination. Since these three viruses possess a common ancestor as demonstrated by phylogeny, recombination most likely occurred in this ancestor. It is important to note that recombinant arenaviruses are able to infect humans (Kosoy et al. 1996). Whether they cause disease in infected individuals is still not clearly established; however, three cases of fatal human infections associated with Whitewater Arroyo virus have been reported in California (CDC 2000).

2.4.4
Evolutive Significance of Interspecies Recombination

The evidence for recombination deduced from the genetic analysis of the genomic S RNA raises major questions concerning the nature of situations that may be conducive to intragenic recombination:

2.4.4.1
Co-infection of the Same Rodent by Two Different Arenaviruses Belonging to Distinct Phylogenetic Lineages

In nature, since arenaviruses can establish chronic infections among their rodent hosts, the more likely scenario for interspecific genome recombination would involve superinfection of a rodent already chronically infected with one arenavirus by a second distinct arenavirus. This hypothesis requires the co-existence of distinct arenaviruses in the same geographic area and this situation is present within several regions of a number of countries. For example, the principal hosts of OLLV, JUNV, and LCMV are rodents of the species *Necromys benefactus* (formerly *Bolomys obscurus;* Wilson and Reeder 2005), *Calomys musculinus* and *Mus musculus*, respectively. These three species and three other common rodent species exist sympatrically in rural regions of Argentina (Mills et al. 1996). Studies of the dynamics of OLLV infection among rodents indicate that dual infections by JUNV and OLLV viruses may occur at low frequency among three species of rodents (*N. benefactus*, *Akodon azarae* , and *M. musculus*) based on comparative IFA titers obtained against specific arenaviral antigens.

Additionally, there is evidence that the principal host for a specific arenavirus can be naturally infected with a different arenavirus associated with a sympatric rodent species. For example, *Sigmodon alstoni* , the principal host of PIRV (lineage A) can naturally be infected with GTOV (lineage B) (Fulhorst et al. 1999b). Moreover, experimental data have shown that immunization of rodents with a virus belonging to a given lineage is poorly protective against infection by viruses belonging to different lineages (Weissenbacher et al. 1975). Consequently, although mixed infections of rodents with distinct arenaviruses have not been reported in the literature, field and experimental data suggest that infections by arenaviruses of different lineages are plausible.

2.4.4.2
Co-infection of One Cell by Two Viruses

Experiments performed in cell cultures have clearly established that co-infection of a single cell by two distinct arenaviruses is possible (Bishop et al. 1980; Rivière et al. 1985; Rivière and Oldstone 1986; Whitton et al. 1988; Lukashevitch 1992). This co-infection could potentially allow the generation of recombinant RNA molecules by template switching of the RNA polymerase. According to this mechanism, the RNA polymerase would jump from one template to another during RNA processing, generating a chimeric RNA molecule including sequences inherited from the two parental strains.

Thus, in summary, our current knowledge concerning the ecology of rodents infected by arenaviruses and the natural circulation of these viruses in the New World, together with experimental data, would suggest that recombination between arenaviruses belonging to different lineages could potentially occur in nature. Furthermore, the recombinant nature of the genome of the 3 arenavirus indigenous to North America (WWAV, TAMV, BCNV) suggests that their ancestor may have been endowed with a selective advantage, facilitating the maintenance and transmission of the recombinant over time. This finding reinforces the fact that future phylogenetic analyses of arenaviruses should be based on complete genomic sequences to allow the identification of recombination and/or reassortment events and therefore a better understanding of the processes of co-speciation and the occurrence of crossing-over or reciprocal recombination.

3
From Enzootic to Epidemic: *Arenavirus* Ecology and Human Health

Persistent infection of the rodent host appears to be a crucial phenomenon in the long-term persistence of the arenaviruses in nature. Infection in the rodent host is associated with a chronic or sporadic viremia and/or viruria and sometimes a life-long shedding of the virus into the environment. The course of the infection is determined by factors such as the age, genetic make-up, immunological resistance, and history of prior infection within the rodent host, but also by the infecting virus strain. Neonatally infected rodents usually become chronic carriers of virus and excrete the virus for a long time (throughout life) in their urine. Virus transmission within rodent populations can occur through three mechanisms: (1) vertical (dam to progeny) transmission, (2) horizontal transmission through direct or indirect contacts, and (3) a balanced combination of both mechanisms. Female rodents infected as neonates with certain arenaviruses (JUNV, MACV) may show reduced fertility or suffer decreased litter sizes (Childs and Peters 1993; Webb et al. 1975); additionally, neonates born to infected dams may experience stunted growth. Accordingly, the persistence of these arenaviruses within a rodent population requires some degree of horizontal transmission. In contrast, other arenaviruses that do not cause infertility, such as certain strains of LCMV (Childs and Peters 1993), can be maintained in a rodent population exclusively by vertical transmission.

Humans usually become infected by arenaviruses through direct contact with infected rodents, including bites, through inhalation of infectious rodent excreta and secreta. The domestic and peridomestic behavior of several species of rodent reservoir hosts is a major contributing factor facilitating viral transmission from rodent to human. Transmission of arenaviruses to humans occurs following recreational or agricultural incursions into environments providing critical habitat for rodent hosts. Additionally, professionals handling infected rodents in the field or laboratory are at increased risk of infection (Sewell 1995). Modifications of the environment driven either by human activities, such as modern farming practices, or ecological changes, such as flooding, have been implicated in the emergence of human disease caused by arenaviruses.

Nine arenaviruses are associated with human diseases. LASV, JUNV, MACV, GTOV, and SABV are known to cause a severe hemorrhagic syndrome, in western Africa, Argentina, Bolivia, Venezuela, and Brazil, respectively (Peters et al. 1996). They are highly infectious, virulent pathogens and all are listed on the Category A Pathogen List (as defined by the CDC); such agents can only be handled in Biosafety Level 4 (BSL-4) laboratories. Infection by LCMV can result in acute central nervous system disease and congenital malformations (Barton and Hyndman 2000; Barton et al. 1993). Very little is known about the health consequences of infection with the other arenaviruses: PICV infection has resulted in numerous seroconversions among humans without any notable clinical significance; FLEV has resulted in two symptomatic laboratory infections and should be regarded as dangerous (F. Pinheiro, unpublished data); TCRV virus has caused a single case of a febrile disease with mild CNS symptomatology (J. Casals, unpublished data) (Peters et al. 1996; Karabatsos 1985; Buchmeier et al. 1974). WWAV has recently been associated with three fatal cases of infection in California (CDC 2000).

3.1 Lymphocytic Choriomeningitis Virus

The first arenavirus to be isolated was LCMV, which was discovered in 1933 during the investigation of an epidemic of St. Louis encephalitis in the USA. In regions where LCMV is known to exist, infection in the two closely related reservoir species hosts, *Mus domesticus* and *M. musculus*, is highly focal (Lehmann-Grube 1971). Studies conducted in Baltimore, Boston, and Washington, DC, revealed a spotty distribution of virus-positive mice in houses (Farmer et al. 1942; Childs et al. 1991, 1992). Similarly, in Germany, much higher infection rates prevail among *Mus* in the west-central region than in the southern or northern portions of the country (Ackermann et al. 1964). Human cases of LCMV infection are most common in autumn. This pattern is the result of peak seasonal population densities of rodents and the movement of house mice into homes and barns with the onset of cold weather. In addition, seasonal variation in infection rates of *Mus* sp. may occur. Situations associated with transmission of virus from infected wild mice to humans include substandard housing such as mobile homes or inner city dwellings, the cleaning of rodent-infested barns or outbuildings, and the autumn entry of wild mice into dwellings. Most human LCMV infections occur among young adults, although persons of all ages have been affected. The mode of transmission in most sporadic human infections is not definitely known; however, experimental and epidemiologic observations implicate aerosols, direct contact with rodents, and rodent bites (in that order) as the most likely vehicles (Enria et al. 1999; Farmer et al. 1942; Hinman et al. 1975). Although most sporadic LCM cases are attributed to contact with infected wild mice, outbreaks of disease have been traced to infected laboratory mice and Syrian hamsters *(Mesocricetus auratus)* (Dykewitz et al. 1992). Individual cases or outbreaks of LCM in the United States and Europe have resulted from exposures to infected pet hamsters (Biggar et al. 1975; Ackermann et al. 1972). Recently, a case of LCMV infection in France was traced back to a population of urban *Mus musculus*; virus isolates were obtained from 60% of the mice trapped in the patient's home (R. Charrel et al).

Although LCMV infection may occur worldwide wherever the house mouse has been introduced, human infection has been conclusively demonstrated only in Europe and the Americas (Lehmann-Grube 1971). LCM cases present most commonly as febrile illnesses with headache and systemic symptoms; leukopenia and thrombocytopenia are usually noted (Peters et al. 1995). After 3–5 days of nonspecific illness, the fever subsides, but it frequently recurs in 2–4 days with several days of even more severe headache. Patients may exhibit meningitis during this second febrile period. In approximately one-third of the cases, cerebrospinal fluid (CSF) exhibits lymphocytic pleocytosis, an elevated protein content, and hypoglycorrhachia. Sometimes there is more severe damage to the central nervous system (CNS) and transient hydrocephalus has been described. Chorioretinitis and congenital hydrocephalus may occur in fetal infections. The second febrile episode, as well as some of the complications of convalescence, have long been thought to represent immunopathologic phenomena, and antibodies detectable by immunofluorescence appear at about this time (Peters et al. 1995). The prevalence of antibody to LCMV is approximately 5% among adults living in large cities of the United States (Childs et al. 1991). Both CNS and congenital infections caused by LCMV may be more common than appreciated and are undoubtedly underdiagnosed (Enria et al 1999; Barton 1996, 2001).

In 2005, LCMV caused an outbreak of infection among four patients who had received solid organ transplants from an infected donor. Severe illness developed in all four patients, three of whom died (CDC 2005). The donor was probably infected from his pet hamster.

3.1.1 South American Arenaviral Hemorrhagic Fever

The clinical picture of the South American arenaviral hemorrhagic fever is almost identical regardless of the virus responsible for the disease. Argentine, Bolivian, and Venezuelan arenaviral hemorrhagic fevers are remarkably similar clinically, and mortality in each is about 15%–30% (Sabattini et al. 1970; Maiztegui et al. 1975; Stinebaugh et al.1966). The disease caused by all three viruses can include neurological symptoms, hemorrhage, and shock; these clinical findings herald a poor prognosis.

3.1.1.1 Argentine Hemorrhagic Fever: Junin Virus

The rodent host reservoir of JUNV is *Calomys musculinus*, a small field rodent of Argentina (Sabattini and Maiztegui 1970). *Calomys* populations reach their highest densities in cornfields and the surrounding weedy fence lines during the austral fall. In the 1950s, a new disease (Argentinean Hemorrhagic Fever, AHF) emerged in the Buenos Aires province of Argentina, a rich farming region, and was associated with intensive deforestation and intensive agricultural practices that considerably increased the contacts between humans and rodents. Most of the infected persons were male agricultural workers engaged in harvesting corn. Transmission from the rodent is by inhalation of infected aerosols produced from rodent excreta or from rodents caught and shredded in mechanical harvesters (Maiztegui 1975). As a consequence, infection with JUNV is strongly seasonal and peaks during the harvest season in autumn. Since the emergence of AHF, a progressive geographic expansion of epidemic outbreaks, occurring a variable intervals, has been observed (Maiztegui 1975; Maiztegui et al. 1986). After its first isolation in 1959, human cases were initially recorded within a 16,000-km^2 area of the rich agricultural pampas north of Buenos Aires province, but AHF progressively expanded to become endemic in a 150,000-km^2 area in southern of Santa Fé, southeastern Cordoba, and northeastern La Pampa provinces (Enria and Feuillade 1998). To date, the human population at risk is estimated to be about 5 million. Several hypotheses were proposed to explain this expansion. Since 1958, cases have been annually recorded, ranging from several hundred to 3,500. An epidemic outbreak of human AHF in southern Santa Fé and northern Buenos Aires provinces was shown to coincide, with a lag of 1–2 months, with the peak density in a rapidly increasing population of *C. musculinus*. The maximum prevalence of JUNV antigen-positive rodents, approximately 25% of adult *C. musculinus*, coincided with peak rodent population density (Mills et al. 1992).

Although human cases present with either neurologic or hemorrhagic manifestations (or a combination of both), molecular studies of JUNV have not associated either syndrome with a particular JUNV genotype (Albarino et al. 1997). Studies of the genetic diversity among JUNV strains circulating in central Argentina demonstrated a high degree of genetic similarity among isolates from the same locale. However, no cluster of related JUNV strains was associated with clinically different forms of AHF (Garcia et al. 2000). Mortality among patients with confirmed AHF was 14%–17% before the routine initiation of immune plasma was implemented (Maiztegui et al. 1979); treatment has reduced the mortality to less than 1%. Introduction of an effective vaccine, using a live-attenuated virus (Candidate#1) (Maiztegui et al. 1998), has decreased the incidence of the AHF to fewer than 100 cases per year (Enria et al. 2002).

3.1.1.2
Bolivian Hemorrhagic Fever: Machupo Virus

The rodent species *Calomys callosus* is the reservoir host of MACV, the agent of Bolivian hemorrhagic fever (BHV) (Johnson et al. 1966). As with JUNV, the dynamics of the rodent population determine the epidemiological features of disease outbreaks among humans (Mercado et al. 1975). In contrast to the rodent host of JUNV, *C. callosus* invades houses during the rainy season, resulting in human cases with identical attack rates among all ages. However, on remote ranches and in fields, adult male patients predominate. A series of outbreaks from 1962 to 1964 in the sparsely populated province of El Beni in northeast Bolivia, involved more than 1,000 patients, 180 of whom died; an increase of rodents invading small towns was coincidentally reported . Transmission was interrupted by a targeted campaign to reduce the rodent population within affected towns.

Bolivian hemorrhagic fever is restricted to the tropical savanna of Beni province and recent investigations have shown that the populations of rodents responsible for the maintenance and transmission of MACV are an independent monophyletic lineage, different from those in other areas of South America (Salazar-Bravo et al. 2002).

The incidence of BHF cases is greatest between April and July (late rainy and early dry season), but the dominant epidemiologic feature is that of small outbreaks in different villages and ranches, with several years of quiescence thereafter. Transmission is thought to occur by aerosols from infected rodents or possibly by contact with food contaminated by infected rodent urine. Most of the recorded infections were acquired by direct contact with *C. laucha* or by aerosol through infected excreta. However, nosocomial transmission of MACV has been clearly demonstrated (Peters et al. 1971; Kilgore et al. 1995). Nosocomial outbreaks have been associated with a single index case who had visited a BHF endemic region. The only recognized hospital-based outbreak resulted in four secondary cases followed by a tertiary case acquired from a necropsy incident; all but one person died. Recently, an epidemic was reported in which seven members of the same family were infected, six of whom died (CDC 1994).

3.1.1.3
Venezuelan Hemorrhagic Fever: Guanarito Virus

In 1989, cases of hemorrhagic fever in the central plains of Venezuela were associated with a new *Arenavirus*, designated Guanarito virus after the region where the first outbreak occurred (Salas et al. 1991). The main affected population was settlers moving into cleared forest areas to practice small-scale agriculture. Since its discovery, GTOV has been responsible for at least 200 cases of VHF. For unknown reasons, the number of reported human cases has spontaneously dropped since 1992, although rodent infection can still be readily demonstrated within and beyond the boundaries of the original endemic zone (Weaver et al. 2001). Natural and experimental data initially suggested that two different rodent species were involved in the transmission cycle of GTOV in nature; the cane rat (*Zygodontomys brevicauda*) and the cotton rat (*Sigmodon alstoni*) (Fulhorst et al. 1999a, 1999b; Tesh et al. 1993). Recently, *Z. brevicauda* has been shown to be the primary reservoir host as it develops a persistent infection with lifelong viruria, accompanied by either low or undetectable levels of antibody. In contrast, the cotton rat has characteristics of an intermediate host infected by spillover of GTOV from cane rats, as it produces neutralizing antibodies and excretes virus for only a limited time.

Research undertaken to better understand the geographic distribution and potential variation in GTOV circulating in the VHF-epidemic area of western Venezuela resulted in the genetic sequencing of 29 isolates of GTOV obtained from rodents and humans (Weaver et al. 2000). Nine genotypes of GTOV were distinguished, all of which, with the exception of the dominant genotype, were restricted to very small geographic areas. All but one of the strains obtained from humans belonged to the dominant genotype. Closely related strains of the dominant GTOV genotype were obtained from a large area covering approximately 75,000 km^2 (Tesh et al. 1993, 1999). A single rodent could be infected by a virus population varying less than 0.5% at the nucleotide level<CHFAN (a low-diversity quasispecies). In contrast, the dominant GTOV strain infecting humans was invariant. Human disease was not associated with a unique genotype restricted to a particular rodent host species. However, overall, the available data are insufficient to conclude whether or not certain genotypes are more pathogenic and/or infectious for humans than others. The limited mobility of rodents in isolated metapopulations could account for the coexistence of independent virus lineages without mixing and competitive exclusion.

3.1.1.4 Brazilian Hemorrhagic Fever: Sabia Virus

Sabia virus has caused a single natural human infection that was fatal, and also two none fatal laboratory infections (Coimbra et al. 1994; Barry et al. 1995). No reservoir host has yet been identified.

3.1.1.5 Lassa Fever: Lassa Virus

Lassa fever is named after a small town in Nigeria, where the first epidemic was described in 1969 (Buckley and Casals 1970). LASV is associated with rodents belonging to the genus *Mastomys* (sometimes referred to as *Praomys*), which are widely distributed in sub-Saharan Africa. In the regions where LASV is endemic, up to 30% of *Mastomys* rodents can carry the virus (Keenlyside 1983). Lassa virus is responsible for an estimated 100,000–300,000 infections and approximately 5,000 deaths annually (McCormick et al. 1987). To date, cases have been reported from Nigeria, Liberia, Sierra Leone, Guinea, Burkina Faso, Ivory Coast, Ghana, Senegal, Gambia, and Mali. Among hospitalized patients, mortality is estimated at 15%–20% (Webb et al. 1986). Serologic surveys suggest that subclinical cases also occur (McCormick et al. 1987). Lassa fever occurs through direct or indirect contact with infected rodents. A number of cases acquired by local residents have been associated with the capture and handling of rodents for consumption (Ter Meulen et al. 1996).

Imported cases of LASV infection among travelers returning from endemic locations have been reported from England, Germany, Japan, the Netherlands, Israel, and the United States.

Nosocomial transmission is a common feature of Lassa fever, and many hospital-based outbreaks have been described (Keenlyside et al. 1983, Fischer-Hoch et al. 1995). However, it is apparent that this aspect of Lassa fever has been overestimated in reports based on infections in hospitals. The additional risk to hospital workers within the endemic zone is not great, as judged by serosurveys, providing that basic hygiene measures are maintained in hospitals dealing with suspected cases (Helmick et al. 1986). Nosocomial cases have been reported only in hospital settings where basic hygiene measures were not enforced. Arenaviruses readily invade the fetus, whether in their natural rodent reservoir, laboratory animals, or humans. Pregnant women infected with LASV often abort and have a high mortality rate; similar observations have been made for Argentinian and Bolivian HFs (Price et al. 1988).

4 Prevention and Control

Prevention of arenaviral disease consists of interrupting the transmission of virus from rodents to humans, from humans to humans, and from infected specimens to laboratory personnel. Strategies for reducing contact between rodents and humans have been effective in the control of outbreaks of BHF; trapping and removal of *C. callosus* in towns reduced the incidence of disease to essentially zero. Rodent intervention strategies have proven more difficult for preventing AHF as conditions under which human exposure occurs are primarily rural and associated with the harvesting of corn. The geographic distribution *C. musculinus* (reservoir host of JUNV) is much wider than *C. callosus* (reservoir of MACV), and Argentinian agricultural practices continue to place workers at risk of exposure to reservoir hosts.

A collaborative effort undertaken by the US and Argentine governments led to the production of a live attenuated Junín virus vaccine named Candid#1. Its efficacy was proven in a double-blind trial in 15,000 agricultural workers at risk to natural infection in Argentina. Subsequently, more than 100,000 people were immunized with JUNV vaccine in Argentina. A prospective study conducted over two epidemic seasons among 6,500 male agricultural workers in Argentina showed that Candid #1 vaccine efficacy was greater or equal to 84%, and no serious adverse effects were detected (Maiztegui et al 1998).

Recent animal protection studies suggest that the JUNV vaccine could be protective against MACV infections as well. However, attenuated JUNV strains do not protect experimental animals against GTOV challenge. Rhesus monkeys (*Cercopithecus aethiops*) challenged with purified inactivated LASV developed humoral antibody responses comparable to that among humans who recovered from Lassa fever. However, these monkeys were not protected when challenged with LASV and died following exposure. A naturally attenuated strain of MOPV from Mozambique protects rhesus monkeys against LASV challenge, but field studies are required to establish the extent and nature of natural human infections with this virus before it can seriously be considered a candidate for human vaccine development. Alternative approaches, including the use of vaccinia virus vectors bearing the LASV GPC or N genes, are being actively investigated and show promising preliminary results.

5 Conclusion

Arenaviruses and their rodent hosts share a common ancient history and the extant diversity of arenaviruses probably evolved through the processes of co-evolution, co-speciation and virus recombination. One can clearly distinguish four major clades of extant arenaviruses which are distributed either in the Old World (including Europe, Africa, and Asia) or the Americas. These observations are congruent with the ancient history of rodents mirroring the ancient paths and spread of *Arenavirus* ancestors. Such a model of co-evolution between parasite and specific hosts appears to apply to other viral groups such as the hantaviruses (Gonzalez 1996a) and Simian immunodeficiency virus (Kuhman et al. 2001). Two new arenaviruses have been recently discovered in Africa: the Morogoro virus isolated from *Mastomys natalensis* in Tanzania, and related to Mopeia virus, and the Kodoko virus detected in pigmy mice (*Mus Nannomys minutoides*) from Guinea. This findings together with the fact that arenavirus have coevolved with their rodent hosts strongly supports that many arenaviruses remain to be discovered not only in Europe, Americas and Africa, but also in Asia and Oceania.

Arenaviruses infect a variety of rodent hosts in which they are often nonpathogenic, whereas several are highly pathogenic for humans, resulting in severe hemorrhagic or neurological syndromes in that accidental host. Since their discovery in the early 1930s, new arenaviruses have been discovered and/or have emerged as human pathogens. As co-evolution and co-speciation occur over a long geological period, recombination appears more likely to occur in the short term and may be potentially most important in giving rise to human pathogenic strains.

References

Ackermann R, Bloedhorn H, Kupper B, Winkens I, Scheid W (1964) Über die Verbreiting des Virus der lymphocytaren Choriomeningitis unter den Mausen in Westdeutschland. I Utersuchungen uberwiegend an Hausmauden *(Mus musculus)*. Zentrabl Bakteriol 194:407–430

Ackermann R, Stille W, Blumenthal W, Helm EB, Keller K, Baldus O (1972) Syrische Goldhamster als Ubertrager von lymphozytarer Choriomeningitis. Dtsch Med Wochenschr 97:1725–1731

Ahmed R, Hahn CS, Somasundaram T, Villarete L, Matloubian M, Strauss JH (1991) Molecular basis of organ-specific selection of viral variants during chronic infection. J Virol 65:4242–4247

Albarino CG, Ghiringhelli PD, Posik DM, Lozano ME, Ambrosio AM, Sanchez A, Romanowski V (1997) Molecular characterization of attenuated Junin virus strains. J Gen Virol 78:1605–1610

Amman BR, Pavlin BI, Albariño CG, Comer JA, Erickson BR, Oliver JB, et al. Pet rodents and fatal lymphocytic choriomeningitis in transplant patients. Emerg Infect Dis [serial on the Internet]. 2007 May [*date cited*]. Available from http://www.cdc.gov/EID/content/13/5/719.htm

Archer AM, Rico-Hesse R (2002) High genetic divergence and recombination in Arenaviruses from the Americas. Virology 20:304:274–281

Armstrong C, Lillie RD (1934) Experimental lymphocytic choriomeningitis of monkeys and mice produced by a virus encountered in studies of the 1933 St Louis encephalitis epidemic. Public Health Rep 49:1019–1027

Banner LR, Lai MM (1991) Random nature of coronavirus RNA recombination in the absence of selection pressure. Virology 185:441–445

Barry M, Russi M, Armstrong L, Geller D, Tesh R, Denbry L, Gonzalez JP, Khan A, Peters CJ (1995) Treatment of a laboratory-acquired Sabia virus infection. N Eng J Med 333:294–296

Barton LL (1996) Lymphocytic choriomeningitis virus: a neglected central nervous system pathogen. Clin Infect Dis 22:197

Barton LL, Hyndman NJ (2000) Lymphocytic choriomeningitis virus: reemerging central nervous system pathogen. Pediatrics 105:E35

Barton LL, Budd SC, Morfitt WS, Peters CJ, Ksiazek TG, Schindler RF, Yoshino MT (1993) Congenital lymphocytic choriomeningitis virus infection in twins. Pediatr Infect Dis J 12:942–946

Beaty BJ, Holterman M, Tabachnick W, Shope RE, Rozhon EJ, Bishop DHL (1981) Molecular basis of bunyavirus transmission by mosquitoes: role of the middle-sized RNA segment. Science 211:1433–1435

Beaty BJ, Miller BR, Shope RE, Rozhon EJ, Bishop DHL (1982) Molecular basis of bunyavirus per os infection of mosquitoes: role of the middle-sized RNA segment. Proc Natl Acad Sci U S A 79:1295–1297

Bergmann M, Garcia-Sastre A, Palese P (1992) Transfection-mediated recombination of influenza A virus. J Virol 66:7576–7580

Biggar RJ, Woodall JP, Walter PD, Haughie GE (1975) Lymphocytic choriomeningitis outbreak associated with pet hamsters: fifty-seven cases from New York State. JAMA 232:494–500

Bishop DH, Beaty BJ, Shope RE (1980) Recombination and gene coding assignments of bunyaviruses and arenaviruses. Ann N Y Acad Sci 354:84–106

Bockstahler LE, Carney PG, Bushar G, Sagripanti JL (1992) Detection of Junin virus by the polymerase chain reaction. J Virol Methods 39:231–235

Bowen MD, Peters CJ, Nichol ST (1996) The phylogeny of New World (Tacaribe complex) arenaviruses. Virology 219:285–290

Bowen MD, Peters CJ, Nichol ST (1997) Phylogenetic analysis of the Arenaviridae: patterns of virus evolution and evidence for cospeciation between arenaviruses and their rodent hosts. Mol Phylogenet Evol 8:301–316

Bowen MD, Rollin PE, Ksiazek TG, Hustad HL, Bausch DG, Demby AH, Bajani MD, Peters CJ, Nichol ST (2000) Genetic diversity among Lassa virus strains. J Virol 74:6992–7004

Bowen MD, Trappier SG, Sanchez AJ, Meyer RF, Goldsmith CS, Zaki SR, Dunster LM, Peters CJ, Ksiazek TG, Nichol ST, the RVF Task Force (2001) A reassortant bunyavirus isolated from acute hemorrhagic fever cases in Kenya and Somalia. Virology 291:185–190

Buckley SM, Casals J (1970) Lassa fever: a new virus disease of man from West Africa. III. Isolation and characterization of the virus. Am J Trop Med Hyg 19:680–691

Buchmeier M, Adam E, Rawls WE (1974) Serologic evidence of infection by Pichinde virus among laboratory workers. Infect Immun 9:821–823

Calisher CH, Tzianabos T, Lord RD, Coleman PH (1970) Tamiami virus, a new member of the TaCaribe group. Am J Trop Med Hyg 19:520–526

Centers for Disease Control and Prevention (1994) Bolivian hemorrhagic fever – El Beni department Bolivia, 1994. Morb Mortal Wkly Rep 43:943–946

Centers for Disease Control and Prevention (2000) Fatal illnesses associated with a New World arenavirus – California, 1999–2000. Morb Mortal Wkly Rep 49:709–711

Centers for Disease Control and Prevention (2005) Lymphocytic choriomeningitis virus infection in organ transplant recipients, Massachusetts Rhode Island, 2005. Morb Mortal Wkly Rep 54:537–539

Charrel RN, de Lamballerie X, Fulhorst CF (2001) The Whitewater Arroyo virus: natural evidence for genetic recombination among Tacaribe serocomplex viruses (family Arenaviridae). Virology 283:161–166

Charrel RN, Feldmann H, Fulhorst CF, Khelifa R, de Chesse R, de Lamballerie X (2002) Phylogeny of New World arenaviruses based on the complete coding sequences of the small genomic segment identified an evolutionary lineage produced by intrasegmental recombination. Biochem Biophys Res Commun 296:1118–1124

Charrel RN, Retornaz K, Emonet S, Noel G, Chaumoitre K, Minodier P, Girard N, Garnier JM, de Lamballerie X. Acquired hydrocephalus caused by a variant lymphocytic choriomeningitis virus. Arch Intern Med. 2006; 166:2044–2046

Childs JE, Peters CJ (1993) Ecology and epidemiology of arenaviruses and their hosts. In: Slavato MS (ed) The arenaviruses. Plenum, New York, pp 331–384

Childs JE, Glass GE, Ksiazek TG, Rossi CA, Oro JG, Leduc JW (1991) Human-rodent contact and infection with lymphocytic choriomeningitis and Seoul viruses in an inner-city population. Am J Trop Med Hyg 44:117–121

Childs JE, Glass GE, Korch GW, Ksiazek TG, Leduc JW (1992) Lymphocytic choriomeningitis virus infection and house mouse *(Mus musculus)* distribution in urban Baltimore. Am J Trop Med Hyg 47:27–34

Clegg JCS, Bowen MD, Buchmeier MJ, Gonzalez JP, Lukashevich IS, Peters CJ, Rico-Hesse R, Romanowski V (2000) Arenaviridae. In: Van Regenmortel MHV, Fauquet CM, Bishop

DHL, Carsten EB, Estes MK, Lemon SM, Maniloff J, Mayo MA, McGeoch DJ, Pringle DR, Wickner RB (eds) Virus taxonomy. Seventh Report of the International Committee for the Taxonomy of Viruses, Academic, New York, pp 633–640

Cummins D (1990) Lassa fever. Br J Hosp Med 43:186–192

Demby AH, Chamberlain J, Brown DW, Clegg CS (1994) Early diagnosis of Lassa fever by reverse transcription-PCR. J Clin Microbiol 32:2898–2903

Desport M, Collins ME, Brownlie J (1998) Genome instability in BVDV: an examination of the sequence and structural influences on RNA recombination. Virology 246:352–361

Desselberger U, Hung T, Follett EA (1986) Genome analysis of human rotaviruses by oligonucleotide mapping of isolated RNA segments. Virus Res 4:357–368

Downs WG, Anderson CR, Spoence L, Aitken THG, Green Hall AH (1963) Tacaribe virus a new agent isolated from *Artibeus* bats and mosquitoes in Trinidad West Indies. Am J Trop Med Hyg 12:640–646

Dropulic LK, Hardwick JM, Griffin DE (1997) A single amino acid change in the E2 glycoprotein of Sindbis virus confers neurovirulence by altering an early step of virus replication. J Virol 71:6100–6105

Dubuisson J, Lustig S, Ruggli N, Akov Y, Rice CM (1997) Genetic determinants of Sindbis virus neuroinvasiveness. J Virol 71:2636–2646

Dykewicz CA, Dato VM, Fisher-Hoch SP, Howarth MV, Perez-Oronoz GI, Ostroff SM, Gary H Jr, Schonberger LB, McCormick JB (1992) Lymphocytic choriomeningitis outbreak associated with nude mice in a research institute. JAMA 267:1349–1353

Emonet S, Retornaz K, Gonzalez J-P, de Lamballerie X, Charrel RN. Mouse-to-human transmission of variant lymphocytic choriomeningitis virus. Emerg Infect Dis [serial on the Internet]. 2007 Mar [*date cited*]. Available from http://www.cdc.gov/EID/content/13/3/472.htm

Engel SR, Hogan KM, Taylor JF, Davis SK (1998) Molecular systematics and paleobiogeography of the South American sigmodontine rodents. Mol Biol Evol 15:35–49

Enria D, Feuillade MR (1998) Argentine haemorrhagic fever (Junin virus-Arenaviridae): a review on clinical, epidemiological, ecological, treatment and preventive aspects of the disease. In: Travassos da Rosa APA, Vasconcelos PFC, Travassos da Rosa JFS (eds) An overview of arbovirology in Brazil and neighboring countries. Instituto Evandro Chaggas Belem Brazil, pp 219–232

Enria D, Bowen M, Mills JN, Shieh WJ, Bausch D, Peters CJ (1999) Arenaviruses. In: Guerrant RL, Walker DH, Weller PF(eds) Tropical infectious diseases: principles pathogens, and practice. Churchill Livingstone; Philadelphia, chapter 111

Enria DA, Barrera Oro JG (2002) Junin virus vaccines. Curr Top Microbiol Immunol 263:239–261

Fischer SA, Graham MB, Kuehnert MJ, Kotton CN, Srinivasan A, Marty FM, Comer JA, Guarner J, Paddock CD, DeMeo DL, Shieh WJ, Erickson BR, Bandy U, DeMaria A Jr, Davis JP, Delmonico FL, Pavlin B, Likos A, Vincent MJ, Sealy TK, Goldsmith CS, Jernigan DB, Rollin PE, Packard MM, Patel M, Rowland C, Helfand RF, Nichol ST, Fishman JA, Ksiazek T, Zaki SR; LCMV in Transplant Recipients Investigation Team. Transmission of lymphocytic choriomeningitis virus by organ transplantation. N Engl J Med. 2006; 354:2235–2249

Fisher-Hoch SP, Tomori O, Nasidi A, Perez-Oronoz GI, Fakile Y, Hutwagner L, McCormick JB (1995) Review of cases of nosocomial Lassa fever in Nigeria: the high price of poor medical practice. BMJ 311:857–859

Fulhorst CF, Bowen MD, Ksiazek TG, Rollin PE, Nichol ST, Kosoy MY, Peters CJ (1996) Isolation and characterization of Whitewater Arroyo virus, a novel North American arenavirus. Virology 224:114–120

Fulhorst CF, Bowen MD, Salas RA, Duno G, Utrera A, Ksiazek TG, De Manzione NM, De Miller E, Vasquez C, Peters CJ, Tesh RB (1999a) Natural rodent host associations of Guanarito and pirital viruses (family Arenaviridae) in central Venezuela. Am J Trop Med Hyg 61:325–330

Fulhorst CF, Ksiazek TG, Peters CJ, Tesh RB (1999b) Experimental infection of the cane mouse *Zygodontomys brevicauda* (family Muridae) with Guanarito virus (Arenaviridae), the etiologic agent of Venezuelan hemorrhagic fever. J Infect Dis 180:966–969

Fulhorst CF, Charrel RN, Weaver SC, Ksiazek TG, Bradley RD, Milazzo ML, Tesh RB, Bowen MD (2001) Geographic distribution and genetic diversity of Whitewater Arroyo virus in the southwestern United States. Emerg Infect Dis 7:403–407

Fulhorst CF, Bennett SG, Milazzo ML, Murray HL Jr, Webb JP Jr, Cajimat MNB, Bradley RB (2002) Bear canyon virus: an arenavirus naturally associated with the California mouse (*Peromyscus californicus*). Emerg Infect Dis 8:717–721

Garcia JB, Morzunov SP, Levis S, Rowe J, Calderon G, Enria D, Sabattini M, Buchmeier MJ, Bowen MD, St Jeor SC (2000) Genetic diversity of the Junin virus in Argentina: geographic and temporal patterns. Virology 272:127–136

Georgescu MM, Balanant J, Macadam A, Otelea D, Combiescu M, Combiescu AA, Crainic R, Delpeyroux F (1997) Evolution of the Sabin type 1 poliovirus in humans: characterization of strains isolated from patients with vaccine-associated paralytic poliomyelitis. J Virol 71:7758–7768

Grande-Perez A, Gomez-Mariano G, Lowenstein PR, Domingo E (2005) Mutagenesis-induced, large fitness variations with an invariant arenavirus consensus genomic nucleotide sequence. J Virol 79:10451–10459

Gritsun TS, Desai A, Gould EA (2001) The degree of attenuation of tick-borne encephalitis virus depends on the cumulative effects of point mutations. J Gen Virol 82:1667–1675

Gunther S, Emmerich P, Laue T, Kuhle O, Asper M, Jung A, Grewing T, Ter Meulen J, Schmitz H (2000) Imported lassa fever in Germany: molecular characterization of a new lassa virus strain. Emerg Infect Dis 6:466–476

Gerrard SR, Li L, Barrett AD, Nichol ST (2004) Ngari virus is a Bunyamwera virus reassortant that can be associated with large outbreaks of hemorrhagic fever in Africa. J Virol 78:8922–8926

Gonzalez JP (1986) Les arénavirus d'Afrique : un nouveau paradigme d'évolution. Bull Institut Pasteur 84:67–85

Gonzalez JP, McCormick JB, Herve JP, Jonhson KM, Georges AJ (1983) An arenavirus isolated from wild-caught rodents in the Central African Republic. Intervirology 19:105–112

Gonzalez JP (1996) Coevolution of rodent and viruses: Arenaviruses and Hantaviruses. *In* New Dimension *in* Parasitology M. Ali Ozcel edit. Acta Parasitologica Turcica. 20, supp. 1:617–638

Gonzalez JP, Bowen M, Nicholl S, Rico-Hesse R (1996) Genetic characterization of Sabiá arenavirus an emerging human pathology, 1996. J Gen Virol 221:218–324

Gonzalez JP, Georges AJ, Kiley MP, Meunier DMY, Peters CJ, McCormick JB (1986) Evolutionary biology of a Lassa virus complex. Med Microbiol Immunol 175:157–159

Gonzalez JP, Sanchez A, Rico-Hesse (1995) Venezuelan molecular phylogeny of Guanarito virus, an emerging human arenavirus. J Am Soc Trop Med Hyg 53:1–6

Hahn CS, Lustig S, Strauss EG, Strauss JH (1988) Western equine encephalitis virus is a recombinant virus. Proc Natl Acad Sci U S A 85:5997–6001

Helmick CG, Webb PA, Scribner CL, Krebs JW, McCormick JB (1986) No evidence for increased risk of Lassa ferver infection in hospital staff. Lancet 2:1202–1205

Hinman AR, Fraser DW, Douglas RG et al (1975) Outbreak of lymphocytic choriomeningitis virus infection in medical center personnel. Am J Epidemiol 101:103

Hirabayashi Y, Oka S, Goto H, Shimada K, Kurata T, Fisher-Hoch SP, McCormick JB (1988) An imported case of Lassa fever with late appearance of polyserositis. J Infect Dis 158:872–875

Hugot JP, Denys CH, Gonzalez JP (2001) Evolution of the Old World Arenaviridae and their rodent hosts: generalized host-transfer or association by descent? Infect Genet Evol 1:22–25

Jarvis TC, Kirkegaard K (1992) Poliovirus RNA recombination: mechanistic studies in the absence of selection. EMBO J 11:3135–3145

Johnson KM, Wiebenga NH, Mackenzie RB, Kuns ML, Tauraso NM, Shelekov A, Webb PA, Justines G, Beye HK (1965) Virus isolations from human cases of hemorrhagic fever in Bolivia. Proc Soc Exp Biol Med 118:113–118

Karabatsos N (1985) International Catalogue of Arboviruses Including Certain Other Viruses of Vertebrates, 3rd edn. American Society of Tropical Medicine and Hygiene, San Antonio

Keenlyside RA, McCormick JB, Webb PA, Smith E, Elliott L, Johnson KM (1983) Case-control study of Mastomys natalensis and humans in Lassa virus-infected households in Sierra Leone. Am J Trop Med Hyg 32:829–837

Kilgore PE, Peters CJ, Mills JN, Rollin PE, Armstrong L, Khan AS, Ksiazek TG (1995) Prospects for the control of Bolivian hemorrhagic fever. Emerg Infect Dis 1:97–100

Kosoy MY, Elliott LH, Ksiazek TG, Fulhorst CF, Rollin PE, Childs JE, Mills JN, Maupin GO, Peters CJ (1996) Prevalence of antibodies to arenaviruses in rodents from the southern and western United States: evidence for an arenavirus associated with the genus Neotoma. Am J Trop Med Hyg 54:570–576

Kuhmann SE, Madani N, Diop OM, Platt EJ, Morvan J, Muller-Trutwin MC, Barre-Sinoussi F, Kabat D (2001) Frequent substitution polymorphisms in African green monkey CCR5 cluster at critical sites for infections by simian immunodeficiency virus SIVagm, implying ancient virus-host coevolution. J Virol 75:8449–8460

Lai MM (1992) RNA recombination in animal and plant viruses. Microbiol Rev 56:61–79

Lecompte E, Ter Meulen J, Emonet S, Daffis S, Charrel RN. Genetic identification of Kodoko virus, a novel arenavirus of the African pigmy mouse (*Mus Nannomys minutoides*) in West Africa. Virology. 2007 Mar 22; [Epub ahead of print]

Lehmann-Grube F (1971) Lymphocytic choriomeningitis virus. Springer, New York

Lisieux T, Coimbra M, Nassar ES, Burattini MN, de Souza LT, Ferreira I, Rocco IM, da Rosa AP, Vasconcelos PF, Pinheiro FP et al (1994) New arenavirus isolated in Brazil. Lancet 343(8894):391–392

Lozano ME, Ghiringhelli PD, Romanowski V, Grau O (1993) A simple nucleic acid amplification assay for the rapid detection of Junin virus in whole blood samples. Virus Res 27:37–53

Lozano ME, Enria D, Maiztegui JI, Grau O, Romanowski V (1995) Rapid diagnosis of Argentine hemorrhagic fever by reverse transcriptase PCR-based assay. J Clin Microbiol 33:1327–1332

Lozano ME, Posik DM, Albarino CG, Schujman G, Ghiringhelli PD, Calderon G, Sabattini M, Romanowski V (1997) Characterization of arenaviruses using a family-specific primer set for RT-PCR amplification and RFLP analysis. Its potential use for detection of uncharacterized arenaviruses. Virus Res 49:79–89

Lukashevich IS (1992) Generation of reassortants between African arenaviruses. Virology 188:600–605

Lunkenheimer K, Hufert FT, Schmitz H (1990) Detection of Lassa virus RNA in specimens from patients with Lassa fever by using the polymerase chain reaction. J Clin Microbiol 28:2689–2692

Maiztegui JI (1975) Clinical and epidemiological patterns of Argentine haemorrhagic fever. Bull World Health Organ 52:567–576

Maiztegui JI, Sabattini MS, Barrera Oro JG (1972) Activity of lymphocytic choriomeningitis virus (LCM) in the endemic area of Argentine hemorrhagic fever (AHF). I. Serological studies in rodents captured in the City of Pergamino. Medicina (B Aires) 32:131–137

Maiztegui JI, Fernandez NJ, de Damilano AJ (1979) Efficacy of immune plasma in treatment of Argentine haemorrhagic fever and association between treatment and a late neurological syndrome. Lancet 2:1216–1217

Maiztegui J, Feuillade M, Briggiler A (1986) Progressive extension of the endemic area and changing incidence of Argentine hemorrhagic fever. Med Microbiol Immunol (Berl) 175:149–152

Maiztegui JI, McKee KT Jr, Barrera Oro JG, Harrison LH, Gibbs PH, Feuillade MR, Enria DA, Briggiler AM, Levis SC, Ambrosio AM, Halsey NA, Peters CJ (1998) Protective efficacy of a live attenuated vaccine against Argentine hemorrhagic fever. AHF Study Group. J Infect Dis 177:277–283

McCormick JB, Webb PA, Krebs JW, Johnson KM, Smith ES (1987) A prospective study of the epidemiology and ecology of Lassa fever. J Infect Dis 155:437–444

McKee KT Jr, Mahlandt BG, Maiztegui JI, Eddy GA, Peters CJ (1985) Experimental Argentine hemorrhagic fever in rhesus macaques: viral strain-dependent clinical response. J Infect Dis 152:218–221

Mercado R (1975) Rodent control programmes in areas affected by Bolivian haemorrhagic fever. Bull World Health Organ 52:691–695

Meyer H, Johnson R, Crawford I, Dascomb H, Rogers N (1960) Central nervous systems syndromes of "viral" etiology: a study of 713 cases. Am J Med 29:334–347

Mills JN, Ellis BA, McKee KT Jr, Ksiazek TG, Oro JG, Maiztegui JI, Calderon GE, Peters CJ, Childs JE (1991) Junin virus activity in rodents from endemic and non endemic loci in control Argentina. Am J Trop Med Hyg 44:589–597

Mills JN, Ellis BA, McKee KT, Calderón GE, Maiztegui JI, Nelson GO, Ksiazek TG, Peters CJ, Childs JE (1992) A longitudinal study of Junín virus activity in the rodent reservoir of Argentine hemorrhagic fever. Am J Trop Med Hyg 47:749–763

Mills JN, Barrera Oro JG, Bressler DS, Childs JE, Tesh RB, Smith JF, Enria DA, Geisbert TW, McKee KT Jr, Bowen MD, Peters CJ, Jahrling PB (1996) Characterization of Oliveros virus, a new member of the Tacaribe complex (Arenaviridae: *Arenavirus*). Am J Trop Med Hyg 54:399–404

Moncayo AC, Hice CL, Watts DW, Travassos da Rosa AP, Guzman H, Russell KL, Calampa C, Gozalo A, Popov VL, Weaver SC, Tesh RB (2001) Allpahuayo virus: a newly recognized arenavirus (Arenaviridae) from arboreal rice rats (Oecomys bicolor and Oecomys paricola) in northeastern Peru. Virology 284:277–286

Murphy FA, Webb PA, Johnson KM, Whitfield SG (1969) Morphological comparison of Machupo virus with lymphocytic choriomeningitis: basis for a new taxonomic group. J Virol 535:541

Musser GG, Carleton MD (1993) Family Muridae In: Wilson DE, Reeder DM (eds) Mammals species of the world. Smithsonian Institution, Washington, DC

Orlich M, Gottwald H, Rott R (1994) Nonhomologous recombination between the hemagglutinin gene and the nucleoprotein gene of an influenza virus. Virology 204:462–465

Park JY, Peters CJ, Rollin PE, Ksiazek TG, Gray B, Waites KB, Stephensen CB (1997) Development of a reverse transcription-polymerase chain reaction assay for diagnosis of lymphocytic choriomeningitis virus infection and its use in a prospective surveillance study. J Med Virol 51:107–114

Parodi AS Greenway DJ, Rugiero HR (1958) Sobre la etiologia del brote epidemico de Junin. El Dia Medico 30:2300–2301

Peters CJ (1995) Arenavirus diseases. In: Porterfield JS (ed) Kass handbook of infectious diseases. Exotic viral infections. Chapman and Hall Medical, New York, pp 227–246

Peters CJ, Kuehne RW, Mercado R, Le Bow RH, Spertzel RO, Webb PA (1974) Hemorrhagic fever in Cochabamba Bolivia, 1971. Am J Epidemiol 99:425–433

Peters CJ, Buchmeier M, Rollin PE, Ksiazek TG (1996) Arenaviruses. In: Fields BN, Knipe DM, Howley PM, Chanock RM, Melnick JL, Monath TP, Roizman R, Straus SE (eds) Fields virology, 3 rd edn. Lippincott-Raven, Philadelphia, pp 1521–1551

Pinheiro FP, Woodal JP, Travassos da Rosa APA, Ravassos da Rosa JF (1977) Studies on arenvairus in Brazil. Medicina B Aires [Suppl 3] 175:197

Price ME, Fisher-Hoch SP, Craven RB, McCormick JB (1988) A prospective study of maternal and fetal outcome in acute Lassa fever infection during pregnancy. BMJ 297:584–587

Riviere Y, Oldstone MB (1986) Genetic reassortants of lymphocytic choriomeningitis virus: unexpected disease and mechanism of pathogenesis. J Virol 59:363–368

Riviere Y, Ahmed R, Southern PJ, Buchmeier MJ, Oldstone MB (1985) Genetic mapping of lymphocytic choriomeningitis virus pathogenicity: virulence in guinea pigs is associated with the L RNA segment. J Virol 55:704–709

Sabattini MS, Maiztegui JI (1970) Fiebre hemorrhagica argentina. Medicina (Buenos Aires). 30 [Suppl]:111

Salas R, de Manzione N, Tesh RB, Rico-Hesse R, Shope RE, Betancourt A, Godoy O, Bruzual R, Pacheco ME, Ramos B et al (1991) Venezuelan hemorrhagic fever. Lancet 338:1033–1036

Salvato M, Borrow P, Shimomaye E, Oldstone MB (1991) Molecular basis of viral persistence: a single amino acid change in the glycoprotein of lymphocytic choriomeningitis virus is associated with suppression of the antiviral cytotoxic T-lymphocyte response and establishment of persistence. J Virol 65:1863–1869

Salvato M, Clegg JCS, Bowen MD, Buchmeier MJ, Gonzalez JP, Lukashevich IS, Peters CJ, Rico-Hesse R, Romanowski V (2005) Arenaviridae. In: Van Regenmortel MHV, Fauquet CM, Bishop DHL, Carsten EB, Estes MK, Lemon SM, Maniloff J, Mayo MA, McGeoch DJ, Pringle DR, Wickner RB (eds) Virus Taxonomy. 8 th ICTV Report of the International Committee for the Taxonomy of Viruses. Academic, New York, pp 633–640

Salazar-Bravo J, Dragoo JW, Bowen MD, Peters CJ, Ksiazek TG, Yates TL (2002) Natural nidality in Bolivian hemorrhagic fever and the systematics of the reservoir species. Infect Genet Evol 1:191–199

Schlaeffer F, Bar-Lavie Y, Sikuler E, Alkan M, Keynan A (1988) Evidence against high contagiousness of Lassa fever. Trans R Soc Trop Med Hyg 82:311

Sewell DL (1995) Laboratory-associated infections and biosafety. Clin Microbiol Rev 8:389–405

Sibold C, Meisel H, Kruger DH, Labuda M, Lysy J, Kozuch O, Pejcoch M, Vaheri A, Plyusnin A (1999) Recombination in Tula hantavirus evolution: analysis of genetic lineages from Slovakia. J Virol 73:667–675

Skinner HH, Knight EH, Buckley LS (1976) The hamster as a secondary reservoir host of lymphocytic choriomeningitis virus. J Hyg (Lond) 76:299–306

Stephensen CB, Blount SR, Lanford RE, Holmes KV, Montali RJ, Fleenor ME, Shaw JF (1992) Prevalence of serum antibodies against lymphocytic choriomeningitis virus in selected populations from two US cities. J Med Virol 38:27–31

Stinebaugh BJ, Schloeder FX, Johnson KM, Mackenzie RB, Entwisle G, De Alba E (1966) Bolivian hemorrhagic fever: a report of four cases. Am J Med 40:217–230

Suzuki Y, Gojobori T, Nakagomi O (1998) Intragenic recombinations in rotaviruses. FEBS Lett 427:183–187

Swanepoel R, Leman PA, Shepherd AJ, Shepherd SP, Kiley MP, McCormick JB (1985) Identification of Ippy as a Lassa-fever-related virus. Lancet 1(8429):639

Ter Meulen J, Lukashevich I, Sidibe K, Inapogui A, Marx M, Dorlemann A, Yansane ML, Koulemou K, Chang-Claude J, Schmitz H (1996) Hunting of peridomestic rodents and consumption of their meat as possible risk factors for rodent-to-human transmission of Lassa virus in the Republic of Guinea. Am J Trop Med Hyg 55:661–666

Tesh RB, Wilson ML, Salas R, De Manzione NM, Tovar D, Ksiazek TG, Peters CJ (1993) Field studies on the epidemiology of Venezuelan hemorrhagic fever: implication of the cotton rat Sigmodon alstoni as the probable rodent reservoir. Am J Trop Med Hyg 49:227–235

Tesh RB, Salas RA, Fulhorst CF, de Manzione N, Duno G, Weaver SC, Utrera A, Parades H, Ellis BA, Mills JN, Bowen MD, Ksiazek TG (1999) Epidemiology of arenaviruses in the Americas. In: Saluzzo JE, Dodet B (eds) Rodent-borne viral Diseases (hantaviruses and arenaviruses). Elsevier, Paris, pp 213–221

Trapido H, Sanmartin C (1971) Pichinde virus: a new virus from Colombia. Am J Trop Med Hyg 20:631–664

Trappier SG, Conaty AL, Farrar BB, Auperin DD, McCormick JB, Fisher-Hoch SP (1993) Evaluation of the polymerase chain reaction for diagnosis of Lassa virus infection. Am J Trop Med Hyg 49:214–221

Van der Heide RM (1982) A patient with Lassa fever from Burkina Faso (formerly Upper Volta), diagnosed in the Netherlands. Ned T Geneesk 126:1–7

Weaver SC, Salas RA, de Manzione N, Fulhorst CF, Duno G, Utrera A, Mills JN, Ksiazek TG, Tovar D, Tesh RB (2000) Guanarito virus (Arenaviridae) isolates from endemic and outlying localities in Venezuela: sequence comparisons among and within strains isolated from Venezuelan hemorrhagic fever patients and rodents. Virology 266:189–195

Weaver SC, Salas RA, de Manzione N, Fulhorst CF, Travasos da Rosa APA, Duno G, Utrera A, Mills JN, Ksiazek TG, Tovar D, Guzman H, Kang W, Tesh RB (2001) Extreme genetic diversity among Pirital virus (Arenaviridae) isolates from Western Venezuela. Virology 285:110–118

Webb PA, Johnson KM, Hibbs JB, Kuns ML (1970) Parana, a new Tacaribe complex virus from Paraguay. Arch Gesamte Virusforsch 32:379–388

Webb PA, Justines G, Johnson KM (1975) Infection of wild and laboratory animals with Machupo and Latino viruses. Bull World Health Organ 52:493–499

Webb PA, McCormick JB, King IJ, Bosman I, Johnson KM, Elliott LH, Kono GK, O'Sullivan R (1986) Lassa fever in children in Sierra Leone West Africa. Trans R Soc Trop Med Hyg 80:577–582

Weissenbacher MC, Coto CE, Calello MA (1975) Cross-protection between Tacaribe complex viruses. Presence of neutralizing antibodies against Junin virus (Argentine hemorrhagic fever) in guinea pigs infected with Tacaribe virus. Intervirology 6:42–49

Whitton JL, Southern PJ, Oldstone MB (1988) Analyses of the cytotoxic T lymphocyte responses to glycoprotein and nucleoprotein components of lymphocytic choriomeningitis virus. Virology 162:321–327

Wilson DE, Reeder DM (2005) Mammal Species of the World, 2nd edn. Johns Hopkins University Press, Baltimore

Worobey M, Rambaut A, Holmes EC (1999) Widespread intras-serotype recombination in natural populations of dengue virus. Proc Natl Acad Sci U S A 96:7352–7357

Wulff H, McIntosh BM, Hammer DB, Johnson KM (1977) Isolation of arenavirus closely related to Lassa virus from *Mastomys natalensis* in South East Africa. Bull World Health Organ 55:441–444

CTMI (2007) 315:289–324
© Springer-Verlag Berlin Heidelberg 2007

Ecological Havoc, the Rise of White-Tailed Deer, and the Emergence of *Amblyomma americanum*-Associated Zoonoses in the United States

C. D. Paddock[1] (✉) · M. J. Yabsley[2]

[1]Division of Viral and Rickettsial Diseases, National Center for Infectious Diseases, Centers for Disease Control and Prevention, Atlanta, GA, USA
cdp9@cdc.gov
[2]Southeastern Cooperative Wildlife Disease Study, College of Veterinary Medicine, University of Georgia, Athens, GA, USA

Abstract Two infectious diseases, and one presumably infectious disease, each vectored by or associated with the bite of the lone star tick (*Amblyomma americanum*), were identified and characterized by clinicians and scientists in the United States during the 1980s and 1990s. These three conditions—human monocytic (or monocytotropic) ehrlichiosis (HME), *Ehrlichia ewingii* ehrlichiosis, and southern tick-associated rash illness (STARI)—undoubtedly existed in the United States prior to this time. However, the near-simultaneous recognition of these diseases is remarkable and suggests the involvement of a unifying process that thrust multiple pathogens into the sphere of human recognition. Previous works by other investigators have emphasized the pivotal role of white-tailed deer (*Odocoileus virginianus*) in the emergence of Lyme disease, human babesiosis, and human granulocytic anaplasmosis. Because whitetails serve as a keystone host for all stages of lone star ticks, and an important reservoir host for *Ehrlichia chaffeensis*, *E. ewingii*, and *Borrelia lonestari*, the near-exponential growth of white-tailed deer populations that occurred in the eastern United States during the twentieth century is likely to have dramatically affected the frequency and distribution of *A. americanum*-associated zoonoses. This chapter describes the natural histories of the pathogens definitively or putatively associated with HME, *E. ewingii* ehrlichiosis, and STARI; the role of white-tailed deer as hosts to lone star ticks and the agents of these diseases; and the cascade of ecologic disturbances to the landscape of the United States that have occurred during the last 200 years that provided critical leverage in the proliferation of white-tailed deer, and ultimately resulted in the emergence of these diseases in human populations.

1 Introduction

> The American white-tailed deer (*Odocoileus virginianus*) is the oldest deer species alive. It is an expert in surviving predation of diverse forms and, like other old North American indigenous mammals, adjusts remarkably well to human activity, to cities, and to agriculture. It is a deer of ecological havoc, a survival virtuoso...
>
> Valerius Geist 1998

Five tickborne infectious diseases—babesiosis, Lyme disease, human monocytic (or monocytotropic) ehrlichiosis (HME), human granulocytic anaplasmosis (HGA), and *Ehrlichia ewingii* ehrlichiosis—were identified and characterized by clinicians and scientists in the United States during a relatively short span of three decades between 1969 and 1999 (Scrimenti 1970; Western et al. 1970; Steere et al. 1978; Maeda et al. 1987; Bakken et al. 1994; Buller et al. 1999). A sixth, as-yet etiologically uncharacterized syndrome, southern tick-associated rash illness (STARI), was also discovered during this period (Schulze et al. 1984; Masters et al. 1994, 1998). The appreciation of these previously unrecognized infections and subsequent discoveries of the varied pathogenic agents that caused these conditions effectively doubled the number of distinct, North American, tick-transmitted diseases and expanded considerably the recognized magnitude of tick-borne infections in the United States. Until the early 1980s, Rocky Mountain spotted fever was the most commonly recognized tick-borne disease in the United States. During 2003, passive surveillance identified approximately 1,100 cases of this disease; however, approximately 320, 360, and

21,300 cases of HME, HGA, and Lyme disease, respectively, were also reported during this same interval (Centers for Disease Control and Prevention 2005). None of these last three diseases had been identified three decades earlier.

It is extremely unlikely that one or more of the pathogens that cause these illnesses arrived in North America during the last half of the twentieth century. For example, DNA of *Borrelia burgdorferi*, the causative agent of Lyme disease, has been detected in archival specimens of New England deer mice and black-legged ticks collected during the 1890s and 1940s, respectively (Persing et al. 1990; Marshall et al. 1994). As outlined in other chapters of this book, multiple factors over time and space contributed to the appreciation of these ecologically and etiologically diverse zoonoses in human populations. Nonetheless, the near-simultaneous recognition of these varied diseases is remarkable and suggests the involvement of a unifying process that thrust these pathogens into the sphere of human recognition.

Several compelling arguments describe the pivotal role of white-tailed deer in the emergence of Lyme disease and babesiosis in the northeastern and upper midwestern United States (Piesman et al. 1979; Wilson et al. 1985; Spielman et al. 1993; Spielman 1994). While this chapter borrows insights provided by these arguments, it focuses primarily on various environmental and ecological imbalances that were introduced to white-tailed deer by a cascade of human interventions during the nineteenth and twentieth centuries and how these combined to create the emergence of three diseases—HME, *E. ewingii* ehrlichiosis, and STARI—each of which is associated with the lone star tick (*Amblyomma americanum*) (Fig. 1).

2 The Natural History of *A. americanum*-associated Zoonoses

The role played by white-tailed deer in the recognition of multiple zoonoses transmitted by *A. americanum* can be linked to several sources of data that implicate deer as the keystone host for lone star tick populations and as an important natural reservoir for the pathogens that cause these diseases.

2.1 White-Tailed Deer as Hosts for *A. americanum*

A. americanum is a widely-distributed, hard tick that obtains its blood meals from a variety of ground-nesting birds and medium-to-large-sized mammals. White-tailed deer support all parasitic stages of *A. americanum* and are

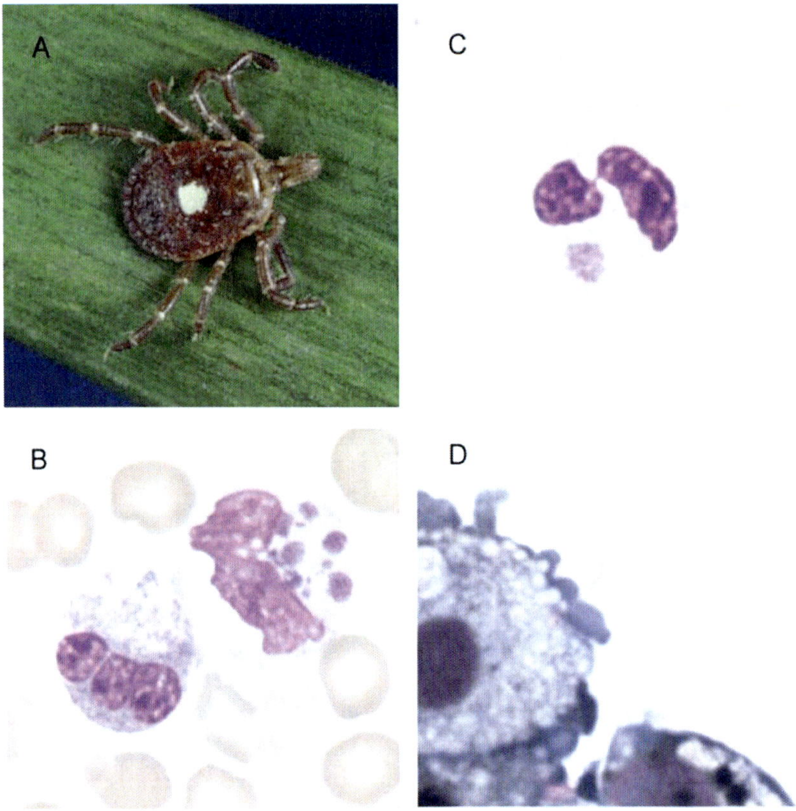

Fig. 1 **A** Adult female lone star tick, *Amblyomma americanum* (photograph provided by Jim Gathany). *A. americanum* is the most frequently encountered human-biting tick in the southeastern and lower midwestern United States (Merten and Durden 2000). **B** Morulae of *Ehrlichia chaffeensis*, the causative agent of human monocytic ehrlichiosis (HME), in the cytoplasm of a mononuclear cell from the peripheral blood of a hospitalized patient (modified Wright's stain). Each morula measures 1.0–6.0 μm in greatest dimension and consists of a cytoplasmic vacuole containing 1 to more than 40 small, coccoid to coccobacillary bacteria that stain dark blue to purple with eosin-azure stains (Paddock and Childs 2003). **C** Morula of *Ehrlichia ewingii* in a neutrophil from the peripheral blood of a patient with *E. ewingii* ehrlichiosis (modified Wright's stain). Morulae of *E. ewingii* are morphologically similar to *E. chaffeensis* but are tropic for neutrophils and occasionally eosinophils of infected hosts (Paddock et al. 2005). **D** *Borrelia lonestari*, the putative agent of southern tick-associated rash illness (STARI), in ISE6 tick cell culture (Giemsa stain). Cultured spirochetes measure 11–25 μm in length and approximately 0.25 μm in width and generally display a flat, wavelike shape with widely variable wavelengths (1.50–2.36 μm) and amplitudes (0.45–0.53 μm). (Varela et al. 2004a)

regarded as the principal wildlife host of lone star ticks (Bishopp and Trembley 1945; Clymer et al. 1970; Bloemer et al. 1986; Kollars et al. 2000). Lone star ticks will perish rapidly of desiccation if isolated from microclimates with high humidity (Hoch et al. 1971). In this context, the abundance of *A. americanum* is influenced primarily by host availability and physiographic variables, which include the degree of ambient moisture, the temperature, the number of daylight hours, and the preferred vegetation type, namely dense understory vegetation in young, second-growth woodland habitats (Hair and Howell 1970; Patrick and Hair 1978). White-tailed deer maintain a dual role in the survival and proliferation of lone star ticks by serving as a preferred food source and as a vehicle for transport and localization within the preferred habitat. In favorable environmental settings, white-tailed deer support enormous numbers of *A. americanum*: in western Kentucky and Tennessee, mean half-body infestations of deer during March through November were as high as 205 adults, 479 nymphs, and 1,150 larvae (Bloemer et al. 1988). As many 2,550 ticks per ear were recorded in an area of Arkansas (Goddard and McHugh 1990).

Environmental and host-related determinants of tick distribution and abundance characteristically vary over time; however, the linkage between the number of white-tailed deer and numbers of lone star ticks has been demonstrated by mathematical models and by deer exclusion studies in various locations. A computer simulation integrating development rates for various stages, fecundity of engorged females, survival of life stages regulated by habitat and climatologic variables, host finding rates, and density-dependent survival rates on hosts demonstrated a linear relationship between the density of deer and of *A. americanum* in a wildlife ecosystem (Mount et al. 1993). Exclusion of white-tailed deer from a 71-ha plot of oak-hickory hardwood forest and reverting fields in western Tennessee during 1985–1988 resulted in a mean percent reduction of larval-, nymphal-, and adult-stage lone star ticks by 88%, 53%, and 51%, respectively, when compared with tick numbers in adjacent control plots where deer were allowed free access during the 4-year interval (Bloemer et al. 1990) (Fig. 2). Similarly, exclusion of white-tailed deer from two approximately 1-ha exclosures in woodland tracts on Fire Island, New York, reduced densities of nymphal-stage *A. americanum* by approximately 48% during the 4 years of post-treatment as compared with pretreatment values (Ginsberg et al. 2002).

2.2
A. americanum as Vectors of Ehrlichiae and Borreliae

Ehrlichia chaffeensis and *E. ewingii* are acquired by *A. americanum* ticks from an infective blood meal from a vertebrate host and are subsequently passed transstadially in the tick vector (Anziani et al. 1990; Ewing et al. 1995). *E. chaffeensis*

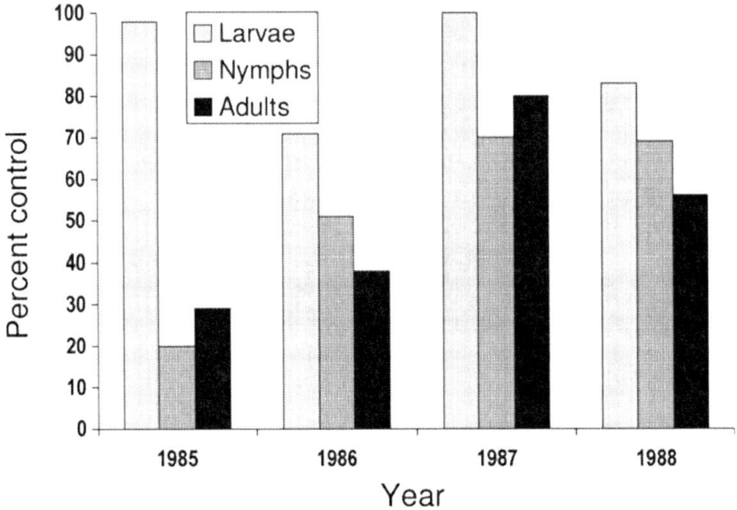

Fig. 2 Percent control of *Amblyomma americanum* larvae, nymphs, and adults (defined as (1–[mean number of ticks per life-stage in deer-excluded plots/mean number of ticks per life-stage in deer-accessible plots] × 100%) during 1985–1988, following exclusion of white-tailed deer from recreational areas at Land Between the Lakes in Kentucky and Tennessee. (Data from Bloemer et al. 1990)

and *E. ewingii* have been detected in adult- and nymphal-stage ticks collected in many southeastern, lower Midwest, and northeastern states (Paddock and Childs 2003; Mixson et al. 2004; Paddock et al. 2005; Schulze et al. 2005; Mixson et al. 2006; Sirigireddy et al. 2006). Because ehrlichiae are not vertically transmitted from adult female ticks to their progeny (Groves et al 1975; Long et al 2003), vertebrate hosts represent important natural reservoirs for *E. chaffeensis* and *E. ewingii*. Infection of *A. americanum* with *B. lonestari* was first reported in 1996 (Barbour et al. 1996; Armstrong et al. 1996) and has been described throughout the range of the lone star tick (Burkot et al. 2001; Stromdahl et al. 2003; Clark 2004; Varela et al. 2004b; Schulze et al. 2005; Taft et al. 2006; Mixson et al. 2006; Schulze et al. 2006). In addition to nymphs and adults, infections have been reported in larval-stage *A. americanum* ticks (Stromdahl et al. 2003), suggesting that transovarial transmission may occur; however, this has not been evaluated experimentally.

Infection prevalences of adult ticks with these agents have been evaluated by using various PCR assays (Table 1). Estimates provided by these studies may not be generalizable over time and space; in addition to extrinsic factors, including

Table 1 Prevalence of infection with *Ehrlichia chaffeensis*, *Ehrlichia ewingii*, and *Borrelia lonestari* in adult lone star ticks in selected areas, as determined by PCR analysis

Agent, state of tick collection	Year(s) of tick collection	No. of ticks tested (% infected)	Reference
E. chaffeensis			
GA	1993–1995	50 (12.0)	Lockhart et al. 1997a
MO	1995	48 (23.0)	Roland et al. 1998
GA	NS	250 (5.2)	Whitlock et al. 2000
CT	1996–1998	106 (7.6)	IJdo et al. 2000
RI	1992	52 (11.5)	IJdo et al. 2000
MO	2000	579 (9.8)	Steiert and Gilfoy 2002
FL	1998	323 (13.6)	Paddock and Childs 2003
NY	1998, 2003	473 (12.5)	Mixson et al. 2004
GA	2001–2003	398 (2.0)	Varela et al. 2004b
NJ	2003	121 (12.3)	Schulze et al. 2005
E. ewingii			
NC	1995, 1998	462 (0.6)	Wolf et al. 2000
FL	1996–1999	121 (1.6)	Sumner et al. 2000
MO	2000	579 (5.4)	Steiert and Gilfoy 2002
TX	NS	66 (7.6)	Long et al. 2004
GA	2001–2003	398 (4.8)	Varela et al. 2004b
NJ	2003	121 (8.2)	Schulze et al. 2005
B. lonestari			
NJ	NS	50 (6.0)	Barbour et al. 1996
NY	NS	318 (3.1)	Barbour et al. 1996
MD	1995	199 (2.0)	Armstrong et al. 1996
AL	1999	19 (10.5)	Burkot et al. 2001
VA	2000	299 (4.3)	Stromdahl et al. 2003
FL	1999–2000	142 (2.8)	Clark 2004
GA	2001–2003	398 (1.0)	Varela et al. 2004b
NJ	2003	121 (9.1)	Schulze et al. 2005

NS Not specified

geographic location, that may influence prevalence estimates, these figures may also vary depending by sample size and DNA detection techniques used by different investigators. For example, a study of *A. americanum* collected from several regions on Long Island, New York, during 2003 revealed *E. chaffeensis* in 0%–27% of adult ticks from five different sampling sites (Mixson et al. 2004). Most studies evaluating individual adult lone star ticks by PCR demonstrate an average prevalence of infection with *E. chaffeensis* of approximately 5%–15% (Lockhart et al. 1997b; Roland et al. 1998; IJdo et al. 2000; Whitlock et al. 2000; Steiert and Gilfoy 2002; Paddock and Childs 2003; Mixson et al. 2004; Varela et al. 2004b; Schulze et al. 2005) and with *E. ewingii* and *B. lonestari* of approximately 1%–10% (Armstrong et al. 1996; Barbour et al. 1996; Wolf et al. 2000; Sumner et al. 2000; Burkot et al. 2001; Steiert and Gilfoy 2002; Stromdahl et al. 2003; Clark 2004; Long et al. 2004; Varela et al. 2004b; Schulze et al. 2005; Mixson et al. 2006; Schulze et al. 2006). Infection prevalences of nymphal-stage ticks are generally lower than prevalences observed in adult *A. americanum* (Paddock and Childs 2003; Mixson et al. 2004). Occasional co-infections of adult lone star ticks with *E. chaffeensis* and *E. ewingii*, *E. chaffeensis* and *B. lonestari*, or *E. ewingii* and *B. lonestari* have been described (Steiert and Gilfoy 2002; Schulze et al. 2005; Mixson et al. 2006). Simultaneous infection of individual adult ticks with two distinct genetic variants of *E. chaffeensis* has also been reported (Mixson et al. 2004).

2.3 White-Tailed Deer as Reservoirs of Ehrlichiae and Borreliae

The current understanding of the epizootiology of HME indicates that white-tailed deer are the principal reservoir host for *E. chaffeensis*. Antibodies reactive with *E. chaffeensis* antigens have been detected at high prevalences in deer populations from many locations in the southeastern and south-central United States (Lockhart et al. 1996; Mueller-Anneling et al. 2000; Yabsley et al. 2003a). Confirmation of deer as reservoirs has been provided by molecular detection and culture isolation from individuals sampled from multiple serologically positive deer populations (Lockhart et al. 1997a, 1997b; Yabsley et al. 2002, 2003a; Arens et al. 2003) (Table 2). However, deer density alone does not represent a significant predictor of risk for HME (Yabsley et al. 2005); instead, densities of *A. americanum* influence the prevalence of infection of *E. chaffeensis* in white-tailed deer, because deer populations are not naturally infected with *E. chaffeensis* unless infested by lone star ticks (Lockhart et al. 1995, 1996; Yabsley et al. 2003a) (Table 3).

Co-infections with *E. chaffeensis* and *E. ewingii* and simultaneous infection with two distinct genetic variants of *E. chaffeensis* in a single white-tailed deer have been described (Yabsley et al. 2002, 2003b). Recent investigations have also

Table 2 Prevalence of infection with *Ehrlichia chaffeensis*, *Ehrlichia ewingii*, and *Borrelia lonestari* in white-tailed deer in selected areas, as determined by PCR analysis

Agent, state of deer collection	Years of deer collection	No. of deer tested (% infected)	Reference
E. chaffeensis			
GA	1993–1995	28 (54.0)	Lockhart et al. 1997a
AR	1996–2001	26 (7.7)	Yabsley et al. 2002
KY	1996–2001	15 (6.7)	Yabsley et al. 2002
NC	1996–2001	9 (22.2)	Yabsley et al. 2002
MO	2000–2001	217 (23.0)	Arens et al. 2003
E. ewingii			
AR	1996–2001	26 (3.8)	Yabsley et al. 2002
KY	1996–2001	15 (6.7)	Yabsley et al. 2002
NC	1996–2001	9 (11.1)	Yabsley et al. 2002
MO	2000–2001	217 (20.3)	Arens et al. 2003
B. lonestari			
AR, FL, GA, KY, LA, MS, NC, SC	1996–2000	80 (8.7)	Moore et al. 2003

demonstrated that primary infection of deer with *E. chaffeensis* does not confer immunologic protection against subsequent infection with a genotypically different strain of *E. chaffeensis* (Varela et al. 2005; Varela-Stokes et al. 2006).

White-tailed deer are the main reservoir responsible for maintenance of the enzootic cycle of *E. chaffeensis* in nature; however, several other vertebrate species are experimentally susceptible, naturally infected, or have evidence of exposure to *E. chaffeensis*. Serologic, molecular, or culture-based evidence of natural infections has been documented for domestic dogs, domestic goats, coyotes, lemurs, rabbits, foxes, and raccoons in the United States (Lockhart et al. 1997b; Davidson et al. 1999; Comer et al. 2000; Dugan et al. 2000; Kocan et al. 2000; Liddell et al. 2003; Yabsley et al. 2004) and in marsh deer (*Blastocercus dichotomus*) in Brazil (Machado et al. 2006).

Comparatively less is known about the natural histories of *E. ewingii* and *B. lonestari*; however, available evidence suggests that deer are also important reservoirs of these two agents. Natural infection of deer with *E. ewingii* has been reported from several locations throughout the distribution of the lone star tick (Yabsley et al. 2002; Arens et al. 2003) (Table 2). Although *E. ewingii* has not been isolated in cell culture, it has been successfully transmitted from naturally infected deer to naïve fawns by blood inoculation (Yabsley et al. 2002). Domestic dogs are

Table 3 Temporal associations between lone star tick infestations and the appearance of antibodies reactive with *Ehrlichia chaffeensis* in white-tailed deer populations in various locations in the United States (from Lockhart et al. 1995; Yabsley et al. 2003b)

Location, year	No. of deer evaluated	Percentage of deer infested with ticks	Percentage of deer with antibodies
Clarke County, GA			
1981	10	0	0
1982	10	0	0
1983	10	10	0
1986	15	47	7
1987	38	87	21
1988	10	80	100
1991	5	100	100
1992	24	100	100
Concordia Parish, LA			
1986	5	0	0
1991	12	67	38
1999	5	100	60
Haywood County, TN			
1989	5	0	0
1994	6	0	0
1998	5	60	20

also common hosts of *E. ewingii* and may represent important natural reservoirs of this agent (Goodman et al. 2003; Liddell et al. 2003; Ndip et al. 2006). White-tailed deer naturally infected with *B. lonestari* have been reported from multiple southeastern states (Moore et al. 2003) (Table 2), and deer have been shown in experiments to be susceptible to infection by inoculation with a culture isolate of this *B. lonestari*, and capable of developing a viable spirochetemia for at least 12 days (Moyer 2005; Moyer et al. 2006). Attempts to infect rodents, domestic dogs, and calves with *B. lonestari* have been unsuccessful (Moyer 2005).

Because white-tailed deer can be naturally infected with multiple, antigenically similar pathogens (e.g., *E. chaffeensis, E. ewingii, A. phagocytophilum*, and a nonspeciated *Anaplasma* sp. [i.e., the "white-tailed deer agent"]) and can also be infected with or exposed to *B. lonestari* and *B. burgdorferi*, the potential for serologic cross-reaction is an important consideration in serologic surveys

(Lockhart et al. 1997b; Yabsley et al. 2002; Arens et al. 2003). More specific serologic tests (e.g., Western blot), molecular-based assays, or culture isolation should be considered when evaluating for various tick-borne infections in white-tailed deer.

No single assay is ideal, because the level of bacteremia may be lower than the level of detection, even by highly sensitive nested PCR assays. As an example, PCR failed to amplify *E. chaffeensis* or *E. ewingii* DNA from whole blood specimens of deer from Jones County, Georgia; however, when blood from these animals was inoculated into naïve fawns, ehrlichiae were later detected in the inoculated fawns (Yabsley et al. 2002). Despite its limitations, PCR has proven to be a useful field surveillance tool, and several studies have used this technique to document the prevalence of infection with *A americanum*-associated ehrlichiae by using molecular assays (Lockhart et al. 1997b; Yabsley et al. 2002; Arens et al. 2003) (Table 2). The limited availability of fresh sterile blood samples, which need to be obtained from deer while these pathogens are in the peripheral circulation, markedly hampers attempts at cell culture isolation of ehrlichiae and Borreliae from wild deer. In addition, white-tailed deer are also nearly ubiquitously infected with a flagellated protozoan parasite (*Trypanosoma cervi*) that often hinders attempts to isolate in culture ehrlichiae and borreliae from naturally infected wild deer. Multiple isolates of *E. chaffeensis* have been obtained in cell culture from wild deer (Lockhart et al. 1997a; Yabsley et al. 2003a); however, *E. ewingii* has not been cultivated from any host, and *B. lonestari* has only recently been isolated from field-collected *A. americanum* (Varela et al. 2004a).

3
Ecological Havoc and White-Tailed Deer Populations

The ability of white-tailed deer to use ecologically disturbed environments to its advantage has contributed considerably to the extraordinary expansion of this animal in the eastern United States during the twentieth century. However, the near-exponential growth of whitetails was not the result of one disastrous human intervention but rather the culmination of various environmental imbalances created during a course of more than 200 years.

3.1
The Fall and Rise of Eastern Forests

As settlers in the United States advanced westward from the Atlantic coast during the 1700s and 1800s, mature forests in the east were felled to provide lumber for local construction and fuel, and for export to Europe. New England forests were harvested particularly for naval stores (e.g., turpentine, tar, and pitch), tannin, ship masts, fences and shingles and as fuel for early industry and domestic purposes; it is estimated that more than 260 million cords of firewood were burned in New England between 1630 and 1800 (Cronon 1983). Large volumes of wood were also consumed to produce charcoal for glassmaking and for smelting iron ore (Spielman 1994). Perhaps more importantly, timbered regions were cleared extensively to provide land for crops and pasturage. Colonial farmers soon recognized that certain tree species were associated with certain types of soil. Hickory, maple, ash, and beech generated rich black humus from centuries of accumulated leaf litter, and settlers identified the presence of these particular trees as indicators of prime agricultural land. Less desirable were the acidic and sandy soils typically associated with hemlock, spruce, and pines (Cronon 1983). In this context, hardwood forests were often the first to disappear to create cultivable acreage. By 1860, woodlands occupied less than 15% of the total land area of New England, having largely been replaced by cleared tracts for farming and agriculture. Farmland comprised approximately 75% of the total land area of Connecticut and New York by 1860 and 1880, respectively (Severinghaus and Brown 1956; Thomson 1977). Deforested landscapes resulted in profound changes in regional microclimate, hydrology, and soil mechanics. Cleared land became sunnier, drier, windier, hotter, and colder (Cronon 1983), changes that are particularly inhospitable to the survival of lone star tick populations.

Vast numbers of eastern farms, fields, and previously harvested forests that were abandoned during the westward expansion of the 1800s and early 1900s became reforested by gradual encroachment of successional trees and shrubs. This transition from farmlands back to forests extended well into the twentieth century. Forest surveys conducted in Virginia in 1940 and 1957 identified an 8.6% increase in forested land in the state during this 17-year period, which occurred almost entirely in agricultural areas that had been abandoned and allowed to revert to second-growth, predominantly hardwood, stands. During this interval, croplands decreased from 6.0 to 3.2 million acres, while hardwood forests increased by 1.4 million acres (Atwood et al. 1965).

Prior to the early twentieth century, the longleaf pine (*Pinus palustris*) dominated much of the forested regions of the southeastern United States. The longleaf forest originally comprised an unbroken belt 100–200 miles wide that covered an estimated 30–60 million acres from southern Virginia to central Florida and westward to central Texas. The longleaf pine was prized in naval architecture for keels, beams, and sideplanks of sailing vessels. It was also valued as structural timber for posts, piles, and joists for bridges, trestles, and warehouses. It was considered a superior wood for wharf construction, and wharves in almost every port from New Orleans to New York were built primarily from longleaf lumber. Longleaf pines were also worked extensively for oleoresin (gum) that was collected and processed to produce turpentine, pitch, and tar (Wahlenberg 1946). In 1880, the annual cut of longleaf pine was estimated at 2 billion board feet and increased steadily to a peak of 13 billion board feet in 1907; by 1946, the longleaf belt was reduced to one-third to one-half of its original area. In extensively harvested regions, longleaf forests were replaced partly or entirely by mixed pines and hardwoods, particularly scrub oak (Wahlenberg 1946). This was accompanied by vigorous growth of formerly suppressed understory flora, creating ecotones comprised of smaller trees and more abundant surface vegetation.

In this context, extensive logging of virgin longleaf pine forests of the Southeast, and the abandonment of farmland in the Northeast, both occurring during the late nineteenth and early twentieth centuries, eventually created extensive tracts of land dominated by young, second-growth woodlands and forests that provided favorable microclimatic conditions for tick survival and an optimum habitat for deer (see below).

3.2 The Fall and Rise of White-Tailed Deer Populations

Prior to and during the early nineteenth century, white-tailed deer were widespread throughout the eastern United States and were important to American Indians and European settlers as an item of trade and as a source of food and clothing. However, unregulated year-round harvests of deer, often aided by packs of dogs, night hunting with fire torches, or hunting from boats, coupled with extensive habitat losses during the mid to late 1800s, led to a dramatic decrease in the number of deer. Deer hunting achieved its zenith with the widespread availability of repeating rifles after the Civil War. In addition, profit motives for market hunters were encouraged by the expansion of the US railway system, which occurred during this same period (Severinghaus and Brown 1956; McCabe and McCabe 1984). By the end of the nineteenth century, an estimated 300,000–500,000 deer remained in North America (Downing 1987). Remnant deer populations were small, isolated, and typically confined to mountainous areas, coastal marshes and swamps, and river bottoms that were inaccessible to hunters.

As early as the mid-1600s, hunting regulations had been established in some areas of the Northeast; however, these laws were not enforced. By the early 1900s, most states had established substantive hunting restrictions to alleviate dramatic population declines. During the mid-1900s, several southeastern and midwestern states began to restock deer populations by translocating large numbers of deer from remnant deer populations. Translocated deer originated primarily from several southeastern states, Wisconsin, and Texas. Increased protection and intensive restocking contributed to a resurgence of white-tailed deer in the United States to an estimated 18 million animals by 1992 (McDonald and Miller 1993) (Fig. 3).

Several biological characteristics of white-tailed deer contribute to rapid and prodigious population growth when food is abundant and natural predators are absent or noncontributory:

1. Relative longevity (6 years or longer)

2. Early reproductive maturation

3. High reproductive rate

4. High fawn survival

5. Social tolerance

6. Relatively indiscriminate food preferences (Leopold et al. 1947; Geist 1998)

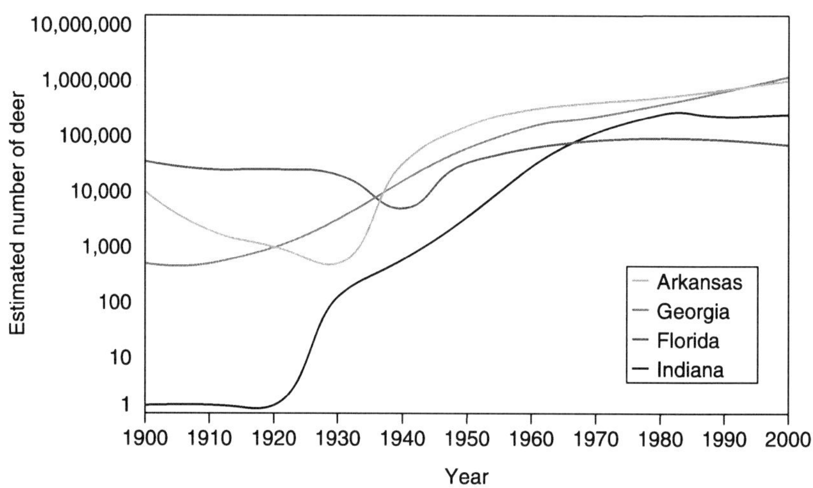

Fig. 3 Approximate number of white-tailed deer in Arkansas, Florida, Georgia, and Indiana during the twentieth century. Precolonial estimates are not available, but deer were widespread and abundant in each of these states. Deer numbers dramatically decreased in Arkansas, Georgia, and Indiana following European settlement of these states and reached the nadir during the late nineteenth century. The principal decrease in Florida deer populations occurred during the 1930s and 1940s following an aggressive deer control program designed to eradicate the tick vector of cattle fever, *Boophilus annulatus*

Whitetails readily consume leaves, twigs, and buds from approximately 100 species of woody plants. Because deer can eat only what they are able to reach—the "browse-line" for white-tailed deer is 6 ft or lower—these animals typically do not flourish in mature forests with sparse understory vegetation. In this context, whitetails thrive best in mosaic habitats where immature, second-growth woods are interspersed with open fields and meadows that provide an ample assortment of accessible foliage (Iker 1983). In addition, white-tailed deer, unlike most other mammalian wildlife species, are notable for their lack of movement from areas with excessive deer densities (i.e., social tolerance) (Leopold et al. 1947).

When situated in environments with abundant low foliage, a white-tailed deer population can potentially double in number every 2 years. For example, the George Reserve in Michigan was stocked with two male and four female whitetails in 1927; within 5 years, the population had increased to an estimated 220 animals. When this same population was thinned to ten deer in 1975, it again increased rapidly to 212 animals by 1980 (McCullough 1984). In Indiana, where deer had been entirely eliminated, 35 whitetails were introduced in 1934; by the early 1980s, that population had multiplied to approximately 100,000 (Iker 1983). The extraordinary growth of white-tailed deer populations is reflected in tabulations of deer-vehicle collisions in the United States. In 1974, a comprehensive listing of road-killed deer compiled by wardens and other game officials amounted to 146,229 animals (Rue 1978). Indiana recorded 34,000 deer kills resulting from automobile collisions in 1 year alone (1987) (Whitaker and Hamilton 1998).

Other ecological disturbances created by humans are likely to have compounded increasing densities of white-tailed deer that occurred during the twentieth century. Natural predators, particularly wolves and cougars, were extirpated from much of the natural range occupied by white-tailed deer. The eastern forest wolf, *Canis lyacon*, was once distributed from Florida to southern Ontario and Quebec, and westward from the Atlantic coast to Oklahoma. In the eastern United States, removal of large carnivores from this region occurred largely during the nineteenth century and often coincided with irruptive growth of deer. Wolves and cougars disappeared from Mount Desert Island in Maine during 1845–1880 and from the Adirondack Mountains in New York between 1882 and 1897. As a result, deer populations expanded considerably in number at these locations (Leopold et al. 1947).

Almost 60 years ago, some wildlife biologists already recognized the problem of deer overabundance in many areas of the United States. In 1947, Aldo Leopold and co-workers wrote, "Prior to the turn of the century, the prevalent population problem in deer was scarcity. Since that time, about a hundred herds of deer . . . have pyramided their numbers to the point of presenting a problem." These authors also mentioned that "there is only one region without deer troubles: the Southeast. Here screw worm and hound dog seem to perform the regulatory functions elsewhere delegated, often without success, to legislatures or conservation commissions. Many parts of the Southeast could support more deer to the advantage of all concerned."

The primary screwworm, *Cochliomyia hominovorax*, caused substantial mortality in domesticated animals and various wildlife species in the southern United States prior to a coordinated control program that used the Sterile Insect Technique during the 1950s and 1960s, which effectively eradicated the screwworm from North America (Krafsur et al. 1987; Baumgartner 1988). The results of this intervention, viewed in context with the observations by Leopold and colleagues (Leopold et al. 1947), might then suggest that human activity aimed at eliminating a deleterious ectoparasite of livestock also eliminated a natural cause of mortality in white-tail deer, particularly in a region of the United States (i.e., the Southeast) that could accommodate greater numbers of these animals.

3.3 Historical Abundance and Range of *A. americanum*

Accurate and quantifiable data that describe *A. americanum* numbers over broad geographic expanses and long intervals of time are limited by the lack of long-term longitudinal studies using controlled methods. However, despite obvious biases and limitations, tick-bite records provide surrogate, albeit crude, regional estimates of lone star population densities. Early twentieth century entomologists commented that in most eastern and southern states, humans were more frequently bitten by *A. americanum* than by any other species of tick (Hooker et al. 1912), and contemporary records seem to support this observation (Merten and Durden 2000). In addition, *A. americanum* was implicated more frequently than any other species in 410 tick-bite records for Air Force personnel from 30 states from 1989–1992 (Campbell and Bowles 1994). The lone star tick accounted for 758 (83%) of 913 ticks removed from 460 persons in Georgia and South Carolina during 1990–1995, and 63 (53%) of 119 ticks recovered from 73 persons in Mississippi during 1990–1999 (Felz et al. 1996; Goddard 2002). In surveys encompassing more restricted geographic areas, the predominance of lone star ticks may be even more pronounced. From a recent study examining the perceived risk of Lyme disease among residents of Gibson Island, Maryland, 1,098 (71%) of 1,556 ticks submitted by residents of during 1994–1996 were *A. americanum* (Armstrong et al. 2001). Although these reports indicate the continuous presence of an aggressive human-biting tick, some anecdotal and prospective evidence indicates that the number of lone star ticks has increased during the last several decades in regions of the southeastern and northeastern United States (Ginsberg et al. 1991; Felz et al. 1996; Ginsberg and Zhioua 1996; Means and White 1997; Mixson et al. 2004; Schulze et al. 2005).

More objective data have documented recent range extensions of the lone star tick within and at the margins of historically established boundaries (Bishopp and Trembley 1945; Cooley and Kohls 1944; Mock et al. 2001). The current distribution of *A. americanum* extends from west-central Texas eastward to the Atlantic Coast, and encompasses the entire southeastern quadrant of the United States, much of the lower Midwest, and parts of coastal New England (Childs and Paddock 2003). Recent studies have identified the appearance of lone star ticks in previously noninfested deer populations from several regions of the southeastern United States during the 1980s (Lockhart et al. 1995; Yabsley et al. 2003b). Importantly, the arrival of *A. americanum* in these populations is clearly associated with subsequent serologic evidence of infection with *E. chaffeensis* or closely related ehrlichiae in these animals (Table 3).

Contemporary range extensions of the lone star tick have become particularly evident in the northeastern United States. In 1754, *A. americanum* became the first North American tick species to be formally described by European naturalists, an event that in all likelihood reflected its relative abundance in the eastern United States during the mid-eighteenth century; however, by 1870 lone star ticks were considered extinct in many parts of New England. Consider the description by New York entomologist Asa Fitch of a "flattened, obovate, chestnut red tick, having a white spot on the end of its scutel, and a whitish ring on its knees:"

> The most common tick of our country, called the wood tick from its inhabiting the woodlands, though formerly abundant throughout the northern and middle states, has now become nearly or quite extinct. The Swedish naturalist Kalm, in passing through the east part of our state 120 years ago, when crossing the Hudson River to Lake Champlain, speaks of the discomfort he experienced from the wood ticks with which the forests there abounded. At this day, along the route he pursued, not one of these insects can probably be found . . . becoming thus extinct with the settlement of the country and the clearing off of its forests. . . In those sections of the country which were settled little over a century ago, tradition still speaks of the annoyances which our American wood ticks were . . . so abundant that if one sits down on the earth or on the trunk of some fallen tree, his clothes and even his body soon gets covered with them (Fitch 1870).

Tick surveys conducted on the southeastern region of Long Island in 1971 identified small, but established, populations of *A. americanum* where none of this species had been recovered during extensive collections approximately 25 years earlier (Collins et al. 1949; Good 1973). Lone star ticks were first documented from Fire Island, New York, in 1988 (Ginsberg et al. 1991) and within several years became the predominant tick from that location (Ginsberg and Zhioua 1996; Ginsberg et al. 2002). Established populations of lone star ticks now exist across Long Island (Mixson et al. 2004).

4 The Emergence of *A. americanum*-Associated Infections in Human Populations

The recognition of *A. americanum*-associated zoonoses can be linked to many factors peculiar to the 1980s and 1990s that occurred independently of the varied environmental disturbances discussed previously. These factors include the development of sensitive and robust molecular diagnostics and the expansion of an immunosuppressed, sentinel patient cohort that was particularly susceptible to the ehrlichioses (Childs and Paddock 2003; Paddock and Childs 2003).

4.1 Human Monocytic Ehrlichiosis

The first documented case of HME occurred in mid-April 1986, when a medical intern at a hospital in Detroit, Michigan, identified unusual intraleukocytic inclusions in a peripheral blood smear of a critically ill patient. The patient, a 51-year-old man, had sustained several tick-bites approximately 2 weeks earlier while planting trees in rural northern Arkansas. Investigators subsequently recognized these inclusions as clusters of bacteria belonging to the genus *Ehrlichia*, a group of organisms previously recognized in the United States solely as veterinary pathogens (Maeda et al. 1987; Fishbein 1990).

During the next several years, clinicians and scientists identified a novel species, *E. chaffeensis*, as a newly recognized agent causing moderately severe to fatal tick-borne disease throughout much of the southeastern, lower midwestern, and mid-Atlantic regions of the United States (Anderson et al. 1991; Fishbein et al. 1994). The identification and characterization of this pathogen was facilitated by isolation of the agent in cell culture (Dawson et al. 1991) and by broadening use of polymerase chain reaction (PCR) technology (Anderson et al. 1992b). During the 1990s, several cases of life-threatening HME were identified among patients with immune systems compromised by neoplasia, corticosteroids, or human immunodeficiency virus (Paddock et al. 2001; Paddock and Childs 2003), and these cases accentuated public health concern regarding *E. chaffeensis*.

Two initial studies that summarized national data for HME during 1986–1997 (742 cases reported by 17 states) (McQuiston et al. 1999) and 1997–2001 (503 cases reported by 23 states) (Gardner et al. 2003) were limited by the lack of a uniform case definition and by inconsistencies in reporting requirements by individual states during the intervals examined. This is reflected by erratic counts in some states (e.g., 54 cases were reported in Virginia during 1986–1997 but only one during 1997–2001). However, some identifiable trends, including consistently high numbers of cases in Arkansas, Missouri, North Carolina, and Oklahoma were identified from these data (McQuiston et al. 1999; Gardner et al. 2003).

Subsequent efforts have been assisted by a uniform case definition for surveillance, which was adopted by state health departments in 1996 and revised in 2000, and by the inclusion of the ehrlichioses in 1999 in the National Electronic Telecommunications System for Surveillance (NETSS). The number of states reporting cases of HME has increased steadily (from three states in 1990 to 48 states by 2003) and the total number of reported cases has risen from 24 cases reported by two states in 1997 to 319 cases reported by 26 states in 2003 (Table 4) (Satalowich 1997; McQuiston et al. 1999; Gardner et al. 2003; Centers for Disease Control and Prevention 2005; Demma et al. 2005). Future estimates of HME incidence are likely to more accurately portray temporal changes in magnitude as the national surveillance system matures. Estimates of regional incidence determined by active surveillance indicate that the frequency of HME may be considerably higher than indicated by passive surveillance in some areas where the disease is endemic. For example, mean incidence rates of 5.2 and 6.8

Table 4 Summary of national case counts and estimated annual incidence of human monocytic ehrlichiosis (HME) in selected states, by year of occurrence, during 1997–2003 (from Satalowich 1997; McQuiston et al. 1999; Gardner et al. 2003; Centers for Disease Control and Prevention 2005; Demma et al. 2005)

	1997	1998	1999	2000	2001	2002	2003
No. of states that report HME (no. reporting >1 case to NETSS)	18 (2)	19 (7)	33 (13)	37 (18)	41 (18)	48 (24)	48(26)
Total reported US cases	24	32	115	196	145	219	319
Estimated annual incidence per million population							
Arkansas	8.7	5.5	8.6	8.2	0.0	6.6	7.0
Missouri	0.0	1.5	8.6	10.5	4.8	8.6	6.0
Maryland	NR	NR	NR	NR	0.4	4.9	9.2
New York	0.0	0.0	0.1	0.2	1.2	0.7	0.6
North Carolina	NR	0.3	1.6	1.2	1.3	2.0	3.2
Oklahoma	NR	NR	3.3	3.5	6.9	3.7	9.4
Tennessee	0.0	0.0	0.2	8.4	3.5	4.8	5.3

NETSS National Electronic Telecommunications System for Surveillance; *NR* HME was not reportable in the given year

per million persons were obtained from passive surveillance in Missouri during 1997–2001 and 2001–2002, respectively (Gardner et al. 2003; Demma et al. 2005); however, active surveillance in southeast Missouri and southwest Illinois during 1997–1999 revealed an incidence of 20–47 cases per million persons (Olano et al. 2003).

4.2 *E. ewingii* Ehrlichiosis

In May 1996, investigators at Washington University Medical Center in St. Louis, Missouri, used a broad-range PCR assay to amplify DNA sequence of *E. ewingii* from a blood sample from an 11-year-old boy from southern Missouri who was assumed to have HME. The child had been exposed to ticks and was subsequently hospitalized with fever, headache, myalgia, and a stiff neck. He had also received a kidney transplant at 27 months of age and was receiving immune-suppressing medications at the time of his illness. During the next 3 years, these same investigators identified other cases of disease caused by *E. ewingii* in two additional immune-suppressed patients and one immune-intact patient. In contrast to findings in patients with HME, morulae were identified in the neutrophils, and occasionally eosinophils, of the patients with *E. ewingii* ehrlichiosis (Buller et al. 1999).

This pathogen had been first identified approximately 25 years earlier as a "new" strain of *Ehrlichia canis* when veterinarians identified morulae in peripheral blood granulocytes of an ill dog from Arkansas in 1970 (Ewing et al. 1971). Investigators subsequently used molecular tools to characterize this ehrlichia as a novel species that they named *E. ewingii* (Anderson et al. 1992a). Following the initial report of human ehrlichiosis caused by *E. ewingii* in 1999 (Buller et al. 1999), cases were identified in Oklahoma and Tennessee in persons co-infected with human immunodeficiency virus (Paddock et al. 2001). Through 2001, 17 patients with *E. ewingii* ehrlichiosis were diagnosed and 12 (70%) had underlying medical conditions causing immune suppression (Paddock et al. 2005).

Cases of disease caused by *E. ewingii* are not identified specifically by NETSS (www.cste.org/ps/2000/2000-id-03.htm). Data examining the relative prevalence of *E. chaffeensis* and *E. ewingii* in canine or deer populations and in lone star ticks in areas where both diseases are endemic suggest that *E. ewingii* occurs in reservoir and vector populations at frequencies similar to or, in some cases, greater than infection with *E. chaffeensis* (Tables 1 and 2) (Yabsley et al. 2002; Steiert and Gilfoy 2002; Arens et al. 2003; Liddell et al. 2003; Long et al. 2004; Varela et al. 2004b; Schulze et al. 2005); however, confirmed cases of disease caused by *E. ewingii* are uncommon relative to cases of HME: investigators at Washington University Medical Center confirmed approximately 200 cases of ehrlichiosis during 1994–2003, of which 89% were caused by *E. chaffeensis* and 11% were caused by *E. ewingii* (Liddell et al. 2003). It has been suggested that *E. ewingii* causes a milder illness than *E. chaffeensis*, particularly in persons without preexisting immune suppression, and that fewer *E. ewingii*-infected patients seek medical attention and confirmatory laboratory evaluation (Paddock et al. 2005).

4.3
Southern Tick-Associated Rash Illness

STARI, also known as southern Lyme disease or as Masters' disease for the physician who identified and described many cases of this illness among patients in southeast Missouri during the late 1980s (Masters et al. 1994, 1998), is a Lyme disease-like condition associated with the bite of *A. americanum* ticks and described in the southeastern and lower midwestern United States. Cases were first documented in the early 1980s (Schulze et al. 1984), and since then more cases have been described from Georgia, Kentucky, Maryland, Missouri, North Carolina, and South Carolina (Masters et al. 1994, 1998; Kirkland et al. 1997; Felz et al. 1999; James et al. 2001; Armstrong et al. 2001; Haddad et al. 2005).

The etiologic agent of STARI has not been definitively identified, although several lines of evidence suggest that a *Borrelia* sp. transmitted by the lone star tick may be a cause of this illness. The clinical presentation of STARI resembles a borreliosis and patients with STARI develop an expanding circular rash at the site of the tick-bite similar to the erythema chronicum migrans rash observed in patients with Lyme disease. Generalized fatigue, headache, and fever may also be present (Kirkland et al. 1997; Masters et al. 1998). *B. burgdorferi*, the causative agent of Lyme disease, has been isolated from rodents and ticks in the southeastern United States (Oliver et al. 1992; Clark 2004); however, the number of confirmed Lyme disease cases in the Southeast is low relative to the number in the Northeast and upper Midwest, and STARI cases are associated with bites of lone star ticks rather than blacklegged ticks (the principal vector of *B. burgdorferi* in the United States) (Schulze et al. 1984; Kirkland et al. 1997; Masters et al. 1998). These observations, and the detection DNA of *B. lonestari* from a rash biopsy specimen from one STARI patient (James et al. 2001), suggest that the etiology of STARI is distinct from *B. burgdorferi*. However, a recent evaluation of 30 STARI patients in Missouri failed to detect *B. lonestari* or *B. burgdorferi* DNA in any of 31 skin biopsy specimens obtained from rash lesions of patients with a clinical diagnosis of STARI; these data suggest that one or more agents other than *B. lonestari* might also contribute to this syndrome (Wormser et al. 2005).

Because the signs and symptoms of STARI closely resemble those of Lyme disease and because the distribution of *A. americanum* and *Ixodes scapularis* are often sympatric, particularly in the mid-Atlantic states, unrecognized cases of STARI may be embedded among cases of presumptively diagnosed Lyme disease (Masters et al. 1994; Armstrong et al. 2001). In this context, an accurate impression of the magnitude of STARI awaits further assessment.

4.4 Other *A. americanum*-Associated Pathogens or Potential Pathogens

Natural infections of lone star ticks with other recognized pathogens and with agents of undetermined pathogenicity have been identified throughout the range of *A. americanum*. These pathogens include *Francisella tularensis* (the causative agent of tularemia) (Hopla and Downs 1953; Calhoun 1954; Hopla 1955), *Coxiella burnetii* (the causative agent of Q fever) (Parker and Kohls 1943; Philip and White 1955), *Rickettsia parkeri* (the cause of a newly recognized, eschar-associated spotted fever rickettsiosis in the United States) (Goddard and Norment 1986), *Rickettsia amblyommii* (a potential agent of spotted fever rickettsiosis) (Burgdorfer et al. 1981; Dasch et al. 1993; Mixson et al. 2006), and lone star virus (an incompletely characterized arbovirus isolated from a lone star tick collected in western Kentucky) (Kokernot et al. 1969).

The most recently discovered bacterium associated with *A. americanum* is the Panola Mountain *Ehrlichia* (PME). This as-yet unnamed *Ehrlichia* species, first identified in lone star ticks collected near Atlanta, Georgia, in 2005, shows close genetic similarity to *Ehrlichia ruminantium*, the agent of heartwater in ruminants (Loftis et al. 2006). The PME has also been detected in *A. americanum* ticks collected in Missouri, and in the blood of naturally infected white-tailed deer in Arkansas, North Carolina, and Virginia (M.J. Yabsely, unpublished observations). *A. americanum* ticks maintain the PME transstadially and are able to transmit this agent to goats and deer in experimental settings; however, the role of the PME as a pathogen of humans requires further investigation (Loftis et al. 2006; M.J. Yabsley, unpublished observations). The impact of various ecological influences on the distribution and abundance of lone star ticks and the resulting frequencies of these agents in human or animal populations has not been explored.

5 Other Zoonoses Associated with White-Tailed Deer

Several investigators, notably Andrew Spielman and co-workers at Harvard University, previously identified the explosive growth of white-tailed deer populations in the United States during the twentieth century as a crucial epizootiological determinant in the emergence of Lyme disease, human babesiosis, and HGA (Piesman et al. 1979; Wilson et al. 1985; Spielman et al. 1993, Spielman 1994;

Thompson et al. 2001). The primary US vector of the pathogens that cause each these diseases is the blacklegged tick, *I. scapularis.* Although deer are an important host for adult blacklegged ticks, the natural histories of *I. scapularis*-associated pathogens are distinct from those described for *A. americanum*-vectored agents in two important ecologic features. First, small rodents, not deer, are the principal hosts for larval- and nymphal-stage *I. scapularis* (Spielman et al. 1993; Spielman 1994). Second, in most regions of the eastern United States, the main vertebrate reservoir host for *B. burgdorferi*, *B. microti*, and *A. phagocytophilum* is the white-footed mouse, *Peromyscus leucopus* (Piesman and Spielman 1982; Donahue et al. 1987; Telford et al. 1996).

Blacklegged ticks can acquire *B. burgdorferi* from experimentally infected deer (Oliver et al. 1992), but disparities between these data, the rarity of recovery of viable spirochetes from deer, and low infection rates of *I. scapularis* ticks collected from whitetails in nature indicate a relative incompetence of white-tailed deer as a reservoir of *B. burgdorferi* (Loken et al. 1985; Telford et al. 1988; Lacombe et al. 1993). White-tailed deer are also refractory to infection with *B. microti* (Piesman et al. 1979). These data suggest that deer serve a minimal role, if any, as reservoirs for some or all of these agents. White-tailed deer are experimentally susceptible to infection with *A. phagocytophilum* (Tate et al. 2005), and a recent study identified molecular evidence of infection with *A. phagocytophilum* in 73 (16%) of 458 deer from 19 states in the southeastern and south-central United States. These studies suggest that white-tailed deer may also be an important sentinel animal for this pathogen (Dugan et al. 2006).

6
Conclusion and Prospectus

> The rapid changes in most environments of the world brought about by the population explosion and socioeconomic events of modern civilization are causing natural enzootics of tickborne infectious agents to change in intensity, distribution, and relation to public health.
>
> Harry Hoogstral 1981

Why were babesiosis, Lyme disease, HME, HGA, *E. ewingii* ehrlichiosis, and STARI not formally described until the last few decades of the twentieth century? Although robust molecular methods were eventually needed to characterize and define the pathogens responsible for each disease, the initial discoveries depended only on astute clinicians and traditional laboratory methods (Western et al. 1970; Fishbein 1990; Bakken 1998), and these resources existed in abundance in the United States for many decades prior to documented recognition of these six tick-borne diseases. The conspicuousness of an expanding, erythematous, targetoid exanthem (i.e., the erythema migrans rash of Lyme disease and STARI) during routine physical examination and the unusual and characteristic appearance of intraerythrocytic babesiae and intraleukocytic ehrlichiae in standard blood smears suggests that descriptions of these tick-borne infections would have appeared earlier and with greater frequency in the medical literature had they been as prevalent in preceding decades as they were during the 1970s and 1980s (Spielman et al. 1993). It can be reasonably assumed that morulae and erythema migrans were identified in a few patients prior to the formal descriptions of the associated disease entities but that a connection of these features to ehrlichiosis or borreliosis was missed or not investigated. The environmental and ecologic imbalances created by human intervention described in this chapter did not create novel tick-borne zoonoses; rather, these events amplified the incidence of the diseases in human populations to a threshold of recognition (Paddock and Childs 2003).

Multiple lines of evidence support the hypothesis that exaggerated growth of white-tailed deer populations provided critical leverage in the emergence of *I. scapularis*- and *A. americanum*-transmitted zoonoses (Spielman et al. 1993; Childs and Paddock 2003; Paddock and Childs 2003). In the case of lone star tick-associated diseases, these changes resulted in (1) expansion of a reservoir pool for ehrlichiae and borreliae, (2) expansion of a keystone host for the vector tick, and (3) range extensions for both tick and pathogen as deer populations were reestablished throughout the eastern United States. Nonetheless, it is also likely that other distinct ecologic disturbances contributed to the emergence of one or more of these diseases.

Several investigators have suggested that rebounding populations of wild turkey (*Meleagris gallopavo*) in the United States might also contribute to recent range extensions of the lone star tick. *A. americanum* has also been called the turkey tick because in its immature stages these ticks are often found attached to *M. gallopavo*, and several studies have identified this bird as an important host of *A. americanum* (Means and White 1997; Kollars et al. 2000; Mock et al. 2001). The fall and rise of wild turkey populations in the eastern United States approximates that of white-tailed deer. Loss of woodland habitat and unrestricted hunting resulted in extirpation of wild turkeys throughout most of their ancestral range. The last recorded observations of native turkeys in Connecticut, New York, and Massachusetts were in 1813, 1844, and 1851, respectively, and by 1907, wild turkeys had also vanished from Kansas, Ohio, Illinois, Indiana, and Iowa (Kennamer et al. 1992). By the early twentieth century, only small populations existed in remote, inaccessible areas. Restoration programs, aided largely by trap-and-transplant programs initiated during the early 1950s, resulted in remarkable population growth and range extensions of wild turkeys. During 1959–1990, the estimated number of eastern wild turkeys

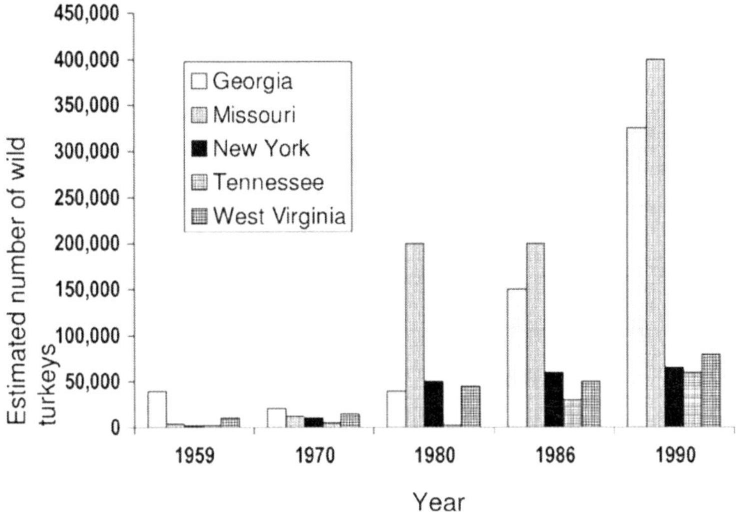

Fig. 4 The estimated number of wild turkeys (*Meleagris gallopavo*) in selected states, 1959–1990 (data from Kennamer et al. 1992). Wild turkeys were nearly extirpated from most of the eastern United States, but populations rebounded considerably during the last half of the twentieth century. This large gallinaceous bird is often a host to larval and nymphal stages of the lone star tick

swelled from approximately 239,000 to over 2,550,000 (Kennamer et al. 1992) (Fig. 4). Increased wild turkey densities have also been suggested as a factor in the recent range extension of *A. americanum* into areas of eastern Kansas (Mock et al. 2001). Population increases of other potential hosts or reservoirs, including coyotes, have also been suggested as contributing to the emergence of *A. americanum*-associated zoonoses (Kocan et al. 2000; Childs and Paddock 2003).

The range of *A. americanum* is increasing, often extending into regions occupied by deer populations not previously infested by lone star ticks (Keirans and Lacombe 1998; Lockhart et al. 1995; Yabsley et al. 2003a). By use of logistic regression modeling, several climatic and landcover variables have been associated with the presence of *E. chaffeensis*-reactive antibodies in deer, a finding that is highly associated with *A. americanum* infestation (Yabsley et al. 2003a, 2005). These models also predict several geographic areas that appear to have suitable tick habitat but where no evidence of ticks or infections of *E. chaffeensis* in deer exists currently (Fig. 5). These regions represent areas of potential spread and should be closely monitored. If *A. americanum* becomes established in these regions, human inhabitants of these areas are placed at risk for disease caused by any of the several pathogens vectored by the lone star tick.

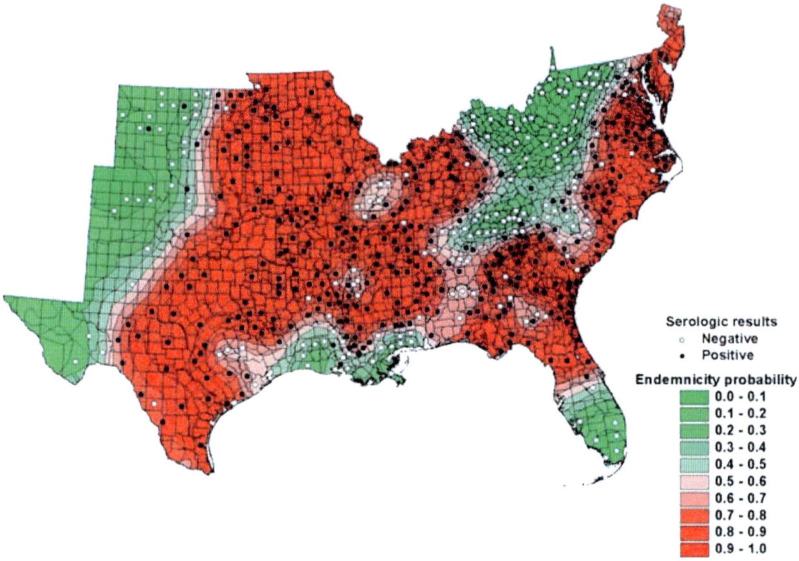

Fig. 5 A Kriging map identifying the endemic probabilities for *Ehrlichia chaffeensis* as determined by geospatial analyses (Yabsley et al. 2005). *Solid circles* represent areas populated by deer with antibodies reactive with *E. chaffeensis*; *open circles* represent areas with seronegative deer (Yabsley et al. 2003b). Increasing probabilities correspond with an increased chance of deer populations that are infected with *E. chaffeensis*

Despite decades of human influence, the natural histories of multiple *A. americanum*-associated diseases in the United States have only recently been unveiled (Childs and Paddock 2003). What will be the prevalence of these pathogens in vector and reservoir populations and the incidence of these diseases in human populations in years to come? It is unlikely that whitetail populations or the incidence of these diseases will continue to climb unrestricted. Valerius Geist, commenting on the recent expansion of whitetails, states that, "This 'weed species' specializes in exploiting opportunities, not at competing for resources through local contests or scrambles." In many aspects, as Geist suggests, the successful adaptation of whitetails to the evolving landscape of the eastern United States parallels the proliferation of weedy plant species that adapt well to disrupted or drastically altered environments. "Weeds" typically flourish because of adverse conditions created by human intervention (e.g., pollution, cultivation, trampling, or herbicide spraying); in this context, "weeds" do not exist in natural environments (Vessel and Wong 1987).

Because disrupted environments require continued intervention to maintain disequilibrium, these landscapes are not stable; for example, a weed-infested lot does not remain weedy indefinitely. Unless continued, the various environmental disturbances and imbalances of the last two centuries that established ideal biotypes for white-tailed deer will not maintain a landscape that allows large numbers of these animals to perpetuate. Over time, whitetail populations can stabilize or diminish as second-growth forests succeed to mature stands. Nonetheless, whitetails have a remarkable propensity to exist in regions despite diminishing food resources; thus downward trends in deer or lone star tick populations are not likely to occur soon. As with white-tailed deer and lone star ticks, the ehrlichioses and STARI are firmly established in North America. Intelligent control and management practices of white-tailed deer populations offer the best hope of stemming further influx of these zoonoses into human populations.

Acknowledgements We thank Robert Holman (CDC), for assistance with NETSS data for HME and the United States Census data for 1997–2003, and Herbert Thompson (CDC) for his insightful comments and thoughtful review of the manuscript.

References

Anderson BE, Dawson JE, Jones DC, Wilson KH (1991) *Ehrlichia chaffeensis*, a new species associated with human ehrlichiosis. J Clin Microbiol 29:2838–2842

Anderson BE, Greene CE, Jones DC, Dawson JE (1992a) *Ehrlichia ewingii* sp nov., the etiologic agent of canine granulocytic ehrlichiosis. Int J Syst Bacteriol 42:299–302

Anderson BE, Sumner JW, Dawson JE, Tzianabos T, Greene CR, Olson JG, Fishbein DB, Olsen-Rasmussen M, Holloway BP, George EH (1992b) Detection of the etiologic agent of human ehrlichiosis by polymerase chain reaction. J Clin Microbiol 30:775–780

Anziani OS, Ewing SA, Barker RW (1990) Experimental transmission of a granulocytic form of the tribe Ehrlichieae by *Dermacentor variabilis* and *Amblyomma americanum* to dogs. Am J Vet Res 51:929–931

Arens MQ, Liddell AM, Buening G, Gaudreault-Keener M, Sumner JW, Comer JA, Buller RS, Storch GA (2003) Detection by PCR and serology of *Ehrlichia* spp. in the blood of wild white-tailed deer in Missouri. J Clin Microbiol 41:1263–1265

Armstrong PM, Rich SM, Smith DR, Hartl DL, Spielman A, Telford SR (1996) A new *Borrelia* infecting lone star ticks. Lancet 347:67–68

Armstrong PM, Brunet LR, Spielman A, Telford SR (2001) Risk of Lyme disease: perceptions of residents of a lone star-tick infested community. Bull World Health Organ 79:916–925

Atwood EL, Lamb JT, Sonenshine DE (1965) A contribution to the epidemiology of Rocky Mountain spotted fever in the eastern United States. Am J Trop Med Hyg 14:831–837

Bakken JS (1998) The discovery of human granulocytotropic ehrlichiosis. J Lab Clin Med 132:175–180

Bakken JS, Dumler JS, Chen SM, Eckman MR, Van Etta LL, Walker DH (1994) Human granulocytic ehrlichiosis in the upper Midwest United States. A new species emerging? J Am Med Assoc 272:212–218

Barbour AG, Maupin GO, Teltow GJ, Carter CJ, Piesman J (1996) Identification of an uncultivable *Borrelia* species in the hard tick *Amblyomma americanum*: possible agent of a Lyme disease-like illness. J Infect Dis 173:403–409

Baumgartner DL (1988) Review of myiasis (Insecta: Diptera: Calliphoridae, Sarcophagidae) of Nearctic wildlife. Wildl Rehab 7:3–46

Bishopp FC, Trembley HL (1945) Distribution and hosts of certain North American ticks. J Parasitol 31:1–54

Bloemer SR, Snoddy EL, Cooney JC, Fairbanks K (1986) Influence of deer exclusion on populations of lone star ticks and American dog ticks (Acari: Ixodidae). J Med Entomol 79:679–683

Bloemer SR, Zimmerman RH, Fairbanks K (1988) Abundance, attachement sites, and density estimators for lone star ticks (Acari: Ixodidae) infesting white-tailed deer. J Med Entomol 25:295–230

Bloemer SR, Mount GA, Morris A, Zimmerman RH, Barnard DR, Snoddy EL (1990) Management of lone star ticks (Acari: Ixodidae) in recreational areas with acaricide applications, vegetative management, and exclusion of white-tailed deer. J Med Entomol 27:543–550

Buller RS, Arens M, Hmiel SP, Paddock CD, Sumner JW, Rikihisa Y, Unver A, Gaudreault-Keener M, Manian FA, Liddell AM, Schmulewitz N, Storch GA (1999) *Ehrlichia ewingii*, a newly recognized agent of human ehrlichiosis. N Engl J Med 341:148–155

Burgdorfer W, Hayes SF, Thomas LA (1981) A new spotted fever group rickettsia from the lone star tick *Amblyomma americanum*. In: Burgdorfer W, Anacker RL (eds) Rickettsiae and rickettsial diseases. Academic, New York, pp 595–602

Burkot TR, Mullen GR, Anderson R, Schneider BS, Happ CM, Zeidner NS (2001) *Borrelia lonestari* DNA in adult *Amblyomma americanum* ticks, Alabama. Emerging Infect Dis 7:471–473

Calhoun EL (1954) Natural occurrence of tularemia in the lone star tick and dogs in Arkansas. Am J Trop Med Hyg 3:360–366

Campbell BS, Bowles DE (1994) Human tick bite records in a United States Air Force population, 1989–1992: implications for tick-borne disease risk. J Wilderness Med 5:405–412

Centers for Disease Control and Prevention (2005) Summary of notifiable diseases—(2003) Morb Mortal Wkly Rep 52:1–85

Childs JE, Paddock CD (2003) The ascendancy of *Amblyomma americanum* as a vector of pathogens affecting humans in the United States. Ann Rev Entomol 48:307–337

Clark K (2004) *Borrelia* species in host seeking ticks and small mammals in northern Florida. J Clin Microbiol 42:576–586

Clymer BC, Howell DE, Hair JA (1970) Animal hosts of economically important ticks in east-central Oklahoma. Ann Entomol Soc Am 63:612–614

Collins DL, Nardy RV, Glasgow RD (1949) Some host relationships of Long Island ticks. J Econ Entomol 42:110–112

Comer JA, Nicholson WL, Paddock CD, Sumner JW, Childs JE (2000) Detection of antibodies reactive with *Ehrlichia chaffeensis* in the raccoon. J Wildl Dis 36:705–712

Cooley RA, Kohls GM (1944) The genus *Amblyomma* (Ixodidade) in the United States. J Parasitol 30:77–111

Cronon W (1983) Changes in the land. Indians, colonists, and the ecology of New England. Hill and Wang, New York

Davidson WR, Lockhart JM, Stallknecht DE, Howerth EA (1999) Susceptibility of red and gray foxes to infection by *Ehrlichia chaffeensis*. J Wildl Dis 35:696–702

Dasch GA, Kelly DJ, Richards AL, Sanchez JL, Rives CC (1993) Western blotting analysis of sera from military personnel exhibiting serological reactivity to spotted fever group rickettsiae. Am J Trop Med Hyg 49 [Suppl 3]:220

Dawson JE, Anderson BE, Fishbein DB, Sanchez JL, Goldsmith CS, Wilson KH, Duntley CW (1991) Isolation and characterization of an *Ehrlichia* sp. from a patient diagnosed with human ehrlichiosis. J Clin Microbiol 29:2741–2745

Demma LJ, Holman RC, McQuiston JH, Krebs JW, Swerdlow DL (2005) Epidemiology of human ehrlichiosis and anaplasmosis in the United States. Am J Trop Med Hyg 73:400–409

Donahue JG, Piesman J, Spielman A (1987) Reservoir competence of white-footed mice for Lyme disease spirochetes. Am J Trop Med Hyg 36:92–96

Downing RL (1987) Success story: white-tailed deer. In: Restoring America's Wildlife. US Dept Interior Fish and Wildlife Service, US Govt Printing Office, Washington, DC, pp 45–57

Dugan VG, Little SE, Beall AD, Stallknecht DE (2000) Natural infection of domestic goats with *Ehrlichia chaffeensis*. J Clin Microbiol 38:448–449

Dugan VG, Yabsley MJ, Tate CM, Mead DG, Munderloh UG, Herron MJ, Stallknecht DE, Little SE, Davidson WR (2006) Evaluation of a prototype *Anaplasma phagocytophilum* surveillance system using white-tailed deer (*Odocoileus virginianus*) as natural sentinels. Vector Borne Zoonotic Dis 6:197–207

Ewing SA, Roberson WR, Buckner RG, Hyat CS (1971) A new strain of *Ehrlichia canis*. J Am Vet Med Assoc 159:1771–1774

Ewing SA, Dawson JE, Kocan AA, Barker RW, Warner CK, Panciera RJ, Fox JC, Kocan KM, Blouin EF (1995) Experimental transmission of *Ehrlichia chaffeensis* (Rickettsiales: Ehrlichieae) among white-tailed deer by *Amblyomma americanum* (Acari:Ixodidae). J Med Entomol 32:368–374

Felz MW, Durden LA, Oliver JH (1996) Ticks parasitizing humans in Georgia and South Carolina. J Parasitol 82:505–508

Felz MW, Chandler FW, Oliver JH, Rahn DW, Schreifer ME (1999) Solitary erythema migrans in Georgia and South Carolina. Arch Dermatol 135:955–960

Fishbein DB (1990) Human ehrlichiosis in the United States. In: Williams JC, Kakoma I (eds) Ehrlichiosis. Kluwer, Amsterdam, pp 100–111

Fishbein DB, Dawson JE, Robinson LE (1994) Human ehrlichiosis in the United States, 1985–1990. Ann Intern Med 120:736–743

Fitch A (1870) Fourteenth report on the noxious, beneficial and other insects of the state of New York. Trans N Y State Ag Soc 30:355–381

Gardner SL, Holman RC, Krebs JW, Berkelman R, Childs JE (2003) National surveillance for the human ehrlichioses in the United States, 1997–2001, and proposed methods for evaluation of data quality. Ann N Y Acad Sci 990:80–89

Geist V (1998) Deer of the world. Their evolution, behavior, and ecology. Stackpole Books, Mechanicsburg, PA

Ginsberg HS, Zhioua E (1996) Nymphal survival and habitat distribution of *Ixodes scapularis* and *Amblyomma americanum* ticks (Acari: Ixodidae) on Fire Island New York USA. Exp Appl Acarol 20:533–544

Ginsberg HS, Ewing CP, O'Connell AF, Bosler EM, Daly JG, Sayre MW (1991) Increased population densities of *Amblyomma americanum* (Acari: Ixodidae) on Long Island, New York. J Parasitol 77:493–495

Ginsberg HS, Butler M, Zhioua E (2002) Effect of deer exclusion by fencing on abundance of *Amblyomma americanum* (Acari: Ixodidae) on Fire Island New York, USA. J Vector Ecol 27:215–221

Goddard J (2002) A ten-year study of tick biting in Mississippi: implications for human disease transmission. J Agromed 8:25–32

Goddard J, McHugh CP (1990) Impact of severe tick infestation at Little Rock AFB, Arkansas on Volant Scorpion military training. Military Med 155:277–280

Goddard J, Norment BR (1986) Spotted fever group rickettsiae in the lone star tick *Amblyomma americanum* (Acari: Ixodidae). J Med Entomol 23:465–472

Good NE (1973) Ticks of eastern Long Island: notes on host relations and seasonal distribution. Ann Entomol Soc Am 66:240–243

Goodman RA, Hawkins EC, Olby NJ, Grindem CB, Hegarty B, Breitschwerdt EB (2003) Molecular identification of *Ehrlichia ewingii* in dogs: 15 cases (1997–2001). J Am Vet Med Assoc 222:1102–1107

Groves MG, Dennis GL, Amyx HL, Huxsoll DL (1975) Transmission of *Ehrlichia canis* to dogs by ticks (*Rhipicephalus sanguineus*). Am J Vet Res 36:937–940

Haddad FA, Schwartz I, Liveris D, Wormser GP (2005) A skin lesion in a patient from Kentucky. Clin Infect Dis 40:429 :475–476

Hair JA, Howell DE (1970) Lone star ticks. Their biology and control in Ozark recreation areas. Oklahoma State University Agricultural Experiment Station Bulletin B 679:1–47

Hoch AL, Barker RW, Hair JA (1971) Measurement of physical parameters to determine suitability of modified woodlots as lone star tick habitat. J Med Entomol 8:725–730

Hoogstral H (1981) Changing patterns of tickborne disease in modern society. Ann Rev Entomol 26:75–99

Hooker WA, Bishopp FC, Wood HP (1912) Some North American ticks. US Bureau Entomol Bull 106:1–204

Hopla CE (1955) The multiplication of tularemia organisms in the lone star tick. Am J Hyg 61:371–380

Hopla CE, Downs CM (1953) The isolation of *Bacterium tularense* from the tick *Amblyomma americanum*. J Kansas Entomol Soc 26:71–72

IJdo JW, Wu C, Magnarelli LA, Stafford KC, Anderson JF, Fikrig E (2000) Detection of *Ehrlichia chaffeensis* DNA in *Amblyomma americanum* ticks in Connecticut and Rhode Island. J Clin Microbiol 38:4655–4656

Iker S (1983) Swamped with deer. Natl Wildl 21:4–11

James AM, Liveris D, Wormser GP, Schwartz I, Montecalvo MA, Johnson BJ (2001) *Borrelia lonestari* infection after a bite by an *Amblyomma americanum* tick. J Infect Dis 183:1810–1814

Keirans JE, Lacombe EH (1998) First records of *Amblyomma americanum Ixodes (Ixodes) dentatus*, and *Ixodes (Ceratixodes) uriae* (Acari: Ixodidae) from Maine. J Parasitol 84:629–631

Kennamer JE, Kennamer M, Brenneman R (1992) History. In: Dickson JG (ed) The wild turkey: biology and management. Stackpole Books, Mechanicsburg, PA, pp 6–17

Kirkland KB, Klimko TB, Meriwether RA, Schriefer M, Levin M, Levine J, MacKenzie WR, Dennis DT (1997) Erythema migrans-like rash illness at a camp in North Carolina: a new tick-borne disease? Arch Intern Med 157:2635–2641

Kocan AA, Levesque GC, Whitworth LC, Murphy GL, Ewing SA, Barker RW (2000) Naturally occurring *Ehrlichia chaffeensis* infection in coyotes from Oklahoma. Emerging Infect Dis 6:477–480

Kokernot RH, Calisher CH, Stannard LJ, Hayes J (1969) Arbovirus studies in the Ohio-Mississippi Basin, 1964–67. Lone star virus, a hitherto unknown agent isolated from the tick *Amblyomma americanum* (Linn.). Am J Trop Med Hyg 18:789–795

Kollars TM, Oliver JH, Durden LA, Kollars PG (2000) Host associations and seasonal activity of *Amblyomma americanum* in Misssouri. J Parasitol 86:1156–1159

Krafsur ES, Whitten CJ, Novy JE (1987) Screwworm eradication in North and Central America. Parasitol Today 3:131–137

Lacombe E, Rand PW, Smith RP (1993) Disparity of *Borrelia burgdorferi* infection rates of adult *Ixodes dammini* on deer and vegetation. J Infect Dis 167:1236–1238

Leopold A, Sowls LK, Spencer DL (1947) A survey of over-populated deer ranges in the United States. J Wildl Mangement 11:162–177

Liddell AM, Stockham SL, Scott MA, Sumner JW, Paddock CD, Gaudreault-Keener M, Arens MQ, Storch GA (2003) Predominance of *Ehrlichia ewingii* in Missouri dogs. J Clin Microbiol 41:4617–4622

Lockhart JM, Davidson WR, Dawson JE, Stallknecht DE (1995) Temporal association of *Amblyomma americanum* with the presence of *Ehrlichia chaffeensis*-reactive antibodies in white-tailed deer. J Wildl Dis 31:119–124

Lockhart JM, Davidson WR, Stallknecht DE, Dawson JE (1996) Site-specific geographic association between *Amblyomma americanum* (Acari: Ixodidae) infestations and *Ehrlichia chaffeensis*-reactive (Rickettsiales: Ehrlichieae) antibodies in white-tailed deer. J Med Entomol 33:153–158

Lockhart JM, Davidson WR, Stallknecht DE, Dawson JE, Howerth EW (1997a) Isolation of *Ehlichia chaffeensis* from wild white-tailed deer (*Odocoileus virginianus*) confirms their role as natural reservoir hosts. J Clin Microbiol 35:1681–1686

Lockhart JM, Davidson WR, Stallknecht DE, Dawson JE, Little SE (1997b) Natural history of *Ehrlichia chaffeensis* (Rickettsiales: Ehrlichieae) in the Piedmont physiographic province of Georgia. J Parasitol 83:887–894

Loftis AD, Reeves WK, Spurlock JP, Mahan SM, Troughton DR, Dasch GA, Levin ML (2006) Infection of a goat with a tick-transmitted *Ehrlichia* from Georgia, U.S.A., that is closely related to *Ehrlichia ruminantium*. J Vector Ecol 31:213–223
Loken KI, Wu CC, Johnson RC, Bey RF (1985) Isolation of the Lyme disease spirochete from mammals in Minnesota. Proc Soc Exp Biol Med 179:300–302
Long SW, Zhang X, Zhang J, Ruble RP, Teel P, Yu XJ (2003) Evaluation of transovarial transmission and transmissibility of *Ehrlichia chaffeensis* (Rickettsiales: Anaplasmataceae) in *Amblyomma americanum* (Acari: Ixodidae). J Med Entomol 40:1000–1004
Long SW, Pound JM, Yu XJ (2004) *Ehrlichia* prevalence in *Amblyomma americanum*, central Texas. Emerging Infect Dis 10:1342–1343
Maeda K, Markowitz N, Hawley RC, Ristic M, Cox D, McDade JE (1987) Human infection with *Ehrlichia canis*, a leukocytic rickettsia. N Engl J Med 316:853–856
Machado RZ, Duarte JM, Dagnone AS, Szabo MP (2006) Detection of *Ehrlichia chaffeensis* in Brazilian marsh deer (*Blatocercus dichotomus*). Vet Parasitol 139:262–266
Marshall WF, Telford SR, Rys RN, Rutledge BJ, Mathiesen D, Malawista SE, Spielman A, Persing DH (1994) Detection of *Borrelia burgdorferi* DNA in museum specimens of *Peromyscus leucopus*. J Infect Dis 170:1027–1032
Masters EJ, Donnell HD, Fobbs M (1994) Missouri Lyme disease: 1989–1992. J Spirochetal Tick-Borne Dis 1:12–17
Masters EJ, Granter S, Duray P, Cordes P (1998) Physician-diagnosed erythema migrans and erythema migrans-like rashes following lone star tick bites. Arch Dermatol 134:955–960
McCabe RE, McCabe TR (1984) Of slings and arrows: an historical perspective. In: Halls LK (ed) White-tailed deer ecology and management. Stackpole Books, Harrisburg, PA, pp 19–72
McCullough DR (1984) Lesson from the George Reserve Michigan. In: Halls LK (ed) White-tailed deer ecology and management. Stackpole Books, Harrisburg, PA, pp 211–242
McDonald JS, Miller KV (1993) A history of white-tailed deer restocking in the United States 1878 to 1992. Research Publication 93–1, The Quality Deer Management Association, Watkinsville, GA
McQuiston JH, Paddock CD, Holman RC, Childs JE (1999) The human ehrlichioses in the United States. Emerging Infect Dis 5:635–642
Means RG, White DJ (1997) New distribution records of *Amblyomma americanum* (L) (Acari: Ixodidae) in New York State. J Vector Ecol 22:133–145
Merten HA, Durden LA (2000) A state-by-state survey of ticks recorded from humans in the United States. J Vector Ecol 25:102–113
Mixson TR, Ginsberg HS, Campbell SR, Sumner JW, Paddock CD (2004) Detection of *Ehrlichia chaffeensis* in adult and nymphal *Amblyomma americanum* (Acari: Ixodidae) ticks from Long Island, New York. J Med Entomol 41:1104–1110
Mixson TR, Campbell SR, Gill JS, Ginsberg HS, Reichard MV, Schulze TL, Dasch GA (2006) Prevalence of *Ehrlichia*, *Borrelia*, and rickettsial agents in *Amblyomma americanum* (Acari: Ixodidae) collected from nine states. J Med Entomol 43:1261–1268

Mock DE, Applegate RD, Fox LB (2001) Preliminary survey of ticks (Acari: Ixodidae) parasitizing wild turkeys (Aves: Phasianidae) in eastern Kansas. J Med Entomol 38:118–121

Moore VA, Varela AS, Yabsley MJ, Davidson WR, Little SE (2003) Detection of *Borrelia lonestari*, putative vector of southern tick-associated rash illness, in white-tailed deer (*Odocoileus virginianus*) from the southeatern United States. J Clin Microbiol 41:424–427

Mount GA, Haile DG, Barnard DR, Daniels E (1993) New version of LSTSIM for computer simulation of *Amblyomma americanum* (Acari: Ixodidae) population dynamics. J Med Entomol 30:843–857

Moyer P (2005) Experimental animal inoculations with *Borrelia lonestari*, putative agent of southern tick-associated rash illness. MS thesis. University of Georgia, Athens, GA

Moyer PL, Varela AS, Luttrell MP, Moore VA, Stallknecht DE, Little SE (2006) White-tailed deer (*Odocoileus virginianus*) develop spirochetemia following experimental infection with *Borrelia lonestari*. Vet Microbiol 115:229–236

Mueller-Anneling L, Gilchrist MJ, Thorne PS (2000) *Ehrlichia chaffeensis* antibodies in white-tailed deer Iowa, 1994 and 1996. Emerging Infect Dis 6:397–400

Ndip LM, Ndip RN, Esemu SN, Dickmu VL, Fokam EB, Walker DH, McBride JE (2005) Ehrlichial infection in Cameroonian canines by *Ehrlichia canis* and *Ehrlichia ewingii*. Vet Parasitol 111:59–66

Olano JP, Masters E, Hogrefe W, Walker DH (2003) Human monocytotropic ehrlichiosis Missouri. Emerging Infect Dis 9:1579–1586

Oliver JH, Stallknecht D, Chandler FH, James AM, McGuire BS, Howerth E (1992) Detection of *Borrelia burgdorferi* in laboratory-reared *Ixodes dammini* (Acari: Ixodidae) fed on experimentally inoculated white-tailed deer. J Med Entomol 29:980–984

Oliver JH, Chandler FW, Luttrell MP, James AM, Stallknecht DE, McGuire BS, Hutcheson HJ, Cummins GA, Lane RS (1993) Isolation and transmission of the Lyme disease spirochete from the southeastern United States. Proc Natl Acad Sci U S A 90:7371–7375

Paddock CD, Childs JE (2003) *Ehrlichia chaffeensis*: a prototypical emerging pathogen. Clin Microbiol Rev 16:37–64

Paddock CD, Folk SM, Shore GM, Machado LJ, Huycke MM, Slater LN, Liddell AM, Buller RS, Storch GA, Monson TP, Rimland D, Sumner JW, Singleton J, Bloch KC, Tang Y, Standaert SM, Childs JE (2001) Infections with *Ehrlichia chaffeensis* and *Ehrlichia ewingii* in persons coinfected with human immunodeficiency virus. Clin Infect Dis 33:1586–1594

Paddock CD, Liddell AM, Storch GA (2005) Other causes of tick-borne ehrlichioses, including *Ehrlichia ewingii*. In: Goodman JL, Dennis DT, Sonenshine DE (eds) Tick-borne diseases of humans. ASM Press, Washington, DC, pp 258–267

Parker RR, Kohls GM (1943) American Q fever: the occurrence of *Rickettsia diaporica* in *Amblyomma americanum* in eastern Texas. Publ Health Rep 58:1510–1511

Patrick CD, Hair JA (1978) White-tailed deer utilization of different habitats and its influence on lone star tick populations. J Parasitol 64:1100–1106

Persing DH, Telford SR, Rys PN, Dodge DE, White TJ, Malawista SE, Spielman A (1990) Detection of *Borrelia burgdorferi* DNA in museum specimens of *Ixodes dammini* ticks. Science 249:1420–1423
Philip CB, White JS (1955) Disease agents recovered incidental to a tick survey of the Mississippi Gulf Coast. J Econ Entomol 48:396–400
Piesman J, Spielman A (1982) *Babesia microti*: infectivity of parasites from ticks for hamsters and white-footed mice. Exp Parasitol 53:242–248
Piesman J, Spielman A, Etkind P, Ruebush TK, Juranek DD (1979) Role of deer in the epizootiology of *Babesia microti* in Massachusetts USA. J Med Entomol 15:537–540
Roland WE, Everett ED, Cyr TL, Hasan SZ, Dommaraju CB, McDonald GA (1998) *Ehrlichia chaffeensis* in Missouri ticks. Am J Trop Med Hyg 59:641–643
Rue LR (1978) The deer of North America. Crown Publishers, New York
Satalowich FT (1997) Tick-borne disease summary: 1996. Missouri Epidemiol 10–12
Schulze TL, Bowen GS, Bosler EM, Lakat MF, Parkin WE, Altman R, Ormiston BG, Shisler JK (1984) *Amblyomma americanum*: a potential vector of Lyme disease in New Jersey. Science 224:601–603
Schulze TL, Jordan RA, Schultze CJ, Mixson T, Papero M (2005) Relative encounter frequencies and prevalence of selected *Borrelia Ehrlichia*, and *Anaplasma* infections in *Amblyomma americanum* and *Ixodes scapularis* (Acari: Ixodidae) ticks from central New Jersey. J Med Entomol 42:450–456
Schulze TL, Jordan RA, Healy SP, Roegner VE, Meddis M, Jahn MB, Guthrie DL (2006) Relative abundance and prevalence of selected *Borrelia* infections in *Ixodes scapularis* and *Amblyomma americanum* (Acari: Ixodidae) from publicly owned lands in Monmouth County, New Jersey. J Med Entomol 43: 1269–1275
Scrimenti RJ (1970) Erythema chronicum migrans. Arch Dermatol 102:104–105
Severinghaus CW, Brown CP (1956) History of the white-tailed deer in New York. N Y Fish Game J 3:129–166
Sirigireddy KR, Mock DC, Ganta RR (2006) Multiplex detection of *Ehrlichia* and *Anaplasma* pathogens in vertebrate and tick hosts by real-time RT-PCR. Ann N Y Acad Sci 1078: 552–556
Spielman A (1994) The emergence of Lyme disease and human babesiosis in a changing environment. Ann N Y Acad Sci 740:146–156
Spielman A, Telford SR, Pollack RJ (1993) The origins and course of the present outbreak of Lyme disease. In: Ginsberg HS (ed) Ecology and environmental management of Lyme disease. Rutgers University Press, New Brunswick, NJ, pp 83–96
Steere AC, Broderick TF, Malawista SE (1978) Erythema chronicum migrans and Lyme arthritis: epidemiologic evidence for a tick vector. Am J Epidemiol 108:312–321
Steiert JG, Gilfoy F (2002) Infection rates of *Amblyomma americanum* and *Dermacentor variabilis* by *Ehrlichia chaffeensis* and *Ehrlichia ewingii* in southwest Missouri. Vector Borne Zoonotic Dis 2:53–60
Stromdahl EY, Williamson PC, Kollars TM, Evans SR, Barry RK, Vince MA, Dobbs NA (2003) Evidence of *Borrelia lonestari* DNA in *Amblyomma americanum* (Acari: Ixodidae) removed from humans. J Clin Microbiol 41:5557–5562

Sumner JW, McKechnie D, Janowski D, Paddock CD (2000) Detection of *Ehrlichia ewingii* in field-collected ticks by using PCR amplification of 16S rRNA gene and *groESL* operon sequences. 15th Meeting of the American Society for Rickettsiology. Captiva Island, FL Abstract 72

Taft SC, Miller MK, Wright SM (2005) Distribution of borreliae among ticks collected from eastern states. Vector-Borne Zoonotic Dis 5:383–389

Tate CM, Mead DG, Luttrell MP, Howerth EW, Dugan VG, Munderloh UG, Davidson WR (2005) Experiemental infection of white-tailed deer with *Anaplasma phagocytophilum*, the etiologic agent of human granulocytic anaplasmosis. J Clin Microbiol 43:3595–3601

Telford SR, Mather TN, Moore SI, Wilson ML, Spielman A (1988) Incompetence of deer as reservoirs of the Lyme disease spirochete. Am J Trop Med Hyg 39:105–109

Telford SR, Dawson JE, Katavolos P, Warner CK, Kolbert CP, Persing DH (1996) Perpetuation of the agent of human granulocytic ehrlichiosis in a deer tick-rodent cycle. Proc Natl Acad Sci U S A 93:6209–6214

Thomson BF (1977) The changing face of New England. Houghton Mifflin, Boston

Thompson C, Spielman A, Krause PJ (2001) Coinfecting deer-associated zoonoses: Lyme disease, babesiosis, and ehrlichiosis. Clin Infect Dis 33:676–685

Varela AS, Luttrell MP, Howerth EW, Moore VA, Davidson WR, Stallknecht DE, Little SE (2004a) First culture isolation of *Borrelia lonestari*, putative agent of southern tick-associated rash illness. J Clin Microbiol 42:1163–1169

Varela AS, Moore VA, Little SE (2004b) Disease agents in *Amblyomma americanum* from northeastern Georgia. J Med Entomol 41:753–759

Varela AS, Stallknecht DE, Yabsley MJ, Moore VA, Howerth EW, Davidson WR, Little SE (2005) Primary and secondary infection with *Ehrlichia chaffeensis* in white-tailed deer (*Odocoileus virginianus*). Vector Borne Zoonotic Dis 5:48–57

Varela-Stokes AS, Stokes JV, Davidson WR, Little SE (2006) Co-infection of white-tailed deer with multiple strains of *Ehrlichia chaffeensis*. Vector Borne Zoonotic Dis 6:140–151

Vessel MF, Wong HH (1987) Natural history of vacant lots. University of California Press, Berkeley

Wahlenberg WG (1946) Longleaf pine, its use, ecology, regeneration, protection, growth and management. Charles Lathrop Pack Forestry Foundation, Washington, DC

Western KA, Benson GD, Gleason NN, Healy GR, Schultz MG (1970) Babesiosis in a Massachusetts resident. N Engl J Med 283:854–856

Whitaker JO, Hamilton WJ (1998) Mammals of the eastern United States, 3rd edn. Cornell University Press, Ithaca NY

Whitlock JE, Fang QQ, Durden LA, Oliver JH (2000) Prevalence of *Ehrlichia chaffeensis* (Rickettsiales: Rickettsiaceae) in *Amblyomma americanum* (Acari: Ixodidae) from the Georgia coast and barrier islands. J Med Entomol 37:276–280

Wilson ML, Alder GH, Spielman A (1985) Correlation between abundance of deer and that of the deer tick *Ixodes dammini* (Acari: Ixodidae). Ann Entomol Soc Am 78:172–176

Wolf L, McPherson T, Harrison B, Engber B, Anderson A, Whitt P (2000) Prevalence of *Ehrlichia ewingii* in *Amblyomma americanum* in North Carolina. J Clin Microbiol 38:2795

Wormser GP, Masters E, Liveris D, Nowakowski J, Nadelman RB, Holmgren D, Bittker S, Cooper D, Wang G, Schwartz I (2005) Microbiologic evaluation of patients from Missouri with erythema migrans. Clin Infect Dis 40:423–428

Yabsley MJ, Varela AS, Tate CM, Dugan VG, Stallknecht DE, Little SE, Davidson WR (2002) *Ehrlichia ewingii* infection in white-tailed deer (*Odocoileus virginianus*). Emerging Infect Dis 8:668–671

Yabsley MJ, Dugan VG, Stallknecht DE, Little SE, Lockhart JM, Dawson JE, Davidson WR (2003a) Evaluation of a prototype *Ehrlichia chaffeensis* surveillance system using white-tailed deer (*Odocoileus virginianus*) as natural sentinels. Vector Borne Zoonotic Dis 3:195–207

Yabsley MJ, Little SE, Sims EJ, Dugan VJ, Stallknecht DE, Davidson WR (2003b) Molecular variation in the variable-length PCR target and 120-kDa antigen genes of *Ehrlichia chaffeensis* from white-tailed deer (*Odocoileus virginianus*). J Clin Microbiol 41:5202–5206

Yabsley MJ, Norton TM, Powell MR, Davidson WR (2004) Molecular and serologic evidence of tick-borne ehrlichiae in three species of lemurs from St. Catherines Island, Georgia USA. J Zoo Wildl Med 35:503–509

Yabsley MJ, Wimberly MC, Stallknecht DE, Little SE, Davidson WR (2005) Spatial analysis of the distribution of *Ehrlichia chaffeensis*, causative agent of human monocytotropic ehrlichiosis, across a multi-state region. Am J Trop Med Hyg 72:840–850

CTMI (2007) 315:325–344
© Springer-Verlag Berlin Heidelberg 2007

Bats, Civets and the Emergence of SARS

L.-F. Wang[1] (✉) · B. T. Eaton[1]

[1]CSIRO Livestock Industries, Australian Animal Health Laboratory, Geelong, Victoria, 3220 Australia
Linfa.Wang@csiro.au

Abstract Severe acute respiratory syndrome (SARS) was the first pandemic transmissible disease of previously unknown aetiology in the twenty-first century. Early epidemiologic investigations suggested an animal origin for SARS-CoV. Virological and serological studies indicated that masked palm civets (*Paguma larvata*), together with two other wildlife animals, sampled from a live animal market were infected with SARS-CoV or a closely related virus. Recently, horseshoe bats in the genus *Rhinolophus* have been identified as natural reservoir of SARS-like coronaviruses. Here, we review studies by different groups demonstrating that SARS-CoV succeeded in spillover from a wildlife reservoir (probably bats) to human population via an intermediate host(s) and that rapid virus evolution played a key role in the adaptation of SARS-CoVs in at least two nonreservoir species within a short period.

1 Introduction

Severe acute respiratory syndrome (SARS) first appeared in mid-November 2002 in Guangdong province in southern China, and continued to spread to more than 30 countries on five continents with 8,098 reported cases and 774 deaths by the end of July 2003, placing it with HIV/AIDS as one of the severe and readily transmissible new diseases to emerge in the twenty-first century (WHO 2004). The high case fatality rate and global spread led to an urgent response by an international network co-ordinated by the World Health Organization (WHO) of the United Nations, which resulted in the rapid identification of the aetiological agent (Drosten et al. 2003; Fouchier et al. 2003; Ksiazek et al. 2003; Kuiken et al. 2003; Peiris et al. 2003). The outbreak was caused by a newly emerged and previously unrecognised coronavirus, now known as the SARS coronavirus (SARS-CoV). The complete genome sequence of SARS-CoV has been determined and the virus is classified within the order *Nidovirales*, family *Coronaviridae*, genus *Coronavirus* (Marra et al. 2003; Rota et al. 2003). From December 16, 2003 to January 8, 2004, four SARS cases were detected in the city of Guanghzou, the capital city of Guangdong province of China (Liang et al. 2004; Wang et al. 2006). None of these cases was fatal or resulted in documented secondary transmission, suggesting the possibility that these sporadic outbreaks were caused by less virulent strains of SARS-CoV.

Coronaviruses are known to infect a variety of avian and mammalian species (Holmes and Lai 2001; see the chapter by Holmes and Drummond, this volume). Before the discovery of SARS-CoV, two human coronaviruses (229E and OC43) were known to cause upper respiratory tract infections that varied in frequency and severity in different disease outbreaks, but were usually mild and self-limited (Holmes and Lai 2001). Since the discovery of SARS-CoV, two new coronaviruses, NL63 (van der Hoek et al. 2004) and HKU1 (Woo et al. 2005), have been isolated from human patients with nonfatal infections. To date, SARS-CoV is the only known coronavirus capable of causing lethal infection in humans. Recently, two groups independently demonstrated that bats in the genus *Rhinolophus* are natural reservoirs of SARS-like viruses (Lau et al. 2005; Li et al. 2005), providing strong evidence that SARS-CoV is indeed a new zoonotic virus with a wildlife origin.

2
Epidemiologic Clues to an Animal Origin

Epidemiological studies of index SARS cases in Guangdong Province provided initial evidence that the agent responsible for the outbreak was zoonotic in origin. Between November 2002 and February 2003, the first cases or clusters of SARS appeared in several independent geographic locations in the Pearl River Delta region in southern Guangdong, and suggested multiple introductions of a virus or similar viruses from a common source. Several of the early cases were reportedly associated with occupations that involved contact with wildlife, including handling, killing and selling wild animals as well as preparing and serving wildlife animal meat in restaurants (Xu et al. 2004). Moreover, a study of early SARS cases (i.e. those with disease onset prior to January 2003) compared to those identified later in the outbreak found that 39% of early-onset cases were food handlers, whereas only 2%–10% of cases between February and April 2003 were associated with this occupation. Also, early-onset cases were more likely to live within walking distance of animal markets than late-onset cases (Xu et al. 2004).

To confirm the initial epidemiologic association of early-onset patients with animal handling, several groups conducted retrospective serologic surveillance in different human populations in Guangdong Province during the outbreak period. In one study by Xu et al. (2004), a total of 1,454 clinically confirmed human cases were analysed covering the period from November 2002 to April 30, 2003. Several observations supported the hypothesis of a wild animal origin for SARS. It was observed that early cases of SARS occurred independently in at least five different well-separated municipalities in Guangdong Province. The study also found that early patients were more likely than later patients to report living near a produce market, but not near a farm, and nine of 23 (or 39%) early patients were food handlers with probable animal contact.

Several studies revealed a higher than normal seroprevalence of SARS-CoV antibodies among wild animal traders. Guan et al. (2003) found that eight of 20 (40%) wild animal traders sampled from a market in Shenzhen, Guandong, in 2004 had anti-SARS-CoV antibodies in comparison to 1 from 20 (5%) vegetable traders from the same market. Yu et al. (2003) analysed serum samples taken on May 4, 2003 from animal traders in three different live animal markets in Guangzhou. Out of 508 animal traders surveyed, 13% had antibodies to SARS-CoV; 72% of traders of masked palm civets (*Paguma larvata*) were seropositive. Interestingly, none of the animal traders had SARS or atypical pneumonia diagnosed during the SARS outbreak in Guangdong, suggesting asymptomatic infection by SARS-CoV or a closely related SARS-like coronavirus. The presence of subclinical infections was corroborated in a separate study conducted by a Hong Kong group (Zheng et al. 2004), who found that 17 of 938 (or 1.7%) adults recruited in 2001 had antibodies to SARS-CoV detected by immunofluorescence and virus neutralisation assays. These findings suggest that a small proportion of healthy individuals in Hong Kong had been exposed to SARS-CoV-related viruses at least 2 years before the SARS outbreak reached Hong Kong in mid-February 2003.

3 Detection of SARS-CoV-Like Viruses among Wildlife

In May 2003, in the middle of the SARS outbreak, a joint team from Hong Kong and Shenzhen sampled a total of 25 animals from seven wild and one domestic animal species from a live animal market in Shenzhen. It was claimed that these animals were sourced from southern China, and that they had been kept in separate storehouses before delivery to the market. The animals remained in the market for a variable period of time and each stall holder had only a few animals of a given species. Animals from different stalls within the market were sampled. Nasal and faecal swabs were collected for PCR and virus isolation and, where possible, blood samples were taken for serology. Among the six masked palm civets sampled, three were PCR-positive, and a SARS-CoV-like virus was isolated from four nasal swabs and one faecal swab (Guan et al. 2003). In addition, a very closely related virus was isolated from the faecal swab of the only raccoon dog (*Nyctereutes procyonoides*) sampled in the study. Two Chinese ferret badgers (*Melogale moschata*) were sampled, and although neither was PCR-positive, one displayed a neutralising antibody titre of 1:160 against SARS-CoV.

Sequencing of PCR products and virus isolates from palm civets and the raccoon dog revealed several important observations. First, the animal SARS-CoVs were almost identical in sequence to SARS-CoVs isolated from human patients, showing a 99.8% sequence identity. Second, the animal SARS-CoVs contained a 29-nt sequence, located in the C-terminal region of the genome immediately upstream from the N gene; this 29-nt sequence was absent from most of the human SARS-CoV isolates. Later it was discovered that human SARS-CoVs isolated during the early phase of the outbreaks contained the 29-nt sequence, suggesting that the deletion event occurred during adaptation of the animal-derived SARS-CoV to its new human host (The Chinese SARS Molecular Epidemiology Consortium 2004).

These data indicated that at least three different wildlife animal species in the Shenzhen market were infected by a coronavirus that is closely related to SARS-CoV. This important discovery provided the first direct evidence that

SARS-CoV existed in animals, and that the virus responsible for the SARS outbreak most likely originated from animals.

4 Multi-directional Transmission of SARS-CoV

Determining the route and direction of transmission is important for the understanding of zoonotic disease emergence and for the control of future outbreaks. For SARS-CoV, there is evidence to suggest that four possible routes of transmission, animal-to-human, animal-to-animal, human-to-human and human-to-animal, occurred during the outbreaks of SARS in 2002–2003 and 2003–2004.

4.1 Animal-to-Human Transmission

When SARS-CoV was identified as the causative agent of the SARS outbreaks, the first question asked was whether this new virus arose from a pre-existing human virus by an evolutionary process which enhanced its virulence or whether it was an animal virus newly introduced into the human population. Retrospective serologic studies indicated that there were no antibodies to SARS-CoV in the human population prior to the SARS outbreak, suggesting that SARS-CoV was not an existing human coronavirus (Ksiazek et al. 2003). Epidemiologic studies, as discussed above, revealed that animal handlers and people working in the food industry had a higher representation than other groups among early SARS patients. Molecular epidemiologic studies confirmed that the earliest genotypes of human SARS-CoV from the 2002–2003 outbreaks were most closely related to those of animal SARS-CoV isolates (Guan et al. 2003; The Chinese SARS Molecular Epidemiology Consortium 2004). During the sporadic outbreaks of 2003–2004, a total of four patients were independently infected with the SARS-CoV (Liang et al. 2004). There was no direct link between any of the four cases and none of the patients had direct or indirect contact history with previously documented SARS cases; all of them had a history of contact with animals. Furthermore, genome sequences of SARS-CoVs from human patients in 2003–2004 were almost identical to those isolated from civets in the market at the same time period, but more divergent from the human SARS-CoVs obtained during the 2002–2003 outbreaks. Taken together, these results demonstrated that animal-to-human transmission was responsible for the introduction of SARS-CoV into the human population.

4.2
Animal-to-Animal Transmission

In the market study conducted by Guan et al. (2003), it was shown, by virus isolation, RT-PCR or serum neutralisation assay, that all of the six masked palm civets were exposed to SARS-CoV. Considering that these animals were sampled at the same time in the same market, but originated from different regions of southern China, it is most likely that some, if not all, of them got infected in the market through animal-to-animal transmission. In the same study, it was shown that the raccoon dog isolate (SZ13) had an S-gene sequence which was identical to that of one of the civet isolates (SZ16) but differed from the other two civet isolates (SZ1 and SZ3) which displayed S-gene sequence variation. This observation strongly indicated the occurrence of inter-species transmission among the animals in the market.

Animal-to-animal transmission has also been demonstrated in experimental situations. Martina et al. (2003) showed that ferrets (*Mustela furo*) and domestic cats (*Felis domesticus*) are susceptible to infection by SARS-CoV and that they can efficiently transmit the virus to previously uninfected animals that are housed with them.

4.3
Human-to-Human Transmission

Numerous epidemiologic studies documented the rapid human-to-human transmission of SARS-CoV, which spread the virus to more than 30 countries in less than 5 months (WHO 2004). One important example was the spread of SARS-CoV from mainland China to Hong Kong by a Chinese doctor attending a conference there. Through the individuals he infected at a Hong Kong hotel, this single human source was mainly responsible for the subsequent spread of SARS to the rest of the world (Tsang et al. 2003; Zhong et al. 2003).

The major routes of SARS CoV transmission are believed to be droplets, aerosols and fomites (Peiris et al. 2004). In general, the average number of secondary cases of infection generated by one infected individual (R_0) was low (see the chapter by Real and Biek, this volume), approximately 2.2–3.7 (Anderson et al. 2004), a figure much lower than the R_0 of influenza, which ranges from 5 to 25. However, for countries with a moderate to large number of cases, superspreading events (SSEs) played a pivotal role in large-scale transmission of the virus. In such circumstances, a few infected individuals caused a much higher number of secondary infections. In addition to the SSE in the Hong Kong hotel, other SSEs occurred in a hospital setting in Hong Kong, an air flight from Hong Kong to Beijing and in healthcare settings in Beijing, Singapore and Toronto

(Anderson et al. 2004). In the SSE in a Beijing hospital, one patient infected 33 out of 74 persons that had close contact with the patient. These secondary cases resulted in a further 43 cases before this chain of transmission subsided (Shen et al. 2004).

4.4 Human-to-Animal Transmission

The exact cause for the rapid transmission of SARS-CoV among the more than 100 residents at the Amoy Gardens apartment block in Hong Kong remains a mystery. Although there have been reports suggesting environmental spread through U-traps contaminated with SARS-CoV in bathrooms, other studies also indicated a possible role played by domestic animals such as rats and cats (Lu and Qu 2004; Martina et al. 2003; Ng 2003). Domestic cats living in the apartment complex were found to be infected with SARS-CoV (Martina et al. 2003), suggesting possible human-to-animal transmission. This notion was supported by the subsequent experimental infection of domestic cats with a human SARS-CoV isolated from a Hong Kong patient (Martina et al. 2003). Experimentally infected cats were asymptomatic, but were able to infect other co-housed cats.

In another potential example of human-to-animal transmission, SARS-CoV was isolated from a pig during a surveillance study in farming villages outside of Tianjin, where a SARS outbreak occurred in the spring of 2003 (Chen et al., 2005). The genome sequence of the pig isolate (designated TJF) revealed it to be closely related to the human isolate BJ01obtained from a patient in Beijing, 120 km from Tianjin, but only distantly related to SARS-CoVs isolated from animals. More importantly, the TJF genome contained the 29-nt deletion, the genetic feature characterising SARS-CoV which circulated among human patients during the later phases of the 2002–2003 outbreaks, but never observed in any of the animal SARS-CoV isolates. The authors concluded that direct human-to-pig transmission was the most likely explanation for these results.

5 Susceptibility of Different Animal Species to Infection by SARS-CoV

During investigations of new zoonotic diseases, it is important to differentiate the roles that different animals may play in distinct stages of disease emergence (see the chapters by Childs et al. and Childs, this volume). It is especially important to distinguish between the reservoir host, which may or may not be responsible for direct pathogen transmission to humans, and the intermediate or amplifying host which introduced the pathogen into the human population. Due to the sudden emergence of SARS, it was extremely difficult to obtain reliable epidemiologic data to pinpoint the source of the outbreak. The vast number of live animals being traded in animal markets in southern China further complicated the investigation process. Experimental animal infection studies therefore became an important component of the SARS-CoV investigation. They provided the proof that SARS-CoV was the causative agent of SARS, helped define the range of animals susceptible to this new virus, elucidated the mechanisms of virus transmission, and established useful animal model(s) for pathogenesis studies and the testing of vaccines and antivirals.

Since the first experimental infection of cynomolgus macaques by Fouchier et al. (2003), rhesus macaques, African green monkeys, cats, ferrets, mice, pigs, hamsters, guinea pigs and civets have also been shown to be susceptible to experimental infection by SARS-CoV (Liang et al. 2005; Martina et al. 2003; Roberts et al. 2005; Subbarao et al. 2004; Weingartl et al. 2004; Wentworth et al. 2004; Wu et al. 2005). Together with the three naturally infected animal species identified in the market study (Guan et al. 2003), more than ten different mammalian species have so far been shown to be susceptible to SARS-CoV. It can be expected that many more susceptible species will be identified in the future.

Rats have also been identified as another potentially susceptible host and may have played a role in the transmission and spread of SARS-CoV in the well-publicised outbreaks of SARS in the Amoy Gardens apartment block in Hong Kong (Ng 2003). Also, the first confirmed SARS case in 2004 in Guangdong was reported not to have had any contact with animals with the exception of rats (Liang et al. 2004). In our inoculation studies, we have obtained serologic evidence to indicate that SARS-CoV was able to replicate asymptomatically in rats (B.T. Eaton, L.-F. Wang et al., unpublished results). It is clear that further studies are required to clarify the role played by rats in the transmission of SARS-CoV.

In contrast, two independent studies conducted in Canada (Weingartl et al. 2004) and the USA (Swayne et al. 2004) indicated that none of the avian species tested, which included chicken, turkey, goose, duck and quail, was susceptible to SARS-CoV infection under laboratory conditions. These findings suggest that domestic poultry were unlikely to be the reservoir or associated with dissemination of SARS-CoV in the animal markets of southern China.

6
The Role of Palm Civets in SARS Outbreak: Natural Reservoir or an Amplifying Host?

In the study by Guan et al. (2003), SARS-CoV-like viruses were isolated from palm civets and a raccoon dog in a live animal market in southern China and serologic evidence indicted that a third species, the Chinese ferret-badger, was also infected by a similar virus. In spite of the diversity of animals susceptible to SARS-CoV-like viruses, subsequent attention focussed on palm civets, probably because of the larger number of these animals being traded in the market. However, despite the initial speculation that palm civets might be the source of SARS-CoV, several studies demonstrated that there was no widespread infection in wild or farmed civets and that infection in this and other species in animal markets was more likely a reflection of an "artificial" market cycle in naïve species than an indication of the natural reservoir of SARS-CoV.

The first clue came from serological surveillance conducted by Tu et al. (2004). In this study, a total of 103 civet serum samples were taken from a number of civet farms and a market in different regions of China. No SARS-CoV antibody was detected in any of the 47 sera taken in June 2003 from two different farms in Hunan and Henan Provinces. The same was true for 28 serum samples obtained in January 2004 from three different farms in Guangdong Province. In contrast, out of the 18 serum samples taken from an animal market in Guangdong during the same period in January 2004, 14 (or 79%) had significant level of neutralising antibodies to SARS-CoV, indicating widespread infection by a virus that is closely related to SARS-CoV.

Molecular analysis was used to investigate the distribution and evolution of SARS-CoV in palm civets and to compare the prevalence of the virus in palm civets in markets and on farms. Following the detection of SARS-CoV in market palm civets at the end of 2003, palm civets were culled in Guangdong Province in an attempt to prevent the potential reemergence of SARS. This provided a unique opportunity for Kan et al. (2005) to sample a relatively large number of animals for molecular epidemiological studies. A total of 91 palm civets and 15 raccoon dogs were sampled in the Xinyuan animal market in Guangzhou in January 2004. The animals were selected from 18 vendors with booths located in four blocks dedicated to the sale of civets and raccoon dogs. PCR analysis indicated that all of the animals sampled were positive and that most animals yielded positive rectal and throat swabs. In the same study, a total of 1,107 palm civets were sampled from 25 farms in 12 provinces from January to September 2004, but none of them was positive when analysed by the same PCR tests. These farms were selected on the basis that they used to sell animals to one of the booths at the Xingyuan animal market or that they claimed to have previously provided more than 80% of their animals to markets in Guangdong province.

In an animal surveillance study conducted in Hong Kong between the summer of 2003 and 2004, Poon et al. (2005) sampled 21 wild trapped palm civets in addition to other mammalian, avian and reptile species. Serological and PCR analyses indicated that none of the animals surveyed was positive for SARS-CoV.

Moreover, when palm civets were experimentally infected with two different strains of human SARS-CoV, one with a 29-nt deletion isolated in Beijing (Qin et al. 2003) and another containing the 29-nt sequence isolated in the early phase of the outbreak from Guangzhou (GZ01), all of the animals developed clinical symptoms including fever, lethargy and loss of aggressiveness (Wu et al. 2005). These results indicated that palm civets were equally susceptible to infection by SARS-CoV with or without the 29-nt sequence.

Taken together, the lack of widespread infection in wild or farmed palm civets and the display of overt clinical symptoms following experimental infection suggest that palm civets are unlikely to be the natural reservoir of SARS-CoV. Instead, the animal's high susceptibility to SARS-CoV and its wide distribution in markets and restaurants made it an ideal amplifying host that is believed to have played an important role in both the major 2002–2003 and sporadic 2003–2004 outbreaks.

7 Identification of Horseshoe Bats as Natural Reservoirs for SARS-Like Viruses

Bats are reservoir hosts of several zoonotic viruses (Calisher et al. 2006), including the Hendra and Nipah viruses, which have recently emerged in Australia and East Asia, respectively (Chua et al. 2000; Murray et al. 1995; Wang and Eaton 2001; see the chapter by Field et al., this volume). They are susceptible and respond asymptomatically to infection with many viruses (Sulkin and Allen 1974; Calisher et al. 2006). These characteristics and the increasing presence of bats and bat products in food and traditional medicine markets in southern China and other Asian countries (Mickleburgh et al. 2002) suggest that bats could be a potential natural reservoir host of SARS-CoV. Recently, two groups have independently reported the presence of SARS-like viruses in different species of horseshoe bats within the genus *Rhinolophus*.

In one study conducted from March to December of 2004, a total of 408 bats representing nine species, six genera and three families from four locations in China (Guangdong, Guangxi, Hubei and Tianjin) were sampled by trapping in their native habitat (Li et al. 2005). Blood, faecal and throat swabs were collected for antibody and PCR analyses. Three communal cave-dwelling species from the genus *Rhinolophus* in the family Rhinolophidae had a high SARS-CoV antibody prevalence: 13 of 46 (28%) in *R. pearsoni* from Guangxi; two of six (33%) in *R. pussilus* from Guangxi; and five of seven (71%) in *R. macrotis* from Hubei. The high seroprevalence and wide distribution of seropositive bats is consistent with the pattern of serology expected from a pathogen's wildlife reservoir host (Hudson et al. 2002).

The serological findings were corroborated by PCR analyses using primer pairs derived from the nucleocapsid (N) and polymerase (P) genes of SARS-CoV. A total of five positive faecal samples were detected, three in *R. pearsoni* from Guangxi and one each in *R. macrotis* and *R. ferrumequinum*, respectively, from Hubei. Genome sequence analysis indicated that SARS-like coronaviruses (SL-CoVs) present in bats have an almost identical genome organisation to those of SARS-CoVs isolated from humans or civets, sharing an overall sequence identity of 92%. The most variable regions were located in the 5′ end of the S gene, which codes for the surface spike protein involved in receptor binding, and in the ORF10-coding region immediately upstream from the N gene, which contains the coding region for putative nonstructural proteins of unknown function (Marra et al. 2003; Rota et al. 2003) and is known to be prone to mutation and deletions of various sizes (Guan et al. 2003; Song et al. 2005; The Chinese SARS Molecular Epidemiology Consortium 2004). When these regions were excluded from the comparison, the sequence identity increased to 94% between SL-CoVs and SARS-CoVs (Li et al. 2005). It was interesting to note that the ORF10-coding region of bat SL-CoVs contained the 29-nt sequence present in civet SARS-CoV isolates and human SARS-CoV isolates from the early phase of the outbreak, but absent from human isolates obtained in the later phases of the outbreak (The Chinese SARS Molecular Epidemiology Consortium 2004). This finding suggests that SL-CoVs and SARS-CoVs may have a common ancestor.

In another study reported by Lau et al. (2005), it was found that 23 (39%) of 59 anal swabs of wild Chinese horseshoe bats (*Rhinolopus sinicus*) contained genetic material closely related to SARS-CoV when analysed by PCR. It was also found that up to 84% of the horseshoe bats examined contained antibodies to a recombinant N protein of SARS-CoV. This study was conducted using wild animals from unpopulated areas of the Hong Kong Special Administration Region of China. Analysis of three full-length genome sequences derived from PCR products revealed similar findings to those reported by Li et al. (2005) in that the bat viruses shared an overall 88% nucleotide and 93% sequence identity to ten human and civet SARS-CoVs isolated from different locations and at different times during the SARS outbreaks, and the major differences were located in the S gene and ORF10-coding region. The bat viruses from Hong Kong also contained the 29-nt sequence in the ORF10 region.

The genetic diversity observed among bat-derived SL-CoVs together with the high prevalence and wide distribution of seropositive bats, as revealed by two independent groups, are consistent with bats being the wildlife reservoir host of SL-CoVs. As shown by Li et al. (2005), comparison of partial sequences from SL-CoVs isolated from three different horseshoe bat species revealed a much higher genetic diversity than those observed among all the reported sequences of civet and human SARS-CoVs. Furthermore, sequence analysis also indicated that human and civet SARS-CoV nestle phylogenetically within the spectrum of SL-CoVs, suggesting that the viruses responsible for the SARS outbreaks were members of this diverse coronavirus group, tentatively named the group 2b coronaviruses or G2b-CoVs (Wang et al. 2006). This notion was strengthened by the comparison of genetic relatedness among the different bat viruses detected in Hong Kong and mainland China. As mentioned above, the overall genome sequence identity between the human or civet SARS-CoV and the bat viruses Rp3 (isolated from *R. pearsoni*) and B24 (isolated from *R. sinicus*) was 92% and 88%, respectively. The sequence identity between Rp3 and B24 is 89%, suggesting that Rp3 has a closer evolutionary relationship to the civet/human isolates than to the B24 isolate of a different bat species.

Further surveillance studies in the region are required to investigate the distribution and diversity of the G2b-CoVs in different bat species, and to find the location and reservoir species of the SARS-CoVs responsible for the SARS outbreaks in 2002–2003 and 2003–2004.

8 Factors Contributing to the Emergence of SARS

Emergence of zoonotic viruses maintained by wildlife reservoir hosts is a complex and poorly understood sequence of events. Childs (2004) and Childs et al., this volume, recognised four transitions in the process by which zoonotic viruses are transmitted and infect other species. Two of these transitions, inter-species contact and cross-species virus transmission (i.e. spillover) are essential and sufficient to cause epidemic emergence. Two other transitions, sustained transmission and virus adaptation within the spillover species, are not required for emergence, but will determine the magnitude and scope of subsequent disease outbreaks. These transition events are discussed below in relation to the potential mechanism of SARS emergence.

8.1 Inter-species Contact and Spillover

There are a number of possibilities for contact between horseshoe bats, the putative reservoir host (H_R), and one or more secondary hosts (H_S). This could happen in the bat's natural habitat and in a variety of other situations. Horseshoe bats are cave-dwelling animals which feed mainly on moths and beetles and may have the opportunity to come into close proximity with other animals which live in or explore caves. It is interesting to note that in the study by Li et al. (2005), serological findings indicated that *Rousettus leschenauti*, a much larger cave-dwelling fruit bat, may also be infected by a closely related G2b-CoV, although at much lower frequency. Contact between bats and other animal species may also arise because bats are used as a source of medicinal components and live bats are among a large number of different animal species that are traded in animal markets. From the studies by Li et al. (2005) and Lau et al. (2005), we know that the main route of excretion of G2b-CoV from naturally infected bats is via faecal shedding. The opportunity of virus transmission between H_R and H_S is therefore further enhanced since direct contact of bats and other animals may not be absolutely required for the virus to pass to a H_S. Live animal trading in China and Asia thus provides the most likely circumstances for inter-species contact.

As revealed in an epidemiology study conducted during the peak of the SARS outbreaks in China, most animal traders handle more than one animal species, thus providing numerous opportunities for animal-to-animal contact. This could happen during transportation, where animal cages are often piled on top of each other, or in the market where more than 100 different animal species can be housed under a single roof simultaneously. Wholesale animal markets or warehouses also offer the possibility of sustained opportunity for inter-species contact because animals may be kept together for an extended time before being sold individually. The notion of inter-species transmission in wholesale or retail markets is supported by the finding in two different studies that G2b-CoVs were detected in civets and raccoon dogs in the market, but not in the farms which claimed to have supplied the animals to the particular markets surveyed (Tu et al. 2004; Kan et al. 2005).

The second transition (i.e. cross-species transmission or spillover) requires not only inter-species contact, but also the susceptibility of H_S animals to the virus. For SARS-CoV, this does not seem to be a major constraint. As discussed above, a large number of mammalian species have been demonstrated to be susceptible to SARS-CoV infection, either under experimental conditions or by natural infection in markets. Spillover is defined as introduction, replication and release of virus from the H_S (Childs 2004). For SARS-CoV, there was ample evidence to suggest that this has happened in more than one H_S species, including civets, raccoon dogs, ferret badgers and humans.

8.2 Sustained Transmission and Virus Adaptation

Three separate surveillance studies indicate that, at least for the civet populations in markets, intra-H_S transmission of SARS-CoV occurred readily (Guan et al. 2003; Tu et al. 2004; Kan et al. 2005). SARS-CoV reactive antibody was found in 79% of civets in January 2004 in one study and 100% of civets tested in another study contained SARS-CoV genomic RNA. Civet trading was banned in May 2003 after the first SARS outbreaks, but was resumed in August 2003. Considering that there was no evidence of widespread infection of SARS-CoV among civet populations on farms and in the wild (Tu et al. 2004; Kan et al. 2005), it can be concluded that re-appearance of the virus in the civet population in markets in late 2003 to early 2004 was a result of separate spillover event(s) which occurred after the resumption of civet trading in August 2003. This would suggest sustainable intra-H_S transmission among civets after spillover events. Similarly, intra-H_S transmission among different human populations was documented in many different cities, especially in Guanghzhou, Hong Kong, Beijing, Singapore and Toronto (Anderson et al. 2004). It is conceivable that such intra-H_S transmission would have been sustained for a much longer period if draconian quarantine measures had not been implemented.

Virus adaptation is the fourth transition considered to be important in determining the scope and magnitude of a disease outbreak after a spillover event (Childs 2004). Several studies demonstrated rapid evolution of the SARS-CoV sequence, especially in the receptor-binding domain (RBD) of the spike protein gene, a location believed to be important for virus adaptation to the different H_S species, i.e. civet and human.

In the first detailed molecular epidemiology study (The Chinese SARS Molecular Epidemiology Consortium 2004), 61 SARS-CoVs derived from early, middle and late phases of the SARS outbreaks in 2003 were analysed by genomic sequencing. It was discovered that genotypes characteristic of each phase could be identified, and that the earliest genotypes were the most similar to those of SARS-CoVs isolated from animals. Moreover, it was shown that while the neutral mutation rate of the viral genome was constant during the different phases of the outbreak, the amino acid substitution rate of the coding region slowed during the course of the outbreak, indicating rapid adaptation to the human host. As expected, the spike protein-coding gene showed the strongest initial responses to host selection pressures.

In a separate study focusing on SARS-CoVs isolated from humans and civets during the 2003–2004 outbreaks, Song et al. (2005) discovered that the ratio of nonsynonymous/synonymous nucleotide substitution in viruses isolated from civets collected 1 year apart and from different geographic locations, was very high. This suggested a rapid process of virus evolution in civets, much like the adaptation process revealed for human SARS-CoV isolates in the first study (The Chinese SARS Molecular Epidemiology Consortium 2004). These results also indicated that civets were not likely to be an H_R, and highlighted their potential role as an H_S involved in transmitting the virus from bats to humans. The authors concluded that major genetic variations in critical genes, particularly the S gene, are essential for the progression from animal-to-human transmission to sustained human-to-human transmission, which eventually led to the first SARS outbreaks in 2002–2003.

The rapid evolution of SARS-CoVs in palm civets in markets in Guangdong was also confirmed by Kan et al. (2005) who analysed a total of 17 animal-derived sequences isolated in January 2004 and compared them to those from animals and humans isolated in 2003. Their study revealed that viruses in palm civets in the live animal markets had undergone a process of evolution that generated viruses with the potential to infect humans. Within the 17 animal-derived sequences, there were three which did not contain any of the novel signature variation residues (SNV) that characterised previously isolated pathogenic viruses. The authors postulated that such viruses were the evolutionary starting point of a process which introduced seven SNVs and caused the substitution of six amino acid residues in the spike protein. The resulting virus jumped to humans and was the cause of the low pathogenic infection of humans in 2003–2004. A further 14 SNVs caused 11 amino acid residue changes and resulted in the high-pathogenicity viruses which were responsible for human infection during early phase of the 2002–2003 outbreaks. Finally, six SNVs with four amino acid changes produced the group of viruses that were responsible for the global epidemic in the middle to late phases of the SARS outbreaks.

The metallopeptidase, angiotensin-converting enzyme 2 (ACE2), has been identified as the functional receptor for SARS-CoV infection (Li et al. 2003). In a comparative study of binding affinity of different S proteins to human and civet ACE2, it was shown that S proteins of SARS-CoVs isolated from civet and the mild human cases in 2004 bind to human ACE2 much less efficiently than the S proteins of SARS-CoV isolated from human patients during 2002–2003 outbreaks (Li et al. 2003). Similar findings were obtained in a separate study by Yang et al. (2005). It was found that the S protein from viruses isolated from a patient in late 2003 and from two civets depended less on the human ACE2 receptor and were markedly resistant to antibody inhibition.

These data demonstrated that SARS-CoVs were successful in both maintaining intra-H_S transmission among at least two different H_S species and in adapting to the new hosts via rapid virus evolution. These attributes made possible the rapid global spread of SARS-CoV to cause the most severe infectious disease outbreak of the twenty-first century.

9
Conclusions

Less than 3 years after the first emergence of SARS, rapid progress has been made in the identification and genetic analysis of the aetiological agent and its molecular epidemiology, the identification of the host receptor and molecular characterisation of the virus–host interaction, and the rapid development of diagnostic assays and vaccine and therapeutic candidates. The recent identification of horseshoe bats as the natural reservoir of this new group of G2b-CoVs will undoubtedly play an important role in facilitating our understanding of SARS emergence and in the prevention of future outbreaks. Bats have been identified or implicated as the natural reservoir host for an increasing number of new and often deadly zoonotic viruses. In addition to the emergence of G2b-CoVs from insectivorous *Rhinolophus* species, Hendra virus, Nipah virus and, most recently, Ebola virus have been shown to have fruit bat reservoir hosts (Chua et al. 2002; Halpin et al. 2000; Leroy et al. 2005; see the chapters by Field et al., and Gonzalez et al., this volume). Bats typically respond asymptomatically to virus infection and display a capacity to permit persistent virus infections (Sulkin and Allen 1974). Their wide distribution and abundant status (one mammalian species in five is a bat) makes them prime candidates for reservoirs of viruses which may, like G2b-CoVs, jump the species barrier and infect humans and other animals. Information on the ecology of bats and the nature of their response to virus infection may not only be scientifically interesting, but may also provide fundamental information on how best to cope with further outbreaks of disease caused by bat-borne viruses (Calisher et al. 2006).

Acknowledgements Work conducted in the authors' group on the identification of the natural reservoir host of SARS-CoV is supported by the Australian Biosecurity Cooperative Research Centre for Emerging Infectious Diseases (Project 1.007R) in collaboration with research activities supported by an NIH/NSF "Ecology of Infectious Diseases" award (no. R01-TW05869) from the John E. Fogarty International Center and the V. Kann Rasmussen Foundation.

References

Anderson RM, Fraser C, Ghani AC, Donnelly CA, Riley S, Ferguson NM, Leung GM, Lam TH, Hedley AJ (2004) Epidemiology, transmission dynamics and control of SARS: the 2002–2003 epidemic. Philos Trans R Soc Lond B 359:1091–1105

Calisher CH, Childs JE, Field HE, Holmes KV, Schountz T (2006) Bats: important reservoir hosts of emerging viruses. Clin Microbiol Rev 19:531–545

Chen WJ, Yan MH, Yang L, Ding BL, He B, Wang YZ, Liu XL, Liu CH, Zhu H, You B, Huang SY, Zhang JG, Mu F, Xiang Z, Feng XL, Wen J, Fang JQ, Yu J, Yang HM, Wang J (2005) SARS-associated coronavirus transmitted from human to pig. Emerg Inf Dis 11:446–448

Childs JE (2004) Zoonotic viruses of wildlife: hither from yon. Arch Virol Suppl 18:1–11

The Chinese SARS Molecular Epidemiology Consortium (2004) Molecular evolution of the SARS coronavirus during the course of the SARS epidemic in China. Science 303:1666–1669

Chua KB, Bellini WJ, Rota PA, Harcourt BH, Tamin A, Lam SK, Ksiazek TG, Rollin PE, Zaki SR, Shieh WJ, Goldsmith CS, Gubler DJ, Roehrig JT, Eaton BT, Gould AR, Olson J, Field H, Daniels P, Ling AE, Peters CJ, Anderson LJ, Mahy BWJ (2000) Nipah virus: a recently emergent deadly paramyxovirus. Science 288:1432–1435

Chua KB, Koh CL, Hooi PS, Wee KF, Khong JH, Chua BH, Chan YP, Lim ME, Lam SK (2002) Isolation of Nipah virus from Malaysian Island flying-foxes. Microbes Infect 4:145–151

Drosten C, Gunther S, Preiser W, van der Werf S, Brodt HR, Becker S, Rabenau H, Panning M, Kolesnikova L, Fouchier RA, Berger A, Burguiere AM, Cinatl J, Eickmann M, Escriou N, Grywna K, Kramme S, Manuguerra JC, Muller S, Rickerts V, Sturmer M, Vieth S, Klenk HD, Osterhaus AD, Schmitz H, Doerr HW (2003) Identification of a novel coronavirus in patients with sever acute respiratory syndrome. New Engl J Med 348:1967–1976

Fouchier RA, Kuiken T, Schutten M, Van Amerongen G, Van Doornum GJ, van den Hoogen BG, Peiris M, Lim W, Stohr K, Osterhaus AD (2003) Aetiology: Koch's postulates fulfilled for SARS virus. Nature 423:240

Guan Y, Zheng BJ, He YQ, Liu XL, Zhuang ZX, Cheung CL, Luo SW, Li PH, Zhang LJ, Guan YJ, Butt KM, Wong KL, Chan KW, Lim W, Shortridge KF, Yuen KY, Peiris JSM, Poon LLM (2003) Isolation and characterization of viruses related to the SARS coronavirus from animals in southern China. Science 302:276–278

Halpin K, Young PL, Field HE, Mackenzie JS (2000) Isolation of Hendra virus from pteropid bats: a natural reservoir of Hendra virus. J Gen Virol 81:1927–1932

Holmes KV, Lai MMC (2001) Coronaviridae: the viruses and their replication. In: Fields BN, Knipe DM, Howley PRM (eds) Fields virology 4th edn., Vol. 1. Lippincott Williams & Wilkins, Philadelphia, pp 1075–1094

Hudson PJ, RizzoliA, Grenfell BT, Heesterbeek H, Dobson AP (2002) The ecology of wildlife diseases. Oxford University Press, Oxford

Kan B, Wang M, Jing H, Xu H, Jiang X, Yan M, Liang W, Zheng H, Wan K, Liu Q, Cui B, Xu Y, Zhang E, Wang H, Ye J, Li G, Li M, Cui Z, Qi X, Chen K, Du L, Gao K, Zhao Y, Zou X, Feng Y, Gao Y, Hai R, Yu D, Guan Y, Xu J (2005) Molecular evolution analysis and geographic investigation of severe acute respiratory syndrome coronavirus-like virus in palm civets at an animal market and on farms. J Virol 79:11892–11900

Ksiazek TG, Erdman D, Goldsmith CS, Zaki SR, Peret T, Emery S, Tong S, Urbani C, Comer JA, Lim W, Rollin PE, Dowell SF, Ling AE, Humphrey CD, Shieh WJ, Guarner J, Paddock CD, Rota P, Fields B, DeRisi J, Yang JY, Cox N, Hughes JM, LeDuc JW, Bellini WJ, Anderson LJ (2003) A novel coronavirus associated with severe acute respiratory syndrome. New Engl J Med 348:1953–1966

Kuiken T, Fouchier RAM, Schutten M, Rimmelzwaan GF, van Amerongen G, van Riel D, Laman JD, de Jong T, van Doornum G, Lim W, Ling AE, Chan PKS, Tam JS, Zambon MC, Gopal R, Drosten C, van der Werf S, Escriou N, Manuguerra JC, Stohr K, Peiris JSM, Osterhaus ADME (2003) Newly discovered coronavirus as the primary cause of severe acute respiratory syndrome. Lancet 362:263–270

Lau SKP, Woo PCY, Li KSM, Huang Y, Tsoi HW, Wong BHL, Wong SSY, Leung SY, Chan KH, Yuen KY (2005) Severe acute respiratory syndrome coronavirus-like virus in Chinese horseshoe bats. Proc Natl Acad Sci U S A 102:14040–14045

Leroy EM, Kumulungui B, Pourrut X, Rouquet P, Hassanin A, Yaba P, Delicat A, Paweska JT, Gonzalez JP, Swanepoel R (2005) Fruit bats as reservoirs of Ebola virus. Nature 438:575–576

Li W, Moore MJ, Vasllieva N, Sui J, Wong SK, Berne MA, Somasundaran M, Sullivan JL, Luzuriaga K, Greeneugh TC, Choe H, Farzan M (2003) Angiotensin-converting enzyme 2 is a functional receptor for the SARS coronavirus. Nature 426:450–454

Li W, Shi Z, Yu M, Ren W, Smith C, Epstein JH, Wang H, Crameri G, Hu Z, Zhang H, Zhang J, McEachern J, Field H, Daszak P, Eaton BT, Zhang S, Wang L-F (2005) Bats are natural reservoirs of SARS-like coronaviruses. Science 310:676–679

Liang GD, Chen QX, Xu JG, Liu YF, Lim W, Peiris JSM, Anderson LJ, Ruan L, Li H, Kan B, Di B, Cheng P, Chan KH, Erdman DD, Gu SY, Yan XG, Liang WL, Zhou DH, Haynes L, Duan SM, Zhang X, Zheng H, Gao Y, Tong SX, Li DX, Fang L, Qin PZ, Xu WB, SARS Diagnosis Working Group (2005) Laboratory diagnosis of four recent sporadic cases of community-acquired SARS, Guangdong Province China. Emerg Infect Dis 10:1774–1781

Liang LC, He C, Lei M, Li SW, Hao YX, Zhu H, Duan Q (2005) Pathology of guinea pigs experimentally infected with a novel reovirus and coronavirus isolated from SARS patients. DNA Cell Biol 24:485–490

Lu ZR, Qu LH (2004) Animal-to-human SARS-associated coronavirus transmission? Emerg Infect Dis 10:959

Marra MA, Jones SJM, Astell CR, Holt RA, Brooks-Wilson A, Butterfield YSN, Khattra J, Asano JK, Barber SA, Chan SY, Cloutier A, Coughlin SM, Freeman D, Girn N, Griffith OL, Leach SR, Mayo M, McDonald H, Montgomery SB, Pandoh PK, Petrescu AS, Robertson AG, Schein JE, Siddiqui A, Smailus DE, Stott JF, Yang GS, Plummer F, Andonov A, Artsob H, Bastien N, Bernard K, Booth T, Bowness D, Czub M, Drebot M, Fernando L, Flick R, Garbutt M, Gray M, Grolla A, Jones S, Feldmann H, Meyers A, Kabani A, Li Y, Normand S, Stroher U, Tipples GA, Tyler S, Vogrig R, Ward D, Watson B, Brunham RC, Krajden M, Petric M, Skowronski DM, Upton C, Roper RL (2003) The genome sequence of the SARS-associated coronavirus. Science 300:1399–1404

Martina BEE, Haagmans BL, Kuiken T, Fouchier RAM, Rimmelzwaan GF, van Amerongen G, Peiris JSM, Lim W, Osterhaus ADME (2003) SARS virus infection of cats and ferrets. Nature 425:915

Mickleburgh SP, Huston AM, Racey PA (2002) A review of the global conservation status of bats. Oryx 36:18–34

Murray K, Selleck P, Hooper P, Hyatt A, Gould A, Gleeson L, Westbury H, Hiley L, Selvey L, Rodwell B (1995) A morbillivirus that caused fatal disease in horses and humans. Science 268:94–97

Ng SKC (2003) Possible role of an animal vector in the SARS outbreak in Amoy Gardens. Lancet 362:570–572

Peiris JSM, GuanY, Yuen KY (2004) Severe acute respiratory syndrome. Nat Med 10:S88-S97

Peiris JSM, Lai ST, Poon LLM, Guan Y, Yam LYC, Lim W, Nicholls J, Yee WKS, Yan WW, Cheung MT, Cheng VCC, Chan KH, Tsang DNC, Yung RWH, Ng TK, Yuen KY (2003) Coronavirus as a possible cause of severe acute respiratory syndrome. Lancet 361:1319–1325

Poon LLM, Chu DKW, Chan KH, Wong OK, Ellis TM, Leung YHC, Lau SKP, Woo PCY, Suen KY, Yuen KY, Guan Y, Peiris JSM (2005) Identification of a novel coronavirus in bats. J Virol 79:2001–2009

Qin E, Zhu QY, Yu M, Fan BC, Chang GH, Si BY, Yang BA, Peng WM, Jiang T, Liu BH, Deng YQ, Liu H, Zhang Y, Wang C, Li YQ, Gan YH, Li XY, Lu FS, Tan G, Cao WC, Yang RF, Wang J, Li W, Xu ZY, Li Y, Wu QF, Lin W, Chen WJ, Tang L, Deng YF, Han YJ, Li CF, Lei M, Li GQ, Li WJ, Lu H, Shi JP, Tong ZZ, Zhang F, Li SG, Liu B, Liu SQ, Dong W, Wang J, Wong GKS, Yu J, Yang HM (2003) A complete sequence and comparative analysis of a SARS-associated virus (isolate BJ01). Chinese Sci Bull 48:941–948

Roberts A, Paddock C, Vogel L, Butter E, Zaki S, Subbarao K (2005) Aged balb/c mice as a model for increased severity of severe acute respiratory syndrome in elderly humans. J Virol 79:5833–5838

Rota PA, Oberste MS, Monroe SS, Nix WA, Campagnoli R, Icenogle JP, Penaranda S, Bankamp B, Maher K, Chen MH, Tong S, Tamin A, Lowe L, Frace M, Derisi JL, Chen Q, Wang D, Erdman DD, Peret TC, Burns C, Ksiazek TG, Rollin PE, Sanchez A, Liffick S, Holloway B, Limor J, McCaustland K, Olsen-Rasmussen M, Fouchier R, Gunther S, Osterhaus AD, Drosten C, Pallansch MA, Anderson LJ, Bellini WJ (2003) Characterization of a novel coronavirus associated with severe acute respiratory syndrome. Science 300:1394–1399

Shen Z, Ning F, Zhou WG, He X, Lin CY, Chin DP, Zhu ZH, Schuchat A (2004) Superspreading SARS events Beijing, 2003. Emerg Infect Dis 10:256–260

Song HD, Tu CC, Zhang GW, Wang SY, Zheng K, Lei LC, Chen QX, Gao YW, Zhou HQ, Xiang H, Zheng HJ, Chern SWW, Cheng F, Pan CM, Xuan H, Chen SJ, Luo HM, Zhou DH, Liu YF, He JF, Qin PZ, Li LH, Ren YQ, Liang WJ, Yu YD, Anderson L, Wang M, Xu RH, Wu XW, Zheng HY, Chen JD, Liang GD, Gao Y, Liao M, Fang L, Jiang LY, Li H, Chen F, Di B, He LJ, Lin JY, Tong SX, Kong XG, Du L, Hao P, Tang H, Bernini A, Yu XJ, Spiga O, Guo ZM, Pan HY, He WZ, Manuguerra JC, Fontanet A, Danchin A, Niccolai N, Li YX, Wu CI, Zhao GP (2005) Cross-host evolution of severe acute respiratory syndrome coronavirus in palm civet and human. Proc Natl Acad Sci U S A 102:2430–2435

Subbarao K, McAuliffe J, Vogel L, Fahle G, Fischer S, Tatti K, Packard M, Shieh WJ, Zaki S, Murphy B (2004) Prior infection and passive transfer of neutralizing antibody prevent replication of severe acute respiratory syndrome coronavirus in the respiratory tract of mice. J Virol 78:3572–3577

Sulkin SE, Allen R (1974) Virus infections in bats. Karger, Basel

Swayne DE, Suarez DL, Spackman E, Tumpey TM, Beck JR, Erdman D, Rollin PE, Ksiazek TG (2004) Domestic poultry and SARS coronavirus, southern China. Emerg Infect Dis 10:914–916

Tsang KW, Ho PL, Ooi GC, Yee WK, Wang T, Chan-Yeung M, Lam WK, Seto WH, Yam LY, Cheung TM, Wong PC, Lam B, Ip MS, Chan J, Yuen KY, Lai KN (2003) A cluster of cases of severe acute respiratory syndrome in Hong Kong. New Engl J Med 348:1977–1985

Tu CC, Crameri G, Kong XG, Chen JD, Sun YW, Yu M, Xiang H, Xia XZ, Liu SW, Ren T, Yu YD, Eaton BT, Xuan H, Wang L-F (2004) Antibodies to SARS coronavirus in civets. Emerg Infect Dis 10:2244–2248

Van der Hoek L, Pyrc K, Jebbink MF, Vermeulen-Oost W, Berkhout RJM, Wolthers KC, Wertheim-van Dillen PME, Kaandorp J, Spaargaren J, Berkhout B (2004) Identification of a new human coronavirus. Nat Med 10:368–373

Wang L-F, Eaton BT (2001) Emerging paramyxoviruses. Infect Dis Rev 3:52–69

Wang L-F, Shi Z, Zhang S, Field H, Daszak P, Eaton BT (2006) Review of bats and SARS. Emerg Infect Dis 12:1834–1840

Weingartl HM, Copps J, Drebot MA, Marszal P, Smith G, Gren J, Andonova M, Pasick J, Kitching P, Czub M (2004) Susceptibility of pigs and chickens to SARS coronavirus. Emerg Infect Dis 10:179–184

Wentworth DE, Gillim-Ross L, Espina N, Bernard KA (2004) Mice susceptible to SARS coronavirus. Emerg Infect Dis 10:1293–1296

World Health Organization (2004) Summary of probable SARS cases with onset of illness from 1 November 2002 to 31 July 2003. http://www.who.int/csr/sars/country/table2003_09_23/en/. Cited 26 February 2007

Woo PCY, Lau SKP, Chu CM, Chan KH, Tsoi HW, Huang Y, Wong BHL, Wong HL, Poon RWS, Cai JJ, Luk WK, Poon LLM, Wong SSY, Guan Y, Peiris JSM, Yuen KY (2005) Characterization and complete genome sequence of a novel coronavirus, coronavirus HKU1, from patients with pneumonia. J Virol 79:884–895

Wu DL, Tu CC, Xin C, Xuan H, Meng QW, Liu YG, Yu YD, Guan YT, Jiang Y, Yin XN, Crameri G, Wang MP, Li CW, Liu SW, Liao M, Feng L, Xiang H, Sun JF, Chen JD, Sun YW, Gu SL, Liu NH, Fu DX, Eaton BT, Wang L-F, Kong XG (2005) Civets are equally susceptible to experimental infection by two different severe acute respiratory syndrome coronavirus isolates. J Virol 79:2620–2625

Xu RH, He JF, Evans MR, Peng GW, Field HE, Yu DW, Lee CK, Luo HM, Lin WS, Lin P, Li LH, Liang WJ, Lin JY, Schnur A (2004) Epidemiologic clues to SARS origin in China. Emerg Infect Dis 10:1030–1037

Yang ZY, Werner HC, Kong WP, Leung K, Traggiai E, Lanzavecchia A, Nabel GJ (2005) Evasion of antibody neutralization in emerging severe acute respiratory syndrome coronaviruses. Proc Natl Acad Sci U S A 102:797–801

Yu D, Li H, Xu R, He J, Lin J, Li L, Li W, Xu H, Huang S, Huang J (2003) Prevalence of IgG antibody to SARS-associated coronavirus in animal traders—Guangdong Province China, 2003. Morb Mort Wkly Rep 52:986–987

Zheng BJ, Guan Y, Wong KH, Zhou J, Wong KL, Young BWY, Lu LW, Lee SS (2004) SARS-related virus predating SARS outbreak Hong Kong. Emerg Infect Dis 10:176–178

Zhong NS, Zheng BJ, Li YM, Poon LLM, Xie ZH, Chan KH, Li PH, Tan SY, Chang Q, Xie JP, Liu XQ, Xu J, Li DX, Yuen KY, Peiris JSM, Guan Y (2003) Epidemiology and cause of acute respiratory syndrome (SARS) in Guangdong People's Republic of China, in February 2003. Lancet 362:1353–1358

CTMI (2007) 315:345–361
© Springer-Verlag Berlin Heidelberg 2007

Poxviruses and the Passive Quest for Novel Hosts

R. L. Regnery (✉)

Poxvirus Program, Division of Viral and Rickettsial Diseases, US Centers for Disease Control and Prevention, Mail-stop G-43, 1600 Clifton Road, Atlanta, GA 30333, USA
rur1@cdc.gov

Abstract Poxviruses are famous, or infamous, as agents of disease introduced into novel host species and between populations of the same species. This discussion concerns selected examples of poxviruses associated with vertebrate infections, i.e., the Chordopoxvirus subfamily of the family Poxviridae. Brief note is made of examples of members of the genera *Leporipoxvirus* and *Parapoxvirus*-like agents that have been recognized to have significant trans-host species impact. The remaining bulk of the discussion involves examples of members of the genus *Orthopoxvirus*, which are known to be (have been) involved with human disease, and their zoonotic origins.

1 Introduction

The family Poxviridae is well represented as infectious agents of almost all animal taxa, including vertebrates and invertebrates (Mayo 2005). Poxviruses replicate within the host cell cytoplasm, poxvirus genomes are double-stranded DNA, and the virions have characteristic morphologic features. Poxviruses appear to be able to participate in occasional exchange of DNA with heterologous sources (Krogh and Shuman 2002), and this ability to share DNA presumably has had significant influence on the evolution of these viruses and their ability to adapt to specific hosts (see the chapter by Holmes and Drummond, this volume). Some poxviruses are exquisitely host-specific and probably pose little threat as sources of novel emergent diseases, whereas certain examples of closely related viruses have potentially wide host ranges and most likely will be the sources of emergent poxvirus infectious disease in the future. A repeating theme of many of the best described scenarios of poxviruses involves the jumping of a virus from one naturally occurring host species to other naïve susceptible host species, often in the context of a novel ecosystem.

2 Leporipoxvirus

The known strains of myxoma virus are members of the *Leporipoxvirus* genus and are native to specific species of rabbits of the genus *Sylvilagus* found in California and extending south into South America (Fenner and Ratcliffe 1965). Epizootics of myxomatosis are thought to be most commonly transmitted mechanically by arthropod vectors. Myxomatosis in the appropriate natural reservoir rabbit species produces disease without apparent effect on the rabbits' fitness as measured by an individual's ability to live and reproduce (Regnery and Miller 1972). A South American strain of myxoma virus was intentionally introduced into the pernicious feral European rabbit (*Oryctologus cunniculus*) population found in Australia (ca. 1950) and was also released with less forethought into the native rabbit populations in Europe. The extreme mortalities of European rabbits that ensued on both continents are landmark events in any discussion of long-term, host species dynamics modification by a viral agent (Fenner and Ratcliffe 1965).

3 Parapoxvirus-Like Virus

A poxvirus with parapoxvirus-like virus morphology was partially described in North American western grey squirrels (*Sciurus grisius*) (Regnery 1975). North American squirrels have successfully been introduced on several occasions into Great Britain (Usher et al. 1992). More recently, evidence for what appears to be a similar poxvirus previously described in North America has been found in feral North American eastern grey squirrels (*Sciurus carolinensis*) introduced into the British Isles, and genome analysis suggests that this virus, although morphologically parapoxvirus-like, is distinct from members of the genus *Parapoxvirus* (Thomas et al. 2003). Although the poxvirus now found in the UK is not apparently severely pathogenic for the introduced grey squirrel, and grey squirrels have a high prevalence of homologous antibodies, the native European red squirrels (*Sciurus vulgaris*) are susceptible to serious infection with the same virus, and few red squirrels appear to live to develop antibodies. Where the distribution of the two squirrel species overlap, the red squirrel's populations are threatened with extinction and the total extinction of the red squirrel on the British Isles is anticipated within the next 20 years. Interestingly, a bacterium of the genus *Bartonella* also appears to have made the eastward journey with the introduced North American grey squirrels to commonly infect British red squirrels; however, there is no evidence that the *Bartonella* species causes fatal disease in either squirrel species (Bown et al. 2002).

4 Orthopoxvirus

Orthopoxviruses, like the leporipoxviruses and parapoxviruses, are well represented as agents of mammalian infections. Naturally occurring orthopoxviruses include variola virus (the now eradicated agent of smallpox), monkeypox virus, ectromelia virus (mousepox virus), camelpox, cowpox, volepox, skunkpox, and raccoonpox viruses. Monkeypox and smallpox (prior to eradication) are recognized to have the potential to cause generalized human infections, typically involving multiple lesions, whereas cowpox and vaccinia infections of otherwise healthy humans typically remain localized to the sites of inoculation. All of the orthopoxviruses share significant DNA sequence homologies (Gubser et al. 2004).

Within recognized species of the genus *Orthopoxvirus*, there are at least two contrasting patterns in host range susceptibility: those viruses with broad host ranges (e.g., vaccinia, cowpox, and monkeypox viruses) and those with limited or single species host ranges (e.g., ectromelia, camelpox, and variola viruses). Viruses currently collectively referred to as cowpox viruses are relatively diverse (perhaps corresponding to multiple preferred rodent life cycles) and are recognized to have the largest recognized orthopoxvirus genomes and the largest set of documented open reading frames (Gubser et al. 2004). And perhaps unlike other models of virus-host interaction, wherein viruses appear to evolve close relationships with single species or even subspecies of the host reservoir, it is an open question whether monkeypox, cowpox, and possibly vaccinia-like viruses continue to persist in nature in part because they may be able to infect multiple reservoir species. Variola virus, on the other hand (prior to eradication), was host species-specific, and although variola virus shares a high percentage sequence homology with cowpox viruses (as well as other orthopoxviruses), the variola virus genome is the smallest recognized, naturally occurring orthopoxvirus genome, suggesting a reductionist evolution of genes as the virus adapted to a single host (humans). Whereas the CHO gene of cowpox virus is associated with the virus's ability to infect in vitro a variety of species' cells, the homologous ectromelia virus gene appears to limit ectromelia virus growth to mouse cells, clearly suggesting that gene specialization contributes to the limited host range of ectromelia virus (Chen et al. 1992). In simplest terms, orthopoxviruses appear to currently reflect two evolutionary pathways, one of retention of host range diversity and another of apparent dependence on a single host species.

4.1 Vaccinia-Like Viruses

Vaccinia virus has been used as a vaccine for the prevention of smallpox, caused by the related variola virus, since the time of Edward Jenner (ca. 1796). However, the origin of vaccinia remains something of a mystery. Vaccinia-like virus infections of water buffalo (*Bubalus bubalis*) are capable of producing primarily localized poxvirus disease in humans in contact with the buffalo, and have been reported in Asia for many years subsequent to the cessation of routine smallpox vaccination (Baxby and Hill 1971; Dumbell and Richardson 1993). It has been proposed that this buffalopox virus represents escaped smallpox vaccine virus used during the eradication of smallpox. Although recent sequence analysis of relatively well-conserved virus envelope genes clearly confirms the close relatedness of buffalopox virus isolate genes and reference vaccinia virus genes (Singh et al. 2006), examples of buffalopox virus isolates have been found to be distinguishable from smallpox vaccine strains of vaccinia by RFLP analysis (Dumbell and Richardson 1993).

Similarly, several isolates of vaccinia-like viruses continue to be made in Brazil, often associated with domestic cattle and the cattlemen that attend the animals (Nagasse-Sugahara et al. 2004; Damaso et al. 2000; da Fonseca et al. 2002; Trindade et al. 2004; de Souza et al. 2003; Schatzmayr et al. 2000). It has likewise been proposed that these isolates may represent examples of escaped Old World vaccinia virus that was used previously as smallpox vaccine (Damaso et al. 2000). Most of the Brazilian vaccinia-like isolates are genetically distinguishable from reference strains of vaccinia. Interestingly, at least one of the Brazilian viruses was isolated from sentinel mice, suggesting arthropod transmission (da Fonseca et al. 2002), and another from a wild rodent of the genus *Oryzomis*, suggesting the existence of an alternate sylvan reservoir species (Fonseca et al. 1998).

Whether or not both the vaccinia-like viruses of Asia and South America are truly escaped vaccine viruses hopefully can be clarified by further detailed genetic analysis. In both geographic examples, it is clear that the viruses have persisted for several years in their presumed novel ecosystems since smallpox vaccination has not been a public health practice in either region since the eradication of smallpox, 20–30 years ago. If vaccine escape is the source of these vaccinia-like viruses, then these viruses may provide valuable models for observing the rates at which orthopoxviruses evolve in the contexts of novel ecologic niches.

The roles that rodents may play in the perpetuation of the vaccinia-like viruses, in addition to the more obvious roles of bovines, are topics worthy of continued investigation. The Brazilian vaccinia-like viruses may be an important precedent for the ability of an Old World orthopoxvirus to establish itself within a novel New World ecologic context (see monkeypox virus discussion below). It will be interesting to follow the fate of the vaccinia virus vaccine vector that has been used to construct modified-live vaccine containing the glycoprotein gene of rabies virus and which has been widely dispersed in the US as a tool to control rabies among wildlife species in the US (see the chapter by Childs, this volume).

4.2
Monkeypox Viruses

Whereas the series of events that must have surrounded the origin of human smallpox will likely be never be fully understood (see variola virus discussion, Sect. 4.3), current events have provided valuable glimpses into evolutionary process of orthopoxviruses with serious pathogenic potential to man. The moniker "monkeypox" originated with the recognition of an orthopoxvirus disease infecting captive nonhuman primates (von Magnus 1959); however, subsequently it has been recognized that monkeypox virus can infect multiple species and in all likelihood primates are not central to the natural disease transmission cycle (Khodakevich et al. 1987a; Meyer et al. 2002; Khodakevich et al. 1987b, 1997; Jezek et al. 1988a; Breman et al. 1980; Hutin et al. 2001a; Arita et al. 1972). The only native African rodent species recovered in the field with monkeypox virus was a moribund rope squirrel (Khodakevich et al. 1986).

In the early 1970s, the original observations that monkeypox virus can result in human disease that closely resembles smallpox generated obvious attention during an era when smallpox was close to being eradicated (Ladnyi et al. 1972; Foster et al. 1972). Subsequently, human monkeypox was relatively well described, especially in the Congo basin of Africa where case fatality ratios for human monkeypox have been variously recorded as 3%–14% in outbreaks and 9.8% among previously smallpox-vaccinated individuals in one prospective study (Jezek et al. 1986b; Hutin et al. 2001). Key epidemiologic differences between variola and monkeypox viruses include not only the host-specific characteristics of variola virus (human only) but also that, whereas variola virus excelled at perpetuation of sustaining transmission between susceptible humans, transmissibility of monkeypox between humans is relatively inefficient. Vaccination with smallpox vaccine (with the related orthopoxvirus, vaccinia virus) is at least partially protective for infection with monkeypox virus. It is generally acknowledged that the *recognition* of on-going human monkeypox in Africa was associated with the eradication of smallpox and the cessation of smallpox vaccination; smallpox vaccination presumably prevented some cases of monkeypox, and prior to the eradication of naturally occurring smallpox, human monkeypox was not recognized and clinically would likely to have been mistaken for smallpox. How long monkeypox has in reality been active as a potentially serious zoonotic disease of humans in Africa will perhaps be impossible to estimate. It is important to note that despite the pathologic similarities between human monkeypox and smallpox, other orthopoxviruses appear to share closer phylogenetic relationships (based on genomic DNA comparisons) with either monkeypox virus or variola than monkeypox virus and variola share with each other (Shchelkunov et al. 2001). Nevertheless, better understanding the components of the close phenotypic similarities between monkeypox and smallpox infections in humans will likely lead to a more complete understanding of the root pathologies associated with serious human orthopoxvirus infections.

Despite the several significant investigations, the natural histories of the monkeypox viruses have remained incompletely understood. Although monkeypox virus has long been recognized to be an example of an orthopoxvirus that can potentially infect a variety of host species, the naturally occurring spectrum of monkeypox infections in African species is unclear. For example, it is unresolved whether African monkeypox viruses sustain transmission within a single, specific, key host species, or whether virus perpetuation continues by virtue of transmission in multiple host species The distribution of the naturally occurring monkeypox, as recognized today, is limited to specific areas of the Congo Basin and coastal Central Africa, and this almost certainly correlates with appropriate reservoir biota (Likos et al. 2005; Peterson et al. 2006; Levine et al. 2007).

The 2003 US monkeypox outbreak renewed interest in analysis of reports of differing relative human pathogenicities of monkeypox virus infections, as well as preliminary observations that, based on limited genomic analysis, there are at least two clades of monkeypox virus isolates that originate in either Western Africa or the Congo Basin (Reed et al. 2004). Recent detailed analysis has demonstrated that genetic differences exist between genes of the two monkeypox virus clades and that some of these genes have been associated with mechanisms of pathogenicity in analogous orthopoxvirus model systems (Likos et al. 2005; Chen et al. 2005). Rigorous case reviews of recent and past human infections suggest significant variations in human pathogenicities associated with the two distinct virus clades; more serious human monkeypox disease is associated with viruses present in the Congo Basin (Likos et al. 2005). That such evolutionary divergence between two monkeypox virus clades exists suggests that there are quite likely at least two distinct cycles of monkeypox virus transmission in Africa, perhaps differentiated by geographic distribution and/or subtle differences within reservoir host species, and/or that monkeypox viruses have evolved from a common ancestor on more than one occasion.

Although monkeypox is typically still regarded primarily as a zoonotic infection, multiple rounds of human-to-human transmission in Africa have been well documented (Jezek et al. 1986a, 1987, 1988b; Learned et al. 2005). Ominously perhaps, the most recent published report of human monkeypox in the Congo basin has also included the most protracted human-to-human case series to date (Learned et al. 2005). The roles of increasing population densities in central Africa and the increased prevalence of immunomodulating disease in the region are likely to be factors that could favor selection for enhanced human-to-human monkeypox virus transmission. Further analysis of the epidemiology and natural histories of monkeypox in Africa, as well as any future importations of monkeypox into novel virus habitats, will potentially offer unique opportunities to study possible early steps in the evolution and divergence of human pathogenic orthopoxvirus disease; similar steps may have occurred in the prehistoric evolution of smallpox-like disease (see Sect. 4.3) and which potentially might occur in the future.

4.3
2003 US Monkeypox Outbreak

The importation and sale of exotic pet species to and within industrialized nations is a frequently overlooked industry. In the US, much of this trade has been legal if largely unregulated: animals destined to be considered exotic pets have administratively neither been considered as agricultural species, nor have most of species incurred scrutiny under legislation designed to curb importation of endangered species. Frequently, large numbers of such animals, destined for the exotic pet trade, have entered the US legally, with no veterinary oversight and no requirement for quarantine. Anecdotal accounts from importers of exotic species suggest that substantial die-off of exotic imported species is expected. Some of these same species have come from areas of the world that also support significant zoonotic disease with recognized host species (e.g., Lassa fever virus).

During the early summer of 2003, human monkeypox was recognized for the first time outside of Africa (CDC 2003a, 2003b, 2003c, 2003d; Reed et al. 2004).

The outbreak resulted in transmission of monkeypox virus to a variety of novel host species, one of which in particular, the prairie dog (*Cynomys* sp.), served as amplification host and vector for 72 confirmed or suspected cases of human disease (Reed et al. 2004). The importation of monkeypox to North American has helped to refocus attention on the natural history of monkeypox in Africa, the events associated with the secondary and tertiary cross-species transmission of virus, and the role of exotic species as vectors of disease to novel host species, including humans.

The reconstruction of events leading to the 2003 US monkeypox outbreak is believed to be relatively complete (Reed et al. 2004) and additionally will be the subject of future publications. Over 500 individual animals were imported from Accra, Ghana, to an animal distributor in Texas who then trans-shipped the animals to other dealerships. Monkeypox virus was isolated from at least three species of imported rodent species; however, which, if any of these species were involved in a natural transmission cycle in Ghana (the origin of the shipment) and which species may have been infected either in captivity in Africa or en route to the US could not be resolved. Infected giant pouched rats (*Cricetomys gambianus*), rope squirrels (*Funisciurus* spp.), and dormice (*Graphiurus murinus*) died after arrival in the US. Despite extensive state and federal trace-back efforts to locate animals from the original shipment, approximately 25% of them remained unaccounted for and the disease status of these animals and the species they represented is likewise unresolved.

Although multiple animals from the original Ghanaian shipment showed evidence of infection, all of the confirmed US human monkeypox infections were associated with contact with infected captive prairie dogs (although some of the same persons may also have had contact with African species). All of the infected prairie dogs were, as far as can be documented, at one time associated with one animal distributor in Illinois. This distributor housed these prairie dogs in the same facility used to house several hundred other animals destined for the exotic pet trade, including African dormice and pouched rats that arrived as a portion of the shipment from Ghana. Mechanisms of transmission between African species and other species (or perhaps secondary infections between prairie dogs reinfecting other species) at the distributor's facility are unclear; however, fomite and arthropod transmission, as well as potential aerosol transmission remain possibilities. Several additional species of animals affiliated with the distributor, not thought to be in direct physical contact with the various African species, have been shown to either be monkeypox virus isolate-positive, monkeypox virus PCR-positive, orthopoxvirus antibody-positive, or a combination of all three (Huston et al. 2007). These included examples of marsupials, insectivores, as well as examples of multiple rodent species. It is unclear how much monkeypox virus mortality was associated with various animal species, prior to recognition of the disease outbreak. However, clearly the prairie dogs attracted the most attention, specifically due to their unique role as vectors of human disease, and to a lesser extent due to their frequently obvious signs of disease, including extensive poxvirus-associated histopathologies and accounts of examples of multiple lesions (Guarner et al. 2004; Langohr et al. 2004).

The underlying reasons why human monkeypox in the US was apparently associated with prairie dog exposure, and not necessarily African rodent species, is not clear; however, there are several factors that may have influenced the observation that prairie dogs appear to have been especially effective vectors of human disease. Firstly, it is clear that prairie dogs are permissive to the growth of virus (Xiao et al. 2005). High titers of virus have been documented in prairie dog urine, feces, and skin tissues by both infectivity (Huston et al. 2007) and histopathology (Guarner et al. 2004; Langohr et al. 2004), suggesting that these infected prairie dogs had the potential to shed virus.

It is also possible that human physical interaction with prairie dogs was significantly different than was interaction experienced with African rodents and perhaps other potential vector species. Anecdotal accounts suggest that one of the attractions that prairie dogs have as captive pets was the rapidity with which prairie dogs habituate to humans and that prairie dogs demonstrate rather fearless captive behavior, which was in the case of the monkeypox outbreak conducive to substantial physical contact with humans. This human physical contact with prairie dogs was reported to include human skin scratching (prairie dogs have substantial claws normally used for digging) and biting by the prairie dogs (Reed et al. 2004; Kile et al. 2005). It is doubtful that similar intimate contact might have been associated with more active wild species, including several of the African species. It should be noted that in addition to direct contact with infected prairie dogs, several other human monkeypox cases were not associated with direct physical prairie dog contact.

Prior to the recognition of an ongoing monkeypox outbreak, there were no special barriers to environmental transmission of virus to North American sylvan species. Imported species were not transported or housed in containment style facilities and animal transactions between human buyers frequently took place at casual pet swap meets. Approximately 25% of the potentially infected prairie dogs as well as animals from the original Ghanaian shipment remain unaccounted for. Prior to the recognition of the human monkeypox outbreak, animals that may have died of monkeypox virus infection were typically not decontaminated prior to disposal, which included disposal into public landfills. Owners of exotic animals and prairie dogs were requested to resist the temptation to release these animals into the wild. Considering these various scenarios, it is reasonable to conclude that opportunities for the escape of monkeypox virus into potential North American reservoir species probably occurred; however, successful introduction would have required large enough populations of susceptible species to have sustained continued transmission. Currently (4 years after outbreak) there is no hard evidence that monkeypox virus did escape or persist within North American sylvan animal populations, but there have been no long-term follow-up studies to validate the absence of such infection, and surveillance for any disease in potential reservoir species would probably be dependent on the reporting of monkeypox in a sentinel human (see the chapters by Childs, by Merianos and by Stallknecht, this volume). So although multiple North American mammal species are susceptible to monkeypox infection and can shed large amounts of virus, it remains unclear whether monkeypox virus has the potential to sustain an introduction, transmission, and subsequent die-off of North American native rodent species, perhaps analogous to what accompanied the effect of introduction of myxomatosis into European rabbit populations in Australia and Europe (see Sect. 2).

Immediately following the June 2003 recognition of a US outbreak of monkeypox associated with importation of animals from Africa, the Centers for Disease Control and Prevention utilized public health quarantine authority to establish a ban on importation of African rodent species (only) and for bushmeat for commercial trade (CDC 2003d). Noting the potential for reimportation of monkeypox virus (or other African rodent-associated human pathogens) in the future, it is likely that this ban will remain in place. On the same day that the ban was placed on importation of African rodent species, the Food and Drug Administration initiated regulatory control of interstate transport of prairie dogs as part of the effort to limit further spread of monkeypox to other susceptible human and nonhuman hosts. Four years after the original importation of monkeypox virus, the long-term continued effect of the ban on interstate transport of prairie dogs on the spread of future monkeypox is not clear. In any case, the combined outbreak response efforts by state and federal partners, together with exotic pet owners, to limit further traffic in infected and potentially infected animals, can reasonably be assumed to have helped limit opportunities for additional human monkeypox and for zoonotic disease introduction into native, highly susceptible, nonhuman North American species. The 2003 US monkeypox importation can optimistically be regarded as a near miss event in the continuing long-term history of poxviruses and their potential exploitation of novel hosts and ecosystems.

Although several human infections associated with the US outbreak were considered serious infections, there were no human fatalities and human disease was generally regarded as less severe than the frequently fatal monkeypox described primarily from the Congo Basin (Reed et al. 2004; Likos et al. 2005). The apparent origin of the monkeypox importation, Ghana, is a region of Africa not previously identified as endemic for human monkeypox (Likos et al. 2005), emphasizing the variation in pathogenic potential of monkeypox viruses in different regions of Africa. Genetic evaluation of the imported monkeypox virus and West African isolates of monkeypox virus, when compared to the genome of monkeypox viruses from the Congo Basin, suggests significant but often subtle differences in genes associated with poxvirus pathogenicity (Likos et al. 2005; Chen et al. 2005).

4.4
Smallpox

The eradication of smallpox (ca. 1970) constitutes one of the greatest achievements of the organized movement to combat disease (Fenner et al. 1988). However, the history of introductions of the agent of smallpox (*Orthopoxvirus variola*) into previously unexposed human populations, even within relatively recent historic times, has been credited with being a major, frequently disastrous, influence on the course of recorded human history (Hopkins 1983; see the chapter by Cleaveland et al., this volume). Attempts to try to deduce earliest evidence for human smallpox are interesting exercises in historic medical deduction (Hopkins 1983; Fenner et al. 1988). Interpretation of historic inferences and clinical descriptions of disease may be best served by recognizing that the viruses responsible for ancestral smallpox were very likely not identical to the *O. variola* that we recognized as the agent of modern-era smallpox, a virus which at least prior to eradication was highly adapted and specialized for the human host.

If there is uncertainty regarding the authentic dates ascribed to sustained transmission of smallpox infections between humans, there is even less certainty about the presumed zoonotic origin and events associated with primal transmission to humans of a zoonotic orthopoxvirus. One of the hallmarks of modern-era smallpox (ca. nineteenth to twentieth centuries) was the virus's ability to maintain continued transmission even within examples of populations with relatively high rates (e.g., 80%) of immunity to smallpox (Fenner et al. 1988). The estimated number of persons typically infected by another smallpox case ($R_0 = 3.5–6$)(Gani and Leach 2002) was lower than that for more highly contagious viruses (e.g., influenza, measles; see the chapter by Real and Biek, this volume); however, this apparently modest infectious nature of smallpox may have contributed to the long-term, slow burn perpetuation of virus transmission to susceptible hosts within finite human populations. Optimal long-term virus survivability does not a priori imply maximal transmissibility. Human-to-human transmission of smallpox, as known to modern medicine, occurred primarily as respiratory droplets over relatively short distances

(Fenner et al. 1988); presumably evolution of effective human host specificity involved selection for such enhanced transmission modes as compared to those more commonly associated with poxvirus zoonotic disease (e.g., direct wound contact). If the ancestral smallpox virus was less professional at measured, sustained, human-to-human transmission, presumably such a virus must have, at some point in its history, adapted to the more optimal transmissibility characteristics for which modern era smallpox was so well known.

The frequently cited skin lesions observed on the mummified corpse of Ramses V appear to be orthopox-like and may very well represent physical evidence of a 3000-year-old smallpox infection (reviewed by Hopkins et al. 2004; Fenner et al 1988); however, without information relating to the responsible virus's identity (e.g., genotype) or transmission characteristics, it is not realistically possible to differentiate Ramses's lesions from those caused by another possibly related orthopoxviruses, e.g., monkeypox. Even if the virus that infected Ramses V was an ancestor of modern-era smallpox, it would be perhaps naïve to expect that the virus of ancient Egypt would not have continued to evolve as a microbial parasite, as human populations continued to increase in density and as opportunities for introduction of virus to naïve populations became more readily possible.

Judging by current experience with human monkeypox virus infections (see Sect. 4), one might expect that during at least the earliest stages of the evolution of smallpox in humans, the prototypic smallpox virus retained zoonotic potential, as has been previously proposed (Hopkins 1983). It is also not unreasonable to consider that the primal zoonotic transmission and subsequent adaptation to growth in humans may have occurred on multiple occasions, as has happened with other zoonotic-turned human-adapted pathogens (Hahn et al. 2000; Wolfe et al. 2005; see the chapter by Childs et al., this volume). Similarly, it would be expected that perhaps additional ancestral variola virus lineages did not survive to the twentieth century. Even as smallpox was being eradicated, it was clear that genetically and clinically divergent lineages of variola virus existed, for example the clearly divergent variola major and minor viruses (Gubser et al. 2004; Fenner et al 1988).

Again, to make an analogy with current monkeypox (see Sect. 4), there is no obvious reason why a zoonotic prototypic smallpox-like virus may not also have had a prolonged, if possibly discontinuous, association with human disease while still being tethered to a nonhuman host species for long-term perpetuation. In the absence of a reliable supply of susceptible, reasonably densely populated human hosts to maintain exclusive human host transmission, an alternate zoonotic host would be a considerable evolutionary advantage for an orthopox virus with potential for exploit humans for hosts. Ironically perhaps, monkeypox virus has outlasted smallpox and probably would be much harder to eradicate. A strict requirement for an alternative, nonhuman reservoir may have been expected to have limited the distribution of a smallpox prototypic virus to correspond with the range of a zoonotic alternate host(s), as is currently the case with monkeypox. However, enhanced human-to-human transmissibility and inability to utilize alternative zoonotic hosts did not necessarily have to happen synchronously. And, as noted above, the ability to sustain ongoing rounds of smallpox within a population might well have selected for viruses that balanced transmissibility with availability of susceptible hosts during a time when human migration and population densities were more limited than in modern times. The timing of the evolution of truly effective human-to-human smallpox transmissibility and sustainability, as well as the independence from a geographically constrained zoonotic reservoir, would have been seminal events in the history of smallpox but were events that may also have occurred later in the development of smallpox-like disease than perhaps is often appreciated. It is interesting to consider that although historic hints suggesting smallpox-like disease can be associated with ancient Egyptian mummies and Indian and Chinese texts dating back at least to 3000 years ago, it is reasonably clear that smallpox did not become an established endemic feature of the European continent until only the sixteenth century, despite reported examples of focal epidemics probably associated with importation (Hopkins et al. 2004). It would be interesting to know if the timing of the advent of endemic European smallpox was simply the consequence of a growing European human population density, or was it also associated with virus evolutionary events relating to enhanced sustainability in human population, perhaps coupled with independence from an unrecognized African or Asian reservoir species? In any case, by the time of smallpox eradication and intensive study in the twentieth century, the variola viruses as we know them had transitioned exclusively to the human host to the exclusion of any potential known zoonotic host.

In terms of emergence of possible novel, naturally occurring pathogenic orthopoxviruses, it seems reasonable that the future parents for human orthopoxvirus disease will most likely be (already are?) examples of the generalist members of the genus (e.g., monkeypox, cowpox, vaccinia-like viruses); the contemporary activities of these generalist viruses deserve our serious attention (see the chapters by Cleaveland et al. and by Holmes and Drummond, this volume).

The more fully we appreciate the evolution of vertebrate host species and their pathogens, the better we will recognize that current era viruses, with which we are most familiar (e.g., variola virus), are probably significantly different from the zoonotic parental viruses that first infected our ancestors. Similarly, the human-adapted viruses of the future will likely be subtly but significantly different from the recognized zoonotic viruses of today. With this awareness, hopefully we may more clearly be able to understand the past and anticipate the future.

Acknowledgements The author wishes to thank Dr. Inger Damon for many thoughtful discussions relative to this manuscript and poxviruses in general.

References

Arita I, Gispen R, Kalter SS, Lim Teong Wah Marennikova SS, Netter R, Tagaya I (1972) Outbreaks of monkeypox and serological surveys in non-human primates. Bull World Health Organ 46:625–631

Baxby D, Hill BJ (1971) Characteristics of a new poxvirus isolated from Indian buffaloes. Arch Gesamte Virusforsch 35:70–79

Bown KJ, Ellis BA, Birtles RJ, Durden LA, Lello J, Begon M, Bennett M (2002) New World origins for haemoparasites infecting United Kingdom grey squirrels (*Sciurus carolinensis*), as revealed by phylogenetic analysis of bartonella infection squirrel populations in England and the United States. Epidemiol Infect 129:647–653

Breman JG, Kalisa R, Steniowski MV, Zanotto E, Gromyko AI, Arita I (1980) Human monkeypox 1970–79. Bull World Health Organ 58:165–182

CDC (1997) Human monkeypox—Kasai Oriental Democratic Republic of Congo February 1996–October 1997. MMWR Morb Mortal Wkly Rep 46:1168–1171

CDC (2003a) Update: multistate outbreak of monkeypox—Illinois, Indiana, Kansas, Missouri, Ohio, and Wisconsin 2003. MMWR Morb Mortal Wkly Rep 52:561–564

CDC (2003b) Update: multistate outbreak of monkeypox—Illinois, Indiana, Kansas, Missouri, Ohio, and Wisconsin 2003. MMWR Morb Mortal Wkly Rep 52:589–590

CDC (2003c) Update: multistate outbreak of monkeypox—Illinois, Indiana, Kansas, Missouri, Ohio, and Wisconsin 2003. MMWR Morb Mortal Wkly Rep 52:616–618

CDC (2003d) Update: multistate outbreak of monkeypox—Illinois, Indiana, Kansas, Missouri, Ohio, and Wisconsin 2003. MMWR Morb Mortal Wkly Rep 52:642–646

Chen N, Li G, Liszewski MK, Atkinson JP, Jahrling PB, Feng Z, Schriewer J, Buck C, Wang C, Lefkowitz EJ, Esposito JJ, Harms T, Damon IK, Roper RL, Upton C, Buller RM (2005) Virulence differences between monkeypox virus isolates from West Africa and the Congo basin. Virology 340:46–63

Chen W, Drillien R, Spehner D, Buller RM (1992) Restricted replication of ectromelia virus in cell culture correlates with mutations in virus-encoded host range gene. Virology 187:433–442

Da Fonseca FG, Trindade GS, Silva RL, Bonjardim CA, Ferreira PC, Kroon EG (2002) Characterization of a vaccinia-like virus isolated in a Brazilian forest. J Gen Virol 83:223–228

Damaso CR, Esposito JJ, Condit RC, Moussatche N (2000) An emergent poxvirus from humans and cattle in Rio de Janeiro State: Cantagalo virus may derive from Brazilian smallpox vaccine. Virology 277:439–449

De Souza TG, da Fonseca FG, Marques JT, Nogueira ML, Mendes LC, Borges AS, Peiro JR, Pituco EM, Bonjardim CA, Ferreira PC, Kroon EG (2003) Aracatuba virus: a vaccinialike virus associated with infection in humans and cattle. Emerg Infect Dis 9:155–160

Dumbell K, Richardson M (1993) Virological investigations of specimens from buffaloes affected by buffalopox in Maharashtra State India between 1985 and 1987. Arch Virol 128:257–267

Fenner F, Henderson DA, Arita I, Jezek Z, Ladnyi ID (1988) Smallpox and its eradication. World Health Organization, Geneva

Fenner F, Ratcliffe FN (1965) Myxomtosis. Cambridge University Press, London

Fonseca FG, Lanna MCS, Campos MA, Kitajima EW, Peres JN, Golgher RR, Ferreira PC, Kroon EG (1998) Morphological and molecular characterization of the poxvirus BeAn 5(8058) Arch Virol 143:1171–1186

Foster SO, Brink EW, Hutchins DL, Pifer JM, Lourie B, Moser CR, Cummings EC, Kuteyi OEK, Eke REA, Titus JB, Smith EA, Hicks JW, Foege WH (1972) Human monkeypox. Bull World Health Organ 46:569–576

Gani R, Leach S (2002) Transmission potential of smallpox in contemporary populations. Nature 414:748–751

Guarner J, Johnson BJ, Paddock CD, Shieh WJ, Goldsmith CS, Reynolds MG, Damon IK, Regnery RL, Zaki SR (2004) Monkeypox transmission and pathogenesis in prairie dogs. Emerg Infect Dis 10:426–431

Gubser C, Hue S, Kellam P, Smith GL (2004) Poxvirus genomes: a phylogenetic analysis. J Gen Virol 85:105–117

Hahn BH, Shaw GM, De Cock KM, Sharp PM (2000) AIDS as a zoonosis: scientific and public health implications. Science 287:607–614

Hopkins DR (1983) Princes and peasants: smallpox in history. University of Chicago Press, Chicago

Hopkins RJ, Kramer WG, Blackwelder WC, Ashtekar M, Hague L, Winker-La Roche SD, Berezuk G, Smith D, Leese PT (2004) Safety and pharmacokinetic evaluation of intravenous vaccinia immune globulin in healthy volunteers. Clin Infect Dis 39:759–766

Hutson CL, Lee KN, Abel J, Carroll DS, Montgomery JM, Olson VA, Li Y, Davidson W, Hughes C, Dillon M, Spurlock P, Kazmierczak JJ, Austin C, Miser L, Sorhage FE, Howell JH, Davis JP, Reynolds MG, Braden Z, Karem KL, Damon IK, Regnery RL (2007) Monkeypox zoonotic associations: Insights from laboratory evaluation of animals associated with the multi-state U.S. outbreak. Am J Trop Med Hyg 76(4): 757–767

Hutin YJ, Williams RJ, Malfait P, Pebody R, Loparev VN, Ropp SL, Rodriguez M, Knight JC, Tshioko FK, Khan AS, Szczeniowski MV, Esposito JJ (2001) Outbreak of human monkeypox, Democratic Republic of Congo 1996 to 1997. Emerg Infect Dis 7:434–438

Jezek Z, Arita I, Mutombo M, Dunn C, Nakano JH, Szczeniowski M (1986a) Four generations of probable person-to-person transmission of human monkeypox. Am J Epidemiol 123:1004–1012

Jezek Z, Grab B, Dixon H (1987) Stochastic model for interhuman spread of monkeypox. Am J Epidemiol 126:1082–1092

Jezek Z, Grab B, Paluku KM, Szczeniowski MV (1988a) Human monkeypox: disease pattern, incidence and attack rates in a rural area of northern Zaire. Trop Geogr Med 40:73–83

Jezek Z, Grab B, Szczeniowski MV, Paluku KM, Mutombo M (1988b) Human monkeypox: secondary attack rates. Bull World Health Organ 66:465–470
Khodakevich L, Jezek Z, Kinzanzka K (1986) Isolation of monkeypox virus from wild squirrel infected in nature. Lancet 1:98–99
Khodakevich L, Szczeniowski M, Manbu MD, Jezek Z, Marennikova S, Nakano J, Messinger D (1987a) The role of squirrels in sustaining monkeypox virus transmission. Trop Geogr Med 39:115–122
Khodakevich L, Szczeniowski M, Nambu, m.D, Jezek Z, Marennikova S, Nakano J, Meier F (1987b) Monkeypox virus in relation to the ecological features surrounding human settlements in Bumba zone Zaire. Trop Geogr Med 39:56–63
Kile JC, Fleischauer AT, Beard B, Kuehnert MJ, Kanwal RS, Pontones P, Messersmith HJ, Teclaw R, Karem KL, Braden ZH, Damon I, Khan AS, Fischer M (2005) Transmission of monkeypox among persons exposed to infected prairie dogs in Indiana in (2003) Arch Pediatr Adolesc Med 159:1022–1025
Krogh BO, Shuman S (2002) A poxvirus-like type IB topoisomerase family in bacteria. Proc Natl Acad Sci U S A 99:1853–1858
Ladnyi ID, Ziegler P, Kima E (1972) A human infection caused by monkeypox virus in Basankusu Territory Democratic Republic of the Congo. Bull World Health Organ 46:593–597
Langohr IM, Stevenson GW, Thacker HL, Regnery RL (2004) Extensive lesions of monkeypox in a prairie dog (*Cynomys* sp.). Vet Pathol 41:702–707
Learned LA, Reynolds MG, Wassa DW, Li Y, Olson VA, Karem K, Stempora LL, Braden ZH, Kline R, Likos A, Libama F, Moudzeo H, Bolanda JD, Tarangonia P, Boumandoki P, Formenty P, Harvey JM, Damon IK (2005) Extended interhuman transmission of monkeypox in a hospital community in the Republic of the Congo 2003. Am J Trop Med Hyg 73:428–434
Levine RS, Peterson AT, Yorita KL, Carroll D, Damon IK, Reynolds MG (2007) Ecological niche and geographic distribution of human monkeypox in Africa. PloS ONE 2:e176
Likos AM, Sammons SA, Olson VA, Frace AM, Li Y, Olsen-Rasmussen M, Davidson W, Galloway R, Khristova ML, Reynolds MG, Zhao H, Carroll DS, Curns A, Formenty P, Esposito JJ, Regnery RL, Damon IK (2005) A tale of two clades: monkeypox viruses. J Gen Virol 86:2661–2672
Mayo MA, Maniloff J, Dusselberger U, Ball LA, Fauquet CM (2005) Virus taxonomy: VIIIth report of the International Committee on Taxonomy of Viruses. Academic Press, New York
Meyer H, Perrichot M, Stemmler M, Emmerich P, Schmitz H, Varaine F, Shungu R, Tshioko F, Formenty P (2002) Outbreaks of disease suspected of being due to human monkeypox virus infection in the Democratic Republic of Congo in 2001. J Clin Microbiol 40:2919–2921
Nagasse-Sugahara TK, Kisielius JJ, Ueda-Ito M, Curti SP, Figueiredo CA, Cruz AS, Silva MM, Ramos CH, Silva MC, Sakurai T, Salles-Gomes LF (2004) Human vaccinia-like virus outbreaks in Sao Paulo and Goias States Brazil: virus detection, isolation and identification. Rev Inst Med Trop Sao Paulo 46:315–322

Peterson AT, Papes M, Reynolds MG, Perry ND, Hanson B, Regnery RL, Hutson CL, Muizniek B, Damon IK, Carroll DS (2006) Native-range ecology and invasive potential of Cricetomys in North America. J Mammology 87:427–432
Reed KD, Melski JW, Graham MB, Regnery RL, Sotir MJ, Wegner MV, Kazmierczak JJ, Stratman EJ, Li Y, Fairley JA, Swain GR, Olson VA, Sargent EK, Kehl SC, Frace MA, Kline R, Foldy SL, Davis JP, Damon IK (2004) The detection of monkeypox in humans in the Western Hemisphere. N Engl J Med 350:342–350
Regnery DC, Miller JH (1972) A myxoma virus epizootic in a brush rabbit population. J Wildl Dis 8:327–331
Regnery RL (1975) Preliminary studies on an unusual poxvirus of the western grey squirrel (*Sciurus griseus griseus*) of North America. Intervirology 5:364–366
Schatzmayr HG, Lemos ER, Mazur C, Schubach A, Majerowicz S, Rozental T, Schubach TM, Bustamante MC, Barth OM (2000) Detection of poxvirus in cattle associated with human cases in the State of Rio de Janeiro: preliminary report. Mem Inst Oswaldo Cruz 95:625–627
Shchelkunov SN, Totmenin AV, Babkin IV, Safronov PF, Ryazankina OI, Petrov NA, Gutorov VV, Uvarova EA, Mikheev MV, Sisler JR, Esposito JJ, Jahrling PB, Moss B, Sandakhchiev LS (2001) Human monkeypox and smallpox viruses: genomic comparison. FEBS Lett 509:66–70
Singh R, Hosammani M, Balamurugan V, Satheesh C, Rasool T, Yadav M (2006) Comparative sequence analysis of envelope protein genes of Indian buffalopox virus isolates. Arch Virol 151:1995–2005
Thomas K, Tompkins DM, Sainsbury AW, Wood AR, Dalziel R, Nettleton PF, Mcinnes CJ (2003) A novel poxvirus lethal to red squirrels (*Sciurus vulgaris*). J Gen Virol 84:3337–3341
Trindade GS, da Fonseca FG, Marques JT, Diniz S, Leite JA, De Bodt S, Van der PY, Bonjardim CA, Ferreira PC, Kroon EG (2004) Belo Horizonte virus: a vaccinia-like virus lacking the A-type inclusion body gene isolated from infected mice. J Gen Virol 85:2015–2021
Usher MB, Crawford TJ, Banwell JL (1992) An American invasion of Great Britain – the case of the native and alien squirrel (*Sciurus*) species. Cons Biol 6:108–115
Von Magnus P (1959) A pox-like disease in cynomologus monkeys. Acta Pathol Microbiol Scand 46:156–176
Wolfe ND, Heneine W, Carr JK, Garcia AD, Shanmugam V, Tamoufe U, Torimiro JN, Prosser AT, LeBreton M, Mpoudi-Ngole E, McCutchan FE, Birx DL, Folks TM, Burke DS, Switzer WM (2005) Emergence of unique primate T-lymphotropic viruses among central African bushmeat hunters. Proc Natl Acad Sci U S A 102:7994–7999
Xiao SY, Sbrana E, Watts DM, Siirin M, da Rosa AP, Tesh RB (2005) Experimental infection of prairie dogs with monkeypox virus. Emerg Infect Dis 11:539–545

CTMI (2007) 315:363–388
© Springer-Verlag Berlin Heidelberg 2007

Ebolavirus and Other Filoviruses

J. P. Gonzalez[1] (✉) · X. Pourrut[1] · E. Leroy[1]

[1]Fundamentals and Domains of Disease Emergence Research Unit, RU178,
Institute for Research Development, IRD, Paris, France
frjpg@mahidol.ac.th

Abstract Since Ebola fever emerged in Central Africa in 1976, a number of studies have been undertaken to investigate its natural history and to characterize its transmission from a hypothetical reservoir host(s) to humans. This research has comprised investigations on a variety of animals and their characterization as intermediate, incidental, amplifying, reservoir, or vector hosts. A viral transmission chain was recently unveiled after a long absence of epidemic Ebola fever. Animal trapping missions were carried out in the Central African rain forest in an area where several epidemics and epizootics had occurred between 2001 and 2005. Among the various animals captured and analyzed, three species of fruit bats (suborder Megachiroptera) were found asymptomatically and naturally infected with Ebola virus: *Hypsignathus monstrosus* (hammer-headed fruit beats), *Epomops franqueti* (singing fruit bats), and *Myonycteris torquata* (little collared fruit bats). From experimental data, serological studies and virus genetic analysis, these findings confirm the importance of these bat species as potential reservoir species of Ebola virus in Central Africa. While feeding bats drop partially eaten fruit and masticated fruit pulp (spats) to the ground, possibly promoting indirect transmission of Ebola virus to certain ground dwelling mammals, if virus is being shed in saliva by chronically and asymptomatically infected bats. Great apes and forest duikers are particularly sensitive to lethal Ebola virus infection. These terrestrial mammals feed on fallen fruits and possibly spats, suggesting a chain of events leading to Ebola virus spillover to these incidental hosts. This chain of events may occur sporadically at different sites and times depending on a combination of the phenology of fruit production by different trees, animal behavior, and various, but as yet still unknown environmental factors, which could include drought. During the reproductive period, infected body fluid can also be shed in the environment and present a potential risk for indirect transmission to other vertebrates.

1 Introduction

In 1976, two geographically isolated epidemics of viral hemorrhagic fever of unknown etiology occurred in Africa, each accompanied by mortality exceeding 50%. The etiological agent causing the outbreaks was found to be a new virus, named Ebola virus. The Ebola virus, together with Marburg virus, a virus of African geographic origin which had been recognized less than 10 years earlier (Martini and Siegert 1971), were subsequently defined as the prototype viruses of a new taxonomic family, *Filoviridae* (Kiley et al. 1982). This name reflects the very unusual morphology of virions observed by electron microscopy, that of a thread or filament (Latin: *filum*; thread). The *Filoviridae* virus family belongs to the order of Mononegavirales characterized by genetic material carried by only one thread of RNA with negative polarity.

Since the emergence of Ebola fever in 1976, many studies have been undertaken to determine the transmission chain of Ebola virus from a hypothetical animal reservoir to humans. This included the search and identification of possible hosts, their characterization as intermediate, incidental and/or amplifying hosts or reservoirs, and finally the search for one or more potential vectors (Monath 1999; Feldmann et al. 2004). This chapter describes the elements of the viral transmission chain, which was described recently after an absence of epidemic activity of Ebola fever for more than 20 years in the specific locales studied. The findings implicate several animal species as playing a central role in the natural maintenance cycle of Ebola virus and in the pathway leading to viral emergence among humans. This pathway involves viral propagation and amplification within the reservoir host and subsequent transmission to intermediate or incidental host(s) capable of sustaining a high incidence of infection accompanied by a high viremia.

2 Ebola Virus and Hosts

2.1 A Variety of Incidental Hosts and an Elusive Reservoir

Due to the nature of its emergence in central Africa, often in areas of inadequate medical infrastructure, studies on Ebola virus have focused on epidemic emergence or resurgence. Intensive research on the origin of this devastating and elusive virus was recently undertaken to find one or more, if any, of its reservoirs. After the first epidemic of 1976 in an area between South Sudan and North Zaire, 3,200 vertebrates and 30,000 insects were collected and tested for the presence of Ebola virus, but no reservoir was identified (Johnson 1976; Arata et al. 1977).

Although a few animal species in Central Africa have been identified with low titers of antibody reactive with Ebola virus antigens, it is only recently that conclusive evidence of active infection with Ebola virus in any wildlife species has been obtained (Table 1). Technological advances in viral diagnostic methods (ELISA, antigen captures, PCR, immunohistochemical labeling of Ebola virus antigens in specific tissues) have made it possible for recent surveys to employ highly sensitive and specific tools to screen for potential reservoir species for zoonotic viruses in ways unavailable to prior investigators (see the chapter by Daniels et al., this volume). Molecular biological methods also permit phylogenetic reconstruction of ancestral viral lineages using sequence data obtained from multiple viral strains isolated at different times; investigators can characterize viruses without isolating each strain and produce models of potential transmission pathways and direction (see the chapter by Holmes and Drummond, this volume). Additionally, prior surveys were conducted during interepidemic phases when there was little or no evidence of Ebola virus activity. During the epidemic outbreaks of Ebola virus occurring between 2001 and 2004 in Gabon and the Republic of Congo, many dead animals were found in the tropical forest areas affected by the epidemic (Leroy et al. 2004). During the 2001–2004 epidemic, 44 carcasses of wild animals were discovered, permitting necropsies to be performed and tissue samples to be collected; samples were transported and analyzed at the high-security laboratory of the Medical Research Centre of Franceville (CIRMF), in Gabon. Sixteen animals (12 gorillas, three chimpanzees, and a

Table 1 Research on the host and reservoir of African Ebola virus

Host type	Test[a]	Habitat	Positive/ total	Origin	Date[b]
Shrew					
Sylvisorex ollula	IFA	F	1/10	CAR	1999
Rodents					
Arvicanthis spp.	IFA	S	9/98	CAR	1979–1983
Mastomys spp.	IFA	F	2/91	CAR	1979–1983
Mastomys spp.	IFA	S	10/265	CAR	1979–1983
Mus spp.	IFA	F	2/54	CAR	1979–1983
Praomys spp	ELISA	F	1/41		1999
Bats					
Hypsignathus monstrosus	ELISA	F	8/32	Gabon, RC	2002–2003
Epomops franquetti	ELISA	F	4/102	Gabon, RC	2002–2003
Myonycteris torquata	ELISA	F	4/58	Gabon, RC	2002–2003
Other mammals					
Cattle	IFA	S	2/108	CAR	1979–1983
Chicken	IFA	S	13/131	CAR	1979–1983
Dog	IFA	F	26/1162	CAR	1979–1983
Donkey	IFA	S	3/13	CAR	1979–1983
Pig	IFA	F	13/80	CAR	1979–1983
Dog	ELISA	F	55/258	Gabon	2002
Nonhuman primates					
Cercopithecus spp.	ELISA	F	1/107	Cameroon, Gabon, RC	1980–2002
Papio spp.	ELISA	F	1/25	Cameroon, Gabon, RC	1980–2002
Mandrillus spp.	ELISA	F	6/215	Cameroon, Gabon, RC	1980–2002
Gorilla gorilla	ELISA	F	2/30	Cameroon, Gabon, RC	1980–2002
Pan troglodytes	ELISA	F	29/225	Cameroon, Gabon, RC	1980–2002

The data presented here are limited to serologic investigations partly done in the Congolese basin rain forest of Central Africa of the Ebola virus enzootic domain (i.e., not including Sudan and Ivory Coast zones which are in different geographical domains). Only animal types with Ebola-Zaire positive antibodies are listed; for more detail please refer to Pourrut et al. 2005; Morvan et al. 2000; Gonzalez et al. 2005

F Forest; *S* Savannah; *CAR* Central African Republic; *RC* Republic of Congo

[a]IFA>1:128, ELISA = screening serum dilution dogs 1:400, nonhuman primates 1:100, bats 1:50

[b]Year of collection

forest duiker) were found positive, by one or more diagnostic test, for Ebola virus infection (E. Leroy, personal communication), demonstrating natural infection and mortality caused by Ebola virus among three species of wildlife. Calculations of relative population size, based on indices of animal presence and abundance in specific locales (excrement, tracks, broken vegetation, nests, etc.), revealed a significant rise in mortality (decline in population size) among certain animal species immediately before and during epidemics of human disease. The populations of gorillas and duikers fell by half between 2002 and 2003 in the Lossi sanctuary (320 km^2), Republic of Congo, and the estimated population size of chimpanzees fell by 88%. Even if these results are approximations (since it is known, for example, that the disappearance of a dominating adult male gorilla causes the break-up of the group and that the dispersed individuals are then difficult to count), they suggested that localized epidemics of Ebola virus can cause significant mortality among certain wildlife species in a very short period of time (Fig. 1). These results complement observations obtained from other studies indicating that significant reductions in populations of gorillas and chimpanzees in the areas of Gabon occur coincidentally with Ebola epidemics affecting humans (Huijbregts et al 2003; Walsh et al. 2003). A previous study conducted in the Taï forest of the Ivory Coast indicated that 11 members of a group of 43 chimpanzees disappeared (a reduction of 26% of the group) during November 1994 when Ebola virus was affecting humans (Formenty et al. 1999).

Sequences of the glycoprotein (GP) gene coding region of the of Ebola virus (the gene considered to be the most variable in Ebola virus genome), obtained from viral RNA extracted from tissue samples obtained from the carcasses of gorillas and chimpanzees, showed that these dead animals were infected by different viral strains. These results indicate that infection of these large primates resulted from simultaneous but independent infections, acquired from an animal reservoir favoring certain environmental conditions (Leroy et al. 2004a). The presence of antibody among chimpanzees sampled before the onset of the first epidemic of Ebola in this area suggests Ebola virus transmission had been occurring prior to the detection of fatal cases (Leroy et al. 2004b).

Observations of the spatial distribution of Ebola virus infection among great ape populations, coupled with reconstruction of the phylogenetic relatedness of viral sequences recovered from different locations, have led to mathematical modeling of the likely spread of Ebola virus. Viral transmission and disease among apes and humans appears to have spread as an epidemic wavefront, originating from a single epidemic epicenter defined by the zone affected by the first Ebola epidemic of 1976 in the Democratic Republic of Congo (RDC), in a northwestern, southeastern direction (Walsh et al. 2003, 2005).

Fig. 1 Victims of the Ebola virus. Carcasses of nonhuman primates infected by Ebola virus discovered in the forest of Gabon and the Republic of Congo. Genetic sequences of the virus were detected in bone tissues (**a** cranium of chimpanzee found approximately 3 weeks after its death), in the skin or the muscle (**b** hand of gorilla found 2 weeks after death of the animal), and in the liver and the spleen (corpse of dead chimpanzee 3–4 days before harvest)

2.2 Animal Species Affected by Ebola Virus

A serologic survey testing 790 samples from 20 species of primates sampled from Cameroon, Gabon, and Republic of Congo (RC), found the prevalence of anti-Ebola IgG, or antibodies reactive to Ebola virus antigens, among chimpanzees to be 12.9% (Leroy et al. 2004b). The results suggested that chimpanzees are in regular contact with the Ebola virus reservoir(s) or infectious virus released into their environment and that some individuals experience nonlethal infections. Furthermore, these data indicate that infection by Ebola virus is ongoing among wildlife during quiescent phases of epidemic disease in humans; Ebola virus may have been enzootic in the forested regions of central Africa for an extended period of time. The presence of anti-Ebola-specific antibodies in other primate species, including five drills (*Papio leucophaeus*), a Western baboon (*Papio papio*), a mandrill (*Papio sphinx*), and a *Cercopithecus* sp., suggests that the circulation of the virus could be much more complex than the simple passage from reservoir to gorillas and chimpanzees. It has been hypothesized that several reservoir species for Ebola virus may exist and epidemiological and virological findings indicate that direct contact with incidentally infected intermediate hosts, such as gorillas and chimpanzees, can lead to sporadic cases and outbreaks of Ebola fever among humans. Additionally, molecular biological data indicate that genetically diverse strains of Ebola virus circulate in nature and can cause fatal disease among wildlife and humans; the potential existence of Ebola virus variants of reduced virulence offers an attractive hypothesis to explain the existence of anti-Ebola specific antibodies among primates.

In addition to Ebola virus spillover to wildlife, domestic dogs have been exposed to and become infected by Ebola virus, as determined by serological studies. At the time of the last Ebola epidemics in Gabon and RC, several dogs were observed consuming the remains of animals fatally infected by Ebola virus. Although the dogs failed to develop any visible clinical signs of disease (Allela et al. 2005), a study undertaken in Gabon after the 2001–2002 epidemics identified anti-Ebola IgG antibodies among dogs; the prevalence of antibodies increased significantly with the proximity of the sample site to the epidemic focus. The prevalence of antibodies to Ebola virus in dogs ranged from 9% in the two largest cities in Gabon, to 15% in the largest town of the epidemic zone, to 25% in rural villages without identifiable human cases, and reached 32% in villages with human Ebola cases directly linked to contact with an infected animal source (Allela et al. 2005). The potential for dogs to survive infection by Ebola virus requires experimental follow-up and, if documented, the potential for infected dogs to shed Ebola virus; by which routes for what duration of time requires elucidation. If domestic dogs prove to be a potential source of human infection the finding could offer one explanation for human epidemics where the original source of exposure has proved elusive.

One can conclude from the above-described investigations and findings that several different taxonomic orders of wild and domestic mammals can be found infected by one or more viral variants of Ebola during epidemics of human disease. Furthermore, there is evidence of ongoing Ebola virus transmission from reservoir host to incidental hosts during interepidemic phases. As the geographic range and preferred habitat of many of the species harboring antibody overlaps, the direction and relative frequency of virus transmission among diverse hosts cannot

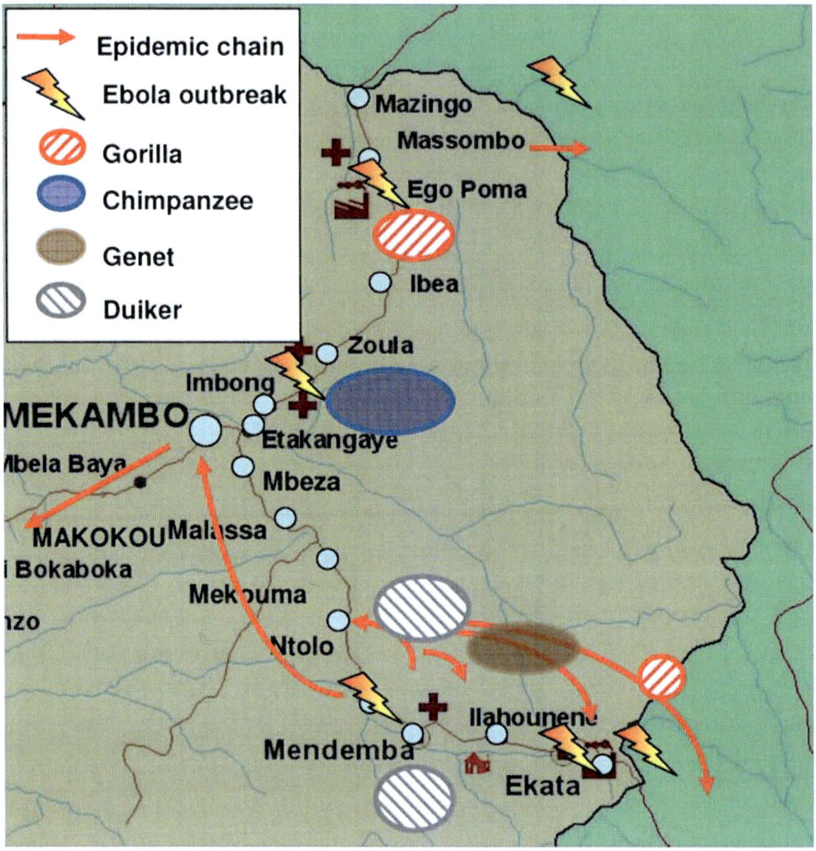

Fig. 2 Areas where infected animals were found during epidemics in Gabon. Territories (*circles*) where infected animals were found are overlapping and thus show the complexity of virus transmission, which can exist between virus host reservoirs, intermediate hosts, or accidental hosts. *Red circles* gorilla; *blue circles* chimpanzee; *black circles* Duikers; *brown circles Genetta* (genets)

be deciphered at the present. Clarifying the individual contribution of incidental hosts in a transmission chain leading to human disease or to a maintenance cycle of different virus variants remains an important, but challenging, endeavor (Fig. 2).

2.3
The Discovery of an Elusive Host: Ebola Virus Reservoirs in Africa

Since 1976, many studies have aimed to identify silently infected but otherwise healthy animal carriers of Ebola virus, but without success (Johnson

1978; Arata and Johnson 1978; Heymann et al. 1980; Gonzalez et al. 1983; Breman et al. 1999; Reiter et al. 1995; Leirs et al. 1999; Formenty et al. 1999). However, analyses carried out on specimens taken from 242 mammals (24 bats, 163 rodents, and 56 insectivorous shrews) captured in the Central African Republic (RCA) in 1998 identified partial Ebola virus genetic sequences in six mice (*Mus setulosus* and *Praomys* sp.) and a shrew (*Sylvisorex ollula*) (Morvan et al. 1999). The absence of a specific serological response, together with the absence of amplifiable total viral sequences, the failure to isolate virus, the nonreproducibility of the results, and the absence of epidemiologic indices favoring the potential role of these animals in Ebola virus epidemics, meant that it was not possible to confirm that these animals were reservoirs of Ebola virus.

Three surveys were recently completed, which targeted the collection of small and medium-sized mammals inhabiting the two forest belts surrounding villages affected by the Ebola epidemics occurring between 2001 and 2005 (Leroy et al. 2005). Animal trapping was initiated a few days after the carcass of an Ebola-infected gorilla was discovered and was limited to the area within 10 km of the carcass site. Over a 3-week period, 1,030 animals were captured and autopsied to obtain tissues for analysis; the laboratory analyses were performed over 4 years (Leroy et al. 2005).

Anti-Ebola IgG was detected in the serum of 16 bats including four *Hypsignathus monstrosus*, eight *Epomops franqueti*, and four *Myonycteris torquata*; no other species of bat or other mammal was seropositive (Fig. 3). Similarly, viral nucleic acid sequences were detected in the tissues from 13 bats; three *H. monstrosus*, five *E. franqueti*, and five *M. torquata* with overlapping domains (Fig. 4) and behavior (Fig. 5). Nucleotide sequences were identified and confirmed as fragments of Ebola virus genes; phylogenetic analyses by Bayesian methods and maximum parsimony identified the greatest similarity was to Ebola virus subtypes found in Zaire (Fig. 6). Although no Ebola virus isolates were obtained, the findings from this study constitute the first virological and biological evidence that certain megachiropteran fruit bats serve as principal reservoir hosts for Ebola virus. Epidemiologic findings collected during previous epidemics suggest contact with fruit bats is relatively common, as these species are a source of bushmeat and the geographic range of the putative reservoir species overlay the known areas of epidemic disease. Additionally, previous studies have documented that a transitory viremia occurs in certain bat species following experimental infection with the Ebola virus (Pourrut et al. 2005; Bergman 1999; Swanepoel et al. 1996). Together, field and experimental findings indicate the plausibility of the hypothesis that certain species of bat serve as the principal reservoir host for at least some variants of Ebola virus.

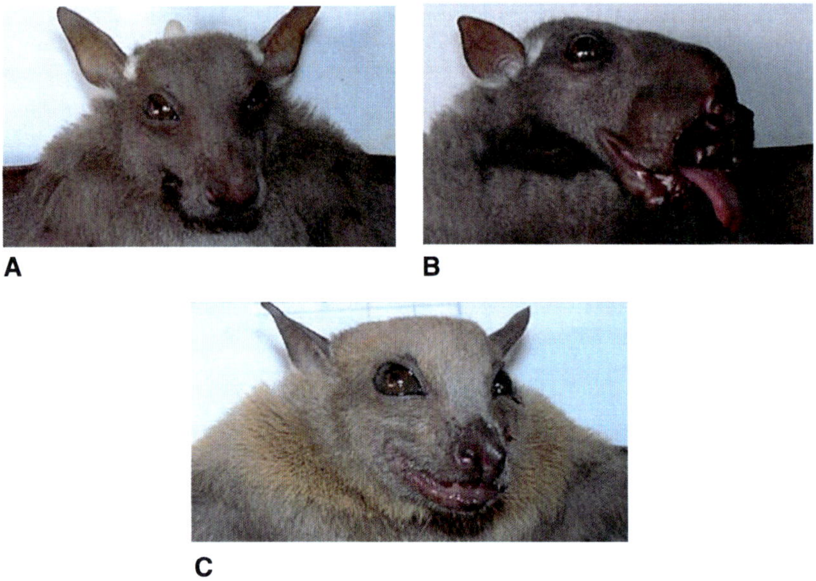

Fig. 3 Three bat species believed to be potential reservoirs of Ebola virus. Photographs of the three species of bats that are potential hosts of Ebola virus: *Hypsignathus monstrosus* (hammer-headed fruit bat), *Epomops franqueti* (singing fruit bat), and *Myonycteris torquata* (little collared fruit bat). These bats are found in the forested areas of central Africa. They have been recorded from Senegal to northern Angola and prefer riverine forests, swamps, mangroves, and palm forests. They play an important role as pollinators of flowers and in dispersing seeds. Their diet consists of fruits, leaves, flowers, nectar, and pollen. *H. montrosus* is the largest bat found in Africa, with males, whose heads are greatly enlarged, significantly larger than females. These three species are nocturnal, roosting during the day in groups of five to 20

It is interesting to note that megachiropteran fruit bats are also reservoirs of Hendra (Halpin et al. 2000) and Nipah (Yob et al. 2001) viruses of the *Paramyxoviridae* family, and Microchiroptera bats are the probable ancestors of all rabies virus variants, serotype 1/genotype 1 of the genus *Lyssavirus* in the family *Rhabdoviridae,* now infecting terrestrial mammals (Badrane and Tordo 2001; Amengal et al. 1997). The *Paramyxoviridae* and *Rhabdoviridae* are the other viral families in the order Mononegavirales and are genetically closely related to the *Filoviridae* (Monath 1999).

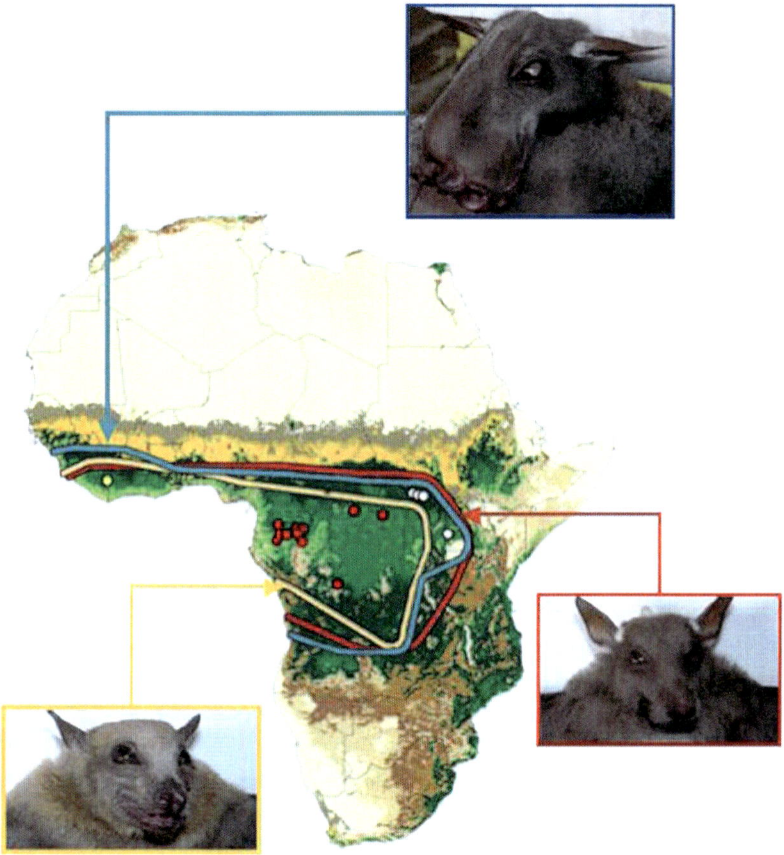

Fig. 4 Distribution of the three bat species that are potential reservoirs of Ebola virus. The *colored lines* mark the limits of distribution of each species (*blue Hypsignathus monstrosus*; *red Epomops franqueti*; *yellow Myonycteris torquata*). It is important to note that the habitats of each species overlap totally

3 Toward Understanding a Complex Natural Cycle and the Origin of Primate Ebola Epidemics

Epidemiological field surveys indicate that mass mortalities of apes and monkey species due to Ebola virus often appear at the end of the dry season (Pinzon et al. 2004), a period when food resources are scarce. Restricted access

Fig. 5 Flight of fruit bats around the canopy trees of equatorial forest in Gabon. The fruit bats live and move in groups of several thousand individuals. This photograph illustrates the large number of contacts that can occur between the large primates and the bats gathered around the same tree to consume the fruits

to a limited number of fruit-bearing trees can lead to spatiotemporal clustering of diverse species of frugivorous animals, such as bats, nonhuman primates, and terrestrial species foraging on fallen or partially eaten fruits or spats. These dense aggregates of different species would increase the probability of contact between infected and susceptible individuals of both reservoir and secondary host species, and promote virus transmission (see the chapter by Real and Biek, this volume). The dry season aggregation of reservoir host species involved in natural maintenance cycles, augmented by incidentally infected secondary hosts serving as sources for intra- and interspecific transmission chains independent of repeated spillover from the reservoir host (see the chapter by Childs et al., this volume), provides an ecological setting for amplifying enzootic transmission of Ebola virus in a manner analogous to draught-induced amplification of enzootic Saint Louis encephalitis virus (SLEV), whereby arthropod vectors and vertebrate hosts are concentrated around a diminished number of water sources (Shaman et al. 2002).

Behavioral and physiological events occurring among bats during and subsequent to the tropical dry season serve to increase the contact rate and types of contacts between individual bats, which can promote transmission of Ebola

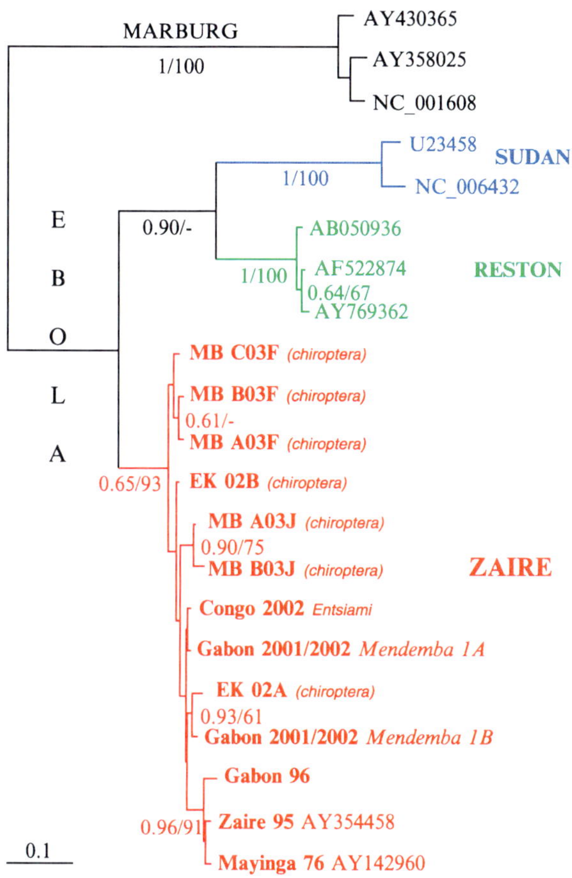

Fig. 6 Zaire Ebola virus sequences detected among fruit bats in Gabon and Republic of Congo. The values indicated below the branches are Bayesian posterior probabilities (*left of the slash*; values below 0.5 are not shown) and the maximum bootstrap percentages obtained with the parsimony method (*right of the slash*; values below less than 50% are not shown). The sequences obtained from bats are indicated as chiroptera (GenBank accession numbers Dq205409–15). The other sequences are viral sequences from symptomatic human cases

virus and increase R_0 (see the chapter by Real and Biek, this volume). In addition to increased competitive interactions driven by unusually high densities of individuals foraging for a common resource of limited availability, breeding activities of megachiropteran fruit bats can involve unusual social behavior as in the case of *H. monstrosus*, where aggregates of males (leks) compete collectively for the attention of females (Hill and Smith 1984); pregnancy can involve physiological changes among female bats that alter immune functions (Langevin and Barclay 1990). Parturition among the African megachiropteran bats occurs throughout the year, although seasonal peaks provide an ample amount of birthing fluids, blood, and placental tissues, potentially containing unusually high titers of Ebola virus, in a medium highly attractive and readily available to scavenging terrestrial mammals (Figs. 7, 8).

4 Other Members of the Filovirus Family

The other two species of virus in the family *Filoviridae* are Marburg virus, which was once regarded as a subtype of African Ebola virus but is now known to be a distinct species, and the Reston subtype of Ebola, limited in distribution to the Philippines and perhaps other regions in southwestern Asia. The

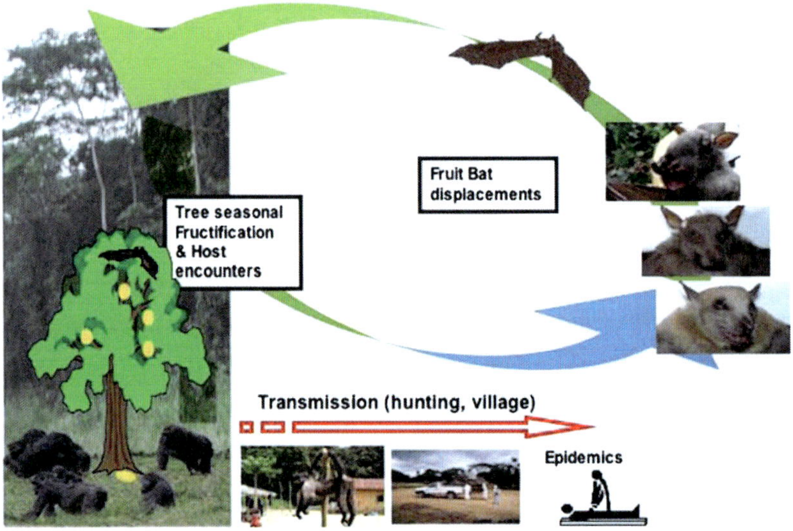

Fig. 7 Hypothetical natural cycle in Central Africa of Ebola viruses. Fruit bats chronically infected with Ebola virus move to consume fruit in the canopy during the fruiting season of certain trees and throw masticated spats and fruit contaminated with Ebola virus-infected saliva. Large monkeys and forest duikers can become infected by eating these fruit and spats on the ground

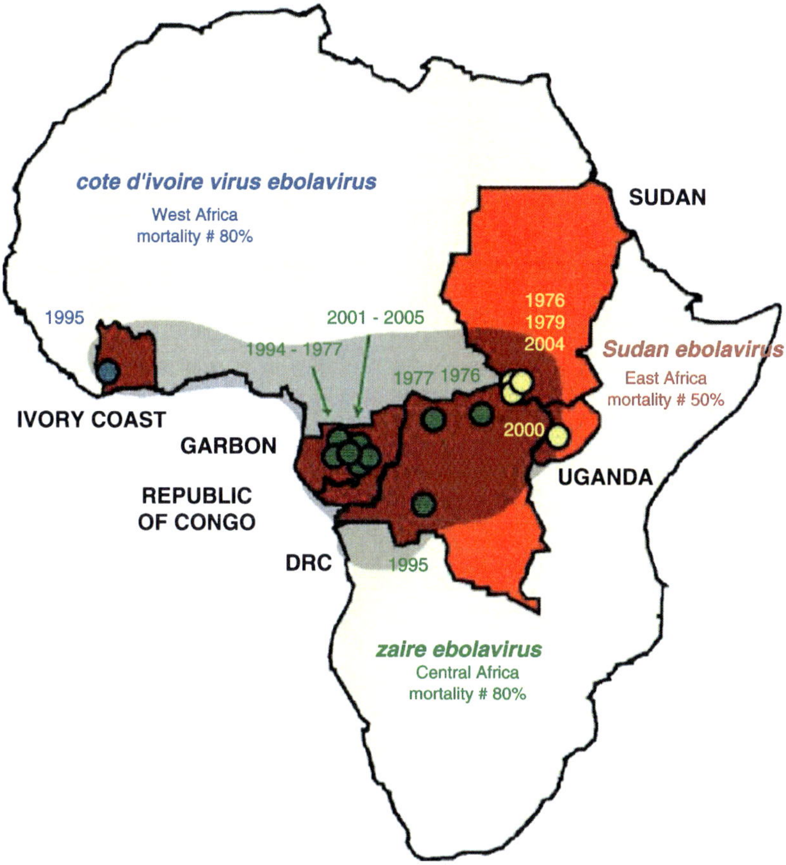

Fig. 8 Geographical distribution of the bat species believed to be potential reservoirs of Ebola virus subtypes and mortality rates of outbreaks due to different Ebola subtypes. The distribution area of bat species appears here in *gray* and represents the complete range of the three species implicated in Ebola virus transmission (see Fig. 4). The death rates of the different epidemics are interesting to note and pose the question of whether different species of bats are associated with each Ebola virus subtype

geographic distribution of Marburg virus is unknown, because knowledge has been restricted to investigations of the patchily distributed and rare outbreaks of human disease occurring in sub-Saharan Africa. The Ebola Reston subtype has been identified among primates in holding facilities in the Philippines and in countries receiving shipments of animals from this island nation. Based on the detection of specific antibodies present among a few humans involved in the care of infected monkeys, the Reston Ebola virus is capable of causing subclinical infection of humans.

4.1 Marburg Virus

The animal reservoir for Marburg virus remains unknown and few secondary hosts have been identified other than humans. The close relationship of Marburg virus to the African Ebola viruses suggests that bats may also be involved in natural maintenance cycles. Several epidemiological observations gleaned from the rare outbreaks of Marburg virus favor this hypothesis. Genetically distinct variants of Marburg virus were obtained during a "single" epidemic affecting gold miners working in mines inhabited by bats. The diversity of Marburg strains infecting these miners suggested multiple independent, but near simultaneous, spillover events requiring close human contact with a relatively common and accessible animal species serving as a reservoir host (Bausch et al. 2003). Although bats met these criteria, demonstration of Ebola virus infection through isolation or detection of viral sequences within tissues derived from field-collected bats or other wildlife has not been reported (Bertherat et al. 1999).

4.2 The Phylogeographic Enigma of Reston Virus

It is suggested that bats may also be the potential reservoir hosts for the subtypes Ebola-Sudan and Ebola-Ivory Coast, and possibly for the Ebola-Reston subtype. If bats serve as reservoir hosts for Ebola virus subtype Reston the species will necessarily be different than the three species of megachiropteran bats implicated as reservoir hosts for the Ebola-Zaire subtype, as the mammalian fauna of the Philippines is distinct from that of Africa, with many species indigenous only to certain islands comprising this nation.

The Reston subtype of Ebola virus is distinguished from the other Ebola subtypes both by its geographical separation and the pathophysiology of infections produced in humans and monkeys. It is the only filovirus found outside Africa, with all strains originating from the Philippines.

The first epidemics due to the Reston subtype among subhuman primates occurred between 1989 and 1990 in colonies of cynomolgus macaques (*Macaca fascicularis*) imported to three American quarantine facilities. The first epidemic was identified in a facility in Reston, Virginia (Geisbert et al. 1992; Jahrling et al. 1996), followed by facilities in Texas and Pennsylvania (Rollin et al. 1999; CDC 1996); all of the affected monkeys had been imported from the same primate breeding center in Manila, Philippines. An additional outbreak occurred in the Manila primate center in 1996, during which 383 of 1,404 macaques died and 85 were diagnosed positive for Ebola infection (Miranda et al. 1999).

The Reston strains induce a primarily respiratory pathology in the macaque, but they do not appear to be pathogenic for humans. A serologic study was carried out in 1990 on 186 animal technicians working in the Reston facility revealed the prevalence of antibody ranged between 6% and 67%, depending on where the technicians worked in the holding facility. The presence of specific antibodies to the Reston Ebola virus and the discovery of high levels of antibodies in three technicians having worked near sick animals infected by Ebola virus-Reston, strongly suggest the occurrence of asymptomatic human infections. Four technicians developed a transitory viremia during this period, including a technician who cut his finger with a lancet during an autopsy on a sick monkey; no symptoms of disease were detected.

In spite of these geographical and pathophysiological differences, the structural and functional characteristics of Ebola-Reston are similar to the three other African Ebola subtypes. The genetic distances between the Reston subtype and the three African subtypes are similar, as are strains within subtypes. Beyond the considerable interest in the genetic basis for the phenotypic differences in the pathogenicity of Ebola-type viruses from Asia and Africa lies the biological mystery of the disjointed distribution of these viruses. Elucidating the evolutionary peregrinations and natural history of the Ebola viruses leading to the bizarre pattern of geographic distribution poses an ongoing biological challenge. Based on the genetic similarity among these viruses, it is reasonable to initially target the search for an animal reservoir host for Ebola-Reston among fruit bats in Asia belonging to the suborder *Megachiroptera*. In addition to identifying animal reservoir hosts within the Philippines, attempts to detect Ebola virus among species on the Asian mainland could provide clues to the origin of these unusual viruses.

5 Conclusions

5.1 Bats, an Underappreciated Reservoir Host for Zoonotic Viruses

Although bats are known to contribute to the epidemiology of a few other human pathogenic parasites (for example, the fungus, *Histoplasma capsulatum* , grows well in bat guano), there are many bat-borne viruses that present major concerns for human health (for review see Calisher et al. 2006). The

Rhabdoviridae family, rabies virus, and rabies-related viruses in the genus *Lyssavirus*, are of obvious public health importance and include Mokola virus, Duvenhage virus, European bat lyssaviruses 1 and 2, and Australian bat lyssavirus. Lagos bat virus and four newly characterized lyssaviruses obtained from Eurasian Microchiroptera, Aravan, Khujand, and West Caucasian bat virus, have not yet been identified as causing human or animal disease (Kuzmin et al. 2003, 2005). Additionally, some viruses in the genus *Flavivirus*, including Japanese encephalitis virus and SLEV (reviewed by Sulkin and Allen 1974; Calisher et al. 2006), have been isolated from bats and are capable of causing epidemics of severe disease among humans and animals. However, of the more than 65 viruses isolated from bats and the numerous other viral infections identified solely on the basis of serological testing of bat sera, there exist few data to assess their risk to human health or to establish details of their maintenance ecology and the significance of bats as reservoir hosts (e.g., Dakar bat virus, Entebbe bat virus, Sokoluk virus, Yokose virus, Jugra virus, and Phnom Penh bat virus).

It is only recently that megachiropteran and microchiropteran bats have achieved notoriety as reservoir hosts of several newly described viruses capable of causing severe disease in humans and other animals. Various *Pteropus* spp. have been shown to be the reservoir hosts of Hendra and Nipah viruses, two novel viruses comprising the newly defined *Henipavirus* genus in the family *Paramyxoviridae*, which have caused severe disease among humans. Initially, Hendra and Nipah virus spillover was from the bat reservoir host to domestic livestock, horses and pigs, respectively, which served as the first secondary host in the transmission chain leading to human infection and disease (see the chapters by Childs et al. and Field et al., this volume). It was only after virus amplification and excretion via the respiratory route from infected livestock that human cases developed (see the chapter by Daniels et al., this volume), although Nipah virus transmission in Bangladesh appears to involve direct human infection from the bat reservoir host or from fruit contaminated by virus followed by human-to-human transmission (see the chapter by Field et al., this volume). In 2005, microchiropteran bats were identified as a reservoir host of the severe acute respiratory syndrome (SARS) coronavirus in Asia (Li et al. 2005). Of special note is the consistent association of bats with viruses containing negative-sense single-stranded RNA and belonging to the order Mononegavirales, lyssaviruses in the family *Rhabdoviridae*, henipaviruses viruses in the family *Paramyxoviridae*, and Ebola virus subtype Zaire in the family *Filoviridae*.

5.2 Bats and Human Disease Emergence

The concept of bats as reservoirs of emerging viral diseases of humans raises questions about an order of mammals for which much remains to be learned about their ecology, taxonomy, and their basic biology. In general, bats possess several unique features that make them notable reservoir hosts (reviewed in Calisher et al. 2006): their capacity to fly and range over long distances when feeding and, in some instances, when migrating; their capacity to entire torpor or hibernation; their tendency to cluster tightly in colonies that may number in the millions (Constantine et al. 1968) or establish large camps (see the chapter by Field et al., this volume); their population structure that can involve seasonal mixing of migratory and nonmigratory metapopulations presenting an opportunity for viral exchange; and the potential for infected bats to become chronic carriers of certain viruses that can be excreted over extended time periods (Sulkin and Allen 1974). Each of these attributes varies from species to species, in part depending on their specific environments.

Questions are arising today on the emergence of the *Filoviridae* within the general context of transmissible diseases of vertebrates. Of particular note is why it has taken more than 20 years to begin to understand the natural transmission cycles of Ebola virus, and why more than 30 years after the first epidemic of Marburg fever in Germany and Yugoslavia, does the animal reservoir for Marburg virus remain unknown? Several factors have hindered the pursuit of the elusive reservoir host for the filoviruses. Epidemics or sporadic cases of filovirus disease have most often occurred in remote locations in countries experiencing varying levels of social upheaval resulting from ongoing armed conflicts or the transient chaos that accompanies disease outbreaks of high mortality linked to hospital-associated transmission. Even at the best of times, the medical and public health infrastructure within central African nations is limited. Animal surveys initiated as a result of investigations into human epidemics have been opportunistic and have not been undertaken at the optimum place or time. Often the identification of a filovirus as the cause of a specific disease outbreak is delayed. Delay results in uncertainty when identifying the index case and, at best, complicates collection of a verbal patient history from surviving relatives or associates. Most animal surveys have biased collection over-representing terrestrial mammals as these species are accessible to trapping and can be purchased as bushmeat. Adequate sampling of bat species has mostly been limited to insectivorous microchiropterans (Liers et al. 1999), as obtaining bats feeding on fruits in the high canopy of trees within dense rainforests is difficult. Animal surveys conducted over short intervals during a single season will most often fail to sample species during critical time periods, such as at the end of the dry season, or when species may be physiologically and immunologically prone to virus infection or excretion, such as during mating and parturition. Finally, as previously mentioned, many surveys relied on technology for viral detection that was vastly inferior to the methods currently available, raising concerns about the reliability of diagnostic procedures of drastically reduced sensitivity.

5.3 Intraspecies and Interspecies Contact and the Risk of Epidemic Initiation

The probability for intra- and interspecific transmission of Ebola virus among species of bats and other wildlife will be influenced by ecological factors such as habitat preferences, roosting sites, and food habits, physiological factors such as age and reproductive status, and virological and immunological factors, such the genotypic and phenotypic variability among Ebola viruses and the potential for viral immunity or cross-immunity. Megachiropteran fruit bats can interact and expose other species to excreted Ebola virus through direct contact, such as inoculating virus present in saliva through bites, or indirect routes, as when discarding partially eaten fruit or spats contaminated with infectious virus. However, once spillover of Ebola virus has been achieved in an initial secondary host species, the emergence of human disease and an epidemic of Ebola hemorrhagic fever may require additional mediating events. If the initial secondary host is human, then a single spillover event may be sufficient to cause human disease, if the Ebola strain is virulent in that host, and the single diseased human may be sufficient to initiate an epidemic, if that individual is treated at a hospital and infects health providers. However, if the initial secondary case is a nonhuman primate or a duiker, then further multiplication of virus, either through repetitive spillover events or sustained transmission within the secondary host population (see the chapter by Childs et al., this volume), may be necessary to amplify the number of infected individuals within that host population such that a susceptible human contacts an infectious primate. Of course, a single infected human does not make an epidemic, but the factors prerequisite to the initiation of an epidemic are in play.

5.4 Ebola Virus and Bats

Knowledge of the natural history of Ebola virus has been clarified by the recent implication of fruit bats as reservoir hosts, and epidemiologic and virologic investigations have further elucidated the varied roles that humans and nonhuman primates and other wildlife can play in the several pathways leading to Ebola hemorrhagic fever emergence and resurgence. The distinction of wildlife species as incidental hosts, dying without further contribution to virus transmission, or significant initial secondary hosts, multiplying the number of infected and infectious individuals necessary to increase the probability that an individual susceptible human, a second secondary host species (see the chapter by Childs, this volume), will achieve contact and become infected by spillover, has provided sufficient details of potential routes of Ebola virus emergence that these elements can now be integrated into mathematical models to investigate transmission pathways.

The terrestrial vertebrates found infected with Ebola virus are highly susceptible to lethal infection and will not retransmit the virus back to the reservoir host (H_R=bats; see the chapter by Childs et al., this volume). Transmission leading to an epidemic/epizootic appears to be unidirectional, with the highly susceptible secondary hosts (H_{S1}=gorillas/chimpanzees) being infected from a common source (directly from the H_R or through contaminated biological products). Once infected, this H_{S1} can sustain transmission to infect additional secondary hosts through intraspecific contact, or transmit the virus to a second incidental host (H_{S2}) through interspecific contact (i.e., from primate to humans).

Such a transmission chain could lead to the wave-like spread of genetically related lineages of Ebola virus (Walsh et al. 2005) or, alternatively, explain outbreaks limited in time and space. Evidence accrued from epidemiological and molecular genetical studies of the genetic distance and phylogenetic origin of isolated strains will provide critical support for these competing hypotheses. Once humans become infected directly through contact with bats or materials contaminated with Ebola virus originating from bats, or, alternatively, through contact (i.e., butchering and consumption) with a previously infected H_{S1}, human-to-human transmission can be sustained for multiple generations.

Fruit bats have a preference for certain fruits that ripen at different times, through which they extract essential nutrients by chewing or sucking the pulp. Bats contaminate fruit with their saliva, potentially imparting any infectious virus present in their saliva, and the masticated fibrous waste or partially eaten fruit is deposited on the ground or left in the tree. As they feed, fruit bats also contaminate the ground with their urine and feces. The contaminated fruit or excreta on the ground or in the tree canopy can be consumed by mammals living in the tree canopy (*Cercopithecus* sp.) or by primates (i.e., gorillas, chimpanzees) or other terrestrial mammals (i.e., duikers) on the ground (H_R-to-H_{S1}transmission). Subsequent intraspecific (H_{S1}-to-H_{S1}) or interspecific (H_{S1}-to-H_{S2}) transmission can occur through contact with infected blood, generally at the time of onset of clinically apparent symptoms within any secondary host species. It is also possible that humans might collect contaminated fruit from the trees or lying on the ground, as has reported with Nipah virus (see the chapter by Field et al., this volume).

5.5
Research Perspectives

Several central questions remain to be addressed in order to understand the dynamics of intraspecific transmission within reservoir host populations and interspecific transmission affecting incidental and secondary hosts. It is essential that the pathogenesis of Ebola virus in the natural reservoir host be investigated to establish the incidence and prevalence of infection, the duration of infectiousness, the mechanisms of virus excretion, the potential for persistent infections to occur and evidence of sporadic shedding of infectious virus, and the influence of physiological or environmental factors (e.g., reproductive status, temperature, etc.) on the pattern of viral maintenance and transmission within H_R populations.

With respect to investigating the role of bats as hosts of pathogens capable of causing diseases of humans and animals, a better understanding of the immune system of bats is critical to delineate responses to infection and to develop improved immunological reagents (see Calisher et al. 2006 for review).

Finally, the origin and evolution of filoviruses and their geographic spread remains totally unknown. The Asian Ebola virus clade hints at the ancient global spread of the Filovirus family, but the enigma posed by our current knowledge of filovirus distribution is begging for answers.

References

Allela L, Boury O, Pouillot R, Delicat A, Yaba P, Kumulungui B, Rouquet P, Gonzalez JP, Leroy EM (2005) Ebola virus antibody prevalence in dogs and human risk. Emerg Infect Dis 11:385–390

Amengual B, Whitby JE, King A, Cobo JS, Bourhy H (1997) Evolution of European bat lyssaviruses. J Gen Virol 78:2319–2328

Arata AA, Johnson B (1977) Approaches towards studies on potential reservoirs of viral haemorrhagic fever in southern Sudan. In: Pattyn SR (ed) Ebola virus haemorrhagic fever. Elsevier/Netherland biomedical, Amsterdam, pp 191–202

Badrane H, Tordo N (2001) Host switching in Lyssavirus history from the Chiroptera to the Carnivora orders. J Virol 75:8096–8104

Bausch DG, Borchert M, Grein T, Roth C, Swanepoel R, Libande ML, Talarmin A, Bertherat E, Muyembe-Tamfum JJ, Tugume B, Colebunders R, Konde KM, Pirad P, Olinda LL, Rodier GR, Campbell P, Tomori O, Ksiazek TG, Rollin PE (2003) Risk factors for Marburg hemorrhagic fever Democratic Republic of the Congo. Emerg Infect Dis 9:1531–1537

Bergmans W (1989) Taxonomy and biogeography of African fruit bats (*Mammalia Megachiroptera*). Beaufortia 39:89–152

Bertherat E, Talarmin A, Zeller H (1999) La République Démocratique du Congo: Entre guerre civile et Virus de Marburg. Le Comité de Cooridantion Scienctifique et Technique International pour l'épideemie de Durba. Med Tropic 59:201–204
Breman JG, Johnson KM, van der Groen G, Robbins CB, Szczeniowski MV, Ruti K, Webb PA, Meier F, Heymann DL (1999) A search for Ebola virus in animals in the Democratic Republic of the Congo and Cameroon: ecologic, virologic, and serologic surveys, 1979–1980. J Infect Dis 179:S139–S147
Calisher CH, Childs JE, Field HE, Holmes KV, Schountz T (2006) Bats: important reservoir hosts of emerging viruses. Clin Microbiol Rev 19:531–545
Centers for Disease Control (1996) Ebola-Reston virus infection among quarantined nonhuman primates–Texas, 1996. Morb Mortal Wkly Rep 45:314–316
Constantine DG, Tierkel ES, Kleckner MD, Hawkins DM (1968) Rabies in New Mexico cavern bats. Pub Health Rep 83:303–316
Feldmann H, Wahl-Jensen V, Jones SM, Stroher U (2004) Ebola virus ecology: a continuing mystery. Trends Microbiol 12:433–437
Formenty P, Boesch C, Wyers M, Steiner C, Donati F, Dind F, Walker F, Le Guenno B (1999) Ebola virus outbreak among wild chimpanzees living in a rain forest of Cote d'Ivoire. J Infect Dis 179 [Suppl 1]:S120–S126
Geisbert TW, Jahrling PB, Hanes MA, Zack PM (1992) Association of Ebola-related Reston virus particles and antigen with tissue lesions of monkeys imported to the United States. J Comp Pathol 106:137–152
Gonzalez JP, McCormick JB, Saluzzo JF, Georges AJ (1983) Les fièvres hémorragiques africaines d'origine virale en République Centrafricaine. Cah ORSTOM Ser Ent Méd Parasit XXI:119–130
Gonzalez JP, Herbreteau V, Morvan J, Leroy EM (2005) Ebola virus circulation in Africa: a balance between clinical expression and epidemiological silence. Bull Soc Pathol Exot 98:210–217
Halpin K, Young PL, Field HE, Mackenzie JS (2000) Isolation of Hendra virus from pteropid bats: a natural reservoir of Hendra virus. J Gen Virol 81:1927–1932
Heymann DL, Weisfeld JS, Webb PA, Johnson KM, Cairns T, Berquist H (1980) Ebola hemorrhagic fever: Tandala 1977–1978. J Infect Dis 142:372–376
Hill JE, Smith JD (1984) Bats: a natural history. University of Texas Press, Austin
Huijbregts B, De Wachter P, Ndong Obiang S, Akou Ella M (2003) Ebola and the decline of gorilla *Gorilla gorilla* and chimpanzee *Pan troglodytes* populations in Minkebe forest, north-eastern Gabon. Oryx 37:437–443
Jahrling PB, Geisbert TW, Jaax NK, Hanes MA, Ksiazek TG, Peters CJ (1996) Experimental infection of cynomolgus macaques with Ebola-Reston filoviruses from the 1989–1990 US epizootic. Arch Virol Suppl 11:115–134
Johnson KM (1978) Ebola haemorrhagic fever in Zaire (1976) Bull World Health Organ 56:271–293
Kiley MP, Bowen ET, Eddy GA, Isaacson M, Johnson KM, McCormick JB, Murphy FA, Pattyn SR, Peters D, Prozesky OW, Regnery RL, Simpson DI, Slenczka W, Sureau P, van der Groen G, Webb PA, Wulff H (1982) Filoviridae: a taxonomic home for Marburg and Ebola viruses? Intervirology 18:24–32

Kuzmin IV, Orciari LA, Arai YT, Smith JS, Hanlon CA, Kameoka Y, Rupprecht CE (2003) Bat lyssaviruses (Aravan and Khujand) from Central Asia: phylogenetic relationships according to N, P, G gene sequences. Virus Res 97:65–79

Kuzmin IV, Hughes GJ, Botvinkin AD, Orciari LA, Rupprecht CE (2005) Phylogenetic relationships of Irkut and West Caucasian bat viruses within the *Lyssavirus* genus and suggested quantitative criteria based on the N gene sequence for lyssavirus genotype definition. Virus Res 111:28–43

Langevin P, Barclay RMR (1990) *Hypsignathus monstrosus*. Mammal Species 357:1–4

Leirs H, Mills JN, Krebs JW, Childs JE, Akaibe D, Woollen N, Ludwig G, Peters CJ, Ksiazek TG (1999) Search for the Ebola virus reservoir in Kikwit Democratic Republic of the Congo: reflections on a vertebrate collection. J Infect Dis 179:S155–S163

Leroy EM, Rouquet P, Formenty P, Souquiere S, Kilbourne A, Froment JM, Bermejo M, Smit S, Karesh W, Swanepoel R, Zaki SR, Rollin PE (2004a) Multiple Ebola virus transmission events and rapid decline of central African wildlife. Science 303:387–390

Leroy EM, Telfer P, Kumulungui B, Yaba P, Rouquet P, Roques P, Gonzalez JP, Ksiazek TG, Rollin PE, Nerrienet E (2004b) A serological survey of Ebola virus infection in central African nonhuman primates. J Infect Dis 190:1895–1899

Leroy EM, Kumulungui B, Pourrut X, Rouquet P, Hassanin A, Yaba P, Delicat A, Paweska JT, Gonzalez JP, Swanepoel R (2005) Fruit bats as reservoirs of Ebola virus. Nature 438:575–576

Li W, Shi Z, Yu M, Ren W, Smith C, Epstein JH, Wang H, Crameri G, Hu Z, Zhang H, Zhang J, McEachern J, Field H, Daszak P, Eaton BT, Zhang S, Wang LF (2005) Bats are natural reservoirs of SARS-like coronaviruses. Science 310(5748):676–679

Mackenzie JS (2005) Emerging zoonotic encephalitis viruses: lessons from Southeast Asia and Oceania. J Neurovirol 1:434–440

Martini GA, Siegert R (1971) Marburg virus disease. Springer, Berlin Heidelberg New York, pp 131

Miranda ME, Ksiazek TG, Retuya TJ, Khan AS, Sanchez A, Fulhorst CF, Rollin PE, Calaor AB, Manalo DL, Roces MC, Dayrit MM, Peters CJ (1999) Epidemiology of Ebola (subtype Reston) virus in the Philippines, 1996. J Infect Dis 179 [Suppl 1]: S115–S119

Monath TP (1999) Ecology of Marburg and Ebola viruses: speculations and directions for the future research. J Infect Dis 179:S127–S138

Morvan JM, Deubel V, Gounon P, Nakoune E, Barriere P, Murri S, Perpete O, Selekon B, Coudrier D, Gautier-Hion A, Colyn M, Volehkov V (1999) Identification of Ebola virus sequences present as RNA or DNA in organs of terrestrial small mammals of the Central African Republic. Microbes Infect 1:1193–1201

Morvan JM, Nakoune E, Deubel V, Colyn M (2000) Forest ecosystems and Ebola virus. B Soc Pathol Exot 93:172–175

Pinzon JE, Wilson JM, Tucker CJ, Arthur R, Jahrling PB, Formenty P (2004) Trigger events: enviroclimatic coupling of Ebola hemorrhagic fever outbreaks. Am J Trop Med Hyg 71:664–674

Pourrut X, Kumulungui B, Wittmann T, Moussavou G, Delicat A, Yaba P, Nkoghe D, Gonzalez JP, Leroy EM (2005) The natural History of Ebola virus in Africa. Microbes Infect 7:1005–1014

Reiter P, Turell M, Coleman R, Miller B, Maupin G, Liz J, Kuehne A, Barth J, Geisbert J, Dohm D, Glick J, Pecor J, Robbins R, Jahrling P, Peters C, Ksiazek T (1999) Field investigations of an outbreak of Ebola hemorrhagic fever Kikwit Democratic Republic of the Congo, 1995: arthropod studies. J Infect Dis 179:S148–S154

Rollin PE, Williams RJ, Bressler DS, Pearson S, Cottingham M, Pucak G, Sanchez A, Trappier SG, Peters RL, Greer PW, Zaki S, Demarcus T, Hendricks K, Kelley M, Simpson D, Geisbert TW, Jahrling PB, Peters CJ, Ksiazek TG (1999) Ebola (subtype Reston) virus among quarantined nonhuman primates recently imported from the Philippines to the United States. J Infect Dis 179 [Suppl 1]:S108–S114

Shaman J, Day JF, Stieglitz M (2002) Drought-induced amplification of Saint Louis encephalitis virus Florida. Emerg Infect Dis 8:575–580

Sulkin SE, Allen R (1974) Virus infections in bats. In: Melnick JL (ed) Monographs in Virolology 8. Karger, Basel

Swanepoel R, Leman PA, Burt FJ (1996) Experimental inoculation of plants and animals with Ebola virus. Emerg Infect Dis 2:321–325

Walsh PD, Abernethy KA, Bermejo M, Beyers R, De Wachter P, Akou ME, Huijbregts B, Mambounga DI, Toham AK, Kilbourn AM, Lahm SA, Latour S, Maisels F, Mbina C, Mihindou Y, Obiang SN, Effa EN, Starkey MP, Telfer P, Thibault M, Tutin CE, White LJ, Wilkie DS (2003) Catastrophic ape decline in western equatorial Africa. Nature 422:611–614

Walsh PD, Biek R, Real LA (2005) Wave-like spread of Ebola Zaire. PLoS Biol 3:e371

Yob JM, Field H, Rashdi AM, Morrissy C, van der Heide B, Rota P, bin Adzhar A, White J, Daniels P, Jamaluddin A, Ksiazek T (2001) Nipah virus infection in bats (order Chiroptera) in peninsular Malaysia. Emerg Infect Dis 7:439–441

CTMI (2007) 315:389–443
© Springer-Verlag Berlin Heidelberg 2007

Pre-spillover Prevention of Emerging Zoonotic Diseases: What Are the Targets and What Are the Tools?

J. E. Childs (✉)

Department of Epidemiology and Public Health and Center for Eco-Epidemiology, Yale University School of Medicine, 60 College Street, P.O. Box 208034 New Haven, CT 06520-8034, USA
jamesechilds@comcast.net

Abstract The uneven standards of surveillance, human- or animal-based, for zoonotic diseases or pathogens maintained and transmitted by wildlife H_Rs, or even domestic species, is a global problem, readily apparent even within the United States, where investment in public health, including surveillance systems, has a long and enviable history. As of 2006, there appears to be little scientific, social, or political consensus that animal-based surveillance for zoonoses merits investment in international infrastructure, other than the fledgling efforts with avian influenza, or targeted nontraditional avenues of surveillance and research. National institutions charged with strategic planning for emerging diseases or intentional releases of zoonotic agents have emphasized improving diagnostic capabilities for detecting human infections, modifying the immune status of human or domestic animals through vaccines, producing better antiviral or antibacterial drugs, and enhancing human-based surveillance as an early warning system. With the possible exception of extensive human vaccination, each of these approaches target post-spillover events and none of these avenues of research will have the slightest impact on reducing the risk of additional emergence of viruses or other pathogens from wildlife. Novel schemes of preventing spillover of human pathogens from animal H_Rs can only spring from improving our understanding of the ecological context and biological interactions of pathogen maintenance among H_Rs. Although the benefit derived from investments to improve surveillance and knowledge of zoonotic pathogens circulating among wildlife H_R populations is uncertain, our experience with HIV and the looming threat of pandemic avian influenza A inform us of the outcomes we can expect by relying on detection of post-spillover events among sentinel humans.

1 Introduction

Individual humans sickened or killed by an unknown infectious cause potentially indicate a zoonotic disease emergence has occurred, but, by themselves, are insufficient to document any instance of emergence. Incident cases of a new zoonotic disease must come to the attention of local authorities and then be the target of clinical, epidemiologic, and microbiologic research prior to any determination that an outbreak was caused by an emerging or reemerging pathogen. Satisfactory fulfillment of Koch's postulates is a daunting process, involving the diagnosis of human disease, i.e., the isolation of the infecting pathogen in cell culture; the molecular and antigenic characterization of pathogens obtained from human or animal tissues; and establishing the novel pathogen's causal role as etiologic agent (Osterhaus et al. 2004).

These endeavors link forever an instance of emergence with a single time point and place, a pinpoint and date on a map [Fig. 2.2 in Institute of Medicine (2003)]. Such an accounting system is necessary, but belies the dynamic ongoing process of disease emergence. As with the invading species that perishes on a foreign shore before being identified and labeled by a knowledgeable biologist, countless cases of zoonotic disease go unrecognized and uncatalogued. These missing data limit comparative analyses of the qualities of successful invading species to the far larger outgroup of pathogens for which there are limited or negative, i.e., not detected, data (Daszak et al. 2000; Cleaveland et al. 2001; Dobson and Foufopoulos 2001; Kolar and Lodge 2001; see the chapter by Cleaveland et al., this volume). Irrespective of the limitations of such studies, coherent trends and suites of plausible traits associated with successfully emerging pathogens have been derived from comparative studies (Dobson and Foufopoulos 2001; Cleaveland et al. 2001; see the chapters by Cleaveland et al. and Holmes and Drummond, this volume), but offer little guidance on how and where to focus attention (but see the chapters by Daszak et al. and Merianos, this volume).

Zoonotic viral emergences surprise even the scientists who are most knowledgeable within a subject area. Witness the identification of a novel *Hantavirus* causing fatal disease in the southwestern United State, after decades of search for pathogenic hantaviruses in the United States (LeDuc et al. 1993), and the discovery of a novel *Lyssavirus* causing a disease indistinguishable from rabies, in supposedly rabies-free Australia (Hooper et al. 1997). Although the process of zoonotic pathogen emergence often begins with identification of a case or cluster of human disease, surveillance and monitoring systems are ill equipped to detect and then characterize the unknown (see the chapters by Merianos and by Stallknecht, this volume).

Once a new zoonotic disease is identified and a case definition is established, the systematic collection of information on incident cases of human disease is used to generate information in a usable form, through appropriate data analytic and publication processes conducted through personnel working through a central repository. When the information is disseminated back to health professionals, from the federal government to individual practitioner level, a surveillance system is established. The country of occurrence, the morbidity and mortality, and the preexisting public health infrastructure, mixed with a good portion of serendipity, influence the likelihood of detecting a newly emerged zoonosis.

2 Disease Detection and Surveillance: Prerequisites to Zoonotic Disease Emergence

Surveillance for zoonotic pathogens is largely based on detecting illness or infection in *Homo sapiens* (see the chapters by Merianos and by Stallknecht, this volume); humans serve as the sentinel species for zoonotic agents maintained in transmission cycles in which, fortunately, they rarely play other than an incidental role as a dead-end host. A variety of surveillance systems and data sources have been successfully, if sometimes unintentionally, employed to monitor existing zoonotic diseases or to detect new diseases (Table 1).

An example of a serendipitous outcome stemming from syndrome-based surveillance for a specific disease occurred in New York City in 2001, with the implementation of a system to detect bioterrorism-related cases of anthrax (Centers for Disease Control and Prevention 2001; Buehler et al. 2003; Paddock et al. 2003). The putative anthrax case definition included a febrile illness accompanied by either a rash or eschar. Rickettsialpox, caused by *Rickettsia akari*, had been an endemic, legally mandated reportable disease in New York City since the mid-1940s (Huebner and Jellison 1947; Huebner et al. 1946), but since the 1980s the median number of annual cases reported was approximately 1 (Paddock et al. 2003). The classical presentation of rickettsialpox includes a fever and one or more eschars at the bite sites produced by the infected mite vector transmitting *R. akari*. Over an 18-month interval, 34 cases of rickettsialpox were diagnosed through the syndromic-based anthrax-surveillance system in New York City; tissue biopsies from patients yielded the first isolates of *R. akari* from the United States in more than 50 years (Koss et al. 2003). Although rickettsialpox was a known entity, anthrax surveillance highlighted the underappreciated level of disease caused by this endemic zoonosis.

Surveillance systems designed to detect and monitor a specific animal disease have also uncovered novel zoonotic pathogens. In the United States, two previously unknown rhabdoviruses have been isolated from dead birds collected for monitoring and forecasting WNV activity (Eidson et al. 2001b, 2001c; Mostashari et al. 2003; Garvin et al. 2004; Travassos da Rosa et al. 2002). While in Australia, laboratory workup of a sick pteropid bat collected in conjunction with Hendra virus (HeV) investigations following an outbreak of disease affecting horses and humans in 1994–1995 (Field et al. 2000, 2004; Halpin et al. 2000) yielded a new *Lyssavirus*, Australian bat lyssavirus (ABL), closely related to rabies virus (Fraser et al. 1996; Gould et al. 1998a). Within months of the isolation of ABL, this virus was demonstrated to be the cause of fatal encephalitis in humans (Gould et al. 1998b); until this time no rabies had been reported from Australia.

Effective, but informal, surveillance systems can be implemented rapidly following the identification of a novel zoonotic disease emergence within

Table 1 Examples of surveillance methods and data sources used to detect and monitor the emergence of zoonotic pathogens causing disease among human sentinels

Surveillance system or data source	Condition monitored
Individual physician	Hantavirus pulmonary syndrome (HPS) in Four Corners region of the United States[a]
Self-reporting of illness	Hot-line telephone reporting of suspect HPS coupled with trace-back for clinical records and samples for diagnostic testing[b]
CDC, Nationally Notifiable Diseases Surveillance System (NNDSS)[c]	West Nile fever and encephalitis[d], human and animal rabies[e], Rocky Mountain spotted fever[f] and others
CDC—Syndrome-based surveillance for anthrax fever, rash or eschar	Rickettsialpox described from New York City during surveillance for anthrax-first isolates of this rickettsia in 50 years[g, h]
EMERGEncy ID NET[i]	Appropriateness of rabies postexposure treatment in sentinel cities given recommendations of ACIP[j]
Automated rumor-tracking web-crawler[k]	Initial cases of severe acute respiratory disease (SARS) in China[l]
Community-based active surveillance, clinical practices and veterinary services	Human and animal rabies in Machakos District, Kenya[m, n]

[a] Duchin et al. 1994
[b] Tappero et al. 1996[3]
[c] Teutsch2000
[d] Centers for Disease Control and Prevention 2002
[e] Krebs et al. 2004
[f] Childs and Paddock 2002
[g] Paddock et al. 2003
[h] Koss et al. 2003
[i] Talan et al. 1998
[j] Moran et al. 2000
[k] A Report of the National Advisory Committee on SARS and Public Health 2003
[l] Heymann and Rodier 2004
[m] Kitala et al. 2000a
[n] Kitala et al. 2000b

countries with a highly developed public health infrastructure. The interplay of factors influencing initial detection and later development of systematic surveillance are illustrated by the outbreak of hantavirus pulmonary syndrome (HPS) in the southwestern United States in May 1993. An Indian Health Service physician noted a temporally and spatially linked cluster of cases of a severe, often fatal, respiratory disease, affecting previously healthy, young-adult Navajo Indians residing on a reservation (Duchin et al. 1994). The physician notified local authorities and subsequently CDC was invited by state officials to help investigate the growing number of fatalities. Testing of patient sera at CDC revealed the presence of antibodies reactive with hantaviral antigens (Ksiazek et al. 1995). Facilitated by epidemiologic knowledge of hantaviruses and hantaviral diseases occurring in Eurasia, rapid progress was made in uncovering the natural history of this mysterious new disease. In a matter of weeks, investigators confirmed the disease was clinically distinct from Eurasian disease (Moolenaar et al. 1995), that the etiologic agent was a new *Hantavirus*, Sin Nombre virus (Nichol et al. 1993), and the reservoir host (H_R; for definition of terminology see the chapter by Childs et al., this volume) was a species of New World rodent, *Peromyscus maniculatus* (Childs et al. 1994).

A relatively crude but effective national surveillance program, capitalizing on media interest in the HPS outbreak, was established by June 1993. Six months later, private citizens or their physicians had reported and submitted clinical specimens for diagnostic testing from 280 persons; 21 confirmed HPS cases were identified from 11 states outside of the four-state region where the initial outbreak was localized (Tappero et al. 1996). This impromptu surveillance system was highly successful in rapidly identifying the widespread geographic distribution and sporadic incidence of HPS cases throughout much of the western United States.

Once a zoonotic disease is characterized, formal, systematic surveillance efforts can be initiated at the state or national level in countries possessing the requisite infrastructure. National surveillance programs coordinated through CDC, with rare exceptions, focus on the systematic collection of data on human disease. National surveillance and the global network for monitoring Influenza A activity among humans is the outstanding example of a system integrating epidemiologic data with the collection and characterization of influenza viral subtypes circulating throughout the world (Centers for Disease Control and Prevention 2004d; Cox et al. 1994). The unquestioned value of the global influenza surveillance program rests with the vaccines produced. Each year's new influenza vaccines are based on determinations of the currently circulating influenza subtypes and divining which subtypes should be incorporated into next season's vaccine cocktail.

A global early warning system to detect zoonotic pathogens transmitted to humans was launched in July 2006 by the UN Food and Agriculture Organization (FAO) and the World Health Organization (WHO) in collaboration with the World Organization for Animal Health (formerly the Office of International Epizooties or OIE) (http://www.who.int/mediacentre/news/new/2006/nw02/en). Specifically mentioned as examples are BSE and SARS; data from infected and diseased humans and animals will be gathered and assessed jointly. Plans to develop a global animal-based influenza surveillance program exist (Centers for Disease Control and Prevention 2004d; Stohr 2003). It remains unclear if animal-based influenza surveillance will extend beyond domestic poultry and livestock to wild waterfowl and shorebird H_Rs, although this latter activity is strongly endorsed (Shortridge et al. 2003; Melville and Shortridge 2004; see the chapter by Webby et al., this volume).

3
Surveillance as Defined by Human and Veterinary Medicine

Surveillance for zoonotic diseases among wildlife, as opposed to domestic animals and livestock, falls through the cracks of both veterinary and human health practices (see the chapter by Stallknecht, this volume). Reviews of animal health monitoring systems mention wildlife disease surveillance only in passing and largely in reference to the difficulties of establishing population estimates (denominator data) for defining rates, such as disease incidence, or the obstacles to developing systematic surveillance programs coordinating with human disease surveillance (Ingram et al. 1975; see the chapters by Daszak et al., Merianos, and by Stallknecht, this volume).

Most regional or state systems collecting information on wildlife diseases are passive surveillance systems. Passive surveillance in the United States, as defined by public health professionals, is the systematic collection of data on human diseases, reportable through legal mandate in most states, obtained within specified time frames on conditions listed by National Notifiable Disease Surveillance System (NNDSS) (Teutsch 2000); data are reported to CDC by electronic submissions via the National Electronic Telecommunications System for Surveillance (NETSS) (Teutsch 2000). International regulations require reporting on quarantinable conditions, such as plague, yellow fever, cholera, and SARS (Teutsch 2000; Centers for Disease Control and Prevention 2002b). Diseases covered by the NNDSS are established through collaborations of the Council of State and Territorial Epidemiologists (CSTE) with the CDC and the nationally reportable diseases are reviewed at 3-year intervals, at which time case definitions are established or modified (Centers for Disease Control and Prevention 1997). By virtue of the population estimates provided by the US Census, human surveillance data collected via NNDSS are population-based. Summary statistics on nationally notifiable disease are published weekly in *Morbidity and Mortality Weekly Report (MMWR)* and summarized in annual reports (Centers for Disease Control and Prevention 2004c).

In contrast, for wildlife and domestic animal diseases, the OIE, situated in Paris, France, determines diseases reportable by its member counties

(Thiermann 2003). The diseases are divided into two lists: List A diseases are of major importance in international trade of animals or animal products and have the potential for very serious and rapid spread irrespective of national borders; List B diseases are of public health importance within counties (Thiermann 2003; http://www.oie.int). Within the United States, mandated reporting of animal diseases varies by state, and voluntary reporting by professionals is a major component of data collection (Salman 2003). At the federal level, information is collected by the Animal and Plant Health Inspection Service (APHIS) of the Department of Agriculture (USDA). Given the lack of accurate population estimates for many domestic animals and livestock, passive veterinary surveillance is not population-based.

3.1
Wildlife-Based Surveillance for Zoonotic Disease: Current Practices

Surveillance for wildlife diseases exists at some level in most developed countries. As with human, surveillance, the infrastructure for receiving, typing, and storing animal specimens and the diagnostic laboratory capacity for establishing diagnoses are minimal prerequisites (see the chapter by Stallknecht, this volume).

Within North America, the Canadian Cooperative Wildlife Health Center (CCWHC), supported by the four Canadian veterinary schools, was established in 1992 to promote nationwide surveillance of wildlife diseases. In Canada, disease detection is carried out by a wide range of professional and voluntary field personnel, including hunters, and specimen diagnosis is conducted at provincial and federal veterinary laboratories. The central repository for data is the CCWHC, which disseminates surveillance information to persons responsible for wildlife programs and policies, and to the public (Leighton et al. 1997).

In the United States, states have often taken the lead in monitoring wildlife diseases, such as WNV among dead birds, arboviral infections among sentinel bird flocks (Mostashari et al. 2003; Eidson et al. 2001a; Komar 2001), and transmissible spongiform encepahlopathy (TSE) associated with elk and white-tailed deer (Williams and Miller 2003). In several states, notably California and Florida, surveillance for arbovirus activity using sentinel flocks of birds have documented trends in the enzootic activity of western equine encephalomyelitis (WEE), St. Louis encephalitis (SLE), and eastern equine encephalomyelitis (EEE) linked to climatic and local weather patterns (Reeves1990; Shaman et al. 2002; Day 2001; Barker et al. 2003).

Surveillance for viruses transmitted from wildlife H_Rs to domestic poultry and livestock, such as avian influenza A, subtypes of which infect and cause disease in humans (Kermode-Scott 2004; Fouchier et al. 2004), is conducted through the USDA. Additionally, the USDA conducts mandated surveillance for zoonotic infections of livestock, such as BSE, anthrax, and bovine tuberculosis (TB) (Anonymous 2004b; Myers et al. 2003).

Regional activities monitoring wildlife diseases, especially among game animals, such as white-tailed deer (*Odocoileus virginianus*), exist through cooperative efforts involving research and educational institutions, state fish and game departments, and hunters. A successful example is the Southeastern Cooperative Wildlife Disease Study (SCWDS) maintained at the University of Georgia, where programs collect regional data on wildlife, ectoparasitic and endoparastic infestations, and microbiologic and serologic evidence of past or current infections. Historical collections and independently funded research programs through SCWDS recently led to the rapid elucidation of the natural history of emerging tick-borne zoonoses caused by bacteria in the genera *Ehrlichia* and *Anaplasma* (Davidson et al. 2001; Little et al. 1998; Lockhart et al. 1996, 1997; see the chapter by Paddock and Yabsley, this volume).

Wildlife disease monitoring in Sweden and Northern Europe has existed since the 1940s, relying heavily on the cooperation and interest of hunters in the collection and submission of samples from game animals (Mörner 2002; Mörner et al. 2002). Surveillance for wildlife diseases in the UK and Ireland has included bovine TB maintained by badgers (see the chapter by Palmer, this volume); current plans call for increased surveillance of wildlife, notably birds for WNV, in England and Wales (Griffin et al. 2005; Gormley and Costello 2003; Crook et al. 2002; Duff et al. 2003; see the chapter by Palmer, this volume).

4 Zoonotic Disease Emergences and Targeted Surveys for Infected Wildlife H_Rs

4.1 Short-Term Surveys Following Zoonotic Disease Emergence

Short-term studies of wildlife H_Rs are the most common survey methods employed in response to specific instances of emergence or spread of zoonotic disease. Following an outbreak of human monkeypox in several US states (Centers for Disease Control and Prevention 2003a; see the chapter by Regnery, this volume), local populations of indigenous North American rodents were captured and examined for infection from areas around animal-holding facilities housing African rodents imported for the pet-trade and implicated as the source of monkeypox virus (Cunha 2004; Check 2004). Native American ground squirrels, coincidentally housed in the same buildings with the African rodents and purchased as pets, were implicated as the source of monkeypox virus transmitted to humans (Guarner et al. 2004; see the chapter by Regnery, this volume). Short-lived studies identifying rabid raccoons were undertaken in Ohio, following the first reported case of raccoon-variant rabies in that state (Stefanak et al. 1999). Testing of trapped and road-killed raccoons helped define the geographic extent of the enzootic area of raccoon rabies in the state in preparation for the deployment of an oral rabies vaccine (ORV) in an effort to prevent the westward expansion of epizootic raccoon rabies into Ohio and west to other states (Kemere et al. 2002; Foroutan et al. 2002; APHIS Wildlife Services Factsheet 2002).

4.2 Long-Term Studies Following Zoonotic Disease Emergence

Long-term prospective studies of zoonotic pathogens circulating within wildlife H_Rs are critical to understanding factors mediating irregular increases and declines within animal populations, which can drive the risk of spillover to humans. The varying population dynamics of zoonotic pathogens and their H_Rs are, in some instances, as with rabies virus, driven by pathogen-induced host mortality (Anderson et al. 1981; Childs et al. 2000; Coyne et al. 1989); the risk of rabies virus spillover to domestic animals is closely, but not perfectly, mirrored by the temporal dynamics within the wildlife H_R (Gordon et al. 2004).

Examples of systematic wildlife disease studies that have exceeded several years in duration are few. One ongoing example is the investigations of the population dynamics of rodent H_Rs and SNV and other hantaviruses in the southwestern United States, which were established in the mid-1990s following the 1993 outbreak of HPS. Replicated and coordinated studies among universities in several states, using similar methodologies for population sampling, virological testing, and data management (Mills et al. 1995), have provided a wealth of information critical for unraveling aspects of the transmission and maintenance of hantaviruses (Mills et al. 1999a, 1999b). The knowledge base established by these efforts allowed increasingly elaborate hypotheses developed from field observations to be tested.

The modalities of hantaviral transmission were assessed by application of microsatellite markers to genetically identify familial relationships among individual mice; related male *P. maniculatus* were more likely to be SNV-infected (Root et al. 2004), providing clues to the chain of transmission events contributing to the male bias in hantaviral infection documented by several descriptive studies (Mills et al. 1999a). Ongoing research is providing clues as to the critical H_R population size required to sustain hantavirus transmission and is exploring the phenomenon of SNV disappearance and reemergence in H_R populations (Calisher et al. 2002), possibly through SNV maintenance within refugia of a special nature (Yates et al. 2002). These ongoing studies spanning more than

6 years, have been sufficient to capture occurrences and effects of environmental drivers, such as El Niño Southern Oscillation (ENSO), which occurs at semi-predictable intervals of approximately 5–10 years (Chen et al. 2004a). ENSO is a principal indicator of global climate which modifies local weather patterns; increasing rainfall associated with ENSO is hypothesized to drive a trophic cascade of events (Polis et al. 2000), ultimately leading to increases in local H_R populations and increased risk of HPS (Glass et al. 2002; Hjelle and Glass 2000). Remote sensing and GIS techniques, coupled to a household-based case–control methodology assessing rodent abundance around residences of HPS cases (Childs et al. 1995), predicted where *P. maniculatus* would be more abundant at future case houses. Analyses of annual satellite images to detect local environmental conditions supportive of rodent HR population growth has proven an effective tool for predicting the qualitative level of risk (low, moderate, high) for HPS over a sizable region of the southwestern US (Glass et al. 2006). Educational recommendations and field trials of rodent-proofing methods were incorporated into the long-term investigations (Glass et al. 1997), to provide readily available control measures in anticipation of increased risk of HPS (Childs et al. 1993).

5
Animal-Based Zoonotic Disease Surveillance: A Horse of Another Color

Animal-based surveillance is a process inherently different from human-based surveillance (Table 2). With the exception of surveillance efforts targeting livestock and poultry, run through the Center for Animal Health Surveillance of the USDA (King 1985), no formal sampling methodology exists for estimating animal population sizes at the regional or continental level (see the chapter by Stallknecht, this volume). Wildlife population estimates at the continental scale are few and generally restricted to tractable populations associated with conservation efforts, with the possible exception of national waterfowl surveys (Butler et al. 1995), or national hunter- or road-killed indices of white-tailed deer populations (Hayne1984).

Targeted ecologic studies directed at species that are endangered or threatened have in several instances provided population-based information complementing the objectives of wildlife disease research. The most notable examples involve species that are relatively easy to observe or for which population-based indices exist, such as carcass, nest, or scat counts (Leroy et al. 2004). Where estimates of animal numbers have been enumerated, the impact of fatal zoonotic viruses indicate certain wildlife species could serve as sentinels for monitoring viral activity; species conservation activities can

Table 2 Key differences in the terms "passive surveillance" and "active surveillance" and methods of data collection as used and defined by veterinary and human health professionals

Surveillance system or manner of data collection	Veterinary health[a]	Human public health[b]
Passive	"The passive collection of data involves the reporting of clinical or subclinical suspect cases to the health authorities by health care professionals at their discretion."	"A passive surveillance system is one in which a health jurisdiction receives disease reports from physicians, laboratories, or other individuals or institutions as mandated by state law."
Key characteristics	Voluntary	Legally mandated, systematically collected within specified time frames, voluntarily reported to CDC
	Not population-based	Specified by state and federal officials within the National Notifiable Disease Surveillance System (NNDSS). Population-based by virtue of the US Census
Active	"An active collection of data for any monitoring and surveillance system (MOSS) is the systematic collection or regular recording of cases of a designated disease or group of diseases for a specific goal of monitoring or surveillance."	"In contrast, an active surveillance system is established when a health department regularly contacts reporting sources (e.g., once per week) to elicit reports, including negative reports (no cases)."
Key characteristics	Not necessarily mandated by law Population-based	Not necessarily mandated by law Population-based Collects negative data

[a] Quoted from Salman (2003)
[b] Quoted from Birkhead and Maylahn (2000)

provide leverage to any additional surveillance investment (see the chapter by Daszak et al., this volume). Examples include great apes killed by Ebola virus (Leroy et al. 2004; Walsh et al. 2003; see the chapters by Gonzales et al., this volume), and rabies induced mortality among African wild dogs (Kat et al. 1995; Gascoyne et al. 1993b; Burrows 1992), and Ethiopian (Whitby et al. 1997; Sillero-Zubiri et al. 1996) and Artic wolves (Ballard and Krausman 1997; Weiler et al. 1995; Chapman 1978). For other wildlife, the lack of population estimates precludes estimation of basic epidemiologic parameters, including rates such as incidence or mortality; these capabilities are beyond those of any existing surveillance system for a wildlife zoonosis.

Novel animal-based surveillance and control programs are being planned for zoonotic agents, such as BSE, SARS-CoV and influenza A subtypes which have realized or potential pandemic importance to humans or domestic animals (http://www.who.int/mediacentre/news/new/2006/nw02/en). The ultimate H_Rs for these agents includes domestic and wild animal species. For example, the H_Rs for influenza A subtype H5NI are among wild waterfowl and shorebirds, and perhaps other avian types, although, domestic chickens and other poultry serve as both the first secondary host (H_{S1}) or intermediate host (H_I)(see the chapter by Childs et al., this volume, for description of terms) and can develop as a novel H_R (see the chapter by Webby et al., this volume). Experts within the WHO and elsewhere, acknowledge a need "... to get rid of the natural reservoir of H5N1, but we need to do it safely" (quote attributed to Klaus Stohr, project leader of WHO's global influenza program; cited in Abbott and Pearson [2004]). However, even rough plans of how such an immense undertaking will be designed and integrated into the countries of greatest significance in Asia are lacking.

5.1
Obstacles to Animal-Based Surveillance

Even when infection within an animal H_R or H_S is relatively detectable, national surveillance programs for monitoring morbidity and mortality among wildlife and establishing the etiologic cause of infection through a system of diagnostic laboratories are rare (see the chapter by Stallknecht, this volume). If the zoonotic agent is a pathogen of domestic livestock, formal surveillance can target abattoirs or production facilities where food animals are processed, as is the major emphasis of BSE surveillance conducted both in the United States by the USDA (Kellar and Lees 2003; Anonymous 2004b) and within European countries (La et al. 2004). Among wildlife, animal rabies is the only disease within the NNDSS for which time-series data of reasonable duration, more than 50 years, quantity and quality has been systematically collected from all US states and territories (Childs et al. 2002).

Animal-based surveillance for pathogens causing emerging zoonotic diseases in humans is often hampered by the lack of clinical signs in infected individuals of the H_R (Table 2). Where zoonotic viruses cause fatal disease among wildlife and domestic animal H_Rs, H_Ss, or H_Is, tracking the spread of these agents is a simpler matter, although this remains a formidable challenge within countries lacking basic surveillance infrastructure. Tracking the spread of influenza A subtype H5N1 of domestic chickens, ducks, and some wild waterfowl in southeastern Asia (Chen et al. 2004b; Li et al. 2004; Lu et al. 2003; see the chapter by Webby et al., this volume), WNV in North America (Garvin et al. 2004; Guptill et al. 2003; Walsh et al. 2003; Larkin 2000), Ebola virus in central Africa (Leroy et al. 2004, 2005; Walsh et al. 2003; see the chapter by Gonzales et al., this volume), and rabies virus in North America, Europe, and southern Africa (Sabeta et al. 2003; Childs et al. 2000; Gordon et al. 2004; see the chapter by Nel and Rupprecht, this volume) has been facilitated by the mortality these viruses cause in wildlife and domestic species.

6
Benefits of Animal-Based Surveillance: Lessons from a Model System for Rabies

National surveillance for animal rabies is a model public health activity. As the CDC is charged with promotion of human health and disease prevention and control, animal-based rabies surveillance data are well integrated into national, state, and local human and veterinary public health programs (Childs et al. 2002). A brief examination of the objectives, types of data collected, and the practical use of the information disseminated through the national animal rabies surveillance program is illustrative of the potential benefits accrued from an animal-based surveillance system.

Surveillance for animal rabies collects information on the current status and level of rabies activity among wildlife and domestic animals at the county level within individual states. Monthly counts of rabid animals, and from some states the tally of negative results, designated to the level of animal species or taxonomic group, are submitted to the CDC (Krebs et al. 2004).

Surveillance information is analyzed, summarized, and disseminated back to the data providers in a timely manner through publications (Krebs et al. 2004) and additional communications, which are updated annually, such as The Compendium of Animal Rabies Prevention and Control (Centers for Disease Control 2005). Surveillance data on animal rabies are sufficiently detailed and accurate to allow human and veterinary health professionals to anticipate levels of rabies activity at the county or regional level, permitting some future planning for preventative activities, including procurement of human vaccine and human rabies immunoglobulin (HRIG) for postexposure treatment of potentially exposed persons (Centers for Disease Control and Prevention 2004b; Advisory Committee on Immunization Practices 1999); increasing vaccination levels of dogs and cats; and initiation of targeted control efforts to vaccinate wildlife using ORV (Krebs et al. 2004; Kemere et al. 2002).

Several species of terrestrial carnivore, raccoons (*Procyon lotor*), red foxes (*Vulpes vulpes*), and striped skunks (*Mephitis mephitis*) serve as H_Rs for particular genetic variants of rabies circulating in the continental US; numerous rabies virus variants are also associated with different species of bats (Messenger et al. 2003). Rabies virus variants can be differentiated by limited sequence analysis or monoclonal antibody methods (Smith et al. 1995) and the enzootic area where rabies variants overlap the geographic range of their terrestrial mammalian hosts can be reasonably determined (Childs et al. 2002). Time-series surveillance data on wildlife rabies, analyzed by statistical algorithms defining and demarcating intervals of increased (epizootic) or diminished (interepizootic or enzootic) rabies activity, provide results concordant with predictions and outcomes based on numerical solutions to mathematical models of the population dynamics of rabies virus within a single H_R species (Childs et al. 2000; Anderson et al. 1981; Coyne et al. 1989). Time-series analyses have defined the temporal dynamics of disease in a wildlife H_R (Childs et al. 2000; Guerra et al. 2003) and demonstrated the close association of this relatively predictable process to the risk of rabies spillover to domestic animals (Gordon et al. 2004). Furthermore, these data can inform epidemiologic simulations and models predicting epizootic rabies spread (Russell et al. 2004, 2005), and have been modified to forecast the savings accrued by preventing rabies spread through the application of ORV (Gordon et al. 2005).

Additionally, local data have provided the raw material to explore formal methodologies for demonstrating and assessing the impact of long-distance translocations (LDTs) of infected animals on the rate and pattern of rabies spread in heterogeneous environments (Smith et al. 2005). The availability of remotely sensed or digitized maps, coupled with GIS-assisted partitioning of landscapes into habitats of varying quality, allow explorations of the impact of landscape heterogeneity on the characteristics of epizootics and the pattern of epizootic wavefront spread (Jones et al. 2003; Smith et al. 2005). Such analyses have been used to assess where remedial prevention activities should be focused when breaches in ORV barriers occur and where active surveillance might be considered as a complement to passive data collection where fine-scale knowledge of the presence of rabies is needed to guide interventions (Russell et al. 2005).

7 Generic and Specific Limitations to Animal-Based Surveillance: Lessons from Rabies

However, rabies surveillance reveals several inherent difficulties to conducting any form of wildlife-based disease surveillance and offers a sobering view of the hurdles to be overcome when considering such programs in other locations for other diseases. Animal rabies surveillance was implemented to provide humans with a measure of rabies risk in their communities and, other than relative species counts over years, there is no information on the incidence or impact of rabies in any animal community. The nature of the human–animal interactions required by an animal-based surveillance system provides a distorted image of rabies as a community process (Fig. 1).

Biases inherent to data collected by animal rabies surveillance at the national level stem from the requirement of human participation in each step of the process culminating in a rabies diagnosis in an animal (Fig. 1a; Gordon et al. 2005). The impact of human demography, measured as absolute population size per county, on the surveillance process is sufficient to account for fully 70% of the variation in total animal specimens tested for rabies (Fig. 1b; Childs et al. 2007). Total county expenditure is almost as strong a predictor, accounting for 65% of the variation in total animal tests performed.

Pathobiologic features of rabies, human behavior, and the expense associated with diagnostic testing of specimens skew the types of animals observed, harvested, and tested for rabies. Medium-to-large-sized mammals are more likely to be observed by humans and reported to wildlife control officials. In a typical surveillance year, small terrestrial mammals, predominantly rodents, but some insectivores, weighing less than 1 kg account for less than 0.5% of the total animals tested and diagnosed as rabid (Real and Childs 2006), although small mammals provide the greatest species diversity and the overwhelming abundance of individuals and biomass of many mammalian communities (Bourliere 1975). Rodents are fully susceptible to rabies infection and are capable of transmitting the virus to other species (Childs et al. 1997; Winkler et al. 1972); in some countries, rodents have been implicated in natural maintenance cycles of the virus (Summa et al. 1987; Verlinde et al. 1975).

A major sampling bias occurs at the level of the rabies diagnostic laboratory where, in an effort to save money on personnel time and diagnostic reagents, rabies testing is typically restricted to specimens from animals directly involved

Fig. 1a, b (Continued) data integrating test outcome with information on the type of animal and date and place of origin produced at the state level and submitted to CDC. **b** Although data on each of the events partitioned in (**a**) are unavailable, a surrogate value of population size is used to measure the importance of human interaction in generating surveillance data, assuming that increasing numbers of humans increase the likelihood of many of the events in (**a**) occurring. There is a strong association between the absolute numbers of humans resident in the smallest surveillance unit (US Census figures), a county within a state, and the total numbers of animals tested for rabies from that surveillance unit. The relationship is a power function in which human population size accounts for 70% of the variance in median total tests conducted for rabies conducted over a decade from 713 counties in a region affected by the raccoon variant of rabies virus. (Adapted from Childs et al. 2006)

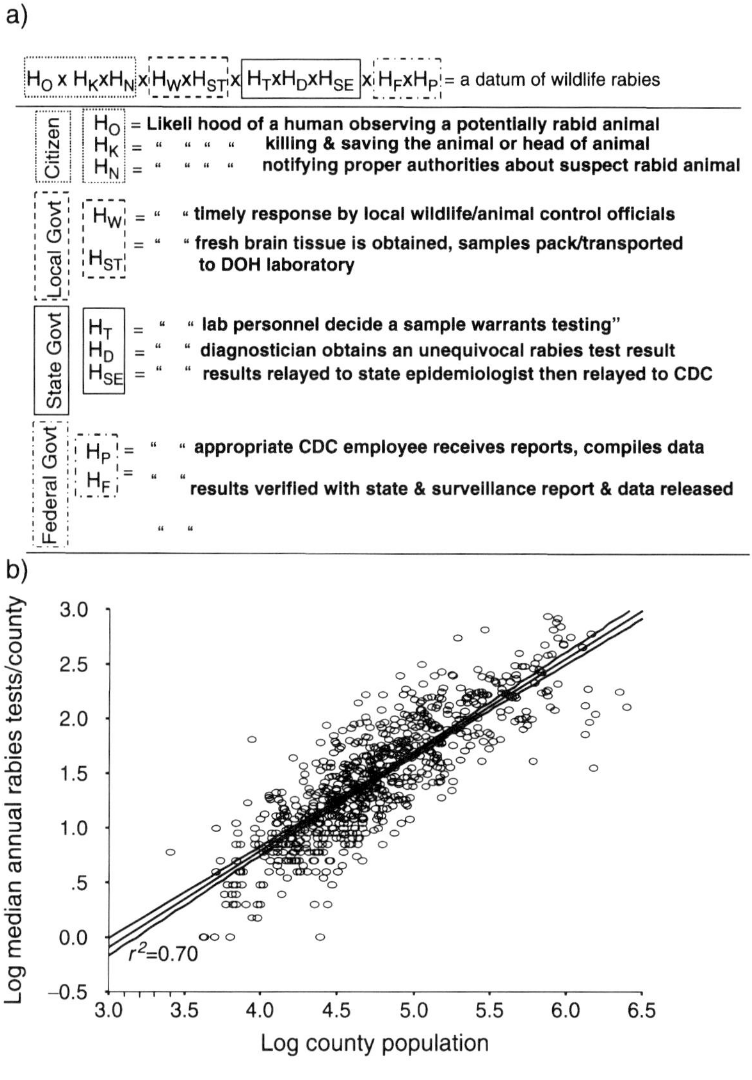

Fig. 1a, b The process of wildlife- and animal-based surveillance is interactive, involving multiple, and frequently independent, interactions between humans and wildlife to generate a single datum captured. Panel **a** depicts some examples of these interactions, which could be assigned a probability if information were available, between private citizens and local and federal agencies in the route to generating a datum on animal rabies. Each process involves some interaction with an animal, a tissue sample taken from the animal, test material derived from the sample, an outcome derived from the sample at the diagnostic laboratory, and the

in the potential exposure of humans or domestic animals to rabies virus (Fig. 1a); other specimens go untested (Torrence et al. 1992; Wilson et al. 1997). Many of these limitations and biases will be generic problems confronting any effort to monitor wildlife species anywhere in the world.

8 From Detection to Intervention: Human-Based Approaches to Zoonotic Disease Control

The most widespread approaches to zoonotic disease control completely ignore the ecology of wildlife and pathogen maintenance and transmission and, therefore, the potential for interrupting pathogen transmission prior to human spillover. Instead, prevention and control strategies focus on defensive measures for the human H_S.

National institutions charged with strategic planning for emerging diseases or intentional releases of zoonotic agents have emphasized improving diagnostic capabilities for detecting human infections, modifying the immune status of human or domestic animals through vaccines, producing better antiviral or antibacterial drugs, and enhancing human-based surveillance as an early warning system (Fauchi 2002; Centers for Disease Control and Prevention 1998). With the possible exception of extensive human vaccination, each of these approaches target post-spillover events and none of these avenues of research will have the slightest impact on reducing the risk of additional emergence of viruses or other pathogens from wildlife.

9 Limitations to Human-Based Intervention Programs for Prevention of Zoonotic Diseases

The current fixation on human vaccines, human diagnostics, human drugs, and human-based surveillance is the legacy of past successes. Landmark achievements for zoonotic disease prevention include vaccines for yellow fever and rabies, and other vaccines of human or veterinary importance exist, or are being developed, for tick-borne encephalitis, Rift-Valley fever, arboviral encephalitides, SARS, Ebola hemorrhagic fever, HPS, and many others (Chang et al. 2004; Cox et al. 2004; Lau 2004; Custer et al. 2003; Matsuoka et al. 2003; Nalca et al. 2003; Warfield et al. 2003; Hjelle 2002; Tomori 2002; Tesh et al. 2002; Stephenson 2001; Monath et al. 2001; Huang et al. 2004). New antiviral drugs can be designed, created, and screened with far better efficiency than at any time in the past and novel candidates and methodologies for improving the delivery of drugs to infected cells are in development (Oxford et al. 2005; Duzgunes et al. 2005; Wu et al. 2005; Pastor-Anglada et al. 2005).

Additionally, traditional measures of case isolation, contact-tracing, and quarantine of exposed persons, banning of public gatherings, or curtailing individual access to international travel have proved highly effective in controlling the spread of zoonotic diseases with pandemic potential, as with SARS (Zhong 2004; Anderson et al. 2004; Speakman et al. 2003; see the chapter by Wang and Eaton, this volume) (Fig. 2). But SARS-CoV is not influenza A. Methods relying on increasing social distance are unlikely to prevent the spread of human-adapted pandemic influenza A (Fraser et al. 2004; Mills et al. 2004). Aerosol transmissibility of influenza virus in the subclinical patient precedes clinical signs by 24 h (Mills et al. 2004; Fraser et al. 2004), unlike the coincidence of clinical disease with the onset of infectiousness with SARS-CoV (Anderson et al. 2004). Influenza A vaccine production capacity and antiviral medication stockpiles to combat influenza spread are insufficient even in wealthy developed countries (Mills et al. 2004). Can we continue to prepare and respond to such pathogens by strictly defensive measures aimed at the human H_S?

So given the proven record of achievement of a medical or technological approach to defending humans from invasion by infectious organisms, is there much to be gained by examining processes, antecedent to human spillover, for potential vulnerabilities and as intervention targets, as a complement to ongoing efforts to improve human-based disease prevention activities? The answer is yes, but a qualified yes. Simply saying we need such systems glosses over the myriad of obstacles in developing programs. Designing and implementing wildlife-based surveillance and targeted interventions will not be achieved in the short term and establishing the infrastructure to support these efforts would be difficult and expensive (see the chapter by Merianos, this volume).

10 From Detection to Intervention: Targets for Wildlife or Domestic Animal Control

The maintenance and transmission cycles of zoonotic viruses within wildlife H_Rs offer many of the same targets for control as do human-based interventions, with the notable exception that population culling can be exploited for control of animal reservoirs, intermediate host populations and arthropod vector species. The ultimate prevention strategy for zoonotic agents affecting humans is to abrogate or greatly reduce cross-species transmission by disrupting transmission and maintenance cycles of zoonotic viruses within the H_R.

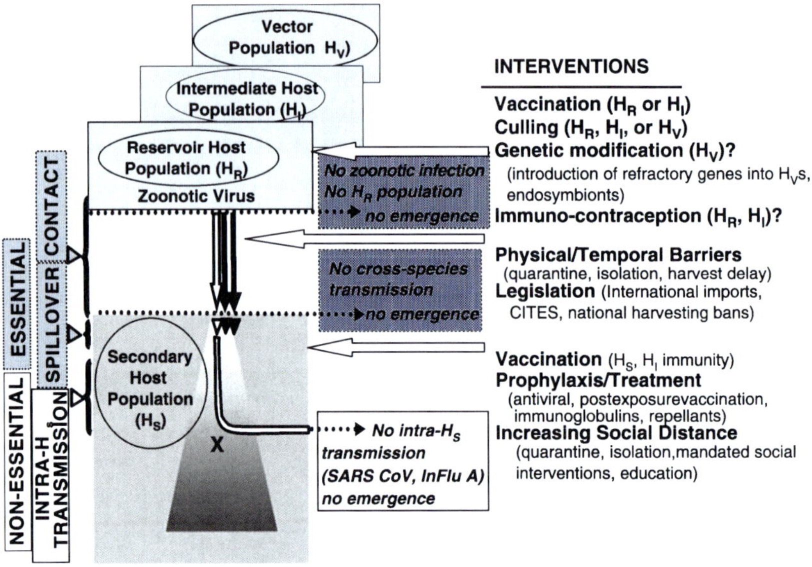

Fig. 2 The targets and types of intervention tools available for preventing spillover of zoonotic pathogens to humans, or for mitigating the impact or spread of zoonotic disease should spillover occur. The *open arrows* leading from the list of intervention types indicate where the intervention acts, either directly at the population level of the reservoir host population (H_R), vector population (H_V), intermediate vertebrate host population (H_I), or at the secondary host population (H_S) assumed to be humans. Other targets are the processes or rates associated with spillover, such as reducing contact between infected H_R and susceptible H_S or between infected H_R and susceptible H_R, or infected H_S and susceptible H_S. The methods employed to reduce host populations are largely restricted to culling (vertebrates) or insecticides (arthropods). Vaccination of host populations is in use for some zoonotic viruses (rabies, influenza A, VEE, etc.) and new vaccines are in development (see Table 3). Animal quarantine, isolation of animals exposed to a pathogen, and legal bans to trade in animals or animal products originating from countries with enzootic disease act to increase social distance and decrease the likelihood of contacts between infected and susceptible hosts. Immunocontraception of H_Rs to reduce population size or genetic modification of H_Vs to render vector populations refractory to infection may play a role in prevention in the future. Human vaccination, treatment, and the prophylactic use of drugs are defensive measures that may prevent or reduce spillover and post-spillover spread, but will not reduce the likelihood of contact between infected H_R and individual humans. (Modified from Childs 2004)

However, rarely has the full force of human scientific creativity and funding been directed at understanding and interrupting vulnerable infectious processes prerequisite to, but intermediate from, the immediate circumstances leading to human infection.

10.1 Culling of Vectors and Wildlife

The most widespread approach to zoonosis control is the culling or killing of individuals of H_Rs, H_Vs, or H_Is, either through selected culling (largely restricted to domestic animals) or indiscriminate population reduction (Wobeser 2002). The most common example of culling is the use of insecticides to control H_V populations and nuisance populations of mosquitoes (Thier 2001; Leprince and Lane 1996; Mount et al. 1996); however, issues related to human and environmental health have limited enthusiasm for this type of control in many circumstances. Culling of wildlife H_R populations has been adopted, or is planned, to curtail transmission of several viral and bacterial zoonotic pathogens to humans or domesticated livestock, although the record of population control as an effective prevention strategy limiting spillover is mixed (Wobeser 2002; see the chapter by Palmer, this volume).

Targeted reduction of specific H_R populations for control of rabies virus variants has been employed in Europe and North America. On both continents, programs have targeted red foxes (Muller 1971; Debbie1991) and in North America raccoon and skunk populations have been targeted (Rosatte et al. 1986; Debbie1991). Efforts are ongoing in Central and South America to reduce vampire bat populations in an effort to curtail the enormous economic loses sustained from vampire-bat transmitted rabies to cattle. Anticoagulants applied topically or systemically by direct inoculation into livestock are the major methods of vampire bat control (Crespo et al. 1979; Fornes et al. 1974; Thompson et al. 1972). However, wildlife culling to control rabies has been deemed largely unsuccessful or unnecessary given the intensive use of ORV to vaccinate susceptible H_Rs (Centers for Disease Control and Prevention 2004b; Macinnes et al. 2001; Aubert 1999b; Brochier et al. 1995; Slate et al. 2005). However, mathematical modeling of different control strategies frequently identifies a combination of vaccination and targeted culling as the optimal strategy for rabies control (Smith and Wilkinson 2003; Anderson et al. 1981).

Culling has recently been halted as a control measure for badgers serving as H_Rs for bovine TB in England (Roper 2003; Gormley and Costello 2003; see the chapter by Palmer, this volume), although in Ireland data suggest badger culling is an effective measure in reducing the incidence of TB in cattle herds (Griffin et al. 2005). The removal of some 20,000 badgers in England from 1975 to 1997 failed to curb bovine TB spread among cattle (Delahay et al. 2003). Vaccination of badgers against TB is now being investigated as a part of an integrated control program that includes targeting specific sites for control and different herd management practices for high-risk regions (White and Benhin 2004).

China initiated culling of live captured and breeding stocks of several species of carnivores, the masked palm civet (*Paguma larvata*), the raccoon dog (*Nyctereutes procyonoides*), and the Chinese ferret badger (*Melogale moschata*), implicated in the transmission of SARS-CoV to humans (Watts 2004; Zhong 2004). The WHO questioned the appropriateness of culling wildlife species (Parry 2004) and it is now appears that wild carnivores are not the actual H_R for SARS-CoV. Current information suggests that bats of the genus *Rhinolophus* are the H_R for ancestral coronaviruses giving rise to SARS-CoV capable of infecting wild carnivores and humans (Li et al. 2005; see the chapter by Wang and Eaton, this volume). Irrespective of the culling of farm-raised animals, the enormous illegal trade in wildlife will continue to stock the wet markets of China, Vietnam, and other southeastern Asian countries, with meat and other animal products from wild carnivores and other wildlife species prized for their culinary and medicinal properties (Bell et al. 2004; Yiming and Dianmo 1998).

10.2
Domestic Livestock and Poultry Culling for Zoonotic Disease Control

Control of emerging zoonotic agents circulating among domestic poultry, livestock, and companion animals is often more finely targeted at specific infected subpopulations or demographic cohorts than methods applied to wildlife. For example, the mass elimination of seropositive dogs in Brazil has been used in control programs for zoonotic visceral leishmaniasis; although evidence suggests dog control has failed to reduce the number of human leishmaniasis cases (Moreira et al. 2004).

Different culling strategies have been used for the control of BSE. Herd culling involves destroying entire herds of cattle from which an index case of BSE originated; birth cohort culling targets the subpopulation of cattle born during a specific interval of time and considered at greatest risk for having acquired BSE before the prohibition of feed containing cattle-derived offal; maternal culling destroys offspring borne to high-risk cows as the risk of vertical transmission of BSE is approximately 10% (Anonymous 2000); a final subpopulation considered to be at high risk, but difficult to identify operationally, is the feeding cohort. In the UK selected culling of birth cohorts (years 1989–1993) and maternal cohorts have been the major methods employed (Donnelly et al. 1997), involving destruction of more than 80,000 animals (Anonymous 2000). France, Portugal, and Ireland have employed mainly herd culling, with some maternal culling in France, with the destruction of approximately 10,000, 6,000, and 15,000 cattle, respectively (Anonymous, 2000). Additional culling methods may be employed as surveillance data accumulate (Calavas et al. 2004). Switzerland and Belgium have adopted both herd and birth cohort culling, with 2,000 and 1,400 animals destroyed as of 2000, respectively (Heim and Murray 2004; Anonymous 2000).

Culling of domestic poultry is the primary means of control for pathogenic influenza A subtypes, some considered to have pandemic potential as human viruses. Millions of chickens and other poultry were killed in Hong Kong in an attempt to prevent the spread of influenza A subtype H5N1 (Watts 2004a; Tam 2002), and in 2004 over 20 million chickens were killed in eight Southeast Asian nations as the threat of a human pandemic looms (Watts 2004a; Abbott and Pearson, 2004; see the chapters by Merianos and Webby et al., this volume). In April 2004, Canada ordered the killing of 19 million chickens and other poultry to contain an outbreak of influenza H7N3; 1 year earlier, the Netherlands culled 30 million chickens to control an outbreak of a related influenza subtype, H7N7 (Stegeman et al. 2004).

Livestock culling resulting in major economic losses accompanied the outbreak of NiV affecting swine and humans in Malaysia in 1997 (Stegeman et al. 2004; Paton et al.1999; see the chapter by Merianos, this volume), where more than 1 million swine were culled (Lam and Chua 2002; Uppal 2000). Nipah virus has since re-emerged in Malaysia, precipitating new rounds of culling (Ahmad 2000). Export bans and culling have enormous economic impacts and emerging zoonotic viruses, such as Influenza H5N1, NiV, WNV, and SARS-CoV, confront the stake holders in a global economy with unprecedented new risks (James 2005; von Overbeck 2003).

10.3 Alternatives to Culling as Population Control

In the future, population reduction by immune contraceptive programs could be used among certain populations of H_Rs or H_Is (Ferro 2002; Miller and Killian 2002; Lurz et al. 2002) (Fig. 2). There are ethical and practical limits as to how culling is, and will be, employed, as populations of game species and other wildlife species considered ecologically and esthetically important will be off limits, even if the species serves as H_R for a zoonotic pathogen. Exceptions occur where species overabundance becomes a nuisance problem or threatens vulnerable environments, as with white-tailed deer (*Odocoileus virgineanus*) in suburban environments or feral horses on barrier islands or federally controlled lands. In such instances, immune contraception may become the population reduction method of choice (Kirkpatrick et al. 1997). Where critical species within a community become environmentally destructive when overabundant, as with elephants within the confines of protected game reserves, controlled culling through hunting could generate income for indigenous peoples, but plans to use immune contraception may present a more acceptable choice (Fayrer-Hosken et al. 1999; Delsink et al. 2002).

10.4 Wildlife Vaccination

The second major approach to zoonotic pathogen control is through vaccination of individuals in the target H_R or H_I populations. Wildlife vaccination is currently limited to few species, although new vaccines are under development (Table 3).

Japanese encephalitis virus (JEV) transmission to humans often requires mosquito vectors which initially obtain a viremic bloodmeal from a swine H_I, alternatively referred to as an amplifying host (Daniels et al. 2002); vaccination of domestic swine to interrupt JEV transmission has been attempted (Daniels et al. 2002; Ueba et al. 1978). Similarly, vaccines for chickens serving as the H_R of influenza A virus subtypes are being employed to remove the intermediary avian host most closely associated with virus transmission to humans (Lee et al. 2004; Ellis et al. 2004b). Intermediate or amplifying vertebrate H_{S1}s, once infected by contact with a H_R, can directly transmit zoonotic viruses to the humans H_{S2}, as occurred with HeV and NiV transmission from pteropid bats initially to horses and swine (Hooper et al. 1998; Selvey et al. 1995; Field et al. 2001; Uppal 2000; see the chapters by Daniels et al. and Field et al., this volume) (Fig. 2). However, the wildlife vaccine with the widest distribution and greatest proven effectiveness is ORV for red foxes and raccoons.

The ORV most commonly in use for rabies control targeting wild carnivores is a recombinant vaccinia virus vaccine expressing the rabies virus glycoprotein gene (V-RG) (Rupprecht et al. 1986 1988); ORV was the first live-recombinant vaccine to be released in the field (Hanlon et al. 1998). The vaccine is distributed in plastic sachets, often covered with a polymer containing additives designed to preferentially attract the target H_R (Linhart et al. 1997, 2002), although nontarget species find these vaccine-laden baits attractive (Olson and Werner 1999).

Millions of ORV doses have been delivered to control red fox rabies in Europe and raccoon rabies in the United States (Aubert 1999a, 1999b; Hanlon and Rupprecht 1998; Slate et al. 2005); ORV has eliminated or reduced red fox rabies in many countries in western Europe (Hanlon and Rupprecht 1998; Aubert 1999b). In the United States, deployment of ORV to reduce enzootic levels of rabies, such as gray fox-associated rabies in Texas (Steelman et al. 2000), or to develop immune barriers to the spread of raccoon variant rabies and coyote/dog variant rabies, in Ohio, West Virginia, and Pennsylvania (the Ohio barrier), and in Texas, respectively (Foroutan et al. 2002; Farry et al. 1998; Slate et al. 2005),

Table 3 Examples of current, planned, or extraordinary interventions by vaccines targeting zoonotic viruses and bacteria among different classes of wildlife or domestic animal hosts

Type of host targeted	Specific example	Vaccine type	Achieved or desired purpose
Reservoir host (H_R)	Oral rabies vaccine (ORV) for red fox, raccoon, coyote, gray foxes[a–d]	Recombinant live vaccinia virus containing rabies glycoprotein gene (V-RG)	Eliminated substantial areas of enzootic red fox rabies in Europe Established an ORV barrier in Ohio to halt westward spread of raccoon-variant rabies virus
	Oral or parenteral vaccination of badgers (*Meles meles*) for TB[e]	Live vaccine, *M. bovis* BCG is main contender	Reduce transmission of TB from a sylvatic badger H_R to cattle
	Oral or parenteral vaccination of *Peromyscus leucopus* for *Borrelia burgdorferi*[f, g]	Recombinant protein A (OspA) from *B. burgdorferi* in *Escherichia coli* and other vectors for oral vaccination of mice	Reduce spirochetal load and prevalence in nymphal ticks to reduce human and domestic animal risk of Lyme disease
	Oral plague vaccine for prairie dogs[h]	Recombinant raccoonpox virus with F1 gene of *Yersinia pestis*	Eliminate plague foci by immunizing H_R among ground dwelling sciurids
Intermediate (H_I) or secondary host (H_S)	Rabies vaccine for domestic animals dogs, cats, ferrets[i]	Killed whole virus	Eliminate the domestic dog as the principal H_R for rabies virus throughout developing world Reduce human exposure to wildlife variants of rabies virus transmitted from a wildlife H_R to a companion animal (H_{S1}) and then to humans (H_{S2})
	Influenza A for chickens, domestic ducks[j, k]	Killed whole virus; recombinant virus vaccine using a Newcastle disease virus with inserted hemagglutinin (HA) gene from avian influenza virus	Prevent emergence and spread of potential pandemic influenza subtypes such as H5N1; prevent domestic poultry mortality from highly pathogenic influenza A viruses transmitted from waterfowl H_R

(*Continued*)

Table 3 Examples of current, planned, or extraordinary interventions by vaccines targeting zoonotic viruses and bacteria among different classes of wildlife or domestic animal hosts—cont'd.

Type of host targeted	Specific example	Vaccine type	Achieved or desired purpose
	Swine vaccinated for JEV[l, m]	Live attenuated Japanese encephalitis virus (JEV); Recombinant vaccine of pseudorabies virus (PRV) expressing NS1 protein of JEV	Prevent human disease from JEV infecting amplifying, intermediate, or secondary host in swine; prevent swine disease caused by JEV
	Horses, other livestock vaccinated for WNV, VEE, WEE, and EEE[n–p]	Inactivated mice brain-derived; live attenuated TC-83; multivalent inactivated VEE, EEE, and WEE viruses	Prevent veterinary losses and remove a H_I for transmission of zoonotic viruses to humans
Secondary host (H_S)	Distemper virus and rabies virus for African wild dog[q, r]	Live attenuated distemper virus and killed distemper and rabies virus vaccines	Protect endangered species from viral diseases introduced by humans through their domestic pets
	Plague vaccine for black-footed ferrets[s]	Recombinant raccoonpox virus with F1 gene of *Yersinia pestis*	Protect endangered and reintroduced species from zoonotic diseases maintained by prairie dog H_R populations
	Whooping cranes vaccinated for EEE[t]	Inactivated EEE virus	Protect endangered species from viral diseases

H_R reservoir host; H_S secondary host; H_I intermediate host

[a–d] Steelman et al. 2000; Fearneyhough et al. 1998; Roscoe et al. 1998; Brochier et al. 1996

[e] Gormley and Costello 2003

[f, g] Luke et al. 1997; Tsao et al. 2004

[h] Mencher et al. 2004

[i] Jenkins et al. 2004

[j k] Ellis et al. 2004b; Swayne et al. 2003

[l m] Xu et al. 2004; Ueba et al. 1978

[n–p] Weaver et al. 2004; Minke et al. 2004; Turell et al. 1999

[q, r] Gascoyne et al. 1993a; Van Heerden et al. 2002

[s] Rocke et al. 2004

[t] Olsen et al. 1997

have established zones where herd immunity is sufficiently high that rabies virus transmission is interrupted.

The Ohio barrier was effective in preventing or reducing raccoon rabies cases west of the vaccination border to a sporadic few, but after 6–7 years of success, a serious breach of the Ohio barrier, 11 km west of the vaccine zone, sparked what appears to be a new epizootic focus (Russell et al. 2005; Anonymous.2004a). Rapid and extensive remedial vaccination was employed and will be essential to contain this new focus from rapidly expanding into a full-blown epizootic (Russell et al. 2005).This long-term approach to rabies control is expensive and demands sustained public commitment (Kemere et al. 2002; Foroutan et al. 2002; Gordon et al. 2005); however, the alternative public health activities required should raccoon rabies become enzootic, are perhaps more expensive and also require sustained support (Gordon et al. 2005).

Although the risk for human exposure to vaccinia virus in ORV exists, relatively few instances of human exposure have been reported (Gordon et al. 2005). In the United States, a case of systemic vaccinia occurred in a pregnant women after she was bitten by her pet dog while trying to remove a vaccine sachet from the dog's mouth (Rupprecht et al. 2001).

10.5 Alternatives to Wildlife Vaccination

If ever fully developed and employed, genetic manipulation of H_V populations, or endosymbionts of H_V populations to establish vector refractoriness to infection by a zoonotic pathogen (Scott et al. 2002; Rasgon et al. 2003; Olson et al. 2002; Blair et al. 2000), will theoretically disrupt the transmission chain leading to human infection (Fig. 2). If refractory gene penetrance into a H_V population is complete, a pathogen could suffer extinction; if partial, the effect would be a mirror image to partial vaccine coverage of humans. Both strategies would reduce the probability of contact (see the chapter by Real and Biek, this volume) between an infectious vector and a susceptible human host, one reducing the proportion or number of infected vectors, the other decreasing the number or proportion of susceptible humans. As yet genetic engineering methods have no proven practical value in zoonotic disease control.

10.6 Quarantine, Isolation, and Legislation

Quarantine of animals arriving into a country from foreign countries, where certain diseases are enzootic, has a long history (Gensini et al. 2004). For example, dogs traveling from the United States to the UK were subject to a 6-month quarantine as part of the UK's rabies prevention law; proof of vaccination and a positive serologic test now suffice (Shaw et al. 2003; Fooks et al. 2002).

National legislation can attempt to reduce within-country movement of species recognized to be H_Rs of zoonotic viruses. Laws pertaining to translocations of rabies H_Rs were passed following the outbreak of a coyote/dog variant of rabies virus in Florida following importation of infected coyotes from Texas (Centers for Disease Control and Prevention 1995). The CDC imposed a ban on the importation of African rodents destined for the US pet trade after the introduction of monkeypox virus and the outbreak of human monkeypox that resulted from transmission of virus through an indigenous North American rodent H_I infected by virus spillover where housed in the same building with the African rodents (Centers for Disease Control 2003b; see the chapter by Regnery, this volume). On the same day as the CDC ban was announced, the Food and Drug Administration initiated regulatory control of interstate transport of prairie dogs in an effort to limit further spread of monkeypox to humans and potentially other susceptible species (see the chapter by Regnery, this volume). In a similar attempt to control the transmission of SARS-CoV, China passed laws prohibiting trade in certain carnivore species following the outbreak of SARS (Zhong 2004).

International laws pertaining to facilitating animal trade, while reducing the risk of exporting diseased animals or animal products, were established by the sanitary and phytosanitary measures, the SPS agreement, coincident with establishment of the World Trade Organization (WTO) in 1994 (Zepeda et al. 2005). The international standards are set by the OIE (OIE 2003). National prohibitions have been instituted by various nations, as exemplified by bans on importing cattle or cattle products from countries where BSE has been detected, listed, and updated on the USDA website (http://www.aphis.usda.gov/lpa/issues/bse/trade/bse_trade_ban_status.html), and bans to importing poultry from countries with enzootic avian influenza (Hall 2004), also listed on the USDA website (http://www.aphis.usda.gov/lpa/issues/ai_us/ai_trade_ban_status.html).

11 Obstacles to Animal-Based Intervention Strategies to Control Zoonotic Disease

11.1 National and International Commitment and Training

Public health professionals have lamented the years of budgetary neglect that have weakened our federal and state infrastructure for conducting surveillance (Bryan et al. 1994). National capacities to collect surveillance data of quality, which can inform prevention and intervention planning, are not developed over a year or even a decade. Any diminishment in support for human-based surveillance activities is a poor prognostic for implementing novel activities, such as designing and implementing regional programs to study zoonotic pathogens within their wildlife H_Rs, as any of these efforts require the same long-term, continuous support.

The United States has already lost much of its capacity to train scientists whose interests span field biology and laboratory sciences; the calls for increased training is a shrill mantra falling on deaf ears (Institute of Medicine1987, 2003, 1992; Centers for Disease Control and Prevention 1994). Even the emergences of SARS-CoV, HIV, WNV, influenza A subtype H5N1, SNV, and NiV have generated little movement toward training, encouraging, or promoting our professional capacity to explore the intricacies by which such pathogens have evolved and are maintained within their wildlife hosts; but by in large, the national response has been a handful of RO1s and a few training grants in vector-borne diseases and disease ecology. Additionally, there has been little success at cross-training of public health and veterinary professionals at the doctoral level; schools of public health tend to have few veterinarians as full-time faculty members, although at the postdoctoral level programs such as the Epidemiologic Intelligence Service (EIS) at CDC recruit veterinarians with each class.

As of July 2006, a joint and coordinated effort to establish an international surveillance network for the monitoring of animals and humans for zoonotic pathogens, or diseases caused by them, has been announced by the WHO and FASO in collaboration with the OIE (http://www.who.int/mediacentre/news/new/2006/nw02/en). The nature of this effort and details concerning program implementation in countries lacking adequate surveillance infrastructure have yet to be announced; any assessment of such a program designed to provide an early warning system for zoonotic pathogen emergence may be years in coming.

11.2
An International Problem with Equivalency in Veterinary Services

The role of veterinary medicine and veterinary epidemiology in support of the SPS agreement is severely hampered by the inequality of services available among nations (Zepeda et al. 2005). Developing nations face an enormous challenge to develop surveillance and monitoring systems, diagnostic laboratories, and the coordinating infrastructure to assure the validity and quality of the process for any domestic animal and livestock disease, much less emerging zoonoses (Zepeda et al. 2005).

11.3 Whose Problem Is It?

The bias toward human-based surveillance and post-spillover treatment of infected humans is firmly institutionalized, and too often the mission-boundaries of federal agencies preclude coordinated advancement toward any integrative policy. As an example of the problems inherent to different federal agencies' ability to cross traditional boundaries to promote integration of human and veterinary epidemiology is illustrated by a report issued by CDC in *Morbidity and Mortality Weekly Reports* in response to the discovery of BSE in cattle in the United States: "The occurrence of BSE in the United States reinforces the need for physicians to be aware of the clinical features of variant Creutzfeldt-Jakob disease (vCJD) and to arrange for brain autopsies in all decedents with suspected or probable CJD to assess the neuropathology of these patients" (Centers for Disease Control 2004a). Although efforts of the USDA to trace the origins of the infected animal were briefly alluded to in this report, the final recommendation focusing on the human consequences of BSE missed an opportunity to re-emphasize the critical component of veterinary surveillance. Perhaps a report, written in collaboration with the USDA, could have highlighted the means by which BSE surveillance in cattle was to be enhanced.

Research focusing on wildlife H_Rs and the human–wildlife interface is most often funded through year-to-year contracts or limited grants to research institutions, which often lack the infrastructure to preserve data, specimens, and, too often, trained investigators for durations exceeding the length of a grant. In addition, if there are no programs in place to disseminate and use the information generated by disparate research efforts, the results from such studies will remain within the confines of some academic journal, rather than translated into recommendations to prevent or reduce the risk of human disease. Currently, any products or recommendations stemming from such studies have little chance of diffusing into the public health culture (Childs 2006, in press).

The same problem exists with theoretical or mathematical approaches to infectious disease epidemiology. Once mathematical models are developed and validated by use of existing data sets (Russell et al. 2004, 2005; Coyne et al. 1989; Childs et al. 2000), the route to integrating insights gleaned from mathematical approaches into public health practice or specific control activities is unclear. Mathematical modeling as an aid to assist policy decisions has come under severe criticism from practicing veterinary professionals operating on the front lines of disease control. The disparate interpretations of the success of mathematical models in forming an effective control policy for an animal-disease disease outbreak are clearly illustrated by postcrisis reviews of the foot-and-mouth-disease (FMD) outbreak in the UK in 2001. Proponents and authors of models saw the utility and predictions of models validated (Woolhouse 2003), while some veterinary practitioners and epidemiologists saw little to no benefit in the models as applied in a real-time crisis (Salman 2004). The serious and widening gulf between mathematical modeling and public health practice requires a systematic and purposeful effort on both sides to bridge these differences (Childs 2006, in press). If communications fail, the danger exists for one class of professional to dismiss the efforts of the other as either irrelevant or hopelessly unsophisticated. Whose problem is it?

11.4
Jumping Zoonoses: The Problems of Long-Distance Translocation

National and international long-range translocations of infected animals have played an extensive role in the emergence of viral zoonoses. The phenomenon is so common that it must be considered in conjunction with any control strategy based on legal restrictions to animal movement, bans to trade in wildlife, or when constructing vaccination barriers to limit pathogen spread.

Instances of transcontinental zoonotic viral spread reinforce the significance of LDTs and the recommendation that contingencies for their occurrence should be included in any strategic plan for zoonotic disease control. In 2002, SARS spread around the world in a matter of months, eventually affecting 27 countries on every populated continent (Heymann 2004). In 2003, monkeypox was introduced into the United States along with a shipment of African rodents destined for the pet trade (Cunha 2004; Centers for Disease Control and Prevention 2003a; see the chapter by Regnery, this volume). In 1999, WNV was recorded in the New World for the first time, introduced into New York City by an infected vector or human host (Lanciotti et al. 1999; Kilpatrick et al. 2005). In 1999, Singapore experienced outbreaks of NiV infection among abattoir workers after importing swine from Malaysia (Chew et al. 2000; see the chapter by Field et al., this volume).

The impact of a within-country LDT is well illustrated by the spread of raccoon rabies from a focus identified in the late 1970s along the Virginia–West Virginia border, a focus likely seeded by the translocation of raccoons incubating rabies from an enzootic region of raccoon-associated rabies virus in the southeastern United States (Nettles et al. 1979). The resulting rabies epizootic, as the disease spread into mid-Atlantic and northeastern states, was one of the most extensive and intensive wildlife epizootics recorded (Childs et al. 2001; Hanlon and Rupprecht 1998). A rabid bat stowaway onboard a ship originating from the west coast of the United States was discovered in Hawaii, which is a rabies-free state (Centers for Disease Control and Prevention 1992); other instances of LDTs of rabid bats, some transcontinental, have been reviewed (Constantine 2003). At a finer scale, quantitatively defined instances of raccoon rabies epizootic foci developing in advance of the epizootic wavefront in Connecticut indicate local translocations influenced the spatial pattern of raccoon rabies spread through that state (Smith et al. 2005). The instance of a rabies virus variant of coyotes/domestic dogs from Texas being introduced into Florida with transported coyotes was described previously (Centers for Disease Control and Prevention 1995).

11.5 Animal Disease Detection and Compensation: How Close Is the Link?

Without adequate compensation for losses accrued through culling or exportation bans, countries attempting to implement animal-based surveillance programs for domestic species, much less wildlife, are likely to encounter problems with voluntary reporting (see the chapter by Merianos, this volume). In some instances, the mere threat of culling, as with swine in areas of Malaysia affected by NiV, can promote epidemic spread as farmers disperse valuable animals to protect their livelihood (Chua 2003; see the chapter by Field et al., this volume). In addition to the enormous economic losses facing individuals whose animals are killed or whose products cannot be sold, the consequences of reporting an outbreak of a new zoonotic disease can be politically unattractive, inviting delays in reporting, as may have occurred with SARS in China (Enserink 2003). Other hidden costs associated with zoonotic disease outbreaks may persist through the burden of surveillance and animal testing (Bradley and Liberski 2004) and the loss to veterinary services (Bennett and Hallam 1998).

11.6 H_R Identification and the Consequences of Getting It Wrong

Before implementation of any control activity, such as culling or vaccination, it is essential that the target species has been accurately and irrefutably identified as the H_R or H_I of importance. Identification of a H_R requires establishing epidemiologic plausibility using definable criteria, such as the temporal and spatial association of putative H_Rs to pathogen spillover, and molecular epidemiologic data linking virus recovered from a H_S to virus circulating among H_Rs (Haydon et al. 2002; Childs 2004). China initiated culling of some species of carnivores and other wildlife intended for human consumption (Watts 2004b), although no SARS-CoV has yet been isolated from wild civets obtained directly from the field (Bell et al. 2004; Guan et al. 2003). In 2005, a putative H_R for coronaviruses ancestral to those isolated and characterized from humans and palm civets was identified among three species of bats of the genus *Rhinolophus* (Li et al. 2005; see the chapter by Wang and Eaton, this volume). Molecular sequencing of SARS-CoV from bats, palm civets, and humans indicates a common ancestor with rapid positive selection for virulent viral subtypes infecting humans and civets (Song et al. 2005; see the chapter by Wang and Eaton, this volume).

Removing carnivores near the top of ecological food chains can have many unforeseen, and in certain circumstances, potentially disastrous, consequences. By diluting, or severing important links in community processes, culling of top-level carnivores can cause changes in species richness and diversity in communities and increases in prey populations (Ostfeld and Holt 2004; Ostfeld and Keesing 2000), including wild rodent H_Rs of other potentially dangerous zoonotic agents, such as *Borrelia burgdorferi* and the arenaviruses and hantaviruses (LoGiudice et al. 2003; Mills and Childs 1998). Use of methods designed to control one species, such as anticoagulants topically applied to cattle to reduce vampire bat populations, can reduce populations of ecologically important species of bats unintentionally dosing themselves when roosting with vampire bats in confined spaces (Mayen 2003; Martinez-Burnes et al. 1997).

12
Priority Zoonoses: The Case for Enhanced Surveillance for HIV and Influenza A

Contrast the purposeful and highly successful surveillance for animal rabies with activities targeting other known or potential pandemic zoonotic threats with wildlife H_Rs. Subtypes of HIV I and HIV II have emerged independently from primate SIVs on at least eight independent occasions (Hahn et al. 2000; B. Hahn, personal communication to JEC). The number of SIVs described among nonhuman primates in Africa, as of 2004, was approximately 40 (Apetrei et al. 2004). Rapid replication, high mutability, and the elevated rates of recombination of lentiviruses (Zhuang et al. 2002; see the chapter by Holmes and Drummond, this volume) virtually assures that new strains of SIV-HIV will make the journey out of Africa. There appears to be little systematic effort to enhance or build the basic infrastructure in regions of West Africa that could begin to conduct surveillance for new emerging HIVs at the human level or monitor the dynamics of transmission of diverse and genetically chimerical SIVs transmitted among nonhuman primates. Detection of spumaviruses among hunters, although uncommon (~1%), signify the extent to which humans are exposed and infected with diverse primate retroviruses (Wolfe et al. 2004). Although some of the countries of importance are war zones and politically unstable, it is unclear that given an improving situation, surveillance for SIVs spilling over to humans would be regarded as a priority among funding institutions concentrating on HIV vaccine trials.

How are we surveilling and preparing for the next pandemic of influenza? Currently influenza A subtype H5N1 has a limited capacity for cross-vertebrate class transmission from birds to mammals, although infection is frequently fatal to humans once spillover succeeds (Guan et al. 2004; Sturm-Ramirez et al. 2004; Claas 2000; Tran et al. 2004; see the chapter by Webby et al., this volume). Monitoring avian H5N1 subtypes has been, and continues to be spotty, and largely limited to domestic poultry in which infection is often fatal to chickens and to a lesser extent ducks (Sturm-Ramirez et al. 2004). Recombination of waterfowl influenza viruses within a domestic duck H_R may have been the origin of highly pathogenic subtypes of H5N1 for chickens (Chen et al. 2004b; Tumpey et al. 2003; Guan et al. 1999, 2000), and successive isolates of H5N1 from domestic ducks over time indicate increasing virulence for mammals (Chen et al. 2004b; Guan et al. 2002a, 2002b). Domestic geese may serve a role as an independent H_R for recombinant wild waterfowl-goose influenza H5N1-subtypes and help drive the rapid evolution of highly pathogenic viruses of ducks and chickens (Webster et al. 2002; Chen et al. 2004b).

Yet the ultimate origin of H5N1 and other influenza subtypes, H7N3 and H9N2, occurring among domestic poultry and representing human threats (Campitelli et al. 2004; Choi et al. 2004), is the diverse species of waterfowl, shorebirds, and possibly other avian types in which these various influenza subtypes circulate, often with minimal morbidity (see the chapter by Webby et al., this volume). Surveillance for influenza subtypes among wild waterfowl and other migratory birds is spotty (Krauss et al. 2004; Campitelli et al. 2004; De Marco et al. 2004; Hatchette et al. 2004) and largely restricted to local or regional populations, as occurs in North America and Italy (Krauss et al. 2004; Slemons et al. 2003; Hatchette et al. 2004; Campitelli et al. 2004; De Marco et al. 2004; Ellis et al. 2004a; Stallknecht et al. 1990). The WHO has proposed establishing an Animal Influenza Network to develop and coordinate research on the ecology and molecular biology of animal influenza viruses and integrate these animal-based activities with the global surveillance program for human influenza (Stohr 2003); presumably emphasis will be placed on wild waterfowl and other migratory birds, in addition to domestic poultry and livestock.

13 Conclusions

The uneven standards of surveillance, human- or animal-based, for zoonotic diseases or pathogens maintained by wildlife H_Rs, or even domestic species (Zepeda et al. 2005), is a global problem, readily apparent even within the United States, where investment in public health, including surveillance systems, has a long and enviable history (Thacker 2000).

As of 2006, there appears to be little scientific, social, or political consensus that animal-based surveillance for zoonoses merits investment in international infrastructure. However, this trend may be changing with the recent announcement of the proposal to develop a global early warning system for certain zoonotic agents or disease to be coordinated by the WHO, FAO, and OIE.

Technologically advanced solutions to addressing vector-borne or zoonotic disease transmission, such as genetic manipulation of mosquitoes or immunocontraception aimed at target vertebrate hosts, may involve good science, but whether these approaches represent good public health is highly debatable (Scott et al. 2002; Furguson et al. 2005). Novel schemes of preventing spillover of human pathogens from animal H_Rs can only spring from improving our understanding of the ecological context and biological interactions of pathogen maintenance among H_Rs.

There are no easy solutions to preventing spillover and there is no reason to expect we will ever predict the wheres and whys of new emergences of zoonotic diseases (see the chapters by Cleaveland et al. and Daszak et al., this volume). Inevitably, the major issue arises of where surveillance and research efforts should focus, and there are many areas worthy of consideration. Where the intent exists to improve global surveillance for specific zoonoses of animals, such as influenza A, every possible effort should be made to bring in new ideas and to set a standard of excellence that will encourage additional forays into these areas. As a speculative example, the ability to genetically modify plants to produce viral antigens of potential vaccine quality (Castle and Dalgleish 2005) may provide a tool to reach wild waterfowl that gather in vast numbers in specific staging areas during migration. Could influenza A subtype H5N1 genes be introduced into corn (Tacket et al. 2004; Lamphear et al. 2004), a favorite food of virtually all waterfowl and poultry, and would such a vaccine immunize sufficient numbers of waterfowl to reduce the susceptible population if widely dispersed among migratory staging areas?

Would there be a payoff from large investments to improve surveillance and knowledge of known or potential zoonotic pathogens circulating among wildlife H_R populations? No one knows, but the alternative is to continue to rely on disease detection among sentinel humans. Our ongoing experience with HIV, the looming threat of pandemic influenza, and the myriad of other zoonotic virus emergences in the last few years inform us of the outcomes we can expect by relying on detection of post-spillover events. Efforts to create a knowledge base of the ecology of zoonotic viruses and other pathogens are not without precedent. A glimpse at the enormous achievements in the field and laboratory by scientists connected to the Rockefeller Foundation Virus Program should convince even skeptical readers of the value of an integrated research approach, without adherence to rigid disciplinary boundaries (Theiler and Downs 1973).

Public health judges its great achievements not by damage control, but permanent prevention or, ultimately, eradication of disease threats. When any zoonotic disease or agent shows up in a human, to a great degree, we have failed; in some notorious instances, such as with HIV, it will already be too late to halt a pandemic's spread. We are aware of the consequences and the difficulties in combating pandemic disease, whether it is HIV in humans or Influenza A subtype H5N1 in domestic poultry. As a conservative measure and complementary strategic approach to defensive planning for disease emergence among humans or domestic animals, more resources and research should be invested on offensive approaches whereby potentially vulnerable points in pre-spillover transmission chains involving animal and vector hosts are identified and interventions are designed and assessed.

References

A Report of the National Advisory Committee on SARS, Public Health (2003) Learning from SARS 1210 Learning from SARS: Renewal of Public Health in Canada: 2003. Health Canada, Ottawa, pp 1–221

Abbott A, Pearson H (2004) Fear of human pandemic grows as bird flu sweeps through Asia. Nature 427:472–473

Advisory Committee on Immunization Practices (1999) Human rabies prevention – United States, 1999. Recommendations of the Advisory Committee on Immunization Practices (ACIP). MMWR Morb Mortal Wkly Rep 48:1–21

Ahmad K (2000) Malaysia culls pigs as Nipah virus strikes again. Lancet 356:230

Anderson RM, Jackson HC, May RM, Smith AM (1981) Population dynamics of fox rabies in Europe. Nature 289:765–771

Anderson RM, Fraser C, Ghani AC, Donnelly CA, Riley S, Ferguson NM, Leung GM, Lam TH, Hedley AJ (2004) Epidemiology, transmission dynamics and control of SARS: the 2002–2003 epidemic. Philos Trans R Soc Lond B Biol Sci 359:1091–1105

Anonymous (2000) Opinion of the Scientific Steering Committee on BSE-related culling in cattle. SSC/14–15 September 2000, pp 1–26

Anonymous (2004a) Rabid raccoon found in Leroy Township. pp 1–2

Anonymous (2004b) USDA plans sharp increase in BSE testing. USDA Release No. (0191) 04 224:1402

Apetrei C, Robertson DL, Marx PA (2004) The history of SIVS, AIDS: epidemiology, phylogeny and biology of isolates from naturally SIV infected non-human primates (NHP) in Africa. Front Biosci 9:225–254:225–254

APHIS Wildlife Services Factsheet (2002) Preventing the westward spread of rabies. USDA, Washington, DC, pp 1–2

Aubert MFA (1999) Costs and benefits of rabies control in wildlife in France. Rev Sci Tech Off Int Epizoot 18:533–543

Ballard WB, Krausman PR (1997) Occurrence of rabies in wolves of Alaska. J Wildl Dis 33:242–245

Barker CM, Reisen WK, Kramer VL (2003) California state mosquito-borne virus surveillance and response plan: a retrospective evaluation using conditional simulations. Am J Trop Med Hyg 68:508–518

Bell D, Roberton S, Hunter PR (2004) Animal origins of SARS coronavirus: possible links with the international trade in small carnivores. Philos Trans R Soc Lond B Biol Sci 359:1107–1114

Bennett RM, Hallam D (1998) Implications of BSE policy for livestock production and veterinary services in the United Kingdom. Vet Rec 142:155–159

Birkhead GS, Maylahn CM (2000) State and local public health surveillance. In: Teutsch SM, Churchill RE (eds) Principles and practice of public health surveillance. Oxford University Press, Oxford, pp 253–286

Blair CD, Adelman ZN, Olson KE (2000) Molecular strategies for interrupting arthropod-borne virus transmission by mosquitoes. Clin Microbiol Rev 13:651–661

Bourliere F (1975) Mammals, small and large: the ecological implications of size. In: Golley FB, Petrusewicz K, Ryszkowski L (eds) Small mammals: their productivity and population dynamics. Cambridge University Press, Cambridge, pp 1–8

Bradley R, Liberski PP (2004) Bovine spongiform encephalopathy (BSE): the end of the beginning or the beginning of the end? Folia Neuropathol 42 Suppl A:55–68

Brochier B, Costy F, Pastoret PP (1995) Elimination of fox rabies from Belgium using a recombinant vaccinia-rabies vaccine: an update. Vet Microbiol 46:269–279

Brochier B, Aubert MF, Pastoret PP, Masson E, Schon J, Lombard M, Chappuis G, Languet B, Desmettre P (1996) Field use of a vaccinia-rabies recombinant vaccine for the control of sylvatic rabies in Europe and North America. Rev Sci Tech 15:947–970

Bryan RT, Pinner RW, Berkelman RL (1994) Emerging infectious diseases in the United States. Improved surveillance, a requisite for prevention. Ann N Y Acad Sci 740:346–361

Buehler JW, Berkelman RL, Hartley DM, Peters CJ (2003) Syndromic surveillance and bioterrorism-related epidemics. Emerg Infect Dis 9:1197–1204

Burrows R (1992) Rabies in wild dogs. Nature 359:277

Butler WI Jr, Stehn RA, Balogh GR (1995) GIS for mapping waterfowl density and distribution from aerial surveys. Wildl Soc Bull 23:140–147

Calavas D, Ducrot C, Baron TG (2004) Past, present and future of bovine spongiform encephalopathy in France. Curr Top Microbiol Immunol 284:51–63

Calisher CH, Root JJ, Mills JN, Beaty BJ (2002) Assessment of ecologic and biologic factors leading to hantavirus pulmonary syndrome. Colorado USA. Croat Med J 43:330–337

Campitelli L, Mogavero E, De Marco MA, Delogu M, Puzelli S, Frezza F, Facchini M, Chiapponi C, Foni E, Cordioli P, Webby R, Barigazzi G, Webster RG, Donatelli I (2004) Interspecies transmission of an H7N3 influenza virus from wild birds to intensively reared domestic poultry in Italy. Virology 20:24–36

Castle D, Dalgleish J (2005) Cultivating fertile ground for the introduction of plant-derived vaccines in developing countries. Vaccine 23:1881–1885

Centers for Disease Control and Prevention (1992) Rabid bat diagnosed in Hawaii. MMWR Morb Mortal Wkly Rep 51:181–185

Centers for Disease Control and Prevention (1994) Addressing emerging infectious disease threats: a prevention strategy for the United States. US Department of Health and Human Services, Atlanta

Centers for Disease Control and Prevention (1995) Translocation of coyote rabies – Florida, (1994) MMWR Morb Mortal Wkly Rep 44:580–587

Centers for Disease Control and Prevention (1997) Case definitions for infectious conditions under public health surveillance. Rep 46. US Department of Health and Human Services, Atlanta

Centers for Disease Control and Prevention (1998) Preventing emerging infectious diseases: a strategy for the 21st century. US Department of Health and Human Services, Atlanta

Centers for Disease Control and Prevention (2001) Update: investigation of bioterrorism-related anthrax and interim guidelines for clinical evaluation of persons with possible anthrax. MMWR Morb Mortal Wkly Rep 50:941–948

Centers for Disease Control and Prevention (2002a) Guidelines for surveillance, prevention and control of West Nile virus. Epidemiol Bull 23:12–14

Centers for Disease Control and Prevention (2002b) Summary of notifiable diseases United States 2000. MMWR Morb Mortal Wkly Rep 49:1–100

Centers for Disease Control and Prevention (2003a) Update: multistate outbreak of monkeypox – Illinois, Indiana, Kansas, Missouri, Ohio, and Wisconsin, 2003. MMWR Morb Mortal Wkly Rep 52:616–618

Centers for Disease Control and Prevention (2003b) Update: multistate outbreak of monkeypox – Illinois, Indiana, Kansas, Missouri, Ohio, and Wisconsin, 2003. MMWR Morb Mortal Wkly Rep 52:642–646

Centers for Disease Control and Prevention (2004a) Bovine spongiform encephalopathy in a dairy cow—Washington State, 2003. MMWR Morb Mortal Wkly Rep 52:1280–1285

Centers for Disease Control and Prevention (2004b) Compendium of animal rabies prevention and control, 2004; National Association of State Public Health Veterinarians Inc (NASPHV). MMWR Morb Mortal Wkly Rep No. RR-9:1–6

Centers for Disease Control and Prevention (2004c) Summary of notifiable diseases—United States, 2002. MMWR Morb Mortal Wkly Rep 51:1–80

Centers for Disease Control and Prevention (2004d) Update: influenza activity – United States and worldwide, 2003–04 season, and composition of the 2004–05 influenza vaccine. MMWR Morb Mortal Wkly Rep 53:547–552

Centers for Disease Control and Prevention (2005) Compendium of animal rabies prevention and control, 2005; National Association of State Public Health Veterinarians Inc. (NASPHV). MMWR Recomm Rep 54:1–8

Chang GJ, Kuno G, Purdy DE, Davis BS (2004) Recent advancement in flavivirus vaccine development. Expert Rev Vaccines. 3:199–220

Chapman RC (1978) Rabies: decimation of a wolf pack in artic Alaska. Science 201:365–367

Check E (2004) Health concerns prompt US review of exotic-pet trade. Nature 427:277

Chen D, Cane MA, Kaplan A, Zebiak SE, Huang D (2004a) Predictability of El Nino over the past 148 years. Nature 428:733–736

Chen H, Deng G, Li Z, Tian G, Li Y, Jiao P, Zhang L, Liu Z, Webster RG, Yu K (2004b) The evolution of H5N1 influenza viruses in ducks in southern China. Proc Natl Acad Sci U S A 101:10452–10457

Chew MH, Arguin PM, Shay DK, Goh KT, Rollin PE, Shieh WJ, Zaki SR, Rota PA, Ling AE, Ksiazek TG, Chew SK, Anderson LJ (2000) Risk factors for Nipah virus infection among abattoir workers in Singapore. J Infect Dis 181:1760–1763

Childs JE (2004) Zoonotic viruses of wildlife: hither from yon. Arch Virol Suppl:1–11

Childs JE (2006) From ecological theory to application. In: Keesing F, Eviner V, Ostfeld RS (eds) Infectious disease ecology. Princeton University Press, in press

Childs JE, Paddock CD (2002) Passive surveillance as an instrument to identify risk factors for fatal Rocky Mountain spotted fever: is there more to learn? Am J Trop Med Hyg 5:450–457

Childs JE, Kaufmann AF, Peters CJ, Ehrenberg RL (1993) Hantavirus infection – southwestern United States: interim recommendations for risk reduction. MMWR Morb Mortal Wkly Rep 42 (RR-11):1–13

Childs JE, Ksiazek TG, Spiropoulou CF, Krebs JW, Morzunov S, Maupin GO, Rollin PE, Sarisky J, Enscore RE, Frey JK, Peters CJ, Nichol ST (1994) Serologic and genetic identification of *Peromyscus maniculatus* as the primary rodent reservoir for a new hantavirus in the southwestern United States. J Infect Dis 169:1271–1280

Childs JE, Krebs JW, Ksiazek TG, Maupin GO, Gage KL, Rollin PE, Zeitz PS, Sarisky J, Enscore RE, Butler JC et al (1995) A household-based, case-control study of environmental factors associated with hantavirus pulmonary syndrome in the southwestern United States. Am J Trop Med Hyg 52:393–397

Childs JE, Colby L, Krebs JW, Strine T, Feller M, Noah D, Drenzek C, Smith JS, Rupprecht CE (1997) Surveillance and spatiotemporal associations of rabies in rodents and lagomorphs in the United States 1985–1994. J Wildl Dis 33:20–27

Childs JE, Curns AT, Dey ME, Real LA, Feinstein L, Bjornstad ON, Krebs JW (2000) Predicting the local dynamics of epizootic rabies among raccoons in the United States. Proc Natl Acad Sci U S A 97:13666–13671

Childs JE, Curns AT, Dey ME, Real AL, Rupprecht CE, Krebs JW (2001) Rabies epizootics among raccoons vary along a North-South gradient in the Eastern United States. Vect Borne Zoonot Dis 1:253–267

Childs JE, Krebs JW, Smith JS (2002) Public health surveillance and the molecular epidemiology of rabies. In: Leitner T (ed) The molecular epidemiology of human viruses. Kluwer Academic, Dordrecht, pp 273–312

Childs JE, Krebs JW, Real LA, Gordon ER (2007) Animal-based national surveillance for zoonotic disease; qualities, limitations and implications of a model system for monitoring rabies. Prev Vet Med 78:246–261

Choi YK, Ozaki H, Webby RJ, Webster RG, Peiris JS, Poon L, Butt C, Leung YH, Guan Y (2004) Continuing evolution of H9N2 influenza viruses in Southeastern China. J Virol 78:8609–8614

Chua KB (2003) Nipah virus outbreak in Malaysia. J Clin Virol 26:265–275

Claas EC (2000) Pandemic influenza is a zoonosis, as it requires introduction of avian-like gene segments in the human population. Vet Microbiol 74:133–139
Cleaveland S, Laurenson MK, Taylor LH (2001) Diseases of humans and their domestic mammals: pathogen characteristics, host range and the risk of emergence. Philos Trans R Soc Lond B Biol Sci 356:991–999
Constantine DG (2003) Geographic translocation of bats: known and potential problems. Emerg Infect Dis 9:17–21
Cox NJ, Brammer TL, Regnery HL (1994) Influenza: global surveillance for epidemic and pandemic variants. Eur J Epidemiol 10:467–470
Cox RJ, Brokstad KA, Ogra P (2004) Influenza virus: immunity and vaccination strategies. Comparison of the immune response to inactivated and live, attenuated influenza vaccines. Scand J Immunol 59:1–15
Coyne MJ, Smith G, McAllister FE (1989) Mathematic model for the population biology of rabies in raccoons in the mid-Atlantic states. Am J Vet Res 50:2148–2154
Crespo RF, Fernandez SS, De Anda L, Velarde FI, Anaya RM (1979) Intramuscular inoculation of cattle with warfarin: a new technique for control of vampire bats. Bull Pan Am Health Organ 13:147–161
Crook PD, Crowcroft NS, Brown DW (2002) West Nile virus and the threat to the UK. Commun Dis Public Health 5:138–143
Cunha BE (2004) Monkeypox in the United States: an occupational health look at the first cases. J Am Assoc Occup Health Nurses 52:164–168
Custer DM, Thompson E, Schmaljohn CS, Ksiazek TG, Hooper JW (2003) Active and passive vaccination against hantavirus pulmonary syndrome with Andes virus M genome segment-based DNA vaccine. J Virol 77:9894–9905
Daniels TJ, Williams DT, Mackenzie JS (2002) Japanese encephalitis virus. In: Morrilla A, Yoon KJ, Zimmerman JJ (eds) Trends in emerging viral infections of swine. Iowa State Press, Ames, IA, pp 249–263
Daszak P, Cunningham AA, Hyatt AD (2000) Emerging infectious diseases of wildlife – threats to biodiversity and human health. Science 287:443–449
Davidson WR, Lockhart JM, Stallknecht DE, Howerth EW, Dawson JE, Rechav Y (2001) Persistent *Ehrlichia chaffeensis* infection in white-tailed deer. J Wildl Dis 37:538–546
Day JF (2001) Predicting St. Louis encephalitis virus epidemics: lessons from recent, and not so recent, outbreaks. Annu Rev Entomol 46:111–138
De Marco MA, Campitelli L, Foni E, Raffini E, Barigazzi G, Delogu M, Guberti V, Di TL, Tollis M, Donatelli I (2004) Influenza surveillance in birds in Italian wetlands (1992–1998): is there a host restricted circulation of influenza viruses in sympatric ducks and coots? Vet Microbiol 98:197–208
Debbie JG (1991) Rabies control of terrestrial wildlife by population reduction. In: Baer GM (ed) The natural history of rabies. CRC Press, Boca Raton, pp 477–484
Delahay RJ, Wilson GJ, Smith GC, Cheeseman CL (2003) Vaccinating badgers (*Meles meles*) against *Mycobacterium bovis*: the ecological considerations. Vet J 166:43–51
Delsink AK, van Altena JJ, Kirkpatrick J, Grobler D, Fayrer-Hosken RA (2002) Field applications of immunocontraception in African elephants (*Loxodonta africana*). Reprod Suppl 60:117–124
Dobson A, Foufopoulos J (2001) Emerging infectious pathogens of wildlife. Philos Trans R Soc Lond B Biol Sci 356:1001–1012

Donnelly CA, Ferguson NM, Ghani AC, Woolhouse ME, Watt CJ, Anderson RM (1997) The epidemiology of BSE in cattle herds in Great Britain. I. Epidemiological processes, demography of cattle and approaches to control by culling. Philos Trans R Soc Lond B Biol Sci 352:781–801

Duchin JS, Koster FT, Peters CJ, Simpson GL, Tempest B, Zaki SR, Rollin PE, Nichol S, Umland ET (1994) Hantavirus pulmonary syndrome: a clinical description of 17 patients with a newly recognized disease. The Hantavirus Study Group. N Engl J Med 330:949–955

Duff JP, Holmes P, Brown I, Gayford P (2003) Surveillance scheme for wildlife disease in England and Wales. Vet Rec 153:538

Duzgunes N, Simoes S, Slepushkin V, Pretzer E, Flasher D, Salem II, Steffan G, Konopka K, Pedroso de Lima MC (2005) Delivery of antiviral agents in liposomes. Methods Enzymol 391:351–373

Eidson M, Komar N, Sorhage F, Nelson R, Talbot T, Mostashari F, McLean R (2001a) Crow deaths as a sentinel surveillance system for West Nile virus in the northeastern United States, 1999. Emerg Infect Dis 7:615–620

Eidson M, Kramer L, Stone W, Hagiwara Y, Schmit K (2001b) Dead bird surveillance as an early warning system for West Nile virus. Emerg Infect Dis 7:631–635

Eidson M, Miller J, Kramer L, Cherry B, Hagiwara Y (2001c) Dead crow densities and human cases of West Nile virus New York State, 2000. Emerg Infect Dis 7:662–664

Ellis TM, Bousfield RB, Bissett LA, Dyrting KC, Luk GS, Tsim ST, Sturm-Ramirez K, Webster RG, Guan Y, Malik Peiris JS (2004a) Investigation of outbreaks of highly pathogenic H5N1 avian influenza in waterfowl and wild birds in Hong Kong in late 2002. Avian Pathol 33:492–505

Ellis TM, Leung CY, Chow MK, Bissett LA, Wong W, Guan Y, Malik Peiris JS (2004b) Vaccination of chickens against H5N1 avian influenza in the face of an outbreak interrupts virus transmission. Avian Pathol 33:405–412

Enserink M (2003) SARS in China. China's missed chance. Science 301:294–296

Farry SC, Henke SE, Beasom SL, Fearneyhough MG (1998) Efficacy of bait distributional strategies to deliver canine rabies vaccines to coyotes in southern Texas. J Wildl Dis 34:23–32

Fauchi AS (2002) NIAID biodefense research agenda for CDC category A agents. NIH Publication No.03–5308:1–54. 2002. US Depaertment of Health and Human Services NIH, http://biodefense.niaid.nih.gov. Cited 27 February 2007

Fayrer-Hosken RA, Bertschinger HJ, Kirkpatrick JF, Grobler D, Lamberski N, Honneyman G, Ulrich T (1999) Contraceptive potential of the porcine zona pellucida vaccine in the African elephant (*Loxodonta africana*). Therio Gen 52:835–846

Fearneyhough MG, Wilson PJ, Clark KA, Smith DR, Johnston DH, Hicks BN, Moore GM (1998) Results of an oral rabies vaccination program for coyotes. J Am Vet Med Assoc 212:498–502

Ferro VA (2002) Current advances in antifertility vaccines for fertility control and non-contraceptive applications. Expert Rev Vaccines 1:443–452

Field H, Young P, Yob JM, Mills J, Hall L, Mackenzie J (2001) The natural history of Hendra and Nipah viruses. Microbes Infect 3:307–314

Field H, Mackenzie J, Daszak P (2004) Novel viral encephalitides associated with bats (Chiroptera) – host management strategies. Arch Virol Suppl 113–121
Field HE, Barratt PC, Hughes RJ, Shield J, Sullivan ND (2000) A fatal case of Hendra virus infection in a horse in north Queensland: clinical and epidemiological features. Aust Vet J 78:279–280
Fooks AR, McElhinney LM, Brookes SM, Johnson N, Keene V, Parsons G, Soldan A (2002) Rabies antibody testing and the UK Pet Travel Scheme. Veterinary Record 150:428–430
Fornes A, Lord RD, Kuns ML, Larghi OP, Fuenzalida E, Lazara L (1974) Control of bovine rabies through vampire bat control. J Wildl Dis 10:310–316
Foroutan P, Meltzer MI, Smith KA (2002) Cost of distributing oral raccoon-variant rabies vaccine in Ohio: 1997–2000. J Am Vet Med Assoc 220:27–32
Fouchier RA, Schneeberger PM, Rozendaal FW, Broekman JM, Kemink SA, Munster V, Kuiken T, Rimmelzwaan GF, Schutten M, van Doornum GJ, Koch G, Bosman A, Koopmans M, Osterhaus AD (2004) Avian influenza A virus (H7N7) associated with human conjunctivitis and a fatal case of acute respiratory distress syndrome. Proc Natl Acad Sci U S A 101:1356–1361
Fraser C, Riley S, Anderson RM, Ferguson NM (2004) Factors that make an infectious disease outbreak controllable. Proc Natl Acad Sci U S A 101:6146–6151
Fraser GC, Hooper PT, Lunt RA, Gould AR, Gleeson LJ, Hyatt AD, Russell GM, Kattenbelt JA (1996) Encephalitis caused by a Lyssavirus in fruit bats in Australia. Emerg Infect Dis 2:327–331
Furguson H, John B, Ng'habi K, Knols BGL (2005) Redressing the sex imbalance in knowledge of vector biology. Trends Ecol Evol 20:202–209
Garvin MC, Tarvin KA, Smith J, Ohajuruka OA, Grimes S (2004) Patterns of West Nile virus infection in Ohio blue jays: implications for initiation of the annual cycle. Am J Trop Med Hyg 70:566–570
Gascoyne SC, King AA, Laurenson MK, Borner M, Schildger B, Barrat J (1993a) Aspects of rabies infection and control in the conservation of the African wild dog (*Lycaon pictus*) in the Serengeti region Tanzania. Onderstepoort J Vet Res 60:415–420
Gascoyne SC, Laurenson MK, Lelo S, Borner M (1993b) Rabies in African wild dogs (*Lycaon pictus*) in the Serengeti region Tanzania. J Wildl Dis 29:396–402
Gensini GF, Yacoub MH, Conti AA (2004) The concept of quarantine in history: from plague to SARS. J Infect 49:257–261
Glass GE, Johnson JS, Hodenbach GA, DiSalvo CLJ, Peters CJ, Childs JE, Mills JN (1997) Experimental evaluation of rodent exclusion methods to reduce hantavirus transmission to humans in rural housing. Am J Trop Med Hyg 56:359–364
Glass GE, Yates TL, Fine JB, Shields TM, Kendall JB, Hope AG, Parmenter CA, Peters CJ, Ksiazek TG, Li CS, Patz JA, Mills JN (2002) Satellite imagery characterizes local animal reservoir populations of Sin Nombre virus in the southwestern United States. Proc Natl Acad Sci U S A 99:16817–16822
Glass GE, Shields TM, Parmenter RR, Goade D, Mills JN, Cheek J, Cook J, Yates TL (2006) Predicted hantavirus risk in 2006 for the southwestern US Occas. Papers Mus Texas Tech Univ 255:1–16

Gordon ER, Curns AT, Krebs JW, Rupprecht CE, Real LA, Childs JE (2004) Temporal dynamics of rabies in a wildlife host and the risk of cross-species transmission. Epidemiol Infect 132:515–524

Gordon ER, Krebs JW, Rupprecht CE, Real LA, Childs JE (2005) Persistence of elevated rabies prevention costs following post-epizootic declines in rates of rabies among raccoons (*Procyon lotor*). Prev Vet Med 68:195–222

Gormley E, Costello E (2003) Tuberculosis and badgers: new approaches to diagnosis and control. J Appl Microbiol 94 Suppl:80S–86S

Gould AR, Hyatt AD, Lunt R, Kattenbelt JA, Hengstberger S, Blacksell SD (1998) Characterisation of a novel lyssavirus isolated from Pteropid bats in Australia. Virus Res 54:165–187

Griffin JM, Williams DH, Kelly GE, Clegg TA, O'boyle I, Collins JD, More SJ (2005) The impact of badger removal on the control of tuberculosis in cattle herds in Ireland. Prev Vet Med 67:237–266

Guan Y, Shortridge KF, Krauss S, Webster RG (1999) Molecular characterization of H9N2 influenza viruses: were they the donors of the "internal" genes of H5N1 viruses in Hong Kong? Proc Natl Acad Sci U S A 96:9363–9367

Guan Y, Shortridge KF, Krauss S, Chin PS, Dyrting KC, Ellis TM, Webster RG, Peiris M (2000) H9N2 influenza viruses possessing H5N1-like internal genomes continue to circulate in poultry in southeastern China. J Virol 74:9372–9380

Guan Y, Peiris M, Kong KF, Dyrting KC, Ellis TM, Sit T, Zhang LJ, Shortridge KF (2002a) H5N1 influenza viruses isolated from geese in Southeastern China: evidence for genetic reassortment and interspecies transmission to ducks. Virology 292:16–23

Guan Y, Peiris JS, Lipatov AS, Ellis TM, Dyrting KC, Krauss S, Zhang LJ, Webster RG, Shortridge KF (2002b) Emergence of multiple genotypes of H5N1 avian influenza viruses in Hong Kong SAR. Proc Natl Acad Sci U S A 99:8950–8955

Guan Y, Zheng BJ, He YQ, Liu XL, Zhuang ZX, Cheung CL, Luo SW, Li PH, Zhang LJ, Guan YJ, Butt KM, Wong KL, Chan KW, Lim W, Shortridge KF, Yuen KY, Peiris JS, Poon LL (2003) Isolation and characterization of viruses related to the SARS coronavirus from animals in southern China. Science 302:276–278

Guan Y, Poon LL, Cheung CY, Ellis TM, Lim W, Lipatov AS, Chan KH, Sturm-Ramirez KM, Cheung CL, Leung YH, Yuen KY, Webster RG, Peiris JS (2004) H5N1 influenza: a protean pandemic threat. Proc Natl Acad Sci U S A 101:8156–8161

Guarner J, Johnson BJ, Paddock CD, Shieh WJ, Goldsmith CS, Reynolds MG, Damon IK, Regnery RL, Zaki SR (2004) Monkeypox transmission and pathogenesis in prairie dogs. Emerg Infect Dis 10:426–431

Guerra MA, Curns AT, Rupprecht CE, Hanlon CA, Krebs JW, Childs JE (2003) Skunk and raccoon rabies in the eastern United States: temporal and spatial analysis. Emerg Infect Dis 9:1143–1150

Guptill SC, Julian KG, Campbell GL, Price SD, Marfin AA (2003) Early-season avian deaths from West Nile virus as warnings of human infection. Emerg Infect Dis 9:483–484

Hahn BH, Shaw GM, De Cock KM, Sharp PM (2000) AIDS as a zoonosis: scientific and public health implications. Science 287:607–614

Hall C (2004) Impact of avian influenza on US poultry trade relations-2002: H5 or H7 low pathogenic avian influenza. Ann N Y Acad Sci 1026:47–53

Halpin K, Young PL, Field HE, Mackenzie JS (2000) Isolation of Hendra virus from pteropid bats: a natural reservoir of Hendra virus. J Gen Virol 81:1927–1932

Hanlon CA, Rupprecht CE (1998) The reemergence of rabies. In: Scheld WM, Armstrong DA, Hughes JM (eds) Emerging infections. Eds. ASM Press, Washington, DC, pp 59–80

Hanlon CA, Niezgoda M, Hamir AN, Schumacher C, Koprowski H, Rupprecht CE (1998) First North American field release of a vaccinia-rabies glycoprotein recombinant virus. J Wildl Dis 34:228–239

Hatchette TF, Walker D, Johnson C, Baker A, Pryor SP, Webster RG (2004) Influenza A viruses in feral Canadian ducks: extensive reassortment in nature. J Gen Virol 85:2327–2337

Haydon DT, Cleaveland S, Taylor LH, Laurenson MK (2002) Identifying reservoirs of infection: a conceptual and practical challenge. Emerg Infect Dis 8:1468–1473

Hayne DW (1984) Population dynamics and analysis. In: halls LK (ed) White-tailed deer: ecology and management. Stackpole Books, Harrisburg, PA, pp 203–210

Heim D, Murray N (2004) Possibilities to manage the BSE epidemic: cohort culling versus herd culling – experiences in Switzerland. Contrib Microbiol 11:186–192

Heymann DL (2004) The international response to the outbreak of SARS in (2003) Philos Trans R Soc Lond B 359:1127–1129

Heymann DL, Rodier G (2004) Global surveillance, national surveillance, and SARS. Emerg Infect Dis 10:173–175

Hjelle B (2002) Vaccines against hantaviruses. Expert Rev Vaccines 1:373–384

Hjelle B, Glass GE (2000) Outbreak of hantavirus infection in the Four Corners region of the United States in the wake of the 1997–1998 El Nino-southern oscillation. J Infect Dis 181:1569–1573

Hooper PT, Lunt RA, Gould AR, Samaratunga H, Hyatt AD, Gleeson LJ, Rodwell BJ, Rupprecht CE, Smith JS, Murray PK (1997) A new lyssavirus-the first endemic rabies-related virus recognized in Australia. Bull Inst Past 95:209–218

Hooper PT, Selleck PW, Gleeson LJ, Daniels PW, Westbury HA, Murray PK (1998) Transmission studies of Hendra virus (equine morbillivirus) in fruit bats, horses and cats. Aust Vet J 76:813–818

Huang Y, Yang ZY, Kong WP, Nabel GJ (2004) Generation of synthetic severe acute respiratory syndrome coronavirus pseudoparticles: implications for assembly and vaccine production. J Virol 78:12557–12565

Huebner RJ, Jellison WL (1947) Rickettsialpox-a newly recognized rickettsial disease. V. Recovery of Rickettsia akari from a house mouse (*Mus musculus*). Pub Health Rep 62:777–780

Huebner RJ, Jellison WL, Pomerantz C (1946) Rickettsialpox – a newly recognized rickettsial disease. IV. Isolation of a rickettsia apparently identical with the causative agent of rickettsialpox from *Allodermanyssus sanguineus*, a rodent mite. Pub Health Rep 61:1677–1682

Ingram DG, Mitchell WR, Martin SW (1975) Animal disease monitoring. Charles C Thomas, Springfield

Institute of Medicine (1987) The US capacity to address tropical infectious disease problems. (1987) National Academy Press, Washington, DC
Institute of Medicine (1992) Emerging infections: microbial threats to health in the United States. National Academy Press, Washington, DC
Institute of Medicine (2003) Microbial threats to health; emergence, detection, and response. The National Academies Press, Washington, DC
James A (2005) The state of veterinary epidemiology and economics. Prev Vet Med 67:91–99
Jones ME, Curns AT, Krebs JW, Childs JE (2003) Environmental and human demographic features associated with epizootic raccoon rabies in Maryland, Pennsylvania, and Virginia. J Wildl Dis 39:869–874
Kat PW, Alexander KA, Smith JS, Munson L (1995) Rabies and African wild dogs in Kenya. Proc R Soc Lond B Biol Sci 262:229–233
Kellar JA, Lees VW (2003) Risk management of the transmissible spongiform encephalopathies in North America. Rev Sci Tech 22:201–225
Kemere PK, Liddel MK, Evangelou P, Slate D, Osmek S (2002) Economic analysis of a large scale oral vaccination program to control raccoon rabies. Clark L (ed) Proceedings Third NWRC Spec Symposium 2002. Ft. Collins, Colorado National Wildlife Research Center. USDA Human Conflicts with Wildlife: Economic Considerations, pp 109–116
Kermode-Scott B (2004) WHO confirms avian flu infections in Canada. BMJ 328:913
Kilpatrick AM, Kramer LD, Campbell SR, Alleyne EO, Dobson AP, Daszak P (2005) West Nile virus risk assessment and the bridge vector paradigm. Emerg Infect Dis 11:425–429
King LJ (1985) Unique characteristics of the national animal disease surveillance system. J Am Vet Med Assoc 186:35–39
Kirkpatrick JF, Turner JW Jr, Liu IK, Fayrer-Hosken R, Rutberg AT (1997) Case studies in wildlife immunocontraception: wild and feral equids and white-tailed deer. Reprod Fertil Dev 9:105–110
Kitala PM, McDermott JJ, Kyule MN, Gathuma JM (2000) Community-based active surveillance for rabies in Machakos District Kenya. Prev Vet Med 44:73–85
Kolar CS, Lodge DM (2001) Progress in invasion biology: predicting invaders. Trends Ecol Evol 16:199–204
Komar N (2001) West Nile virus surveillance using sentinel birds. Ann N Y Acad Sci 951:58–73
Koss T, Carter EL, Grossman ME, Silvers DN, Rabinowitz AD, Singleton J Jr, Zaki SR, Paddock CD (2003) Increased detection of rickettsialpox in a New York City hospital following the anthrax outbreak of 2001: use of immunohistochemistry for the rapid confirmation of cases in an era of bioterrorism. Arch Dermatol 139:1545–1552
Krauss S, Walker D, Pryor SP, Niles L, Chenghong L, Hinshaw VS, Webster RG (2004) Influenza A viruses of migrating wild aquatic birds in North America. Vect Borne Zoonot Dis 4:177–189
Krebs JW, Mandel EJ, Swerdlow DL, Rupprecht CE (2004) Rabies surveillance in the United States during 2003. J Am Vet Med Assoc 225:1837–1849

Ksiazek TG, Peters CJ, Rollin PE, Zaki S, Nichol S, Spiropoulou C, Morzunov S, Feldmann H, Sanchez A, Khan AS, Mahy BWJ, Wachsmuth K, Butler JC (1995) Identification of a new North American hantavirus that causes acute pulmonary insufficiency. Am J Trop Med Hyg 52:117–123

La BC, Calavas D, Abrial D, Morignat E, Ducrot C (2004) Estimating the trend of the French BSE epidemic over six birth cohorts through the analysis of abattoir screening in 2001 and 2002. Vet Res 35:299–308

Lake County General Health District News Release, July 27, OH

Lam SK, Chua KB (2002) Nipah virus encephalitis outbreak in Malaysia. Clin Infect Dis 34 [Suppl 2]: S48–S51

Lamphear BJ, Jilka JM, Kesl L, Welter M, Howard JA, Streatfield SJ (2004) A corn-based delivery system for animal vaccines: an oral transmissible gastroenteritis virus vaccine boosts lactogenic immunity in swine. Vaccine 22:2420–2424

Lanciotti RS, Roehrig JT, Deubel V, Smith J, Parker M, Steele K, Crise B, Volpe KE, Crabtree MB, Scherret JH, Hall RA, Mackenzie JS, Cropp CB, Panigrahy B, Ostlund E, Schmitt B, Malkinson M, Banet C, Weissman J, Komar N, Savage HM, Stone W, McNamara T, Gubler DJ (1999) Origin of the West Nile virus responsible for an outbreak of encephalitis in the northeastern United States. Science 286:2333–2337

Larkin M (2000) Hunting and logging linked to emerging infectious diseases. Lancet 356:1173

Lau YL (2004) SARS: future research and vaccine. Paediatr Respir Rev 5:300–303

LeDuc JW, Childs JE, Glass GE, Watson AJ (1993) Hantaan (Korean hemorrhagic fever) and related rodent zoonoses. In: Morse SS (ed) Emerging viruses.Oxford University Press, New York, pp 149–158

Lee CW, Senne DA, Suarez DL (2004) Effect of vaccine use in the evolution of Mexican lineage H5N2 avian influenza virus. J Virol 78:8372–8381

Leighton FA, Wobeser GA, Barker IK, Daoust PY, Martineau D (1997) The Canadian Cooperative Wildlife Health Centre and surveillance of wild animal diseases in Canada. Can Vet J 38:279–284

Leprince DJ, Lane RS (1996) Evaluation of permethrin-impregnated cotton balls as potential nesting material to control ectoparasites of woodrats in California. J Med Entomol 33:355–360

Leroy EM, Rouquet P, Formenty P, Souquiere S, Kilbourne A, Froment JM, Bermejo M, Smit S, Karesh W, Swanepoel R, Zaki SR, Rollin PE (2004) Multiple Ebola virus transmission events and rapid decline of central African wildlife. Science 303:387–390

Leroy EM, Kumulungui B, Pourrut X, Rouquet P, Hassanin A, Yaba P, Delicat A, Paweska JT, Gonzalez JP, Swanepoel R (2005) Fruit bats as reservoirs of Ebola virus. Nature 438:575–576

Li W, Shi Z, Yu M, Ren W, Smith C, Epstein JH, Wang H, Crameri G, Hu Z, Zhang H, Zhang J, McEachern J, Field H, Daszak P, Eaton BT, Zhang S, Wang LF (2005) Bats are natural reservoirs of SARS-like coronaviruses. Science 310:676–679

Li KS, Guan Y, Wang J, Smith GJ, Xu KM, Duan L, Rahardjo AP, Puthavathana P, Buranathai C, Nguyen TD, Estoepangestie AT, Chaisingh A, Auewarakul P, Long HT, Hanh NT, Webby RJ, Poon LL, Chen H, Shortridge KF, Yuen KY, Webster RG, Peiris JS (2004) Genesis of a highly pathogenic and potentially pandemic H5N1 influenza virus in eastern Asia. Nature 430:209–213

Li W, Shi Z, Yu M, Ren W, Smith C, Epstein JH, Wang H, Crameri G, Hu Z, Zhang H, Zhang J, McEachern J, Field H, Daszak P, Eaton BT, Zhang S, Wang LF (2005) Bats are natural reservoirs of SARS-like coronaviruses. Science 310:676–679

Linhart SB, King R, Zamir S, Naveh U, Davidson M, Perl S (1997) Oral rabies vaccination of red foxes and golden jackals in Israel: preliminary bait evaluation. Rev Sci Tech 16:874–880

Linhart SB, Wlodkowski JC, Kavanaugh DM, Motes-Kreimeyer L, Montoney AJ, Chipman RB, Slate D, Bigler LL, Fearneyhough MG (2002) A new flavor-coated sachet bait for delivering oral rabies vaccine to raccoons and coyotes. J Wildl Dis 38:363–377

Little SE, Stallknecht DE, Lockhart JM, Dawson JE, Davidson WR (1998) Natural coinfection of a white-tailed deer (*Odocoileus virginianus*) population with three *Ehrlichia* spp. J Parasitol 84:897–901

Lockhart JM, Davidson WR, Stallknecht DE, Dawson JE (1996) Site-specific geographic association between *Amblyomma americanum* (Acari: Ixodidae) infestations and *Ehrlichia chaffeensis*-reactive (Rickettsiales: Ehrlichieae) antibodies in white-tailed deer. J Med Entomol 33:153–158

Lockhart JM, Davidson WR, Stallknecht DE, Dawson JE, Howerth EW (1997) Isolation of *Ehrlichia chaffeensis* from wild white-tailed deer (*Odocoileus virginianus*) confirms their role as natural reservoir hosts. J Clin Microbiol 35:1681–1686

LoGiudice K, Ostfeld RS, Schmidt KA, Keesing F (2003) The ecology of infectious disease: effects of host diversity and community composition on Lyme disease risk. Proc Natl Acad Sci U S A 100:567–571

Lu X, Cho D, Hall H, Rowe T, Sung H, Kim W, Kang C, Mo I, Cox N, Klimov A, Katz J (2003) Pathogenicity and antigenicity of a new influenza A (H5N1) virus isolated from duck meat. J Med Virol 69:553–559

Luke CJ, Huebner RC, Kasmiersky V, Barbour AG (1997) Oral delivery of purified lipoprotein OspA protects mice from systemic infection with Borrelia burgdorferi. Vaccine 15:739–746

Lurz PW, Shirley MD, Shirley MD, Rushton SP (2002) Evaluation of immunocontraception as a publicly acceptable form of vertebrate pest species control: the introduced grey squirrel in Britain as an example. Environ Manage 30:342–351

Macinnes CD, Smith SM, Tinline RR, Ayers NR, Bachmann P, Ball DG, Calder LA, Crosgrey SJ, Fielding C, Hauschildt P, Honig JM, Johnston DH, Lawson KF, Nunan CP, Pedde MA, Pond B, Stewart RB, Voigt DR (2001) Elimination of rabies from red foxes in eastern Ontario. J Wildl Dis 37:119–132

Martinez-Burnes J, Lopez A, Medellin J, Haines D, Loza E, Martinez M (1997) An outbreak of vampire bat-transmitted rabies in cattle in northeastern Mexico. Can Vet J 38:175–177

Matsuoka Y, Chen H, Cox N, Subbarao K, Beck J, Swayne D (2003) Safety evaluation in chickens of candidate human vaccines against potential pandemic strains of influenza. Avian Dis 47:926–930

Mayen F (2003) Haematophagous bats in Brazil, their role in rabies transmission, impact on public health, livestock industry and alternatives to an indiscriminate reduction of bat population. J Vet Med B Infect Dis Vet Public Health 50:469–472

Melville DS, Shortridge KF (2004) Influenza: time to come to grips with the avian dimension. Lancet Infect Dis 4:261–262

Mencher JS, Smith SR, Powell TD, Stinchcomb DT, Osorio JE, Rocke TE (2004) Protection of black-tailed prairie dogs (*Cynomys ludovicianus*) against plague after voluntary consumption of baits containing recombinant raccoon poxvirus vaccine. Infect Immun 72:5502–5505

Messenger SL, Rupprecht CE, Smith JS (2003) Bats, emerging virus infections, and the rabies paradigm. In: Kunz TH, Fenton MB (eds) Bat ecology. University of Chicago Press, Chicago, pp 622–679

Miller LA, Killian GJ (2002) In search of the active PZP epitope in white-tailed deer immunocontraception. Vaccine 20:2735–2742

Mills CE, Robins JM, Lipsitch M (2004) Transmissibility of 1918 pandemic influenza. Nature 432:904–906

Mills JN, Childs JE (1998) Ecological studies of rodent reservoirs: their relevance for human health. Emerg Infect Dis 4:529–537

Mills JN, Childs JE, Ksiazek TG, Peters CJ, Velleca WM (1995) Methods for trapping and sampling small mammals for virologic testing. US Department of Health and Human Services, Atlanta

Mills JN, Ksiazek TG, Peters CJ, Childs JE (1999a) Long-term studies of hantavirus reservoir populations in the southwestern United States: a synthesis. Emerg Infect Dis 5:135–142

Mills JN, Yates TL, Ksiazek TG, Peters CJ, Childs JE (1999b) Long-term studies of hantavirus reservoir populations in the southwestern United States: rationale, potential, and methods. Emerg Infect Dis 5:95–101

Minke JM, Siger L, Karaca K, Austgen L, Gordy P, Bowen R, Renshaw RW, Loosmore S, Audonnet JC, Nordgren B (2004) Recombinant canarypoxvirus vaccine carrying the prM/E genes of West Nile virus protects horses against a West Nile virus-mosquito challenge. Arch Virol Suppl: 221–230

Monath TP, Arroyo J, Miller C, Guirakhoo F (2001) West Nile virus vaccine. Curr Drug Targets. Infect Disord 1:37–50

Moolenaar RL, Dalton C, Lipman HB, Umland ET, Gallaher M, Duchin JS, Chapman L, Zaki SR, Ksiazek TG, Rollin PE, Nichol S, Cheek JE, Butler JC, Peters CJ, Breiman RF (1995) Clinical features that differentiate hantavirus pulmonary syndrome from three other acute respiratory illnesses. Clin Infect Dis 21:643–649

Moran GJ, Talan DA, Mower W, Newdow M, Ong S, Nakase JY, Pinner RW, Childs JE (2000) Appropriateness of rabies postexposure prophylaxis treatment for animal exposures. Emergency ID Net Study Group. JAMA 284:1001–1007

Moreira ED Jr, Mendes DSV, Sreenivasan M, Nascimento EG, Pontes de CL (2004) Assessment of an optimized dog-culling program in the dynamics of canine Leishmania transmission. Vet Parasitol 122:245–252

Mörner T (2002) Health monitoring and conservation of wildlife in Sweden and Northern Europe. In: Gibbs EPJ, Bokma BH (eds) The domestic animal/wildlife interface: issues for disease control conservastion sustainable food and emerging diseases. Ann N Y Acad Sci New York, pp 34–38

Mörner T, Obendorf DL, Artois M, Woodford MH (2002) Surveillance and monitoring of wildlife diseases. Rev Sci Tech 21:67–76

Mostashari F, Kulldorff M, Hartman JJ, Miller JR, Kulasekera V (2003) Dead bird clusters as an early warning system for West Nile virus activity. Emerg Infect Dis 9:641–646
Mount GA, Biery TL, Haile DG (1996) A review of ultralow-volume aerial sprays of insecticide for mosquito control. J Am Mosq Control Assoc 12:601–618
Muller J (1971) The effect of fox reduction on the occurrence of rabies. Observations from two outbreaks of rabies in Denmark. Bull Off Int Epizoot 75:763–776
Myers TJ, Rhorer MD, Clifford J (2003) USDA options for regulatory changes to enhance the prevention and control of avian influenza. Avian Dis 47:982–987
Nalca A, Fellows PF, Whitehouse CA (2003) Vaccines and animal models for arboviral encephalitides. Antiviral Res 60:153–174
Nettles VF, Shaddock JH, Sikes RK, Reyes CR (1979) Rabies in translocated raccoons. Am J Pub Health 69:601–602
Nichol ST, Spiropoulou CF, Morzunov S, Rollin PE, Ksiazek TG, Feldmann H, Sanchez A, Childs JE, Zaki S, Peters CJ (1993) Genetic identification of a hantavirus associated with an outbreak of acute respiratory illness. Science 262:914–917
OIE (2003) Terrestrial Animal Health Code 2003.Paris World Organization for Animal Health (Office International des Epizooties). http//www.oie.int. Cited 28 February 2007
Olsen GH, Turell MJ, Pagac BB (1997) Efficacy of eastern equine encephalitis immunization in whooping cranes. J Wildl Dis 33:312–315
Olson CA, Werner PA (1999) Oral rabies vaccine contact by raccoons and nontarget species in a field trial in Florida. J Wildl Dis 35:687–695
Olson KE, Adelman ZN, Travanty EA, Sanchez-Vargas I, Beaty BJ, Blair CD (2002) Developing arbovirus resistance in mosquitoes. Insect Biochem Mol Biol 32:1333–1343
Osterhaus ADME, Fouchier RAM, Kuiken T (2004) The aetiology of SARS: Koch's postulates fulfilled. Philos Trans R Soc Lond B 359:1082
Ostfeld RS, Holt RD (2004) Are predators good for your health? Evaluating evidence for top-down regulation of zoonotic disease reservoirs. Front Ecol Environ 2:13–20
Ostfeld RS, Keesing F (2000) Biodiversity and disease risk: the case of Lyme disease. Conserv Biol 14:722–728
Oxford JS, Balasingam S, Chan C, Catchpole A, Lambkin R (2005) New antiviral drugs, vaccines and classic public health interventions against SARS coronavirus. Antivir Chem Chemother 16:13–21
Paddock CD, Zaki SR, Koss T, Singleton J Jr, Sumner JW, Comer JA, Eremeeva ME, Dasch GA, Cherry B, Childs JE (2003) Rickettsialpox in New York City: a persistent urban zoonosis. Ann N Y Acad Sci 990:36–44
Parry J (2004) WHO queries culling of civet cats. BMJ 328:128
Pastor-Anglada M, Cano-Soldado P, Molina-Arcas M, Lostao MP, Larrayoz I, Martinez-Picado J, Casado FJ (2005) Cell entry and export of nucleoside analogues. Virus Res 107:151–164
Paton NI, Leo YS, Zaki SR, Wong MC, Lee KE, Ling AE, Chew SK, Ang B, Rollin PE, Ksiazek TG, Auchus AP, Umapathi T, Sng I, Lee CC, Lim E, Kurup A, Lam MS, Wong SY (1999) Outbreak of Nipah virus infection among abattoir workers in Singapore: description of a new infectious disease. Lancet 354:1253–1256

Polis GA, Sears AL, Huxel GR, Strong DR, Maron J (2000) When is a trophic cascade a trophic cascade? Trends Ecol Evol 15:473–475

Rasgon JL, Styer LM, Scott TW (2003) *Wolbachia*-induced mortality as a mechanism to modulate pathogen transmission by vector arthropods. J Med Entomol 40:125–132

Real LE, Childs JE (2006) Rabies. In: Collinge SK, Ray C (eds) Disease ecology: community structure and pathogen dynamics. Oxford University Press, Oxford, pp 168–185

Reeves WC (1990) Epidemiology and control of mosquitoe-borne arboviruses in California, 1943–1987. California Mosquito and Vector Control Association Inc., Sacramento

Rocke TE, Mencher J, Smith SR, Friedlander AM, Andrews GP, Baeten LA (2004) Recombinant F1-V fusion protein protects black-footed ferrets (*Mustela nigripes*) against virulent *Yersinia pestis* infection. J Zoo Wildl Med 35:142–146

Root JJ, Black WC, Calisher CH, Wilson KR, Beaty BJ (2004) Genetic relatedness of deer mice (*Peromyscus maniculatus*) infected with Sin Nombre virus. Vector Borne Zoonot Dis 4:149–157

Roper TJ (2003) Ecology: badger cull culled. Nature 426:782–783

Rosatte RC, Pybus MJ, Gunson JR (1986) Population reduction as a factor in the control of skunk rabies in Alberta. J Wildl Dis 22:459–467

Roscoe DE, Holste WC, Sorhage FE, Campbell C, Niezgoda M, Buchannan R, Diehl D, Niu HS, Rupprecht CE (1998) Efficacy of an oral vaccinia-rabies glycoprotein recombinant vaccine in controlling epidemic raccoon rabies in New Jersey. J Wildl Dis 34:752–763

Rupprecht CE, Wiktor TJ, Johnston DH, Hamir AN, Dietzschold B, Wunner WH, Glickman LT, Koprowski H (1986) Oral immunization and protection of raccoons (*Procyon lotor*) with a vaccinia-rabies glycoprotein recombinant virus vaccine. Proc Natl Acad Sci U S A 83:7947–7950

Rupprecht CE, Hamir AN, Johnston DH, Koprowski H (1988) Efficacy of a vaccinia-rabies glycoprotein recombinant virus vaccine in raccoons (*Procyon lotor*). Rev Infect Dis 10 [Suppl 4]: S803–S809

Rupprecht CE, Blass L, Smith K, Orciari LA, Niezgoda M, Whitfield SG, Gibbons RV, Guerra M, Hanlon CA (2001) Human infection due to recombinant vaccinia-rabies glycoprotein virus. N Engl J Med 345:582–586

Russell CA, Smith DL, Waller LA, Childs JE, Real LA (2004) A priori prediction of disease invasion dynamics in a novel environment. Proc R Soc Lond B Biol Sci 271:21–25

Russell CA, Smith DL, Childs JE, Real LA (2005) Predictive spatial dynamics and strategic planning for raccoon rabies emergence in Ohio. PLoS Biol 3:1–7

Sabeta CT, Bingham J, Nel LH (2003) Molecular epidemiology of canid rabies in Zimbabwe and South Africa. Virus Res 91:203–211

Salman MD (2003) Surveillance and monitoring systems for animal health programs and disease surveys. In: Salman MD (ed) Animal disease surveillance and survey systems. Iowa State Press, Ames, IA, pp 3–13

Salman MD (2004) Controlling emerging diseases in the 21st century. Prev Vet Med 62:177–184

Scott TW, Takken W, Knols BG, Boete C (2002) The ecology of genetically modified mosquitoes. Science 298:117–119

Selvey LA, Wells RM, McCormack JG, Ansford AJ, Murray K, Rogers RJ, Lavercombe PS, Selleck P, Sheridan JW (1995) Infection of humans and horses by a newly described morbillivirus. Med J Aust 162:642–645
Shaman J, Day JF, Stieglitz M (2002) Drought-induced amplification of Saint Louis encephalitis virus Florida. Emerg Infect Dis 8:575–580
Shaw SE, Lerga AI, Williams S, Beugnet F, Birtles RJ, Day MJ, Kenny MJ (2003) Review of exotic infectious diseases in small animals entering the United Kingdom from abroad diagnosed by PCR. Vet Rec 152:176–177
Shortridge KF, Peiris JS, Guan Y (2003) The next influenza pandemic: lessons from Hong Kong. J Appl Microbiol 94 Suppl:70S–79S
Sillero-Zubiri C, King AA, Macdonald DW (1996) Rabies and mortality in Ethiopian wolves (*Canis simensis*). J Wildl Dis 32:80–86
Slate D, Rupprecht CE, Rooney JA, Donovan D, Lein DH, Chipman RB (2005) Status of oral rabies vaccination in wild carnivores in the United States. Virus Res 111:68–76
Slemons RD, Hansen WR, Converse KA, Senne DA (2003) Type A influenza virus surveillance in free-flying, nonmigratory ducks residing on the eastern shore of Maryland. Avian Dis 47:1107–1110
Smith DL, Waller LA, Russell CA, Childs JE, Real LA (2005) Assessing the role of long-distance translocation and spatial heterogeneity in the raccoon rabies epidemic in Connecticut. Prev Vet Med 71:225–240
Smith GC, Wilkinson D (2003) Modeling control of rabies outbreaks in red fox populations to evaluate culling, vaccination, and vaccination combined with fertility control. J Wildl Dis 39:278–286
Smith JS, Orciari LA, Yager PA (1995) Molecular epidemiology of rabies in the United States. Sem Virol 6:387–400
Song HD, Tu CC, Zhang GW, Wang SY, Zheng K, Lei LC, Chen QX, Gao YW, Zhou HQ, Xiang H, Zheng HJ, Chern SW, Cheng F, Pan CM, Xuan H, Chen SJ, Luo HM, Zhou DH, Liu YF, He JF, Qin PZ, Li LH, Ren YQ, Liang WJ, Yu YD, Anderson L, Wang M, Xu RH, Wu XW, Zheng HY, Chen JD, Liang G, Gao Y, Liao M, Fang L, Jiang LY, Li H, Chen F, Di B, He LJ, Lin JY, Tong S, Kong X, Du L, Hao P, Tang H, Bernini A, Yu XJ, Spiga O, Guo ZM, Pan HY, He WZ, Manuguerra JC, Fontanet A, Danchin A, Niccolai N, Li YX, Wu CI, Zhao GP (2005) Cross-host evolution of severe acute respiratory syndrome coronavirus in palm civet and human. Proc Natl Acad Sci U S A 102:2430–2435
Speakman J, Gonzalez-Martin F, Perez T (2003) Quarantine in severe acute respiratory syndrome (SARS) and other emerging infectious diseases. J Law Med Ethics 31:63–64
Stallknecht DE, Shane SM, Zwank PJ, Senne DA, Kearney MT (1990) Avian influenza viruses from migratory and resident ducks of coastal Louisiana. Avian Dis 34:398–405
Steelman HG, Henke SE, Moore GM (2000) Bait delivery for oral rabies vaccine to gray foxes. J Wildl Dis 36:744–751
Stefanak M, Vaughn KA, Shaheen JF (1999) Positive raccoon-strain rabies cases in Mahoning County Ohio, 1997. J Public Health Manage Pract 5:33–34

Stegeman A, Bouma A, Elbers AR, de Jong MC, Nodelijk G, de KF, Koch G, van Boven M (2004) Avian influenza A virus (H7N7) epidemic in The Netherlands in 2003: course of the epidemic and effectiveness of control measures. J Infect Dis 190:2088–2095
Stephenson JR (2001) Genetically modified viruses: vaccines by design. Curr Pharma Biotechnol 2:47–76
Stohr K (2003) The WHO Global Influenza Program and its Animal Influenza Network. Avian Dis 47:934–938
Sturm-Ramirez KM, Ellis T, Bousfield B, Bissett L, Dyrting K, Rehg JE, Poon L, Guan Y, Peiris M, Webster RG (2004) Reemerging H5N1 influenza viruses in Hong Kong in 2002 are highly pathogenic to ducks. J Virol 78:4892–4901
Summa ME, Carrieri ML, Favoretto SR, Chamelet EL (1987) Rabies in the State of Sao Paulo: the rodents question. Rev Inst Med Trop Sao Paulo 29:53–58
Swayne DE, Suarez DL, Schultz-Cherry S, Tumpey TM, King DJ, Nakaya T, Palese P, Garcia-Sastre A (2003) Recombinant paramyxovirus type 1-avian influenza-H7 virus as a vaccine for protection of chickens against influenza and Newcastle disease. Avian Dis 47:1047–1050
Tacket CO, Pasetti MF, Edelman R, Howard JA, Streatfield S (2004) Immunogenicity of recombinant LT-B delivered orally to humans in transgenic corn. Vaccine 22:4385–4389
Talan DA, Moran GJ, Mower WR, Newdow M, Ong S, Slutsker L, Jarvis WR, Conn LA, Pinner RW (1998) EMERGEncy ID NET: an emergency department-based emerging infections sentinel network. Ann Emerg Med 32:703–711
Tam JS (2002) Influenza A (H5N1) in Hong Kong: an overview. Vaccine 20 [Suppl 2]: S77–S81
Tappero J, Khan A, Pinner R, Wenger J, Graber J, Armstrong L, Holman R, Ksiazek T, Khabbaz R (1996) Utility of emergency, telephone-based national surveillance for hantavirus pulmonary syndrome. JAMA 275:398–400
Tesh RB, Arroyo J, Travassos da Rosa AP, Guzman H, Xiao SY, Monath TP (2002) Efficacy of killed virus vaccine, live attenuated chimeric virus vaccine, and passive immunization for prevention of West Nile virus encephalitis in hamster model. Emerg Infect Dis 8:1392–1397
Teutsch SM (2000) Considerations in planning a surveillance system. In: Teutsch SM, Churchill RE (eds) Principles and practice of public health surveillance. Oxford University Press, Oxford, pp 17–29
Thacker SB (2000) Historical development. In: Teutsch SM, Churchill RE (eds) Principles and practice of public health. Eds. Oxford University Press, Oxford, pp 1–16
Theiler M, Downs WG (1973) The Arthropod-borne viruses of vertebrates: an account of the Rockefeller Foundation Virus Program, 1951–1970. Yale University Press, New Haven
Thier A (2001) Balancing the risks: vector control and pesticide use in response to emerging illness. J Urban Health 78:372–381
Thiermann AB (2003) The OIE role in control of List A and emerging diseases. Dev Biol (Basel) 114:1–4

Thompson RD, Mitchell GC, Burns RJ (1972) Vampire bat control by systemic treatment of livestock with an anticoagulant. Science 177:806–808

Tomori O (2002) Yellow fever in Africa: public health impact and prospects for control in the 21st century. Biomed 22:178–210

Torrence ME, Jenkins SR, Glickman LT (1992) Epidemiology of raccoon rabies in Virginia, 1984 to 1989. J Wildl Dis 28:369–376

Tran TH, Nguyen TL, Nguyen TD, Luong TS, Pham PM, Nguyen VC, Pham TS, Vo CD, Le TQ, Ngo TT, Dao BK, Le PP, Nguyen TT, Hoang TL, Cao VT, Le TG, Nguyen DT, Le HN, Nguyen KT, Le HS, Le VT, Christiane D, Tran TT, Menno de J, Schultsz C, Cheng P, Lim W, Horby P, Farrar J (2004) Avian influenza A (H5N1) in 10 patients in Vietnam. N Engl J Med 350:1179–1188

Travassos da Rosa AP, Mather TN, Takeda T, Whitehouse CA, Shope RE, Popov VL, Guzman H, Coffey L, Araujo TP, Tesh RB (2002) Two new rhabdoviruses (Rhabdoviridae) isolated from birds during surveillance for arboviral encephalitis, northeastern United States. Emerg Infect Dis 8:614–618

Tsao JI, Wootton JT, Bunikis J, Luna MG, Fish D, Barbour AG (2004) An ecological approach to preventing human infection: vaccinating wild mouse reservoirs intervenes in the Lyme disease cycle. Proc Natl Acad Sci U S A 101:18159–18164

Tumpey TM, Suarez DL, Perkins LE, Senne DA, Lee J, Lee YJ, Mo IP, Sung HW, Swayne DE (2003) Evaluation of a high-pathogenicity H5N1 avian influenza A virus isolated from duck meat. Avian Dis 47:951–955

Turell MJ, Ludwig GV, Kondig J, Smith JF (1999) Limited potential for mosquito transmission of genetically engineered, live-attenuated Venezuelan equine encephalitis virus vaccine candidates. Am J Trop Med Hyg 60:1041–1044

Ueba N, Kimura T, Nakajima S, Kurimura T, Kitaura T (1978) Field experiments on live attenuated Japanese encephalitis virus vaccine for swine. Biken J 21:95–103

Uppal PK (2000) Emergence of Nipah virus in Malaysia. Ann N Y Acad Sci 916:354–357

Van Heerden J, Bingham J, van Vuuren M, Burroughs RE, Stylianides E (2002) Clinical and serological response of wild dogs (*Lycaon pictus*) to vaccination against canine distemper, canine parvovirus infection and rabies. J South Afr Vet Assoc 73:8–12

Verlinde JD, Li-Fo-Sjoe E, Versteeg J, Dekker SM (1975) A local outbreak of paralytic rabies in Surinam children. Trop Geogr Med 27:137–142

Von Overbeck J (2003) Insurance and epidemics: SARS, West Nile virus and Nipah virus. J Insur Med 35:165–173

Walsh PD, Abernethy KA, Bermejo M, Beyers R, De Wachter P, Akou ME, Huijbregts B, Mambounga DI, Toham AK, Kilbourn AM, Lahm SA, Latour S, Maisels F, Mbina C, Mihindou Y, Obiang SN, Effa EN, Starkey MP, Telfer P, Thibault M, Tutin CE, White LJ, Wilkie DS (2003) Catastrophic ape decline in western equatorial Africa. Nature 422:611–614

Warfield KL, Bosio CM, Welcher BC, Deal EM, Mohamadzadeh M, Schmaljohn A, Aman MJ, Bavari S (2003) Ebola virus-like particles protect from lethal Ebola virus infection. Proc Natl Acad Sci U S A 100:15889–15894

Watts J (2004a) Asian nations step up action to curb spread of avian influenza. Outbreak is spreading at an unprecendented speed WHO says, and nowhere can be considered safe. Lancet 363:373

Watts J (2004b) China culls wild animals to prevent new SARS threat. Lancet 363:134

Weaver SC, Ferro C, Barrera R, Boshell J, Navarro JC (2004) Venezuelan equine encephalitis. Annu Rev Entomol 49:141–174

Webster RG, Guan Y, Peiris M, Walker D, Krauss S, Zhou NN, Govorkova EA, Ellis TM, Dyrting KC, Sit T, Perez DR, Shortridge KF (2002) Characterization of H5N1 influenza viruses that continue to circulate in geese in southeastern China. J Virol 76:118–126

Weiler GJ, Garner GW, Ritter DG (1995) Occurrence of rabies in a wolf population in northeastern Alaska. J Wildl Dis 31:79–82

Whitby JE, Johnstone P, Sillero-Zubiri C (1997) Rabies virus in the decomposed brain of an Ethiopian wolf detected by nested reverse transcription-polymerase chain reaction. J Wildl Dis 33:912–915

White PC, Benhin JK (2004) Factors influencing the incidence and scale of bovine tuberculosis in cattle in southwest England. Prev Vet Med 63:1–7

Williams ES, Miller MW (2003) Transmissible spongiform encephalopathies in non-domestic animals: origin, transmission and risk factors. Rev Sci Tech 22:145–156

Wilson ML, Bretsky PM, Cooper GH Jr, Egbertson SH, Van Kruiningen HJ, Cartter ML (1997) Emergence of raccoon rabies in Connecticut, 1991–1994: spatial and temporal characteristics of animal infection and human contact. Am J Trop Med Hyg 57:457–463

Winkler WG, Schneider NJ, Jennings WL (1972) Experimental rabies infection in wild rodents. J Wildl Dis 8:99–103

Wobeser G (2002) Disease management strategies for wildlife. Rev Sci Tech 21:159–178

Wolfe ND, Switzer WM, Carr JK, Bhullar VB, Shanmugam V, Tamoufe U, Prosser AT, Torimiro JN, Wright A, Mpoudi-Ngole E, McCutchan FE, Birx DL, Folks TM, Burke DS, Heneine W (2004) Naturally acquired simian retrovirus infections in central African hunters. Lancet 363:932–937

Woolhouse ME (2003) Foot-and-mouth disease in the UK: what should we do next time? J Appl Microbiol 94 Suppl: 126S–130S

Wu CJ, Huang HW, Liu CY, Hong CF, Chan YL (2005) Inhibition of SARS-CoV replication by siRNA. Antiviral Res 65:45–48

Xu G, Xu X, Li Z, He Q, Wu B, Sun S, Chen H (2004) Construction of recombinant pseudorabies virus expressing NS1 protein of Japanese encephalitis (SA14-2) virus and its safety and immunogenicity. Vaccine 22:1846–1853

Yates TL, Mills JN, Parmenter CA, Ksiazek TG, Parmenter RR, Castle V, Calisher CH, Nichol ST, Abbott JC, Young ML, Morrison BJ, Beaty BJ, Dunnum JL, Baker RJ, Salazar-Bravo J, Peters CJ (2002) The ecology and evolutionary history of an emergent disease: Hantavirus pulmonary syndrome. Bioscience 52:989–998

Yiming L, Dianmo L (1998) The dynamics of trade in wildlife across the Guangxi border between China and Vietnam during 1993–1996 and its control strategies. Biodivers Conserv 7:895–914

Zepeda C, Salman M, Thiermann A, Kellar J, Rojas H, Willeberg P (2005) The role of veterinary epidemiology and veterinary services in complying with the World Trade Organization SPS agreement. Prev Vet Med 67:125–140

Zhong N (2004) Management and prevention of SARS in China. Philos Trans R Soc Lond B Biol Sci 359:1115–1116

Zhuang J, Jetzt AE, Sun G, Yu H, Klarmann G, Ron Y, Preston BD, Dougherty JP (2002) Human immunodeficiency virus type 1 recombination: rate, fidelity, and putative hot spots. J Virol 76:11273–11282

CTMI (2007) 315:445–461
© Springer-Verlag Berlin Heidelberg 2007

Impediments to Wildlife Disease Surveillance, Research, and Diagnostics

D. E. Stallknecht (✉)

Southeastern Cooperative Wildlife Disease Study, Department of Population Health, College of Veterinary Medicine, The University of Georgia, Athens, GA 30605, USA
dstall@vet.uga.edu

Abstract There is a recognized need for increased wildlife disease surveillance and research related to understanding the epidemiology and control of emerging wildlife and zoonotic diseases. Although both passive and active surveillance strategies can and have been effectively used with wildlife, some unique problems are often encountered. These can include limitations related to case acquisition and under-reporting, difficulty in designing sampling strategies that adequately represent the population of interest, the lack of properly validated diagnostic tests, problems related to data interpretation due to missing or inaccurate denominator data, and the lack of an existing wildlife surveillance infrastructure. Many of these same problems are often encountered in field research, which can be further complicated by the complexity and scale of the natural systems in which this work takes place. Although such studies may be difficult, there are numerous examples of success and our understanding of wildlife and wildlife-related zoonotic and emerging disease continues to grow.

1 Introduction

Although there is a recognized need for wildlife disease surveillance and research related to understanding the epidemiology of both wildlife and emerging zoonotic diseases, the challenges associated with such work can be daunting. Such challenges not only relate to the practicality of case, sample, and field data collection, but also to the interpretation of field data and the validation of field observations through experimental studies. This chapter is not meant to discourage such work. Rather, it is hoped that it will provide some guidance in eliminating or negotiating potential problems associated with an increasing need for surveillance and epidemiologic studies involving wildlife populations.

Wildlife disease surveillance and research, relative to emerging diseases, are linked in a progression of activity, moving from disease discovery to the implementation of disease control measures. This process can start with the detection of a unique disease through a diagnostic case submission, as was the case with chronic wasting disease (CWD) (Williams and Young 1980); from unexplained population declines, as occurred with amphibian declines and the resulting discovery of chytridiomycosis (Berger et al. 1998); or through field epidemiologic studies supporting investigations of newly described zoonotic (or domestic animal) diseases, as in the cases of Sin Nombre virus (Ksiazek et al. 1995) and Hendra virus (Young et al. 1996). From this point, traditional epidemiologic goals follow, relative to describing and understanding natural history and epidemiology and development of control measures. It is important to note that this epidemiologic process cannot stand alone, and needs ongoing support from a diversity of disciplines dedicated to improved diagnostics and our basic understanding of pathogenesis and molecular biology.

2 Wildlife Disease Surveillance

Epidemiologic surveillance is defined as the ongoing, systematic, and continuous collection, analysis, and interpretation of health data (Toma et al. 1990). For an infectious disease surveillance system to be effective, it must provide a high probability of capturing an infected individual as soon as possible and must incorporate diagnostic technologies that maximize the probability of detecting the agent (Thurmond 2003). These requirements often are difficult to fulfill in relation to wildlife diseases or zoonotic diseases involving wildlife populations and must be viewed in the context of the varied objectives and value that can be associated with wildlife surveillance systems. Objectives can be time-sensitive, if related to a need for an immediate response to disease detection, such as would occur with human exposure to a rabid wild animal or the detection of West Nile virus (WNV) as part of a dead bird surveillance program. In contrast, objectives may relate to more long-term goals, such as understanding causes of mortality and morbidity in wildlife populations, defining spatial and temporal disease patterns and species susceptibility, evaluating diagnostic techniques, or problem definition. In addition, wildlife disease surveillance provides a unique opportunity to obtain native biological materials, such as wild type field isolates, that may be fundamental to pathogen discovery or understanding pathogen phylogenetics and disease emergence. For these reasons, there is no single formula for a successful wildlife surveillance program, and individual approaches should be carefully constructed to meet defined objectives under the very real constraints of practicality.

2.1
Case and Sample Acquisition

As with public and domestic animal health surveillance systems, case and sample material can be obtained passively through a utilization of existing data sources and infrastructure, or actively through investigator-driven data collection designed to meet specific information needs. The first obstacle that will be encountered with either approach is the acquisition of representative cases or samples.

2.1.1
Passive Surveillance Systems

The central problem associated with a wildlife disease surveillance system that is dependent on diagnostic submissions relates to detection of naturally occurring mortality and morbidity, that is, the identification and submission of a case or cases to the diagnostic facility. Wildlife case submissions are dependent on a complex of interrelated natural and decision-making outcomes. For a case to reach a diagnostic laboratory, it must persist in the environment, it must be detected, it must be reported in a timely manner, and it must be delivered to that facility. This chain, of often low probability events, is generally dependent on initial detection by the public. It can be time-sensitive and species-dependent, and often there is little or no incentive for either reporting or submission.

The extent of potential underreporting is poorly documented but can occur even with large and very abundant wildlife species such as white-tailed deer (*Odocoileus virginianus*). For example, during a focal hemorrhagic disease outbreak in white-tailed deer in Missouri, a mortality rate of approximately 8% was estimated based on observed mortality in 100 radio-monitored animals (Berringer et al. 2000). During the course of this outbreak, not a single case of mortality of morbidity was reported by the public and under normal circumstances this outbreak would have remained undetected.

There is little available information relating to potential impediments or biases associated with case submissions, and such information is limited to a small number of studies on carcass persistence and the probability of carcass detection. Carcass persistence can be species-dependent, especially in relation to size or gender (in the case of sexual dimorphisms such as occurs in the plumage of birds) and can be surprisingly short-term (Wobeser 1994). For example, it has been estimated that more than 75% of passerine bird carcasses may be naturally removed from the field within 24 h (Wobeser 1992). Even with persistence in the environment, mortality can be very difficult to detect, as is evident from a controlled study of waterfowl carcass detection in a Texas wetland where only 12% of placed birds were detected in active searches (Stutzenbaker et al. 1986). Potential problems relative to reporting bias (the probability that a found case will be reported to the diagnostic laboratory) also need to be considered, but currently, there is no information available to help one navigate through this unknown. As previously stated, the detection and reporting of wildlife mortality and morbidity is dependent on public interest, and this is likely controlled by a variety of interacting variables ranging from perceived value of individual species to a perceived need for such a submission. This perceived need may be enhanced during large-scale mortality events as opposed to observations involving individual mortality. As evident from the large number of wild bird submissions relating to the introduction of WNV into the United States, some of these potential problems can be overcome by broad-based or directed public education. Effective public case reporting associated with the emergence of *Mycoplasma gallisepticum* in house finches through a created feeder watch program also exemplifies the potential utility of such a public-supported approach (Dondt et al. 1998).

Submission related to morbidity can also be problematic as the actual recognition of disease, capture, and containment of the moribund animal represents additional links in the submission chain. However, such animals are being submitted to a growing number of wildlife rehabilitation centers, and with the incorporation of appropriate diagnostics, such cases could greatly enhance surveillance efforts (Kelly and Sleeman 2003). The initial detection of *Mycoplasma gallisepticum* in house finches was attributable to cases submitted to wildlife rehabilitation centers (Ley et al. 1996).

The potential value of passive surveillance should not be diminished by these impediments, as these systems provide an ideal setting for disease discovery, especially in relation to emerging wildlife diseases. Some important examples of such discoveries in the United States alone include CWD (Williams and Young 1980), bovine tuberculosis in white-tailed deer in Michigan (Schmitt et al. 1997), avian vacuolar myelinopathy in bald eagles (*Haliaeetus leucocephalus*) and American coots (*Fulica americana*) (Thomas et al. 1998), adenovirus hemorrhagic disease in black-tailed deer (*Odocoileus hemionus columbianus*) in California (Woods et al. 1996), and WNV in North American birds (Steele et al. 2000).

2.1.2
Active Surveillance

Active surveillance, in this discussion, is defined as investigator-driven data collection designed to meet specific information needs. Such needs may relate to large-scale questions, such as determining the presence (or absence) of a pathogen in a given species or geographic area, or they can be very focused and directed at individual populations. In the latter case, the desired data most often relates to defining specific risk factors associated with disease or antibody prevalence. Traditionally, active surveillance has relied heavily on a cross-sectional study design. This design is well suited for wildlife disease studies, as it requires only a point sample (single capture or sample collection) in which data on potential risk factors and disease and infection status are collected. However, sample collection, while under the control of the investigator, can provide a major obstacle to this type of study both in relation to cost and feasibility. Ideally, samples should include adequate numbers for pathogen and antibody detection, reliable prevalence estimates, or statistical analysis. Samples should also be representative of the study population in question and incorporate relevant biological, spatial, and temporal variables. These ideal conditions are not often achieved and sample collection usually represents a compromise associated with availability. For example, readily available sources of convenience samples such as hunter-killed animals, animals captured for marking or banding studies, animals killed by vehicles on highways, or removed nuisance animals are often targeted for surveillance. Finally, it should be noted that with some wildlife species, either related to population numbers, biology, or a lack of capture technology, reasonable sampling through direct capture methods may be extremely difficult. In this case, indirect sampling methods, such as fecal sampling for avian influenza virus (AIV) (Kawaoka et al. 1988), may have utility.

As with all epidemiologic studies, a potential problem associated with wildlife sampling relates to samples being representative of the population. With hunter-killed samples, sample collection is temporally and spatially restricted, and available animals may not be representative of either the age or gender structure of the population. Although temporal and age effects may be somewhat reduced in serologic-based studies, provided that an antibody response persists, they still may effect results. Spatially, variation in local population density may not be adequately incorporated into the sampling scheme. Finally, there may be an inherent sampling bias associated with the capture and submission of species or individual animals. For example, house finches affected with *M. gallisepticum* are lethargic and often have an affinity for bird feeders. Animal killed on the highways may be predisposed to this outcome based on other problems affecting behavior or condition. Animals sampled from hunters are often selected based on purely nonrandom criteria such as antler development or as directed by specific hunting regulations.

The quality of basic biological data collected in such studies also can be problematic. Age is often a critical variable in these types of studies, and these data are often unavailable or incomplete due to a lack of defined age criteria for many wildlife species. Even when available, these age criteria often allow only gross categorization limited to classifying an individual to only juvenile and adult categories. In such a case, the resulting lack of precision in the adult age class determinations can cause serious problems when comparing prevalence estimates between or within populations. Location data relative to point of capture and sampling also can be problematic, especially in species that exhibit extended home ranges or migratory behaviors. These potential problems should not be viewed as insurmountable, but it is extremely important to consider them in study design, data analysis, and the interpretation of results.

Another type of active investigation that is commonly utilized is the outbreak investigation. Outbreak investigations are usually driven by high-mortality and morbidity events and are subject to the same impediments described for active and passive surveillance.

2.2 Availability and Applicability of Diagnostic Tests

The availability of reliable diagnostic assays is often a problem, and even when available, data are often lacking in relation to species-specific test performance. Diagnostic tests can be grouped into two broad categories: those designed for antibody detection and those designed for pathogen and nucleic acid detection. In the latter case, these can be further divided into direct detection through visualization (applicable to such things as ecto- and endo-parasites), culture, immune capture or staining, and nucleic acid detection.

Two major considerations relative to serologic test performance and interpretation include field specificity and sensitivity. An understanding of test sensitivity and specificity is fundamental to data interpretation, but these estimates may not adequately describe the field performance of the assay. In most cases, estimates of test performance are derived from acute phase samples, and from experimental infections that may or may not include the exact species under study. The persistence of antibodies following infection most often will be unknown and will vary with test formats. All serologic tests need to be supported with species-specific positive and negative control serum samples, and these basic samples may be difficult to acquire. In addition, some test formats, such as the indirect ELISA, rely on species-specific antisera, which also may be unavailable.

As for specificity, most serologic tests are experimentally evaluated with a panel of known antisera for related and/or unrelated pathogens. While such evaluations are extremely important, it is unlikely that even the best of these evaluations will cover the multitude of potential pathogens (some unknown) that may be encountered in the field. In short, every attempt should be made to validate serologic test performance prior to field utilization and this can best be achieved through a combination of long-term experimental studies supported by reliable field data (integrated diagnostics). Serologic assays have a major application to both wildlife disease surveillance and research, but results, especially in cases with low prevalence estimates, should be highly scrutinized and always approached with some caution.

For pathogen detection from field-collected samples, similar problems with test sensitivity and specificity can exist. In this case, target degradation presents an additional influence on test sensitivity. This can be especially true in the case of delayed diagnostic submissions. Another source of potential variation relates to target tissue, as optimum tissues or samples for pathogen detection may vary between species or between systems. A recent example of such variation relates to the detection of pseudorabies virus in feral swine, which can be routinely isolated only from genital swabs rather than the tonsil or nasal swabs that prove effective with domestic swine (Romero et al. 2001). In addition, the ability to detect a pathogen may be influenced by species-specific variation in pathogenesis as in the case of WNV where viral loads vary greatly between bird species (Komar et al. 2003). Such species-related variation may be especially important when utilizing diagnostic tools with limited sensitivity such as the Vec Test (Vec Test, Medical Analysis Systems, Camarillo, CA), which is currently being utilized for rapid WNV diagnosis (Stone et al. 2004). Potential problems related to specificity, as with serologic testing, can occur with cross-reactive antigens in immune-based diagnostics such as IHC and in sequence similarities with PCR-based diagnostics. An example of such a problem was reported in a study of *Ehrlichia* in white-tailed deer where a PCR protocol designed for the human granulocytic agent (now *Anaplasma phagocytophila*) resulted in a similar molecular weight amplicon that was produced by the presence of a recently discovered *Anaplasma* sp. of deer (Little et al. 1998). With PCR-based diagnostics, especially those targeting highly conserved regions such as 16S rDNA, amplicon identification through sequence confirmation should always be included. As with the validation of serologic methods, results from representative experimental infections can greatly assist in data interpretation. Multiple test formats such as utilizing culture and PCR or multiple PCR assays designed for different targets also can minimize problems relative to diagnostic test performance.

2.3 Considerations Regarding Data Interpretation

2.3.1 The Missing Denominator

In wildlife disease surveillance, whether through a passive or active system, the lack of reliable population data may represent the major impediment to data interpretation. It is a misconception that population information is readily available for our free-ranging wildlife species. In fact, even for very abundant and economically important species, such as white tailed deer, these estimates often represent nothing more than an educated guess. In addition, population data often are in the form of an index rather than a true estimate. An example of this can be seen with avian population data generated by the Christmas bird counts in North America. Such indices can be used to demonstrate trends, but give no information on population numbers or density. In the case of density, the spatial distribution of the population (which can be greatly influenced by behavior and habitat quality and diversity), may greatly influence this potentially important variable. This elusive information can present a major difficulty in the interpretation of mortality and morbidity data. In addition, it provides a problem when comparing results over time or between different populations, and unfortunately this often is not considered in such comparisons.

2.3.2 Negative Data Are Important and Will Be Greatly Underrepresented in the Literature

The acquisition of negative data certainly does not represent the goal of most wildlife disease surveillance programs, but these data can be extremely important to our understanding of wildlife and zoonotic diseases. Although negative data often are not published and tend to get lost, this information can be extremely important in reservoir determination and in risk assessments. For example, there have been many isolates of AIV recovered from species of shorebirds utilizing Delaware Bay during spring migration, and this has been interpreted to indicate that shorebirds represent a major reservoir for these viruses (Kawaoka et al. 1988). This perspective changes when the negative data from these species are examined, and in fact, the high prevalence of AIV infection in shorebirds at Delaware Bay currently is unique, and is probably driven by a high infection rate in only one species, the ruddy turnstone (*Arenaria interpres*) (Stallknecht 1997). These perspectives are very different, and such differences can greatly influence subsequent study design, interpretation of field data, and the development of disease management options.

2.3.3 Problems with Problem Definition

Important goals of wildlife research and surveillance are to identify risk factors, transmission and maintenance cycles, and in the case of wildlife disease, population impacts. The complexity of natural ecosystems coupled with anthropogenic impacts within these systems results in many interactive and sequential events that can lead to the emergence of both zoonotic and wildlife disease. For this reason, such events can be very difficult to define and predict. An example of this can be seen with the emergence of *M. gallisepticum* in house finches. This disease emergence probably represents the culmination of a series of events that included the release of the house finch to the eastern United States as a result of newly imposed federal legislation that restricted an attempt to commercialize this species as a pet; the establishment of this invasive species within the eastern US as a result of major habitat changes related to urbanization and the growing popularity of bird feeding; and an opportunity for this expanding population to eventually interact with *M. gallisepticum* -infected birds, presumably infected free-ranging or commercial poultry (Fischer et al. 1997). This entire chain of events covers close to 60 years and includes many seemingly unrelated factors.

Population impacts associated with new diseases also can be problematic as exemplified by recent attempts to document WNV effects on wild avian populations. Although it is well established that WNV has caused significant mortality in wild birds in North American, detection of population impacts have ranged from negligible (Caffrey and Peterson 2003), inconclusive (Hochachka et al. 2004), to locally extreme (Yaremych et al. 2004), depending on data sources, scale, species, and study design.

2.4 Surveillance Infrastructure

As previously stated, passive surveillance systems related to wildlife diseases have inherent problems with detectability and reporting. These problems are exacerbated by a relative scarcity (worldwide) of diagnostic and research labs dedicated to wildlife disease. Diagnostic submissions are a basic approach to wildlife disease discovery and are critical to the detection of emerging wildlife diseases. Diagnostic-based surveillance for rabies and WNV in the United States demonstrates the potential gains associated with wildlife surveillance when an adequate infrastructure is in place. As previously stated, however, the success of these surveillance networks relates to a targeted approach and the support of a diagnostic infrastructure dedicated to detection of these specific diseases. The WNV dead bird surveillance program provides a relevant example of both the success and potential limitations of such a targeted approach. The goal of this surveillance clearly was related to public health, and the early detection of WNV in birds was effectively used as an indicator of impending human risk (Guptill et al. 2003). The surveillance system that evolved primarily utilized corvids and rapid diagnostic tools, such as PCR and immunohistochemistry, which were limited to WNV detection. In retrospect, although much was gained from this program, a wealth of potential data relating to other avian diseases and even the presence of other arboviruses associated with these samples was not generated due to restricted submissions and diagnostics that efficiently targeted WNV.

3 Wildlife Disease Research

For the purpose of this discussion, wildlife disease research has been categorized into two broad categories. The first involves field epidemiology. The second approach relates to experimental infections directed at improving an understanding of pathogenesis and/or the validation of diagnostic tests. With few exceptions, most of the previously discussed impediments to wildlife surveillance also have application to wildlife research.

3.1 Epidemiologic Studies

Standard epidemiologic approaches, such as cohort studies and ecological studies, have been used effectively in the study of wildlife diseases. A unique aspect and challenge to epidemiologic studies involving wildlife, however, relates to the need to integrate the collection of both disease and basic biological data. Unlike human and domestic animal populations, population and basic biological data are often lacking and must be actively collected. Another challenge, related to the practicality of field wildlife research, involves sample collection, including both animal capture and availability of samples that do not compromise the health of the captured animal. Capture options often are limited and may be extremely demanding in both time and financial expenditures. Diagnostic options also may be limited depending on conditions under which an animal is sampled.

3.1.1
Biological Considerations

Population data needed in an epidemiologic study may include population size, density, age structure, sex ratio, recruitment and attrition, home range, habitat utilization, and species composition, among others. Such information often is critical to understanding pathogen transmission and maintenance within wildlife populations. The natural history of AIV in wild ducks and geese in North America provides a relevant example. The highest prevalence of AIV infection in ducks (which sometimes exceeds 30%) is associated with mallards (*Anas platyrhynchos*) in late summer and early fall (Hinshaw et al. 1980). This temporal relationship is driven by behavior and population structure and is related to the gathering of large numbers of juvenile birds (that year's reproduction) at a premigration staging area. In contrast, AIV prevalence in blue-winged teal (*Anas discors*) populations is relatively low at this time of year (Stallknecht et al. 1990). This species-related difference in prevalence probably is linked to migratory behavior, as early migrating teal have already arrived on wintering habitats in August and September and escape these annual epidemics on the staging areas.

Biological traits related to the species or population can also represent a relevant tool for hypothesis testing in epidemiologic studies and should always be considered in study design. For example, the detection of AIV from mottled ducks (*Anas fulvigula*), which is a nonmigratory resident species in the southern United States, documented AIV transmission on waterfowl wintering areas (Stallknecht et al. 1990). The detection of a pathogen or antibodies in a wildlife species in conjunction with knowledge related to home range (as measured directly through mark-recapture or inferred through results of previous studies) can also be used to effectively delineate the point of transmission to often small areas. Home ranges can be very restricted, especially with small rodents, and in Hantaviurs-related studies, this biological attribute has been used effectively to document risk factors such as specific habitat types and population dynamics (Calisher et al. 2001) and gene flow within these populations (Root et al. 2003).

The need for this biological data can be met through the integration of standard wildlife biology techniques into field epidemiology. Both are field-oriented and population-based sciences and the extensive degree of overlap between these disciplines allow for very effective collaborations.

3.1.2 Temporal and Spatial Variation

The possibility of temporal and spatial variation always needs to be considered in both study design and data interpretation and such variation can provide very relevant information regarding pathogen maintenance and transmission mechanisms. Both seasonal and secular variation related to changes in climatic, vector and host population dynamics, or host population immunity are commonly encountered in epidemiologic studies involving wildlife but often are not adequately addressed due to the need for extended long-term studies. Likewise, spatial distribution is especially important when the targeted disease or infection is highly clustered. An example of the importance of both temporal and spatial variation is evident with the natural history of vesicular stomatitis virus (VSV) on Ossabaw Island, Georgia, which is the only known focus of this virus in the United States. This focus was initially discovered through serologic testing of feral swine, and as with other vector-borne diseases, VSV transmission is seasonal and occurs in late spring and summer (Stallknecht et al. 1987). This seasonality results in variation in antibody prevalence estimates in young pigs that can range in a given year from 0% to 90%, depending on the season of sampling. In addition, there is a strong spatial dependency as the vector *Lutzomyia shannoni* is associated with maritime hardwood habitats. Even on this very small island, VSV antibody prevalence estimates from pigs can range from 3% to 67% depending on the forest type from which they are sampled (Comer et al. 1993). This localization also exemplifies the value of incorporating wildlife surveillance into traditional public and domestic animal health surveillance networks. Without surveillance and research directed at wildlife, this focus of VSV would not be known.

3.1.3 Scale and Regional Variation in Causal Relationships

One of the primary goals of wildlife research is to identify risk factors and causal relationships related to pathogen distribution, seasonal or spatial transmission patterns, or pathogen transmission to domestic animal or human populations. It is important to consider, however, that these relationships may vary over the range of many wildlife species and this is especially true for species with extended distributions. For example, the prevalence of antibodies to *Ehrlichia chaffeensis* in white-tailed deer is dependent on different climatic and land use variables over the range of this species in the Southeasten United States (Yabsley et al. 2005). With this same species, the prevalence of hemorrhagic disease, caused by related orbiviruses in the epizootic hemorrhagic disease and bluetongue serogroups, varies between regions and depending on location can be primarily driven by vector distribution and abundance, acquired herd immunity, or innate resistance possibly related to long-term virus–host co-evolution (Gaydos et al. 2002). Variation in causal relationships over time or space, whether related to host susceptibility, pathogen virulence, or environmental factors, should not be viewed as a major impediment; such differences can provide valuable insight into the recognition of potential mechanisms for disease emergence.

3.2 Experimental Studies

Controlled experimental studies are a critical component to wildlife disease research, especially related to understanding pathogenesis and in the development and validation of diagnostic methods. These studies, however, are not without their own set of difficulties. Practical impediments include animal procurement, potentially unusual husbandry and housing requirements, artificial infection routes or doses, and cost associated with the need for extended studies. For these reasons, most experimental studies, especially with larger wildlife species, rely on relatively small sample sizes. These problems can be further complicated by unknown exposure histories if field-caught animals are used in such studies, and if unrecognized variables, such as age or genetics, are associated with clinical response.

Despite these potential shortcomings, experimental studies are of primary importance in understanding pathogenesis, validating diagnostics, and providing a necessary perspective for interpreting data. However, it should be recognized that such results are subject to many potential problems associated with the artificial and often uncontrolled nature of this type of work and experimental results may not be entirely representative of field events. An example of this can be seen in comparing WNV viremia associated with experimental infections (Komar et al. 2003) and field infections (Allison et al. 2004) in rock pigeons (*Columba livia*). In this case, overall viremia titers obtained from these studies generally were consistent and indicated a low potential for this species to act as an efficient amplifying host for this virus. In contrast, several high titer outliers were detected in the field study, and it was proposed that this could have been related to co-infection with pigeon paramyovirus (a variable not included in the experimental studies). These results demonstrate the need to cross-validate and question both experimental and field data, and in fact, this process often leads to the discovery of many additional relationships that are relevant to pathogenesis and epidemiology.

4 Conclusions

As initially stated, the goal of this chapter is to provide some guidance in eliminating and negotiating potential problems associated with conducting and interpreting results from surveillance and epidemiologic studies involving wildlife. This work often is difficult but extremely important. Some general guidelines to consider in planning such activities are:

a. Develop an integrated plan. Gaining an understanding of natural history, epidemiology, or the factors related to disease emergence can be complicated and will require ongoing work. This understanding can be achieved only through a variety of surveillance approaches that are supported by both field epidemiology and laboratory research.

b. Archive. Disease emergence often is not associated with a "new" pathogen or wildlife reservoir. Much can be achieved in the short term if retrospective samples (serum, tissue, or nucleic acid) are available. Such historic samples can be fundamental to understanding the risk factors associated with the emergence event.

c. Maintain quality control. Understand the limitations of the data and strive to improve and validate results. This can be facilitated through integrating different but supporting diagnostic tools and by incorporating experimental studies into long-range plans.

d. Take care in the interpretation. As in all scientific endeavors, there is a constant need to question results, and in most cases, such questioning will not only provide guidance for future research but also may provide much insight into sometimes complex systems.

e. Do not be restricted by traditional approaches. Surveillance and research involving wildlife present unique challenges that may require unique approaches. The biology of the species under study is of major importance and biological attributes can be utilized for unique and efficient experimental and surveillance designs.

f. Do not be intimidated. Although such studies are difficult, examples of success are everywhere and our understanding of wildlife and wildlife-related zoonotic and emerging disease has greatly increased. Enter into this arena with an understanding that no individual study is perfect; it is the collective product that counts.

References

Allison AB, Mead DG, Gibbs SEJ, Hoffman DM, Stallknecht DE (2004) West Nile virus viremia in wild rock pigeons. Emerg Infect Dise 10:2252–2255

Berger L, Speare R, Daszak P, Green DE, Cunningham AA, Goggin CL, Slocombe R, Ragan MA, Hyatt AD, McDonald KR, Hines HB, Lips KR, Marantelli G, Parks H (1998) Chytridiomycosis causes amphibian mortality associated with population declines in the rain forests of Australia and Central America. Proc Natl Acad Sci U S A 95:9031–9036

Berringer J, Hanson LP, Stallknecht DE (2000) An epizootic of hemorrhagic disease in white-tailed deer in Missouri. J Wildl Dis 36:588–591

Calisher CH, Mills JN, Sweeney WP, Choate JR, Sharp DE, Canestorp KM, Beaty BJ (2001) Do unusual site-specific population dynamics of rodent reservoirs provide clues to the natural history of hantaviruses? J Wildl Dis 37:280–288

Caffrey C, Peterson CC (2003) West Nile virus may not be a conservation issue in the northeastern United States. American Birds (103 rd Christmas Bird Count). 14–21

Comer JA, Kavanaugh DM, Stallknecht DE, Ware GO, Corn JL, Nettles VF (1993) Effect of forest type on the distribution of Lutzomia shannoni (Diptera: Psychodidae) and vesicular stomatitis virus on Ossabaw Island GA. J Med Entomol 30:555–560

Dhondt AA, Tessaglia DL, Slothower RL (1998) Epidemic mycoplasma conjunctivitis in house finches from eastern North America. J Wildl Dis 34:265–280

Fischer JR, Stallknecht DE, Luttrell MP, Dhondt AA, Converse KA (1997) Mycoplasma conjunctivitis in wild songbirds: An example of the spread of a new contagious disease in a mobile population. J Emerg Dis 3:69–72

Gaydos JK, Davidson WR, Howerth EW, Murphy M, Elvinger F, Stallknecht DE (2002) Innate resistance to epizootic hemorrhagic disease virus serotypes 1 and 2 in white-tailed deer. J Wildl Dis 38:713–719

Guptill SC, Julian KG, Campbell GL, Price SD, Marfin AA (2003) Early-season avian deaths from West Nile virus as warnings of human infection. Emerg Infect Dis 9:483–484

Hinshaw VS, Webster RG, Turner B (1980) The perpetuation of orthomyxoviruses and paramyxoviruses in Canadian waterfowl. Can J Microbiol 26:622–629

Hochachka WM, Dhondt AA, McGowan KG, Kramer LD (2004) Impact of West Nile virus on American crows in the northeastern United States, and its relevance to existing monitoring programs. Ecol Health 1:60–68

Kawaoka Y, Chambers TM, Sladen WL, Webster RG (1988) Is the gene pool of influenza viruses in shorebirds and gull different from that in wild ducks? Virology 163:247–250

Kelly TR, Sleeman JM (2003) Morbidity and mortality of red foxes (*Vulpes vulpes*) and gray foxes (*Urocyon cineroargenteus*) admitted to the Wildlife Center of Virginia, 1993–2001. J Wildl Dis 39:467–469

Komar N, Langevin S, Hinten S, Nemeth N, Edwards E, Hettler D, Davis B, Bowen R, Bunning M (2003) Experimental infection of North American birds with the New York 1999 strain of West Nile virus. Emerg Infect Dis 9:311–322

Ksiazek TG, Peters CJ, Rollin PE, Zaki S, Nichol S, Spiropoulou C, Morzunov S, Feldmann H, Sanchez A, Khan AS, Mahy BWJ, Wachsmuth K, Butler JC (1995) Identification of a new North American hantavirus that causes acute pulmonary insufficiency. Am J Trop Med and Hygiene 52:117–123

Ley DH, Berkhoff JE, Mclaren JM (1996) *Mycoplasma gallisepticum* isolated from house finches (*Carpodacus mexicanus*) with conjunctivitis. Avian Dis 40:480–483

Little SE, Stallknecht DE, Lockhart JM, Dawson JE, Davidson WR (1998) Natural coinfection of a white-tailed deer (*Odocoileus virginianus*) population with three Ehrlichia species. J Parasitol 84:897–901

Romero CH, Meade PN, Shultz JE, Chung HY, Gibbs EP, Hahn EC, Lollis G (2001) Venereal transmission of pseudorabies viruses indigenous to feral swine. J Wildl Dis 37:289–296

Root JJ, Black WC, Calisher CH, Wilson KR, Mackie RS, Schountz T, Mills JN, Beaty BJ (2003) Analysis of gene flow among populations of deer mice (*Peromyscus maniculatus*) at sites near hantavirus pulmonary syndrome case-patient residences. J Wildl Dis 39:287–298

Schmitt SM, Fitzgerald SD, Cooley TM, Bruinging-Fann CS, Sulivan L, Berry D, Carlson T, Minnis RB, Payer JB, Sikarskie J (1997) Bovine tuberculosis in free-ranging white-tailed deer from Michigan. J Wildl Dis 33:749–758

Stallknecht DE (1997) Ecology and epidemiology of avian influenza viruses in wild bird populations: Waterfowl, shorbirds, pelicans, cormorants, etc. Proceedings of the fourth international symposium on avian Influenza. pp 61–69

Stallknecht DE, Fletcher WO, Ericson GA, Nettles VF (1987) Antibodies to vesicular stomatitis New Jersey type virus in wild and domestic sentinel swine. Am J Epidemiol 125:1058–1065

Stallknecht DE, Shane SM, Zwank PJ, Senne DA, Kearney MT (1990) Avian influenza viruses from migratory and resident ducks of coastal Louisiana. Avian Dis 34:398–405

Steele KE, Linn MJ, Schhoepp RJ, Komar N, Geisbert TW, Manduca RM, Calle PP, Raphael BL, Clippinger TL, Larson T, Smith J, Lanciotti RS, Panella NA, McNamara TS (2000) Pathology of fatal West Nile virus infections in native and exotic birds during the 1999 outbreak in New York City New York. Vet Pathol 37:208–224

Stone WB, Okoniewski JC, Therrien JE, Kramer LD, Kauffman EB, Eidson M (2004) VecTest as diagnostic and surveillance tool for West Nile virus in dead birds. Emerg Infect Dis 10:2175–2181

Stutzenbaker CD, Brown K, Lobpries D (1986) Special Report: an assessment of the accuracy of documenting waterfowl die-offs in a Texas coastal marsh. In: Feierabend JS, Russel AB (eds) Lead poisoning in wild waterfowl. National Wildlife Federation, Washington, DC, pp 88–95

Thomas NJ, Meteyer CU, Sileo L (1998) Epizootic vacuolar myelinopathy of the central nervous system of bald eagles (*Haliaeetus leucocephalus*) and American coots (*Fulica americana*) Vet Pathol 35:479–487

Thurmond MC (2003) Conceptual foundations for infectious disease surveillance. Journal of Veterinary Diagn Invest 15:501–514

Toma B, Vaillancourt J-P, Dufour B, Eloit M, Moutou F, Marsh W, Benet JJ, Sanaa M, and Michel P (1999) Dictionary of Veterinary Epidemiology Iowa State University Press Ames

Williams ES, Young S (1980) Chronic wasting disease of captive mule deer: A spongiform encephalopathy. J Wildl Dis 16:89–98
Wobeser GA (1994) Investigation and management of disease in wild animals. Plenum Press, New York
Wobeser G, Wobeser AG (1992) Carcass disappearance and estimation of mortality in a simulated die-off of small birds. J Wildl Dis 28:548–554
Woods LW, Swift PK, Barr BC, Horzinek MC, Nordhausen RW, Stillian MH, Patton JF, Oliver MN, Jones KR, MacLachlan NJ (1996) Systemic adenovirus infection associated with high mortality in mule deer (*Odocoileus hemionus*) in California. Vet Pathol 33:125–132
Yaremych SA, Warner RE, Mankin PC, Brawn JD, Raim A, Novak R (2004) West Nile virus and high death rate in American crows. Emerg Infect Dis 10:709–711
Yabsley MJ, Wimberly MC, Dugan VG, Little SE, Stallknecht DE, Davidson WR (2005) Spatial analysis of the distribution of *Ehrlichia chaffeensis*, causitive agent of human monocytotropic ehrlichiosis across a multi-state region. Am J Trop Med Hyg 72:840–850
Young PL, Halpin K, Selleck PW, Field H, Gravel JL, Kelly MA, MacKenzie JS (1996) Seologic evidence for the presence in Pteropus bats of a paramyxovirus related to equine morbillivirus. Emerg Infect Dis 2:239–240

CTMI (2007) 315:463–475
© Springer-Verlag Berlin Heidelberg 2007

Collaborative Research Approaches to the Role of Wildlife in Zoonotic Disease Emergence

P. Daszak[1] (✉) · J. H. Epstein[1] · A. M. Kilpatrick[1] · A. A. Aguirre[2] · W. B. Karesh[3] · A. A. Cunningham[4]

[1]Consortium for Conservation Medicine, 460 West 34th Street, 17th Floor, New York, NY 10001, USA
daszak@conservationmedicine.org
[2]Wildlife Trust, 460 West 34th Street, New York, NY 10001, USA
[3]Wildlife Conservation Society, 2300 Southern Blvd, Bronx, NY 10460, USA
[4]Institute of Zoology, Regent's Park, London, NW1 4RY, UK

Abstract Emerging infectious diseases are a key threat to public health and the majority are caused by zoonotic pathogens. Here we discuss new collaborative approaches to understanding the process of zoonotic disease emergence that link veterinary medicine, public health, and ecological approaches: conservation medicine and one health. We demonstrate how studies on the underlying drivers of disease emergence (bushmeat hunting, wildlife trade, and deforestation) can provide ways to model, predict, and ultimately prevent zoonotic disease emergence and spread.

1 Introduction

Emerging infectious diseases (EIDs) are a significant threat to global public health, and around 75% of these are caused by zoonotic pathogens (Taylor et al. 2001; Woolhouse and Gowtage-Sequeria 2005). Zoonotic emerging diseases cause significant mortality (e.g., HIV-1 and -2), threaten, or have caused pandemic spread (e.g., SARS coronavirus, Nipah virus, Avian influenza virus), or threaten global health due to high case fatality rates and no available vaccines or therapies (e.g., Ebola virus, Nipah virus, Hendra virus) (Chua et al. 2000; Guan et al. 2003; Hahn et al. 2000; Klenk 1999; Subbarao et al. 1998). The underlying causes of zoonotic disease emergence are often the same processes that threaten wildlife populations and are largely environmental (e.g., agricultural expansion), human behavioral (e.g., increased travel and trade), or demographic (e.g., migration into new regions) (Morens et al. 2004; Smolinski et al. 2003). These changes alter the contact rates among humans, wildlife, and domestic animals, and provide a bridge for pathogens to move into new host populations (Daszak et al. 2001). Zoonotic pathogens that move across this bridge are either microbes already established as infecting humans, or those with no prior ability but able to evolve and adapt to this new host (Daszak et al. 2000; Morse 1993b). In this review, we discuss strategies for investigating the process of disease emergence from wildlife. We highlight this process by providing examples of key emerging zoonoses and make a case for increased integration of wildlife biologists, epidemiologists, veterinarians, ecologists, microbiologists and others in dealing with the global threat of emerging zoonoses.

2 Local Spillover, Global Emergence

The dynamics of disease emergence from wildlife are complex, involve an array of anthropogenic factors, and a diverse assemblage of known and unknown viruses, fungi, bacteria, and other pathogens. Anthropogenic factors bring human and domestic animal populations into increasing contact with wildlife reservoirs of zoonotic pathogens. For example, in Malaysia, intensive management of pig production in farms located in fruit bat habitat led to the spillover of Nipah virus, a paramyxovirus for which these bats serve as a reservoirs (Chua et al. 2000). Similarly, logging routes carved into primary forest have provided easier access to hunters in search of animals to eat or sell. The trade in bushmeat, which brings wild animals from geographically disparate habitats into contact with each other and with people, has led to the spillover of several important zoonotic viruses including HIV, Ebola, and SARS (Hahn et al. 2000; Karesh et al. 2005; Leroy et al. 2004; Li et al. 2005). As human populations continue to increase, so do these anthropogenic pressures on wildlife habitat and populations. The result is likely to be continued spillover of new zoonotic pathogens into human populations, and perhaps even an increase in spillover rates, reflecting increases in these anthropogenic drivers of emergence.

Each spillover event is not necessarily a threat to global health per se. The process of emergence of pandemic pathogens, such as HIV-1, occurs in a series of stages (Hahn et al. 2000; Wolfe et al. 2005a). First, there usually is a series of initial spillover events, only some of which result in virus replication in the new, human, host. In some cases, viral pathogens spill over into domestic animals before reaching humans (e.g., Nipah virus, SARS) (Chua et al. 2000; Li et al. 2005). The spillover process has been termed viral traffic (Morse 1993a) or viral chatter (Wolfe et al. 2004b) for viruses' and is the initial phase of invasion of a pathogen into a new population described by disease ecologists (Anderson and May 1986). The next stage is the persistence of new viral pathogens in the human population (Anderson and May 1986). This occurs only if the zoonotic pathogen is able to be transmitted successfully from person to person, a characteristic which may occur naturally (e.g., the recent small-scale persistence of Nipah virus in Bangladesh; Hsu et al. 2004). It may also be a product of evolution from an ancestral nonhuman animal virus to a human-adapted strain, such as occurred when SIV_{CPZ} entered the human population to become HIV-1 (Hahn et al. 2000) or the initial emergence of measles virus from a morbillivirus of domestic animals (Dobson and Carper 1996). Finally, the spread phase of emergence (Anderson and May 1986) occurs when local chains of transmission link into denser human populations or populations that are well connected through sex or through needle-sharing by intravenous drug users (HIV) or through increased travel (SARS).

Just as the pressures that foster spillover have increased, there has been a significant increase over the past few decades in international trade and travel, with a resulting increased potential for the last, pandemic phase of emergence. Between 1986 and 1999, the global GDP per capita increased by an average of around 2.5% p.a., while an index of global air travel increased by 5% p.a. (The Boeing Company 2000) (Fig. 1). Likewise, as the demand for air travel has doubled during this period, the number of kilometers of new routes developed has increased in direct proportion (The Boeing Company 2000). Air travel industry projections suggest that, during the next 20 years, the air travel share of GDP will rise steadily, with air traffic growing by 4.0% annually, two percentage points faster than global mean annual growth in GDP (The Boeing Company 2002). This expansion will represent a doubling of global air traffic that is

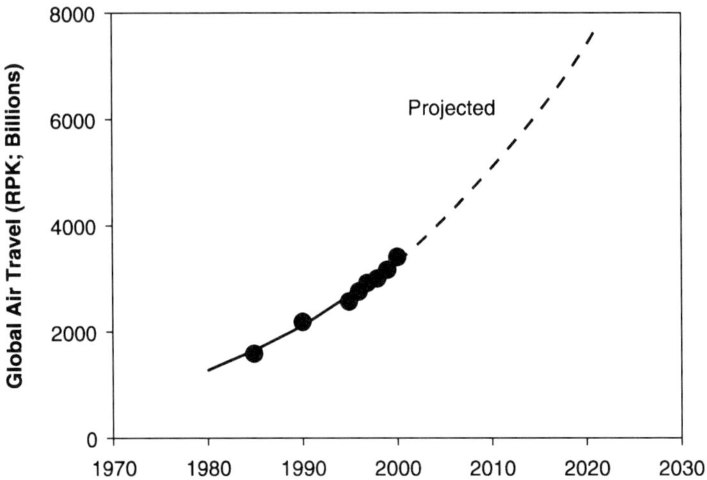

Fig. 1 Recent, and projected future, expansion of global air travel. Data are based on an index of air travel, revenue passenger kilometers (RPK), which describes the number of people traveling annually in relation to the number of kilometers traveled. (Data obtained from The Boeing Company, 2000)

unlikely to be mirrored by a similar expansion in disease surveillance budgets (particularly for wildlife EIDs). The impact of these trends on pathogen spread is likely to be significant, with a resulting growth in the ability of local spillover events to become pandemic outbreaks.

3 Fusing Ecology with Medical Sciences

Despite a large and growing literature on emerging zoonoses, there is still a dearth of research on the *process* of disease emergence. However, the association of disease emergence with anthropogenic environmental changes and demographic or human behavioral changes (Morse 1993b; Patz et al. 2004) may provide a fruitful way to examine this phenomenon (Aguirre et al. 2002; Daszak et al. 2000; Karesh and Cook 2005). For example, research that measures the rate of these anthropogenic or demographic changes and how they affect pathogen dynamics in wildlife would provide a way to assess the risk of spillover to people. Understanding relative or actual risks of spillover could provide a tool to focus resources on high-risk regions (e.g., areas of recent deforestation) or high-risk behaviors (e.g., bushmeat hunting or agricultural changes). This may ultimately allow public health programs to predict and prevent zoonotic disease emergence.

The key to such approaches is a detailed, mechanistic understanding of the processes that drive disease emergence. Here we provide case-study examples in which ecologists have obtained a detailed understanding of the driver of an emergence event, and in which there have been varying degrees of collaboration between ecologists and medical scientists to study the process of disease emergence. These collaborations are especially fruitful areas of future research.

3.1 Bushmeat Hunting and Disease Emergence

In tropical forests, the building of roads to support logging or mining operations, or to connect villages and towns provides access to wildlife and is often associated with increased demand for, and access to, bushmeat. This process is thought to have led to spillover of simian immunodeficiency viruses and the emergence of HIV-1 and -2 (Hahn et al. 2000). Ebola hemorrhagic fever virus outbreaks in humans have repeatedly been linked to the handling of infected great apes (Leroy et al. 2004). With the expansion of human populations, there has been an increase in demand for bushmeat, particularly in locations where alternative sources of protein have been scarce (Bulte and Horan 2002; de Merode et al. 2004; Fa et al. 2003). It is therefore likely that the risk of future disease emergence through this process will also increase.

The process of deforestation, building of logging roads and expanding demand for bushmeat is complex, but there is a great deal of research that has quantified these factors. For example, researchers have analyzed the diversity of wildlife hunted, the weight of meat extracted from forests, its monetary and nutritional value, and the seasonal and interannual dynamics of hunting (Bulte and Horan 2002; de Merode et al. 2004; Fa et al. 2002a, 2002b). Anthropologists have studied how incentives for hunting wildlife vary from region to region. In Central and West Africa, hunters are often local village members hunting for subsistence or trade to local or regional markets (Wolfe et al. 2004a), whereas in other regions luxury, familiarity, medicinal value, tradition, prestige, taste preference, or subsistence drive the process (Wilkie et al. 2005). Ecologists have also measured the impact of hunting on the populations of some species (Bowen-Jones and Pendry 1999; Maisels et al. 2001; Plumptre et al. 1999). Combining these series of data would allow an assessment of how hunting increases as a logging road is built, and how this affects the dynamics of wildlife populations.

The final critical link of how these changes will affect the risk of pathogen spillover requires only three additional elements: measurements of background diversity of pathogens in hunted wildlife; assessment of whether these pathogens will be able to replicate in people; and measurement of the type of contact and rate of contact between hunted species and/or their meat, and people. Studies focused on these issues have already begun. For example, studies of exposure to nonhuman primate viruses in Cameroon bushmeat hunters have revealed new spillover events (Wolfe et al. 2005a, 2005b, 2005c). Studies of the patterns of spillover have identified those pathogen groups most likely to move from nonhuman to human hosts (Taylor et al. 2001; Woolhouse and Gowtage-Sequeria 2005). However, to truly fuse wildlife research with medical research on emerging disease will require a greater degree of collaboration among diverse disciplines. For example, few studies have used molecular techniques to survey hunted wildlife or other zoonotic reservoirs for novel pathogens, a study which would require collaboration among molecular biologists and wildlife biologists, and veterinarians. The wildlife mortality and monitoring network established in Central Africa is one example (Rouquet et al. 2005). Similarly, ecologists have used mathematical models to describe pathogen dynamics in wildlife (Dobson and Foufopoulos 2001; Hudson et al. 2002), but few studies have used these approaches to predict patterns of spillover from bushmeat. Expanding collaboration in the face of new zoonotic threats (e.g., H5N1 avian influenza) is likely to increase capacity to understand these complex processes.

3.2
Wildlife Trade and Disease Emergence

In 2002, SARS coronavirus emerged in humans in China (Drosten et al. 2003). The epidemiological risk factors of the first cases in Southeast China were proximity to live animal (wet) markets and working in the restaurant industry (Xu et al. 2004). Virus isolation and genome sequence data suggested a role for masked palm civets (*Paguma larvata*); however, infection seemed to be limited to those animals in the marketplace, as opposed to on farms or in the wild (Xu et al. 2004). Preliminary surveillance had suggested involvement of other species of small mammal traded for food in these markets; however, the natural reservoir for SARS-CoV remained unknown (Guan et al. 2003). Further work has demonstrated that *Rhinolophus* spp. bats are the wildlife reservoir of SARS-like coronaviruses, and it has been suggested that the trade in these animals for food initially led to spillover to other wet-market species and humans (Lau et al. 2005; Li et al. 2005). Bats, civets, and other mammals are traded in large numbers in Chinese wet markets, but the origins of these animals are often many hundreds of miles from their point of sale. With an increasing demand for wild animal meat as China's economy grows, pathogen spillover from this wildlife trade is likely to be a continuing problem.

Wildlife trade also leads to the movement of animals over great distances and has caused the emergence of viruses in areas significantly outside their natural range (Karesh et al. 2005). For example, in 2003, Monkeypox emerged in the United States through the importation of Gambian giant rats from Africa (Sejvar et al. 2004). Pathogens may also be introduced into new regions in animals inadvertently carried with traded goods (Cook and Karesh 2005). West Nile virus emerged in the USA in 1999, with the first cluster of cases occurring in Queens, New York, close to a large international air and sea port (Lanciotti et al. 1999). However, it is unknown if this virus was introduced within birds shipped into New York for the pet trade or for livestock production, or whether it was carried within a mosquito in an airplane (Kilpatrick et al. 2004, 2006a). The highly pathogenic H5N1 strain of avian influenza has spread within Southeast Asia and Europe via the trade in poultry and via migrating birds, as well as through the trade in birds for pets (Kilpatrick et al. 2006c). A pair of H5N1-infected crested hawk eagles was confiscated by authorities at Brussels airport after being illegally smuggled into the country from Southeast Asia (Van Borm et al. 2005) and several mesias in a group of ornamental birds imported from Taiwan were found to be infected with the H5N1 virus while in quarantine in the UK (http://www.defra.gov.uk/animalh/diseases/notifiable/disease/ai/pdf/ai-epidemrep111105.pdf).

The model of collaboration between wildlife ecologists and medical scientists described in the bushmeat studies above suggests some simple but valuable strategies to predict and prevent zoonotic disease spillover and spread. For example, the spread of West Nile virus throughout the continental USA and into South America has led to concern over its potential spread to regions with endangered bird species that may be at risk of extinction by this pathogen. In Hawaii, over one-third of endemic bird species have been driven to extinction or endangered status by introduced avian malaria and pox (Van Riper et al. 1986). In the Galapagos islands, endemic Darwin's finches exist as separate species on individual islands, and are greatly threatened by disease introduction (Wikelski et al. 2004). Two recent studies have collated data on the average number of mosquitoes transported on airplanes and ships, the number and identity of migratory birds, pet birds and poultry imported onto these islands, and the number of people visiting these islands (Kilpatrick et al. 2004, 2006a). Using these data and simple mathematical models, it is possible to identify the most likely pathways of introduction of this zoonotic disease to the islands (mosquitoes carried by airplanes), and therefore to advise policy to reduce the risk of pathogen introduction. This approach can easily be expanded using global data on wildlife trade to predict the risk of introduction of new or known zoonoses into new regions.

3.3 Urban Sprawl, Fragmentation, and Zoonotic Disease Emergence

Logging in tropical regions is paralleled by the process of urban sprawl in developed countries, and the removal of natural wildlife habitat. The impact of this process on wildlife diversity, ecology, and habitat quality has been studied extensively by ecologists (Johnson and Klemens 2005). The impact of this process on human health has also been well-studied, but largely regarding the impact of pollution and stress on human health and welfare. However, recent work by disease ecologists has shown a strong connection between urban sprawl, habitat fragmentation, the loss of biodiversity and increased risk of zoonotic disease spillover to people. Lyme disease emergence is facilitated by urban sprawl and associated fragmentation that reduce the diversity of mammal communities to a pair of highly competent reservoirs, the white-footed mouse (*Peromyscus leucopus*) and the eastern chipmunk (*Tamias striatus*) (LoGiudice et al. 2003). In the northeastern USA, the now endemic West Nile virus (WNV) is transmitted within a diverse assemblage of birds, mammals, and mosquitoes (Marra 2004). However, recent analyses of WNV dynamics across an urban-to-rural land use gradient has shown that the bulk of mosquitoes become infected by feeding on American robins (*Turdus migratorius*) (Kilpatrick et al. 2006b), a common suburban species. The risk of WNV infection at these sites is higher in urban and suburban habitat than in heavily forested habitat (A.M. Kilpatrick, unpublished observations).

4 A Call for Cross-disciplinary Collaboration

During the last few years, interest in the emergence of zoonotic diseases has grown. There has been a series of new programs developed by funding agencies in the USA, Europe, Canada, and Australia and national agencies working on public health and wildlife health in the USA, Canada, and Europe. Global organizations including the World Health Organization (WHO), the Food and Agriculture Organization (FAO), and the Office Internationale des Epizooites (OIE), also known as the World Organization for Animal Health, have recently recognized the significance of wildlife as reservoirs for zoonotic diseases. Also, new journals have been published to deal with an increasing output of research on emerging zoonotic diseases (e.g., *Vector-borne and Zoonotic Diseases*) and on ecological research in zoonotic emergence (e.g., *EcoHealth*).

The bulk of research on emerging zoonotic diseases in the USA and Europe continues to be concerned with developing new vaccines and drug therapies and surveillance in the human population. However, there might be more cost-effective and efficient ways of addressing this growing phenomenon. We have shown here that there is an increasing understanding of the ecological processes that underlie disease emergence from wildlife to people. We have highlighted examples of new, collaborative, and interdisciplinary approaches to emerging zoonotic diseases that are necessary to develop this new understanding into a focused surveillance and research approach that will ultimately allow for prediction and prevention of zoonotic disease spread. There has been a growing interest in such collaboration from veterinary researchers, through the new fields of "Conservation Medicine" (Aguirre et al. 2002) and "One Health" (Karesh and Cook 2005). There have also been calls for expansion of these initiatives from the National Research Council of the USA (Womack et al. 2005). However, there remains an urgent need to expand the connections and collaborations among veterinary researchers, microbiologists, public health researchers, and ecologists. We propose the following measures to encourage this approach:

1. Fostering collaboration among the disciplines. In particular, linking ecological approaches with laboratory advances in pathogen surveillance. This should be encouraged through education (undergraduate, postgraduate, and professional) and in research.
2. Encouraging studies to discover new pathogens in wildlife. This will provide a critical link toward predicting the risk of zoonotic disease spillover from wildlife. It will require the development of testing protocols and expansion of laboratory support in countries with high vertebrate biodiversity (those with a likely high biodiversity of potentially zoonotic agents).
3. Supporting studies that address the underlying drivers of emergence. Understanding how anthropogenic environmental changes and sociological or demographic factors affect the risk of disease emergence is likely to provide more cost-effective and ultimately more sustainable mechanisms to mitigate these threats.

References

Aguirre AA, Ostfeld RS, Tabor GM, House C, Pearl MC (2002) Conservation medicine: ecological health in practice. Oxford University Press, New York

Anderson RM, May RM (1986) Philos Trans R Soc Lond B 314:533–570

The Boeing Company (2000) Current market outlook 2000: Into the next century. The Boeing Company, Seattle

The Boeing Company (2002) Current market outlook 2002. The Boeing Company, Seattle

Bowen-Jones E, Pendry S (1999) The threat to primates and other mammals from the bushmeat trade in Africa, and how this threat could be diminished. Oryx 33:233–246

Bulte EH, Horan RD (2002) Does human population growth increase wildlife harvesting? An economic assessment. J Wildl Manage 6:574–580

Chua K, Bellini W, Rota P, Harcourt B, Tamin A, Lam S, Ksiazek T, Rollin P, Zaki S, Shieh W-J, Goldsmith C, Gubler D, Roehrig J, Eaton B, Gould A, Olson J, Field H, Daniels P, Ling A, Peters C, Anderson L, Mahy B (2000) Nipah virus: a recently emergent deadly paramyxovirus. Science 28:1432–1435

Cook RA, Karesh WB (2005) Ebola SARS, and other diseases that imperil people and animals. In: Guynup S (ed) State of the wild. Island Press, Washington, pp 131–138

Daszak P, Cunningham AA, Hyatt AD (2000) Emerging infectious diseases of wildlife: threats to biodiversity and human health. Science 28:443–449

Daszak P, Cunningham AA, Hyatt AD (2001) Anthropogenic environmental change and the emergence of infectious diseases in wildlife. Acta Tropica 7:103–116

De Merode E, Homewood K, Cowlishaw G (2004) Biological conservation. 11:573–581

Dobson A, Foufopoulos J (2001) Emerging infectious pathogens of wildlife. Philos Trans R Soc Lond B Biol Sci 35:1001–1012

Dobson AP, Carper ER (1996) Infectious diseases and human population history. Biosci 4:115–126

Drosten C, Gunther S, Preiser W, van der Werf S, Brodt HR, Becker S, Rabenau H, Panning M, Kolesnikova L, Fouchier RAM, Berger A, Burguiere AM, Cinatl J, Eickmann M, Escriou N, Grywna K, Kramme S, Manuguerra JC, Muller S, Rickerts V, Sturmer M, Vieth S, Klenk HD, Osterhaus A, Schmitz H, Doerr HW (2003) Identification of a novel coronavirus in patients with severe acute respiratory syndrome. New Engl J Med 34:1967–1976

Fa JE, Juste J, Burn RW, Broad G (2002a) Bushmeat consumption and preferences of two ethnic groups in Bioko Island, West Africa. Human Ecol 3:397–416

Fa JE, Peres CA, Meeuwig J (2002b) Bushmeat exploitation in tropical forests: an intercontinental comparison. Conserv Biol 1:232–237

Fa JE, Currie D, Meeuwig J (2003) Bushmeat and food security in the Congo Basin: linkages between wildlife and people's future. Environ Conserv 3:71–78

Guan Y, Zheng BJ, He YQ, Liu XL, Zhuang X, Cheung CL, Luo SW, Li PH, Zhang LJ, Guan YJ, Butt K M, Wong K L, Chan KW, Lim W, Shortridge KF, Yuen KY, Peiris JSM, Poon LLM (2003) Isolation and characterization of viruses related to the SARS coronavirus from animals in southern China. Science 30:276–278

Hahn BH, Shaw GM, de Cock KM, and Sharp PM (2000) AIDS as a zoonosis: scientific and public health implications. Science 28:607–614

Hsu VP, Hossain MJ, Parashar UD, Mohammed MA, Ksiazek TG, Kuzmin I, Niezgoda M, Rupprecht C, Bresee J, Breiman RF (2004) Emerg Infect Dis 1:2082–2087

Hudson P J, Rizzoli A, Grenfell BT, Heesterbeek H, Dobson AP (2002) The ecology of wildlife diseases. Oxford University Press, Oxford
Johnson EA, Klemens MW (2005) Nature in fragments: the legacy of sprawl. Columbia University Press, New York
Karesh WB, Cook RA (2005) The Human-Animal Link, One World – One Health. Foreign Affairs 8:8–50
Karesh WB, Cook RA, Bennett EL, Newcomb J (2005) Wildlife trade and global disease emergence. Emerg Infect Dis 1:1000–1002
Kilpatrick AM, Gluzberg Y, Burgett J, Daszak P (2004) Ecohealth 1:205–209
Kilpatrick AM, Daszak P, Goodman SJ, Rogg H, Kramer LD, Cedeño V, Cunningham AA (2006a) Predicting pathogen introduction: West Nile virus spread to Galaipagos. Conserv Biol 20:1224–1231
Kilpatrick AM, Daszak P, Jones MJ, Marra PP, Kramer LD (2006b) Proc R Soc Lond B 273:2327–2333
Kilpatrick AM, Chmura AA, Gibbons DW, Fleischer RC, Marra PP, Daszak P (2006c) Predicting the global spread of H5N1 avian influenza. Proc Natl Acad Sci U S A 103:19368–19373
Klenk H-D (1999) Marburg and Ebola viruses. Current topics in microbiology and immunology. Springer, Berlin Heidelberg New York
Lanciotti RS, Roehrig JT, Deubel V, Smith J, Parker M, Steele K, Crise B, Volpe KE, Crabtree MB, Scherret JH, Hall RA, MacKenzie JS, Cropp CB, Panigrahy B, Ostlund E, Schmitt B, Malkinson M, Banet C, Weissman J, Komar N, Savage HM, Stone W, McNamara T, Gubler DJ (1999) Origin of the West Nile virus responsible for an outbreak of encephalitis in the northeastern United States. Science 28:2333–2337
Lau SKP, Woo PCY, Li KSM, Huang Y, Tsoi H-W, Wong BHL, Wong SSY, Leung S-Y, Chan K-H, Yuen K-Y (2005) Severe acute respiratory syndrome coronavirus-like virus in Chinese horseshoe bats. Proc Natl Acad Sci U S A 10:14040–14045
Leroy EM, Rouquet P, Formenty P, Souquiere S, Kilbourne A, Froment J-M, Bermejo M, Smit S, Karesh W, Swanepoel R, Zaki S R, Rollin PE (2004) Multiple Ebola virus transmission events and rapid decline of central African wildlife. Science 30:387–390
Li W, Shi Z, Yu M, Ren W, Smith C, Epstein JH, Wang H, Crameri G, Hu Z, Zhang H, Zhang J, McEachern J, Field H, Daszak P, Eaton BT, Zhang S, Wang L-F (2005) Bats are natural reservoirs of SARS-like coronaviruses. Science 31:676–679
LoGiudice K, Ostfeld RS, Schmidt KA, Keesing F (2003) Proc Natl Acad Sci U S A 10:567–571
Maisels F, Keming E, Kemei M, Toh C (2001) The extirpation of large mammals and implications for montane forest conservation: the case of the Kilum-Ijim Forest, North-west Province, Cameroon. Oryx 3:322–331
Marra PP, Griffing S, Caffrey CL, Kilpatrick AM, McLean RG, Brand C, Saito E, Dupuis AP, Kramer LD, Novak R (2004) Bioscience 5:393–402
Morens DM, Folkers GK, Fauci AS (2004) The challenge of emerging and re-emerging infectious diseases. Nature 43:242–249
Morse SS (1993a) Emerging viruses. Oxford University Press, New York
Morse SS (1993b) Examining the origins of emerging viruses. In: More SS (ed) Emerging viruses. Oxford University Press, New York, pp 10–28

Patz JA, Daszak P, Tabor GM, Aguirre AA, Pearl M, Epstein J, Wolfe ND, Kilpatrick AM, Foufopoulos J, Molyneux D, Bradley DJ (2004) Unhealthy landscapes: policy recommendations on land use change and infectious disease emergence. Environ Health Perspec 11:1092–1098

Plumptre A, McNeilage A, Hall J (1999) Am J Phys Anthropol 108:224

Rouquet P, Froment J M, Bermejo M, Kilbourn A, Reed P, Karesh WB, Kumulungui B, Yaba P, Deilcat A, Rollin PE, Leroy EM (2005) Wild animal mortality monitoring and human Ebola outbreaks, Gabon and Republic of Congo, 2001–2003. Emerg Infec Dis 1:283–290

Sejvar JJ, Chowdary Y, Schomogyi M, Stevens J, Patel J, Karem K, Fischer M, Kuehnert MJ, Zaki SR, Paddock CD, Guarner J, Shieh WJ, Patton JL, Bernard N, Li Y, Olson VA, Kline RL, Loparev VN, Schmid DS, Beard B, Regnery RR, Damon IK (2004) Human monkeypox infection: a family cluster in the midwestern United States. J Infect Dis 19:1833–1840

Smolinski MS, Hamburg MA, Lederberg J (2003) Microbial threats to health: emergence detection, and response. National Academies Press, Washington, DC

Subbarao K, Klimov A, Katz J, Regnery H, Lim W, Hall H, Perdue ML, Swayne DE, Bender C, Huang J, Hemphill M, Rowe T, Shaw M, Xu X, Fukuda K, Cox N (1998) Characterization of an avian influenza A (H5N1) virus isolated from a child with a fatal respiratory illness. Science 27:393–396

Taylor LH, Latham SM, Woolhouse MEJ (2001) Risk factors for human disease emergence. Philos Trans R Soc Lond B 35:983–989

Van Borm S, Thomas I, Hanquet G, Lambrecht N, Boschmans M, Dupont G, Decaestecker M, Snacken R, van den Berg T (2005) Highly pathogenic H5N1 influenza virus in smuggled Thai eagles, Belgium. Emerg Infect Dis 1:702–705

Van Riper C, Van Riper SG, Goff LM, Laird M (1986) The epizootiology and ecological significance of malaria in Hawaiian land birds. Ecol Monogr 5:327–344

Wikelski M, Foufopoulos J, Vargas H, Snell H (2004) Ecology and Society

Wilkie DS, Starkey M, Abernethy K, Effa EN, Telfer P, Godoy R (2005) Role of prices and wealth in consumer demand for bushmeat in Gabon, central Africa. Conserv Biol 1:268–274

Wolfe ND, Daszak P, Kilpatrick AM, Burke DS (2005a) Bushmeat hunting, deforestation, and prediction of zoonoses emergence. Emerg Infect Diss 1:1822–1827

Wolfe ND, Heneine W, Carr JK, Garcia AD, Shanmugam V, Tamoufe U, Torimiro J N, Prosser AT, LeBreton M, Mpoudi-Ngole E, McCutchan FE, Birx DL, Folks TM, Burke DS, Switzer WM (2005b) Emergence of unique primate T-lymphotropic viruses among central African bushmeat hunters. Proc Natl Acad Sci U S A 10:7994–7999

Wolfe N D, Heneine W, Tiffany A, Garcia A D, Wright A, Carr JK, Tamoufe U, Torimiro JN, Prosser A, LeBreton M, Mpoudi-Ngole E, McCutchan FE, Birx DL, Folks TM, Burke DS, Switzer WM (2005c) Identification of a novel simian foamy virus infection in a Central African originating from a mona monkey (*Cercopithecus mona*). AIDS Res Human Retroviruses 2:476

Wolfe ND, Prosser AT, Carr JK, Tamoufe U, Mpoudi-Ngole E, Torimiro JN, LeBreton M, McCutchan FE, Birx DL, Burke DS (2004a) Emerg Infect Dis 1:2094–2099

Wolfe ND, Switzer WM, Folks TM, Burkes DS, Heneine W (2004b) Simian retroviral infections in human beings – reply. Lancet 36:139–140
Womack JE, Anderson LC, Bull LS, Capen CS, Cheville NF, Daszak P, Dodds WJ, Doyle MP, Franz DR, Shadduck JA, Shaw DH, Swayne DE, Tolwani RJ (2005) Critical needs for research in veterinary science. National Academies Press, Washington, DC
Woolhouse MEJ, Gowtage-Sequeria S (2005) Host range and emerging and reemerging pathogens. Emerg Infect Dis 1:1842–1847
Xu RH, He JF, Evans MR, Peng GW, Field HE, Yu DW, Lee CK, Luo HM, Lin WS, Lin P, Li LH, Liang WJ, Lin JY, Schnur A (2004) Epidemiologic clues to SARS origin in China. Emerg Infect Dis 1:1030–1037

CTMI (2007) 315:477–508
© Springer-Verlag Berlin Heidelberg 2007

Surveillance and Response to Disease Emergence

Angela Merianos* (✉)

Alert and Response Operations, Department of Epidemic and Pandemic Alert and Response, World Health Organization, 20 Avenue Appia, 1211 Geneva 27, Switzerland
merianosa@who.int

Abstract New and emerging infectious diseases affect humans, domestic animals, livestock and wildlife and can have a significant impact on health, trade and biodiversity. Of the emerging infectious diseases of humans, 75% are zoonotic, with wildlife being an increasingly important source of inter-species transmission. Recent animal health

*Formerly a Professor in Curtin University.

emergencies have highlighted the vulnerability of the livestock sector to the impact of infectious diseases and the associated risks to human health. Outbreaks resulting from wildlife trade have resulted in enormous economic losses globally. On a global level, the human health sector lags behind the animal health sector in the assessment of potential threats, although substantive differences exist among countries in the state of national preparedness planning for emerging diseases. The lack of surveillance data on emerging zoonoses from many developing countries means that the burden of human, livestock and wildlife disease is underestimated and opportunities for control interventions thereby limited. In the context of emerging zoonoses, comprehensive risk assessments are needed to identify the animal–human and animal–animal interfaces where transmission of infectious agents occurs and the feasibility of risk reduction interventions. The impact of emerging diseases can be minimised through a well-prepared and strong public health system and similar systems developed by the livestock, wildlife and food safety sectors. National animal disease emergencies, especially those that spill over to affect human health, require a whole-of-government approach for effective disease containment. As it is highly likely that zoonoses and animal diseases with the potential to affect human health will continue to emerge, surveillance and response systems for emerging zoonotic diseases will need to be strengthened and maintained at national and international levels. Applied research, linked across the human, livestock and wildlife sectors, is needed to inform preparedness planning and the development of evidence-based approaches to zoonotic disease prevention and control.

1 The Impact of Emerging Zoonoses

Emerging infectious diseases are defined as diseases that have recently increased in incidence or geographic range, recently moved into new host populations, recently been discovered or are caused by newly evolved pathogens (Lederberg et al. 1992; Smolinski et al. 2003). New and emerging infectious diseases affect humans, domestic animals, livestock and wildlife and can have a significant impact on health (WHO 2001), trade and biodiversity (Daszak et al. 2001). Of the emerging infectious diseases of humans, 75% are zoonotic, with wildlife being an increasingly important source of inter-species transmission (Daszak et al. 2001; Taylor et al. 2001; see the chapter by Cleaveland et al., this volume).

Massive global increases in demand for food of animal origin associated with population growth, income growth, urbanisation and devolution in global agriculture, is having a profound effect on health, livelihoods and environments. These factors are contributing to the exacerbation of public health and environmental problems, pressure on food production and distribution, and the illegal transport and trade in livestock, food products and people. The livestock sector represents almost half of the world's agricultural economy. Recent animal health emergencies have highlighted the vulnerability of the livestock sector to the impact of infectious diseases and the associated risks to human health (FAO/OIE 2004). Outbreaks resulting from wildlife trade have resulted in enormous economic losses globally (Karesh et al. 2005). In addition, the world is seeing unprecedented levels of international travel that has facilitated the spread of infectious diseases. The International Civil Aviation Organization estimates that air travel among its 185 members will reach two billion passengers annually by 2005 (ICAO Circular 2005).

The management of emerging zoonoses in humans requires a public health response closely linked to control measures in livestock animals and wildlife and that takes the complex interconnections among species into full account (Wildlife Conservation Society 2005; see the chapters by Childs and Daszak et al., this volume). Considerable resources by the agricultural and animal health sectors go into modelling risk and the economic impact of crises in consumer confidence resulting from animal diseases or infected animal products (University of Sydney FAH Report 2005). On a global level, the human health sector lags behind the animal health sector in the assessment of potential threats, although substantive differences exist among countries in their national preparedness planning for emerging diseases. Until recently, little attention has been given to determining the direct and indirect costs of human disease outbreaks, including morbidity and excess mortality, health service delivery costs, public health expenditure, the psychosocial impact on affected individuals, families and communities, the economic impact on travel, tourism and the insurance industry, and loss of confidence in governments and health services.

The economic burden of emerging zoonoses often falls disproportionately on the rural sector and the poor because of their greater risk of exposure to diseases of livestock and wildlife and pre-existing urban–rural socioeconomic inequalities. The health and socioeconomic impact of zoonoses are increasingly being felt most particularly, although not exclusively, by developing countries (Seimenis 1998). The lack of surveillance data on emerging zoonoses from many developing countries means that the burden of human, livestock and wildlife disease is underestimated and opportunities for control interventions thereby limited (see the chapters by Childs, by Nel and Rupprecht, and by Stallknecht, this volume).

1.1 The AIDS Epidemic

Most of the emerging infectious diseases identified since the mid-1990s have been caused by viruses. The AIDS epidemic caused by the human immunodeficiency virus, HIV, is one of the most destructive pandemics in human history (UNAIDS 2005). Since its recognition in 1981, AIDS has killed over 25 million people and an estimated 40.3 million people were living with HIV/AIDS in 2005. HIV emerged from at least two nonhuman primate reservoirs in Africa in the 1950s (Hahn et al. 2000). There are currently 33 nonhuman primates known to harbour their own unique simian immunodeficiency virus (SIV) strains (Kalish et al. 2005) and primate bush meat has a high prevalence of SIV (Peeters et al. 2002). In a study of 16 SIV isolates from five different primate species, 12 were able to infect human monocyte-derived macrophages, while 11 showed replication in human peripheral blood mononuclear cells, although the authors state that cell tropism does not necessarily predict virus pathogenicity in vivo (Grimm et al. 2003). Hunters in sub-Saharan Africa continue to be exposed to SIV during hunting and butchering nonhuman primates such as chimpanzees and sooty mangabeys, or by keeping wild primates as pets (Kalish et al. 2005; see the chapter by Daszak et al., this volume). Such spillover events have implications for the safety of the blood supply through the genesis of new HIV strains that are not detected by current HIV tests (Kalish et al. 2005).

More recently, Nipah virus, severe acute respiratory syndrome (SARS) and highly pathogenic avian influenza (A/H5N1) have also highlighted the importance of emerging zoonoses and their impact on health and economic development (see the chapters by Field et al., Wang and Eaton, and Webby et al., this volume).

1.2 Nipah Virus

Nipah virus, a henipavirus (Field et al. 2001), was first diagnosed in Malaysia in 1999 (Chua et al. 1999) and has caused serious disease in humans and livestock in Malaysia, Singapore (Paton et al. 1999; Tambyah et al. 2001), Bangladesh (ICDDR,B 2003, 2004a, 2004b, 2005) and India (Chadha 2006; see the chapter by Field et al., this volume). Outbreaks of Nipah virus encephalitis have been characterised by high mortality in humans. Transmission to humans is primarily through contact with infected pigs (Chua et al. 1999, 2000), although recent outbreak investigations in Bangladesh and India provide evidence for limited human-to-human transmission (WHO 2004a; Hsu et al. 2004; ICDDR,B 2004a; Chadha 2006), transmission via ingestion of food products contaminated with the saliva or urine (Enserink 2000) of Old World fruit bats (*Pteropodidae*) (WHO 2004a; ICDDR,B 2005) and/or contact transmission (WHO 2004a; ICDDR,B 2005) in environments contaminated by fruit bats. Pteropid bats are considered the natural reservoirs of Nipah virus (Eaton et al. 2006; Field et al. 2001; Chua et al. 2002). *Pteropus* species are distributed from Madagascar through the Indian subcontinent to south-eastern Asia and Australia and as far east as the Cook Islands in the Pacific (Chua et al. 2002). Serological evidence of Nipah virus infection in pteropid bats has also been found in Cambodia, although there are no reported outbreaks of Nipah encephalitis in humans (Reynes et al.

2005). Additional work remains to be done to improve our understanding of risk factors for transmission to humans and livestock, the disease ecology of Nipah virus and the geographic distribution of the reservoir species.

The cost of the outbreak to Malaysia is estimated at overUS $500 million. Over one million pigs were destroyed for outbreak control (US $97 million), control activities cost US $136 million, 36,000 jobs were lost, and there were 257 cases of encephalitis including 105 deaths (Nor and Ong 2000; FAO/APHCA 2002).

Four outbreaks of Nipah virus have occurred in the same region of Bangladesh from 2001 to 2005, all occurring between January and April but each attributed to different exposure factors (Hsu et al. 2004; WHO 2004a; ICDDR,B 2003 2004a 2004b 2005). Genotyping virus from the Bangladesh outbreaks have showed a 95% homology with isolates from the Malaysian outbreak in 1999. In outbreaks in the Meherpur district (2001) and Faridpur district (2004), direct contact with the secretions of ill patients is thought to have played a role in transmission of the disease (Hsu et al. 2004; ICDDR,B 2004b; WHO 2004a). In the outbreak in Naogaon district (2003), cases were associated with exposure to a herd of pigs (ICDDR,B 2004a). In Goalanda, Rajbari district, nine of the 12 Nipah cases were boys under 19 years who climbed trees where fruit bats fed overnight (WHO 2004a). Contamination is thought to have occurred while eating the same fruits, although whether infection was a result of ingestion or contact transmission was not determined. In January 2005, 12 cases of Nipah virus were reported in Basail Upazila, Tangail District, of whom 11 died (92%) (ICDDR,B 2005). The only significant exposure associated with illness was drinking raw date palm juice, which is consumed within a few hours of collection. Date palm juice potentially contaminated with the saliva and/or urine of *Pteropus giganteus*, the species of fruit bat widely distributed throughout Bangladesh, is considered the most likely source of transmission.

1.3 Severe Acute Respiratory Syndrome

Severe acute respiratory syndrome caused by the SARS coronavirus (SARS-CoV) emerged in the Guangdong province of the People's Republic of China in November 2002 (see the chapter by Wang and Eaton, this volume). The Himalayan masked palm civet (*Paguma larvata*) is considered the source of infection in humans (Chinese SARS Molecular Epidemiology Consortium 2004; Guan et al. 2003; Kan et al. 2005). One reservoir of SARS-like coronaviruses closely related to those responsible for the SARS outbreak is now known to be cave-dwelling bats in the genus *Rhinolophus* (Chinese horseshoe bats) (Li et al. 2005). These viruses, termed SARS-like coronaviruses, display greater genetic variation than SARS-CoV isolated from humans or from civets. The SARS epidemic demonstrated that even in well-resourced countries, the initial response to SARS was hindered by inadequate disease surveillance systems, poor communication and information sharing, and insufficient public health capacity. Unprecedented levels of international travel and trade enabled the rapid spread of SARS within and between continents. In global terms, SARS was a small epidemic resulting in just over 8,000 cases and 774 deaths (WHO 2004b). However, SARS severely challenged the capacity of curative and preventive health services, including the ability of public health services in unaffected countries to investigate suspected cases of SARS. The epidemic temporarily reduced consumer confidence in Asia, costing Asian economies US $11–18 billiion and resulting in estimated losses of 0.5%–2% of total output according to official macroeconomic data, and economic impact studies from international financial institutions, industry associations, and public policy research institutions (US General Accounting Office 2004). SARS had significant, but temporary, negative effects on a variety of economic activities, especially travel and tourism even in unaffected countries. Tourism fell by 9.7% in the Asia Pacific region as a direct result of SARS (Department of Tourism, Industry and Resources 2004).

1.4 Highly Pathogenic Avian Influenza

Human cases of A/H5N1 avian influenza were first reported in Hong Kong in 1997, when it infected 18 people with six deaths (Tam 2002). The World Organization for Animal Health (OIE) received the first report of its re-emergence on 12 December 2003 from the Republic of Korea. The disease spread rapidly within South-East Asia (Cambodia, China, Indonesia, Japan, the Lao People's Democratic Republic, Malaysia, the Republic of Korea, Thailand and Viet Nam), had infected domestic poultry flocks and wild birds in Russia, Kazakhstan and Mongolia by July 2005, and Romania, Croatia and Turkey by October 2005, confirming the westward spread of the virus (FAO 2005; see the chapter by Webby et al., this volume). Evidence shows that the A/H5N1 virus is now enzootic in many parts of Asia and is spreading rapidly in Europe. One estimate of the direct costs to the agricultural sector in Cambodia, Thailand and Viet Nam is of the order of US $560 million (McLeod 2005). Genetic analyses of isolates from Mongolia, Kazakhstan, Romania, Russia and Turkey show a close genetic relationship to wild bird isolates from the Qinghai Lake outbreak, China. Outbreaks of A/H5N1 have recurred despite aggressive control measures, including the culling of millions of poultry since December 2003. At the time of writing in January 2006, human cases with an overall fatality rate around 50% have been reported in Cambodia, China, Indonesia, Thailand, Turkey and Viet Nam (see the chapter by Webby et al., this volume).

A/H5N1 avian influenza has also proven highly pathogenic to wading birds and a number of terrestrial mammals. In 2004, concurrent with outbreaks of avian influenza in poultry, a total of 147 of 441 tigers (*Panthera tigris*) and two leopards (*P. pardus*) kept in the zoo in Suphanburi, Thailand, died after an acute respiratory illness with high fever or were euthanised to prevent possible spread to other zoo animals. The aetiological agent was subsequently confirmed as A/H5N1 avian influenza. The animals had been fed raw chicken carcasses that were contaminated with the A/H5N1 virus (Keawcharoen et al. 2004). Cases occurring 12 days after the tigers were last fed raw poultry were attributed to tiger-to-tiger transmission (Thanawonghuwech et al. 2005). During the outbreak, there were also anecdotal reports of fatal A/H5N1 virus infection in domestic cats, previously thought to be resistant to influenza A infections (ProMed mail 2004). When cats ($n=3$) were experimentally infected with A/H5N1 virus isolated from a fatal human case in Vietnam (A/Vietnam/1194/04), they exhibited respiratory symptoms, diffuse alveolar damage and excreted virus at 3 days post-infection (Kuiken et al. 2004; see the chapter by Webby et al., this volume). Three control cats inoculated with a human A/H3N2 virus isolate from a human (A/Netherlands/18/94) showed no evidence of infection or disease. The study also demonstrated that cats could be infected with A/H5N1 virus both by horizontal transmission and by feeding on virus-infected birds (Kuiken et al. 2004). There is considerable concern that other carnivores may also be susceptible to infection through eating infected poultry or infected wild birds.

Almost all human infections can be linked to contact with infected poultry, but instances of inefficient human-to-human transmission may have occurred in several family clusters in Vietnam (Tran et al. 2004), and possibly in Thailand (Ungchusak et al. 2005) and Indonesia (WHO 2007). The risk of further human cases continues, as do opportunities for a human-adapted pandemic strain to emerge following a recombination event. Kuiken et al. concluded that cats might also enable the adaptation of A/H5N1 to mammals, thereby increasing the risk of a human influenza pandemic (Kuiken et al. 2004). More recently, concerns have been raised that inappropriate vaccination of poultry to control the disease may lead to asymptomatic transmission among birds and spread of the virus between farms from poor biosecurity during vaccination campaigns (Parry 2005).

These scenarios highlight the importance of controlling avian influenza in livestock as far as possible to prevent human infections, and the need for strong collaboration between the animal and human health sectors. The United Nations Food and Agriculture Organization (FAO)/OIE regional animal laboratory network will be closely linked to the World Health Organization (WHO) Global Influenza Programme (WHO 2004c) to allow rapid sharing of virus samples and assessment of changes in A/H5N1 strains circulating in animal populations suggestive of increasing resistance to antiviral drugs or which may diminish the effectiveness of the human prototype H5 vaccines currently under development.

Although there is considerable epidemiological uncertainty about the extent of an influenza pandemic, it is expected to be more damaging in human health, social and economic development terms than previous public health emergencies. The Asian Development Bank has modelled the economic impact on Asia of a relatively mild influenza pandemic of 1 year's duration and with an attack rate of 20% and a case fatality ratio of 0.5%. The scenario is far less severe than the pandemic of 1918 but probably more severe than the pandemics of 1957 and 1968. The model puts the potential cost to the region at between US $99.2 billion and $282.7 billion in lost consumption, trade and investment, with an additional $14.2 billion lost through staff incapacity and death (Bloom et al. 2005).

1.5 Transmissible Spongiform Encephalopathies

The transmissible spongiform encephalopathies (TSEs) are a group of fatal neurodegenerative diseases of humans and other mammalian species (WHO 2003a). Although the pathogenesis of TSEs is incompletely understood, most researchers believe the aetiological agent is a prion, the misfolded form of a normal cellular protein designated PrP^{Sc}, that acquires infectivity. TSEs are genetically determined, sporadic or acquired from exposure to TSE-contaminated materials. The accumulation of PrP^{Sc} in the brain is a hallmark of most forms of TSE.

Scrapie of sheep and goats and bovine spongiform encephalopathy (BSE) are serious livestock diseases that have resulted in significant losses to livestock producers through death or destruction of affected animal populations. Both are subject to eradication programs (Ramasamy 2004) in affected countries and import restrictions in unaffected countries.

Scrapie has been known to infect sheep for at least 250 years (WHO 2003a) and is not transmissible to humans. Its infective nature was first described in 1935 following transmission studies in sheep that involved the intraocular inoculation of a healthy ewe with infected sheep spinal cord tissue. Disease surveillance, herd depopulation and selective breeding programs were proving successful control measures until recently. There is a well-established association between sheep prion protein genotype and the risk of death from scrapie (Baylis et al. 2004). Certain genotypes are associated with susceptibility to the disease and others with resistance. The intensified surveillance of scrapie in the European

Union, together with the improvement of PrP^{Sc} detection techniques, has led to the discovery of a growing number of atypical scrapie cases. In 2002, researchers in Germany, Portugal and France identified a variant form of scrapie that appears to infect sheep of the genotype ARR/ARR purposefully bred in Europe as a lineage resistant to scrapie (LeDur et al. 2005; Roden et al. 2006). The prion proteins of the variant form accumulate in different parts of the brain, have different biochemical properties and produce a spectrum of disease that differs slightly from traditional scrapie. Inoculation of transgenic mice expressing ovine PrP with material from three sheep homozygous for the resistant PrP(ARR) allele efficiently transmitted the disease to the mice. These observations suggest that a previously unrecognised infectious TSE agent infects sheep flocks (LeDur et al. 2005) and may have important implications in terms of scrapie control and public health.

The appearance of BSE resulted in an explosive epidemic of fatal encephalopathy in cattle herds in Britain. BSE has been causally linked to variant Creutzfeldt-Jakob disease (vCJD) in humans (Bruce et al. 1997; Collinge et al. 1997). BSE has had profound effects on the livestock industry, animal and human food safety, the international requirements for import risk assessments and certification of freedom of disease. The history of BSE is a cautionary tale of the unanticipated and unintended impact of new technologies and production practices introduced by the livestock industry on human and wildlife health. BSE also highlights the various economic, social and political costs and impacts resulting from disease prevention and large-scale control strategies.

BSE was first reported in British cattle in November 1986, and by September 2005 183,850 confirmed cases had been reported to the OIE (OIE 2005a). Mathematical modelling indicates that the epidemic began in the mid-1970s and that approximately one million cattle must have been infected and entered the food supply. Current evidence supports the hypothesis that BSE originated from the recycling of cattle infected with a scrapie-like agent derived from either sheep or cattle in feed containing rendered meat and bonemeal. Changes to the rendering process from the 1970s to the early 1980s appear to have reduced the inactivation of PrP^{Sc} and enabled propagation of the agent. BSE became a notifiable disease in the UK in June 1988, and soon afterwards, a ban on the feeding of ruminant-derived protein to ruminants became mandatory. The ban was extended to specified high-risk bovine offals (SBOs) for human consumption in November 1989 based on the infectivity of tissues of scrapie-infected sheep, and in September 1990 SBOs were prohibited for use in feed for all animals and birds in the UK. The BSE epidemic in Britain peaked at the end of 1992 when 37,280 incident cases were detected and then declined rapidly, although a small number of cases continued to occur (Enserink 2005; OIE 2005a). In 2004, 343 cases were reported in Britain and just over 150 in 2005. BSE in animals born after the ruminant feed ban have been attributed to exposure to contaminated feed after the ban, maternal transmission or other unidentified routes of transmission. In October 2004, French researchers confirmed a TSE in a goat slaughtered in 2002 that could not be distinguished from BSE on the mouse bioassay which takes 2 years to complete (Europa 2004). One additional goat tested positive of 140,000 goats examined from April 2002 to January 2005.

Because of the global export of cattle and cattle-derived products, BSE has since been reported on a smaller scale from all 25 EU countries with the exception of Sweden (Grist 2005), and in Israel (Nitzan-Kaluski and Leventhal 2003), Japan (Yamakawa et al. 2003) and most recently from Canada (Coulthart et al. 2003) and the United States (Larkin 2002). In some of these countries, BSE-affected cattle were detected even after a probabilistic risk assessment integrating release, exposure and consequence assessments indicated a negligible probability that BSE was introduced and established (Morley et al. 2003). Materials potentially contaminated with the BSE agent had been distributed around the world through the trade in live cattle and cattle by-products before export bans and import risk assessments were put into place. These products include a range of high-risk materials, some masked by trading patterns that have included processing and re-exportation of hazardous products. The occurrence of BSE in cattle in Europe and elsewhere raised new concerns about the precautions needed to ensure the safety of the international trade of cattle and cattle products. Many countries still have no monitoring systems or insensitive surveillance in place for BSE and may not have the financial and response capacity to eliminate BSE should cases occur. From 2001 to 2004, abattoir-based testing of asymptomatic cattle for BSE in European Union countries cost €1.6 million per BSE case detected, with an overall cost of approximately €1.6 billion (Enserink 2005).

In March 1996, ten cases of a newly recognised variant of Creutzfeldt-Jakob disease (CJD), the most commonly recognised form of human TSE, were reported in the United Kingdom. The new form was designated variant-CJD. Consumption of BSE-infected beef products, particularly mechanically recovered meat, is the most likely route of transmission in humans. These cases were temporally and geographically linked to outbreaks of BSE, making an aetiological link highly likely. Several different PrP^{Sc} types in humans have been identified, each associated with a different clinical phenotype of CJD. Strain-typing experiments have shown that the vCJD agent is different from that causing sporadic CJD but similar to the BSE agent. Humans that are homozygous (methionine/methionine) at codon 129 are more susceptible to both variant and sporadic CJD. All but one of the cases of vCJD to date has been homozygous at codon 129; the single heterozygous (methionine/valine) case was infected via a blood transfusion and demonstrated for the first time that codon 129-heterozygous individuals are susceptible to vCJD infection

(Peden et al. 2004). Speculation continues on whether cases of vCJD with very long incubation periods will occur among individuals heterozygous or homozygous (valine/valine) at codon 129 who were exposed to high-risk beef products before the bans.

Since 2003, two cases of vCJD in the UK were attributed to infections via the transfusion of red cells from donors who later died of vCJD (Peden et al. 2004; Llewelyn et al. 2004). A substantial body of animal data have also demonstrated that TSEs can be transmitted through blood (Ironside and Head 2003), even when the donor is in the subclinical phase of disease (Houston et al. 2000). Epidemiological studies of lymphoreticular system tissues have shown a low, but measurable, carrier state in vCJD. PrP^{Sc} has been found in appendix, spleen, tonsil and lymph nodes of patients with vCJD, and in this regard differs to other human TSEs (Hill et al. 1999). TSEs are highly resistant to the sterilisation and equipment reprocessing techniques that readily destroy bacterial and viral pathogens and have radically changed the practice of infection control during surgical and invasive diagnostic procedures. The widespread distribution of PrP^{Sc} throughout the lymphoid and central nervous systems raises concerns about the risk of transmission through surgical and ophthalmological procedures (Dunstan and Alpers 2005). The appearance of vCJD has also challenged the safety of the blood supply and organ donation. Changes in surgical practices, such as the use of disposable equipment for common procedures and the need to destroy or quarantine expensive equipment that would previously have been reprocessed for use, have resulted in considerable costs to health care systems around the world.

From 1986 to 2003, 37 cases of TSEs occurred in 37 zoo animals involving 12 species, including the ungulate species *Tragelaphus strepsiceros* (greater kudu) and wild-captive *Felidae* (cheetah, tiger and lion). In 1990, the first case of feline spongiform encephalopathy in a domestic cat was reported in the UK, with 91 reports by September 2001. Exposure to uncooked infected bovine materials is assumed to be the source of transmission in the felids. The ongoing risk of interspecies transmission of TSEs needs careful assessment (Ramasamy 2004) in view of the experimental evidence that tissues from subclinically infected animals (Race and Chesebro 1998) can be infectious to other species.

Historically TSEs have only affected wildlife in small numbers. Transmissible mink encephalopathy is associated with exposure through feed contaminated with a TSE agent (Williams and Miller 2003). Chronic wasting disease (CWD) of mule deer and elk, first discovered in Wyoming and Colorado in the 1980s, has been spreading across the United States and Canada, raising concerns about the risk of transmission to free-ranging cervids that may lead to losses in biodiversity (Daszak et al. 2001) and that threatens the viability of game farming industries (Williams and Miller 2003). CWD is thought to be spread orally, either through direct contact among animals or via environmental contamination.

Current TSE risk assessments (Grist 2005) acknowledge the importance of generic uncertainties in the following areas: the prevalence levels of TSE-infected individuals in animal and human populations; whether a threshold dose of prions is required to initiate infection; whether ingested prions accumulate in an individual over time; the dose of prions required to overcome the species barrier for interspecies transmission to occur; the nature of prion transportation and longevity in the environment; and whether genetic heterozygosity will lead to a second wave of vCJD of very long incubation periods. These and other unanswered questions raise concerns that the lifting or loosening of BSE control measures and reductions in research funding recently announced by the European Union is premature, and that long-term vigilance is required to prevent a resurgence of disease and to monitor the effects of emerging TSE variants (Enserink 2005).

1.6 Wildlife Zoonoses

Emerging infectious diseases of wildlife such as Ebola virus and West Nile virus, which have resulted in spillover events to humans and livestock, are a threat to animal welfare and biodiversity (Daszak et al. 2001; Pourrut et al. 2005; see the chapter by Daszak et al., this volume). Others, such as chronic wasting disease in elk and deer, may result in transmission to humans through the consumption of game meats. The outbreak of monkeypox in pet owners and handlers (including a veterinarian) in the USA in 2003, highlighted the importance of wildlife species in zoonotic disease and the extent of the international trade in wildlife species (Guarner et al. 2004; CDC 2003). The source of the outbreak was traced to the legal importation of exotic rodent reservoirs of monkeypox from Ghana in West Africa (see the chapter by Regnery and Damon, this volume). Native pet prairie dogs housed near some of these rodents in a distributor's premises became infected, and the subsequent multi-state distribution and sale of the prairie dogs resulted in human infections.

2 Minimising the Impact of Emerging Zoonoses

Preparedness planning for disease emergence usually involves some form of risk assessment to assess the likelihood of infection and disease, and the impact on susceptible populations. In the context of emerging zoonoses, comprehensive risk assessments are needed to identify the animal–human and animal–animal interfaces where transmission of infectious agents occurs and risk reduction interventions are feasible (see the chapter by Cleaveland et al., this volume).

As wildlife is important in the epidemiology of many, if not most, zoonoses, wildlife should be taken into account in the risk analysis framework (Kruse et al. 2004). Health risk assessments for emerging zoonotic diseases should be undertaken whenever possible in the context of developmental projects that have ecological impacts and are likely to bring people into greater contact with wildlife (see the chapter by Daszak et al., this volume).

Assessing the risk of spillover events (Daszak et al. 2000) among species requires an understanding of the behaviour and ecology of emerging pathogens and the complex interactions between the agent, its natural reservoir(s), the behaviour of humans or animals susceptible to infection, and the ecosystems in which they interact. It is becoming increasingly apparent that bats are the reservoirs for a number of pathogenic viruses (Calisher et al. 2006; Field et al. 2004; see the chapter by Field et al., this volume), including rabies (Warrell and Warrell 2004; see the chapter by Nel and Rupprecht, this volume), the Australian bat lyssavirus (Fraser et al. 1996; Field et al. 1999, 2004; Gould et al. 2002; Warrell and Warrell 2004), henipaviruses (Eaton et al. 2006), SARS-like coronaviruses (Li et al. 2005), and Ebola virus (Leroy et al. 2005; see the chapter by Gonzalez et al., this volume), and are considered candidate reservoirs for Marburg virus (Leroy et al. 2005; see the chapter by Gonzalez et al., this volume). Other taxa may also prove to have co-evolved with a variety of viruses pathogenic for humans and animals (Peterson et al. 2004). For some emerging zoonoses, limited knowledge of these relationships, especially for wildlife diseases, makes the risk assessment of spillover particularly difficult (Polley 2005), thereby also limiting our ability to design interventions that will reduce opportunities for interspecies transmission.

Data to inform risk assessments, especially in less developed countries, are often lacking or unreliable, and some risk models have therefore extrapolated the results obtained from data collected in developed countries (FAO 2004). Accordingly, differences between countries and regions in the risk parameters used to develop the model need to be considered in designing and implementing surveillance and diagnostic systems for emerging diseases and risk reduction strategies. Some of these data are routinely collected or arise from research conducted in the human health, agriculture and wildlife sectors. In some countries, national livestock databases designed to increase the safety and traceability of livestock products are potentially valuable sources of data and are being used to strengthen veterinary epidemiology and economic analysis (James 2005). Livestock data which can be used for epidemiological purposes include movement records, animal health program data, quality assurance schemes, production records and breeding records.

Insufficient work has gone into collating and triangulating data from these various sources to build an integrated and dynamic picture of the evolution of emerging zoonoses. The potential applications of integrated human, livestock and wildlife data include developing a better understanding of the descriptive epidemiology of emerging zoonoses, improved risk and decision analysis, and mathematical models to inform policy development and disease control management in all sectors. Using cartographic and geostatistical methods during epidemiological investigations can provide real-time quantitative data for identifying and tracking the geospatial spread of infectious diseases (Lai et al. 2004).

3 Mechanisms for Surveillance and Response to Emerging Zoonoses

Factors that drive disease emergence in human, livestock and wildlife populations are increasingly the result of human activity, and include changes to global ecology and climate, land use, animal husbandry and food production practices, air travel and the globalisation of trade (see the chapter by Childs et al., this volume). The impact of emerging diseases can be minimised through a well-prepared and strong public health system and similar systems developed by the livestock, wildlife and food safety sectors. To respond to emerging zoonoses effectively, preparedness plans, early warning systems and response capacity must be strengthened and implemented in a coordinated way across all sectors.

To meet the global challenge that emerging disease outbreaks present, the International Health Regulations (IHR) (WHO 2005a; Merianos and Peiris 2005) provide a legal framework for the international public health response to control cross-boundary infectious diseases. The purpose and scope of the revised IHR "are to prevent, protect against, control and provide a public health response to the international spread of disease in ways that are commensurate with, and restricted to, public health risks and which avoid unnecessary interference with international traffic and trade." The IHR (2005) explicitly recognise the need for intersectoral and multidisciplinary cooperation in managing risks of potential international public health importance. The IHR include a decision algorithm to assist countries in determining whether an outbreak or other unusual disease event may constitute a threat to international public health. National health authorities are required to report to the World Health Organization in the event of the following: smallpox, wild type poliovirus, human influenza (new subtype) and SARS; any event of potential international public health concern; and known epidemic-prone diseases that have the potential to spread internationally or threaten trade (e.g. cholera, plague, viral haemorrhagic fevers and West Nile fever).

In 2000, the WHO Department of Communicable Diseases Surveillance and Response in Geneva, Switzerland, initiated the formation of the Global Outbreak Alert and Response Network (GOARN) (WHO 2000), which provides the operational and technical response arm for the control of global outbreaks. Since April 2000, GOARN has played a key role in providing support to outbreak investigations in countries seeking assistance. Technical cooperation includes the provision of multidisciplinary field teams to assist in outbreak investigation and control, laboratory diagnosis and verification, clinical case management, and the delivery of vaccines and other therapeutic agents, equipment and logistics. Recent GOARN responses to diseases of zoonotic origin include multiple outbreaks of SARS and highly pathogenic avian influenza (A/H5N1) in humans, Ebola and Marburg haemorrhagic fevers, Nipah virus disease, plague and Rift Valley fever.

In response to the profound effects of emerging zoonoses such as Nipah virus, SARS and human cases of influenza A/H5N1 in the Asia Pacific Region, countries of the region in collaboration with the WHO South-East Asia and Western Pacific Regional Offices have adopted the *Asia Pacific Strategy for Emerging Diseases* (WHO 2005b). The Strategy aims to minimise the health, economic and social impact of emerging diseases through a targeted program of capacity building for public health surveillance and outbreak response in accordance with the core requirements of the IHR. Similar strategies are being implemented through a variety of public health networks in other WHO regions. Reducing the risk of diseases acquired from animals is a key objective of the Asia Pacific Strategy, which describes a broad, multinational, and multisectoral approach over the medium to long term. Success in the prevention and control of emerging zoonoses will require close collaboration between local and national health, agriculture, wildlife and food safety authorities in parallel with risk reduction activities involving international organisations such as WHO, FAO and OIE.

The quality of pathogen surveillance in animals varies greatly among countries and typically does not include wildlife (Kuiken et al. 2005; see the chapters by Childs et al. and by Stallknecht, this volume). The Terrestrial Animal Health Code (2005) (OIE 2005b) aims to assure the sanitary safety of international trade in terrestrial animals and their products through health measures to be used by national veterinary authorities to avoid the transfer of agents pathogenic for animals or humans, while avoiding unjustified sanitary barriers. The Terrestrial Code states that "countries shall make available to other countries, through the OIE, whatever information is necessary to minimise the spread of important animal diseases and to assist in achieving better worldwide control of these diseases". The Terrestrial Code lists procedures for the international reporting of diseases, ethical rules for international trade, certification and animal welfare, the principles of import risk analysis, and the organisation of import and export procedures. There are a large number of notifiable animal diseases under international surveillance: anthrax, bovine spongiform encephalopathy, bovine tuberculosis, brucellosis, Crimean Congo haemorrhagic fever, highly pathogenic avian influenza, hydatid disease, Japanese encephalitis, leptospirosis, Nipah virus encephalitis, Q fever, Rift Valley fever, *Salmonella enteritidis* and *S. typhimurium* in poultry, screwworm, trichinellosis, tularaemia, and West Nile fever have the potential to cause human disease. National disease control requirements under the Terrestrial Code identify the need for a formal and ongoing system for detecting and investigating outbreaks of disease, procedures for the rapid collection and transport of clinical specimens, laboratory investigation guidelines for diagnostic quality assurance and a system for recording, managing and analysing diagnostic and surveillance data. In addition, the Terrestrial Code makes recommendations for the standardised monitoring of antimicrobials used in animal husbandry to evaluate usage patterns by animal species, antimicrobial class, potency and type of use in order to evaluate antimicrobial use and detect the emergence of resistance. Antimicrobial resistance may also have implications for antimicrobial efficacy in human health and in wildlife.

Agricultural pests and diseases may spread across borders or be introduced through travel, trade and the illegal trafficking of animals. Infectious agents can cause disease control emergencies, especially in developing countries with limited response capacity, and may result in major economic losses. On occasion, extensive emergency operations with international assistance become necessary particularly if detection and response are delayed. In 1994, the FAO established an Emergency Prevention System (EMPRES) for Transboundary Animal and Plant Pests and Diseases (FAO 2005) in order to minimise the risk of such emergencies developing. EMPRES has four main components – early warning, early reaction, co-ordination and applied research – and all are integral to preparedness planning for emerging infectious diseases.

All countries should participate in regional, and where possible, global surveillance and diagnostic networks for human, livestock and wildlife health, and enable the sharing of information to characterise risk, prevent disease spread, and enhance control efforts. To be most effective, preparedness planning for emerging zoonoses requires a whole-of-government approach, clear command, control and coordination structures across the health, agriculture and wildlife sectors, and appropriate funding of the human health and veterinary services for their disease alert and response operations. Opportunities for shared training and involvement in multi-sectoral outbreak simulations should be identified to test operational communications, networking and partnerships, and to identify gaps in preparedness across the various sectors.

Countries should define the criteria (trigger points) for declaring a national animal disease emergency and initiating whole-of-government action. The availability of human, material and financial resources, including technical expertise and surge capacity, should be assessed as part of preparedness planning for emerging diseases and linkages formed with regional and global networks, such as a the Global Outbreak Alert and Response Network, that can provide emergency support to affected countries. Relevant local trigger points for alert and response should be defined as part of emergency preparedness planning by all human and veterinary health services.

4 Elements of Early Warning and Response Systems for Emerging Zoonoses

Early warning and response systems for emerging zoonoses require effective cross-jurisdictional, intersectoral and interdisciplinary collaboration. Early warning systems have been implemented at sub-national, national, regional and global levels. Networking, and linking individuals and agencies, will be key factors in building and sustaining surveillance and response capacity against existing and emerging disease threats. These activities can also provide the support needed in the areas where key capacities, such as diagnostics, do not currently exist or are under-resourced and require development.

Areas of expertise considered critical to improve detection, monitoring and investigation of emerging infections include field epidemiology, clinical and veterinary sciences, laboratory diagnostics, field ecology (mammalogy and entomology), behavioural science, medical anthropology, risk communication, social mobilisation (behaviour change communication) and other related disciplines.

4.1 Early Warning Systems

Early warning systems are based predominantly on epidemiological surveillance in the form of event-based and case-based activities. Event-based surveillance is purposely designed to detect unusual or unexpected disease events such as disease clusters or unexplained deaths (Merianos and Peiris 2005; WHO 2005a; see the chapter by Childs, this volume). Case-based surveillance provides information on individual cases of disease. Both lead to improved awareness and knowledge of the distribution of disease or infection and, depending on the completeness and quality of the data collected, might permit forecasting the evolution of an outbreak. Development, strengthening and implementation of early warning and response functions within integrated national disease surveillance systems are critical steps in building the core capacities for surveillance and response under the IHR (2005). Similar guidance is provided to detect, investigate and control outbreaks of disease in domestic animals, livestock (OIE 2005b; FAO 2005) and wildlife.

Mortality surveillance—the investigation of unusual mortality—should be an integral part of early warning systems for public health, domestic animals, livestock and wildlife. Wild bird mortality has provided early indications of highly pathogenic avian influenza infection (Sturm-Ramirez et al. 2004; Liu et al. 2005) and West Nile virus (McLean et al. 2002). West Nile virus occurs over a broad geographic range and in diverse vertebrate hosts and vector species. Until recently, there were few reports of deaths in wild birds and a small number of cases of equine encephalitis. Mortality in domestic birds was first reported in Israel in 1997 (Banet-Noach et al. 2003), and encephalitis was reported in horses in Italy in 1998 (Cantile et al. 2000) and France in 2000 (Murque et al. 2001). In 1999, West Nile virus caused an outbreak of encephalitis in humans in the New York area concurrent with cases of equine encephalitis and deaths in crows and other native and exotic bird species. A mortality surveillance system for the rapid detection of West Nile virus was implemented as an integrated response between wildlife health and public health agencies (McLean et al. 2002). The death of nonhuman primates has been associated with outbreaks of Ebola haemorrhagic fever (Rouquet et al. 2005). Wild animal outbreaks began before each of the five human Ebola outbreaks in the forest zone between Gabon and Republic of Congo. All human Ebola virus outbreaks from 2001 to 2003 in that area resulted from handling infected wild animal carcasses. Through the establishment of an animal mortality monitoring network, health authorities were twice alerted to the imminent risk of a human Ebola outbreak weeks before they occurred (Rouquet et al. 2005).

Supporting effective surveillance are the routine clinical, laboratory and epidemiological information systems that can provide valuable baseline data and are often the sources of data that help identify and track unusual disease events. Such data sources include outpatient, hospital-based and public health and animal health records, hospital mortality data, the laboratory accession system used for specimen tracking, and data on the use of pharmaceuticals. Routinely collected data can support surveillance activities and may be the only ongoing data for general mortality surveillance in the veterinary field.

4.2 Risk-Based Surveillance

Targeted surveillance of high-risk settings and populations can provide cost-effective early warning of infection. Risk settings include farms, slaughterhouses, livestock and wildlife markets, hospitals, laboratories, international borders and hubs for international travel and trade. High-risk occupations include health care workers, laboratory staff, veterinarians, primary producers, cullers, stock transporters and chicken catchers, abattoir workers, hunters, and distributors of animals, especially wildlife. Serological surveillance of high-risk populations, including baseline serology for occupation risk groups, health monitoring, and methods for identifying disease in vaccinated animals (such as monitoring unvaccinated sentinel animals and laboratory investigations that can discriminate vaccinated from infected animals), can provide important information on background rates of infection and disease, the size and distribution of susceptible and immune populations and species, and the effectiveness of control measures such as immunisation. Molecular epidemiology, especially when combined with human networking and animal movement data, allows tracing of disease transmission pathways and the identification of pathogen maintenance cycles (James 2005). The ability to differentiate vaccine-induced and wild antigens and antibodies has profound implications for epidemiological surveillance and disease control policy.

Major hubs of wildlife trade provide practical surveillance and control opportunities, especially if there is a supportive regulatory framework in place (Karesh et al. 2005). Air travel statistics have been used to model the importance of international travel hubs in the spread of epidemic-prone diseases in humans (Bauch et al. 2005; see the chapter by Daszak et al., this volume).

The effectiveness of existing local and national human and animal disease surveillance systems to detect known and novel zoonoses should be routinely evaluated to identify gaps and weaknesses. Astute clinicians and veterinarians are often the first to detect unusual disease events and are an integral part of the surveillance system for emerging diseases. Building awareness, knowledge and skills of clinicians in both sectors about emerging zoonoses will improve their early detection.

Effective wildlife surveillance is often limited by funding constraints, which necessitates optimisation of study design, sampling methodology and diagnostic methods; these are potential areas of applied research.

4.3 Improving Pathogen Identification

Laboratory diagnosis is an essential component of disease surveillance, both for the routine confirmation of diseases and for rapid determination of the aetiological agent during outbreaks (WHO 2005a). Laboratory surveillance systems are particularly useful for the detection of rare zoonotic infections that have spilled over into humans, domestic animals or livestock.

Laboratory assistance on-site to support outbreak investigations has proven very useful in emerging disease outbreaks. The use of new technologies for field use, such as rapid diagnostic tests, robust and portable nucleic acid-based technologies and multi-pathogen microarrays for the detection of known pathogens and their virulence factors (Burton et al. 2005; Sergeev et al. 2004) have greatly reduced the time taken to arrive at a definitive diagnosis during outbreaks.

There is an urgent need to strengthen linkages between national clinical and veterinary reference laboratories with regional and international laboratory networks that support verification and quality assurance and can provide diagnostic services for emerging and dangerous pathogens when necessary. These networks can also collaborate in the development of rapid diagnostic tests, including point of care tests, for surveillance purposes and test their performance under field conditions.

The WHO has been active in strengthening global laboratory networks to ensure that all countries have access to technical expertise for pathogen identification, reference and verification in humans, internal and international quality assurance, logistical assistance in the form of equipment, supplies and transport, and access to appropriate levels of biocontainment. Diagnostic and molecular biological capacity of OIE/FAO Reference Laboratories and Collaborating Centres are also being strengthened, and technology transfer is provided to National Agricultural Research Systems through the established system of networks of national and regional laboratories (FAO/OIE 2004).

4.4
Improving Information Management for the Early Detection of Emerging Diseases

Effective surveillance for emerging zoonoses requires the exchange of information among public health authorities, veterinary services and the wildlife sector. Timely analysis of surveillance data are needed to identify, track and manage threats to public health, the livestock industry and to wildlife, and to support evidence-based interventions for control. Information management should include systems to support the alert and event confirmation functions of early warning systems. All sectors should aim to improve or develop information systems for epidemic intelligence, verification status, laboratory investigations and field operations. Wherever possible, these systems should be integrated so that critical information for decision making is readily available. In addition, mechanisms and communication technologies that facilitate the rapid exchange of epidemic intelligence across the health, livestock and wildlife sectors as required should be implemented and tested as part of emergency preparedness. Because information of zoonotic disease occurrence in animals is important to public health officials, WHO, FAO and OIE developed GLEWS, the Global Early Warning and Response System for Major Animal Diseases, including Zoonoses, to combine information from each organization so that outbreaks can be detected earlier and the coordination of response to emerging zoonoses improved (WHO/FAO/OIE 2004).

5 Control Measures

5.1 General and Threat-Specific Control Measures

Decreasing contact among species through community education, legislation and regulation or direct intervention is considered a practical approach to reducing the risk of emerging zoonoses (Karesh et al. 2005; see the chapter by Real and Biek, this volume). Following an outbreak of A/H5N1 in Hong Kong in 1997 that resulted in 18 cases and six deaths, control measures aimed at reducing exposure of humans to potential H5-infected poultry were instituted and included culling of all poultry in Hong Kong, the segregation of waterfowl and chickens, the introduction of import control measures for chickens and waterfowl and central slaughtering of waterfowl (Tam 2002). Following illness caused by influenza A/H9N2 (G1) strain in two children in Hong Kong in 1999, closing down retail poultry markets for 1 day per month and subsequent exclusion of quail from live bird markets reduced the rate of A/H9N2 avian influenza virus (especially the G1 strain) in market birds (Kung et al. 2003).

In addition to health monitoring for occupational exposure to dangerous pathogens, evidence-based protective measures for high-risk groups, such as vaccination and the use of personal protective equipment, should be applied wherever possible. However, in some situations the groups at highest risk of animal-to-human transmission of infectious diseases are poorly defined and may require specific prevention interventions that are culturally and socially acceptable. A/H5N1 infections in women and children exposed to infected poultry through activities such as slaughtering, defeathering and/or handling sick or dead birds is an important example.

Activities to prevent and control zoonotic diseases must also recognise the local cultural and economic factors that influence the patterns of human–animal and animal–animal interactions, and the ecological changes associated with land usage and animal husbandry practices that increase the frequency and intensity of human exposure to animal reservoirs of disease.

5.2 Improving Pathogen Containment in Laboratories and Biosafety in the Field

"Laboratory biosafety" describes the containment principles, technologies and practices that are implemented to prevent unintentional exposure to pathogens and toxins or their accidental release (WHO 2004d). "Laboratory biosecurity" describes the institutional and personal security measures designed to prevent the loss, theft, misuse, diversion or intentional release of pathogens and toxins (WHO 2004a). Effective biosafety systems depend on well-formulated laboratory policies, optimal work practices, appropriate containment equipment and inventory controls, personnel risk assessments and effective management. Managing risks in the laboratory is dependent on both biosafety and biosecurity.

Breaches in laboratory biosafety and biosecurity have resulted in individual cases or outbreaks of disease caused by dangerous pathogens (Heymann et al. 2004). The three laboratory-associated outbreaks of SARS after transmission had ceased in July 2003 are a salient lesson. These incidents were attributed to breaches in laboratory biosafety and resulted in one or more cases of SARS: Singapore (WHO 2003b; Report of the Review Panel on New SARS Case and Biosafety; Lim et al. 2004), Taipei (WHO 2003c) and Beijing (WHO 2004e, 2004f). Fortunately only one of these incidents resulted in secondary transmission outside of the laboratory. The last incident was a cluster of nine cases, one of whom died, in three generations of transmission affecting family and hospital contacts of a laboratory worker.

All countries have an ongoing responsibility to develop, implement and monitor national standards to protect specimens, pathogens and toxins from accidental release or misuse. Biosafety also includes the measures put in place to protect laboratory staff and others involved in the diagnostic chain: appropriate training, health monitoring, the use of appropriate personal protective equipment, procedures for the investigation of spills and other incidents, and the laboratory equipment and engineering of the physical environment needed to reduce risks. The US Office of Health and Safety has developed a security plan based on facility risk assessments (Richmond and Nesby-O'Dell 2002). According to that plan, the key elements of laboratory security are systematic site reviews of physical security, data security, employee security, access controls to laboratory and animal areas, procedures for agent inventory and accountability; controls on shipping or transfer and receiving of select agents, incident and injury policies and emergency response plans, and a mechanism to investigate and address breaches in security. Preventive measures such as the immunisation of staff against vaccine-preventable diseases, and protocols for post-exposure prophylaxis where applicable, should also be written into laboratory management plans.

The responsibility for biosafety and biosecurity begins at the point of collection of clinical specimens, whether in a clinical setting, for research purposes or as part of a field investigation. New or poorly characterised infectious diseases such as emerging zoonoses pose particular difficulties for biosafety risk assessments. When knowledge of the pathogenic agent is insufficient to perform an appropriate risk assessment, for example, with clinical specimens or epidemiological samples collected in the field, a precautionary approach should be adopted during specimen manipulation. Standard precautions, especially handwashing, should always be followed and barrier protection used (gloves, gowns, eye protection) when handling clinical specimens. When dealing with poorly understood pathogens, additional (transmission-based) precautions and the use of special protective equipment, such as high-efficiency respirators, are recommended.

Decisions about the level of biocontainment required should consider available epidemiological data (morbidity and mortality data, suspected route of transmission, other outbreak investigation data) and the geographical origin of the specimen. Both human health laboratories and animal facilities are designated according to a risk assessment and the risk group of the microorganisms under investigation, as Biosafety Level (BSL) 1, 2, 3 or 4 (WHO 2004d). At Biosafety Level 3, manipulation of all potentially infectious material must be conducted within a biological safety cabinet or other primary containment device. The maximum containment laboratory – Biosafety Level 4 – is designed for work with dangerous pathogens. The WHO Laboratory Biosafety Manual recommends that any activities which require virus culture or manipulation involving the growth or concentration of a pathogen should be carried out in a BSL3 facility while routine diagnostic procedures (such as serology, haematology and biochemistry) or the manipulation of inactivated agents can be conducted under BSL2 conditions. Aerosol-generating procedures must be carried within a class 2 biological safety cabinet within a BSL3 laboratory and the operator should follow strict transmission-based precautions, including the use of appropriate personal protective equipment.

Concerns have been raised that there is a lack of standardisation in biosafety policy, practice and monitoring of the current levels of biocontainment within and between countries (Mackenzie and Olowokure, in press). Differences in requirements exist between animal and human laboratory biocontainment requirements within the United States, and between the US, British, Australian, European, Canadian and WHO guidelines (Mackenzie and Olowokure, in press). Accreditation of laboratories does not occur in many developing countries that handle dangerous pathogens. A set of international standards would assist in assuring conformity with good operating procedures and standards of biosafety and biosecurity. International standards are also required for quality assurance, building engineering, laboratory management, staff training, health monitoring of laboratory staff, and incident investigation and management in the event of accidental breakage, spills and other potentially hazardous events (Mackenzie and Olowokure in press).

6 Applied Research

Global efforts are underway to develop a comprehensive research agenda on the determinants of inter-species transmission of disease for policy development and evidence-based prevention and control activities. Key areas of research include:

- The environmental, ecological and climatic factors which facilitate the emergence, maintenance and transmission of zoonoses, including deforestation, developmental projects, global warming, urban ecology, the dynamics of inter-species transmission of infectious diseases between wild and domestic animals and between animals and humans.
- The evolutionary changes of pathogenic infectious agents that result in increased infectivity, virulence or transmissibility and mechanisms of pathogen dispersal.
- The human, livestock and wildlife host factors that facilitate the emergence of infections and their spread and the protective factors resulting in resistance to disease, including genetic analysis.
- New diagnostic tools and surveillance technologies that can support rapid and accurate diagnosis under field conditions. Technologies that have proven particularly useful in the study of emerging zoonoses include remote sensing and global information systems.
- Improved mathematical models of transmission dynamics to improve our ability to predict future disease outbreaks.
- Improved case management and the development of new vaccines and other therapeutic modalities for the treatment and prevention of emerging zoonoses.
- The social inequalities and behavioural factors that influence the distribution of emerging diseases, their course and the populations that are most affected.
- The impact of disease control strategies on affected populations, including the costs, benefits, incentives and disincentives of participation in control measures in order to frame effective interventions.

- The effectiveness of intervention methods used by public health, agriculture and wildlife sectors to prevent, mitigate and control emerging zoonotic diseases, and the risks and benefits for other sectors.
- Economic evaluation of historical outbreaks and modelling of future outbreaks of zoonotic disease.
- Development of more powerful study designs and sampling methodologies, and diagnostic methods, to optimise wildlife surveillance.

7 Conclusions

As it is highly likely that zoonoses and animal diseases with the potential to affect human health will continue to emerge, surveillance for zoonotic diseases will need to be strengthened and maintained at national and international levels. Surveillance, laboratory capability, knowledge, skills and technology transfer, and communications along with adequate funding for all these aspects are key elements when developing capacity to detect and respond to emerging diseases. Applied research is another critical component that is often under-funded, with evident funding shortfalls in the wildlife sector.

Viral zoonoses are the most common diseases to have emerged in the last four decades. Recognition of the importance of wildlife as a reservoir of zoonoses is increasing, although in most countries, the resources provided to wildlife research and conservation management remain limited. An expanded research agenda in the factors leading to disease emergence integrated across the human health, livestock and wildlife sectors is needed to inform risk assessments and preparedness planning for the prevention and control of zoonoses. Cost-effective prevention, investigation and control strategies necessitate an interdisciplinary and multi-sectoral approach within countries and internationally.

References

Banet-Noach C, Malkinson M, Brill A, Samina I, Yadin H, Weisman Y, Pokamunski S, King R, Deubel V, Stram Y (2003) Phylogenetic relationships of West Nile viruses isolated from birds and horses in Israel from 1997 to 2001. Virus Genes 26:135–141

Bauch CT, Lloyd-Smith JO, Coffee MP, Galvani AP (2005) Dynamically modelling SARS and other newly emerging respiratory illnesses: past, present, and future. Epidemiology 16:791–801

Baylis M, Chihota C, Stevenson E, Goldmann W, Smith A, Sivam K, Tongue S, Gravenor MB (2004) Risk of scrapie in British sheep of different prion protein genotype. J Gen Virol 85:2735–2740

Bloom E, De Wit V, Carangal-San, Jose MJ (2005) ERD Policy Brief Series No. 42. Potential economic impact of an avian flu pandemic on Asia. Asian Development Bank November (2005) http://www.adb.org/Documents/EDRC/Policy_Briefs/PB042.pdf. Cited 9 January 2006

Bruce ME, Will RG, Ironside JW, McConnell I, Drummond D, Suttie A, McCardle L, Chree A, Hope J, Birkett C, Cousens S, Fraser H, Bostock CJ (1997) Transmissions to mice indicate that the new variant CJD is caused by the BSE agent. Nature 389:498–501

Burton JE, Oshota OJ, North E, Hudson MJ, Polyanskaya N, Brehm J, Lloyd G, Silman NJ (2005) Development of a multi-pathogen oligonucleotide microarray for detection of Bacillus anthracis. Mol Cell Probes 19:349–357

Calisher CH, Childs JE, Field HE, Holmes KV, Schountz T (2006) Bats: Important reservoir hosts of emerging viruses. Clin Microbiol Rev 19:531–545

Cantile C, Di Guardo G, Eleni C, and Arispici M (2000) Clinical and neuropathological features of West Nile virus equine encephalomyelitis in Italy. Equine Vet J 32:31–35

Centres for Disease Control and Prevention (2003) Update: Multistate outbreak of monkeypox—Illinois, Indiana, Kansas, Missouri, Ohio, and Wisconsin, 2003. MMWR Morb Mortal Wkly Rep 52:561–564

Chadha MS (2006) Nipah Virus-associated encephalitis outbreak Siliguri India. Emerg Infect Dis12:235–240

Chinese SARS Molecular Epidemiology Consortium (2004) Molecular evolution of the SARS coronavirus during the course of the SARS epidemic in China. Science 303:1666–1669

Chua KB, Goh KJ, Wong KT, Kamarulzaman A, Tan PS, Ksiazek TG, Zaki SR, Paul G, Lam SK, Tan CT (1999) Fatal encephalitis due to Nipah virus among pig-farmers in Malaysia. Lancet 354:1257–1259

Chua KB, Bellini WJ, Rota PA, Harcourt BH, Tamin A, Lam SK, Ksiazek T, Rollin P, Zaki S, Shieh., W-J, Goldsmith C, Gubler D, Roehrig J, Eaton B, Gould A, Olson J, Field H, Daniels P, Ling A, Peters C, Anderson L, Mahy B (2000) Nipah virus: a recently emergent deadly paramyxovirus. Science 288:1432–1435

Chua KB, Koh CL, Hooi PS, Wee KF, Khong JH, Chua BH, Chan YP, Lim ME, Lam SK (2002) Isolation of Nipah virus from Malaysian Island flying-foxes. Microbes Infect 4:145–151

Collinge J, Hill AF, Desbruslais M, Joiner S, Sidle KCL, Gowland I (1997) The same prion strain causes vCJD, BSE. Nature 389:48–50

Coulthart MB, Mogk R, Rancourt JM, Godal DL, Czub S (2003) Prion protein gene sequence of Canada's first non-imported case of bovine spongiform encephalopathy (BSE). Genome 46:1005–1009

Daszak P, Cunningham AA, Hyatt AD (2000) Emerging infectious diseases of wildlife—Threats to biodiversity and human health. Science 287:443–449

Daszak P, Cunningham AA, Hyatt AD (2001) Anthropogenic environmental change and the emergence of infectious diseases in wildlife. Acta Trop 78:103–116

Department of Tourism Industry and Resources (2004) Tourism Overview August (2004) http://www.industry.gov.au/assets/documents/itrinternet/TOURISMoverview Aug042005020812(3813)pdf. Cited 7 July 2005
Dunstan RA, Alpers MP (2005)Variant Creutzfeldt-Jakob disease: implications for the health care system. Aust N Z J Public Health 29:308–312
Eaton BT, Broder CC, Middleton D, Wang L-F (2006) Hendra and Nipah viruses: different and dangerous. Nature Rev Microbiol 4:23–35
Enserink M (2000) Malaysian researchers trace Nipah virus outbreaks to bats. Science 289:518–519
Enserink M (2005) After the crisis: More questions about prions. Science 310:1756–1758
Europa (2004) Commission submits French Research Findings on TSE in a goat to Expert Panel. Brussels, 28 October 2004 http://europa.eu.int/rapid/pressReleasesAction-do?reference=IP/04/1324&format=HTML&aged=0&language=EN&guiLanguage=en. Cited 9 January 2006
FAO (2003) Emergency Prevention System (EMPRES) for Transboundary Animal and Plant Pests and Diseases http://www.fao.org/EMPRES/default.htm. Cited 5 December 2005
FAO (2004) Guiding principles for highly pathogenic avian influenza surveillance and diagnostic networks in Asia. FAO Expert meeting on surveillance and diagnosis of avian influenza in Asia. Bangkok Thailand, 21–23 July 2004
FAO (2005) FAO Avian Influenza Disease Emergency News. Update on the Avian Influenza situation (as of 12/11/2005) – Issue no. 36. Avian influenza spread into Europe Situations update (July–October 2005)
FAO/APHCA (2002) Manual on the diagnosis of Nipah virus infection in animals. Food and Agriculture Organization (FAO) and the Animal Production and Health Commission for Asia and the Pacific (APHCA) RAP publication no. 2002/01, FAO, Rome. http://www.fao.org/docs/eims/upload//195002/AVIbull036.pdf. Cited 5 December 2005
FAO/OIE (2004) The Global Framework for the Progressive Control of Transboundary Animal Diseases. Food and Agriculture Organization (FAO) and the World Organisation for Animal Health (OIE) http://www.fao.org/ag/againfo/subjects/en/health/diseases-cards/cd/documents/GF-TADs24May(2004)pdf (accessed 9 January 2006)
Field H, McCall B, Barrett J (1999) Australian bat lyssavirus infection in a captive juvenile black flying fox. Emerg Infect Dis 5:438–440
Field H, Young P, Yob JM, Mills J, Hall L, Mackenzie J (2001) The natural history of Hendra and Nipah viruses. Microbes Infect 3:307–314
Field H, Mackenzie J, Daszak P (2004) Novel viral encephalitides associated with bats (Chiroptera) - host management strategies. Arch Virol Suppl 18:113–121
Fraser GC, Hooper PT, Lunt RA, Gould AR, Gleeson LJ, Hyatt AD, Russell GM, Kattenbelt JA (1996) Encephalitis caused by a Lyssavirus in fruit bats in Australia. Emerg Infect Dis 2:327–331
Gould AR, Kattenbelt JA, Gumley SG, Lunt RA (2002) Characterisation of an Australian bat lyssavirus variant isolated from an insectivorous bat. Virus Res 89:1–28

Grimm TA, Beer BE, Hirsch VM, Clouse KA (2003) Simian immunodeficiency viruses from multiple lineages infect human macrophages: implications for cross-species transmission. J Acquir Immune Defic Syndr 32:362–369

Grist EPM (2005) Transmissible spongiform encephalopathy risk assessment: The UK experience. Risk Analysis 25:519–531

Guan Y, Zheng BJ, He YQ, Liu XL, Zhuang ZX, Cheung CL, Luo SW, Li PH, Zhang LJ, Guan YJ, Butt KM, Wong KL, Chan KW, Lim W, Shortridge KF, Yuen KY, Peiris JS, Poon LL (2003) Isolation and characterization of viruses related to the SARS coronavirus from animals in southern China. Science 302:276–278

Guarner J, Johnson BJ, Paddock CD, Shieh W-J, Goldsmith CS, Reynolds MG, Damon IK, Regnery RL, Zaki SR (2004) Monkeypox transmission and pathogenesis in prairie dogs. Emerg Infect Dis 10:426–431

Hahn BH, Shaw GM, De Cock KM, Sharp PM (2000) AIDS as a zoonosis: scientific and public health implications. Science 287:607–614

Heymann DL, Aylward RB, Wolff C (2004) Dangerous pathogens in the laboratory: from smallpox to today's SARS setbacks and tomorrow's polio-free world. Lancet 363:1566–1568

Hill AF, Butterworth RJ, Joiner S, Jackson G, Rossor MN, Thomas DJ, Frosh A, Tolley N, Bell JE, Spencer M, King A, Al-Sarraj S, Ironside JW, Lantos PL, Collinge J (1999) Investigation of variant Creutzfeldt-Jakob disease and other human prion diseases with tonsil biopsy samples. Lancet 353:183–189

Houston F, Foster JD, Chong A, Hunter N, and Bostock CJ (2000) Transmission of BSE by blood transfusion in sheep. Lancet 356:999–1000

Hsu VP, Hossain MJ, Parashar UD, Ali MM, Ksiazek TG, Kuzmin I, Niezgoda M, Rupprecht C, Bresee J, Breiman RF (2004) Nipah virus encephalitis re-emergence Bangladesh. Emerg Infect Dis 10:2082–2087

ICDDR,B (2003) Outbreaks of encephalitis due to Nipah/Hendra-like viruses, western Bangladesh. Health Sci Bull 1:1–6

ICAO (2004a) Outlook for air transport to the year (2005) International Civil Aviation Organization Circular 270-AT/111

ICDDR,B (2004b) Person-to-person transmission of Nipah virus during outbreak in Faridpur District, 2004. Health Science Bull 2:5–9

ICDDR,B (2005) Nipah virus outbreak from date palm juice. Health Science Bull 3:1–5

Ironside JW, Head MW (2003) Variant Creutzfeldt-Jakob disease and its transmission by blood. J Thrombol Haemost 1:1479–1486

James A (2005) The state of veterinary epidemiology and economics. Prevent Vet Med 67:91–99

Kalish ML, Wolfe ND, Ndongmo CB, McNicholl J, Robbins KE, Aidoo M, Fonjungo PN, Alemnji G, Zeh C, Djoko CF, Mpoudi-Ngole E, Burke DS, Folks TM (2005) Central African hunters exposed to simian immunodeficiency virus. Emerg Infect Dis 11:1928–1930

Kan B, Wang M, Jing H, Xu H, Jiang X, Yan M, Liang W, Zheng H, Wan K, Liu Q, Cui B, Xu Y, Zhang E, Wang H, Ye J, Li G, Li M, Cui Z, Qi X, Chen K, Du L, Gao K, Zhao YT, Zou XZ, Feng YJ, Gao YF, Hai R, Yu D, Guan Y, Xu J (2005) Molecular evolution analysis and geographic investigation of severe acute respiratory syndrome

coronavirus-like virus in palm civets at an animal market and on farms. J Virol 79:11892–11900
Karesh WB, Cook RA, Bennett EL, Newcomb J (2005) Wildlife trade and global disease emergence. Emerg Infect Dis 11:1000–1002
Keawcharoen J, Oraveerakul K, Kuiken T, Fouchier RAM, Amonsin A, Payungporn S, Noppornpanth S, Wattanodorn S, Theamblooniers A, Tantilertcharoen R, Pattanarangsan R, Arya N, Ratanakorn P, Osterhaus ADME, Poovorawan Y (2004) Avian influenza H5N1 in tigers and leopards. Emerg Infect Dis 10:2189–191
Kruse H, Kirkemo A-M, and Handeland K (2004) Wildlife as a source of zoonotic infections. Emerg Infect Dis12:2067–2072
Kuiken T, Rimmelzwaan G, van Riel D, van Amerongen G, Baars M, Fouchier R, Osterhaus A (2004) Avian H5N1 influenza in cats. Science 308:241
Kuiken T, Leighton IFA, Fouchier RAM, LeDuc JW, Peiris JSM, Schude A, Stohr K, Osterhaus AD (2005) Pathogen surveillance in animals. Science 309:1680–1681
Kung NY, Guan Y, Perkins NR, Bissett L, Ellis T, Sims L, Morris RS, Shortridge KF, Peiris JS (2003) The impact of a monthly rest day on avian influenza virus isolation rates in retail live poultry markets in Hong Kong. Avian Dis 47:1037–1041
Lai PC, Wong CM, Hedley AJ, Lo SV, Leung PY, Kong J, and Leung GM (2004) Understanding the spatial clustering of severe acute respiratory syndrome (SARS) in Hong Kong. Environ Health Perspect 112:1550–1556
Larkin M (2002) First US case of variant Creutzfeldt-Jakob disease reported. Lancet Infect Dis 2:715
Lederberg J, Shope RE, Oakes SC (1992) Emerging infections: microbial threats to health in the United States. Institute of Medicine. National Academy Press, Washington, DC
Le Dur A, Beringue V, Andreoletti O, Reine F, Lai TL, Baron T, Bratberg B, Vilotte JL, Sarradin P, Benestad SL, Laude H (2005) A newly identified type of scrapie agent can naturally infect sheep with resistant PrP genotypes. Proc Natl Acad Sci U S A 102:16031–16036
Leroy EM, Kumulungui B, Pourrut X, Rouquet P, Hassanin A, Yaba P, Delicat A, Paweska JT, Gonzalez JP, Swanepoel R (2005) Fruit bats as reservoirs of Ebola virus. Nature 438:575–576
Li W, Shi Z, Yu M, Ren W, Smith C, Epstein JH, Wang H, Crameri G, Hu Z, Zhang H, Zhang J, McEachern J, Field H, Daszak P, Eaton BT, Zhang S, Wang LF (2005) Bats are natural reservoirs of SARS-like coronaviruses. Science 310:676–679
Lim PL, Kurup A, Gopalakrishna G, Chan KP, Wong CW, Ng LC, Se-Thoe SY, Oon L, Bai X, Stanton LW, Ruan Y, Miller LD, Vega VB, James L, Ooi PL, Kai CS, Olsen SJ, Ang B, Leo YS (2004) Laboratory-acquired severe acute respiratory syndrome. New Engl J Med 250:1740–1745
Liu J, Xiao H, Lei F, Zhu Q, Qin K, Zhang XW, Zhang XL, Zhao D, Wang G, Feng Y, Ma J, Liu W, Wang J, Gao GF (2005) Highly pathogenic H5N1 influenza virus infection in migratory birds. Science 309:1206
Llewelyn GA, Hewitt PE, Knight RS, Amar K, Cousens S, MacKenzie C, Houston F (2004) Possible transfusion of variant Creutzfeldt-Jakob disease by blood transfusion. Lancet 363:417–421

Mackenzie JS, Olowokure B (2006) Biocontainment and Biosafety Issues Related to SARS-Coronavirus. In: SARS: The inside story on stopping an epidemic Western Pacific Regional Office Manila (in press)
McLean RG, Ubico SR, Bourne D, Komar N (2002) West Nile virus in livestock and wildlife. Curr Topics Microbiol Immunol 267:271–308
McLeod R (2005) The socioeconomic impacts of emerging infectious diseases in Asia, with a focus on the Greater Mekong Sub-Region. Asian Development Bank, 200. In: Bloom E, De Wit V, Carangal-San Jose MJ (eds) ERD Policy Brief Series No. 42. Potential economic impact of an Avian Flu pandemic on Asia. Asian Development Bank November 2005. http://www.adb.org/Documents/EDRC/Policy_Briefs/PB042.pdf. Cited 9 January 2006
Merianos A, Peiris M (2005) International Health Regulations (2005) Lancet Comment. Lancet 366:1249–1251
Morley RS, Chen S, Rheault N (2003) Assessment of the risk factors related to bovine spongiform encephalopathy. Rev Sci Tech 22:157–178
Murgue B, Murri S, Zientara S, Durand B, Durand JP, Zeller H (2001) West Nile outbreak in horses in southern France, 2000: the return after 35 years. Emerg Infect Dis 7:692–696
Nitzan-Kaluski D, Leventhal A (2003) Bovine spongiform encephalopathy in Israel: implications for human health. Israel Med Assoc J 5:662–665
Nor M, Ong B (2000) The Nipah virus outbreak and the effect on the pig industry in Malaysia. In: Proceedings of the 16th International Pig Veterinary Congress, Ocean Grove, pp 548–550
OIE (2005a) Number of cases of bovine spongiform encephalopathy (BSE) reported in the United Kingdom. http://www.oie.int/eng/info/en_esbru.htm. Cited 27 February 2007
OIE (2005b) Terrestrial Animal Health Code. http://www.oie.int/eng/publicat/en_code.htm. Cited 12 December 2005
Parry J (2005) Vaccinating poultry against avian flu is contributing to spread. Br Med J 331:1223
Paton NI, Leo YS, Zaki SR, Auchus AP, Lee KE, Ling AE, Chew SK, Ang B, Rollin PE, Umapathi T, Sng I, Lee CC, Lim E, Ksiazek TG (1999) Outbreak of Nipah-virus infection among abattoir workers in Singapore. Lancet 354:1253–1256
Peden AH, Head MW, Ritchie DL, Bell JE, Ironside JW (2004) Preclinical vCJD after blood transfusion in a PRNP codon 129 heterozygous patient. Lancet 364:527–529
Peeters M, Courgnaud V, Abela B, Auzel P, Pourrut X, Bibollet-Ruche F, Loul S, Liegeois F, Butel C, Koulagna D, Mpoudi-Ngole E, Shaw GM, Hahn BH, Delaporte E (2002) Risk to human health form a plethora of simian immunodeficiency viruses in primate bushmeat. Emerg Infect Dis 8:451–457
Peterson AT, Carroll DS, Mills JN, Johnson KM (2004) Potential mammalian filovirus reservoirs. Emerg Infect Dis 10:2073–2081
Polley L (2005) Navigating parasite webs and parasite flow: Emerging and re-emerging parasitic zoonoses of wildlife origin. Int J Parasitol 35:1279–1294
Pourrut X, Kumulungui B, Wittmann T, Moussavou G, Delicat A, Yaba P, Nkoghe D, Gonzalez JP, Leroy EM (2005) The natural history of Ebola virus in Africa. Microbes Infect 7:1005–10014

ProMED Mail (2004) Avian influenza H5N1, mammals—East Asia. Archive number 2004(0221)056:1 February 2004. www.promedmail.org. Cited 27 February 2007

Race R, Chesebro B (1998) Scrapie infectivity found in resistant species. Nature 392:770

Ramasamy I (2004)The risk of accidental transmission of transmissible spongiform encephalopathy: identification of emerging issues. Public Health 118:409–420

Review Panel on New SARS Case and Biosafety (2003) Biosafety and SARS Incident in Singapore September 2003 Report. http://www.moh.gov.sg/corp/sars/pdf/Report_SARS_Biosafety.pdf. Cited 27 February 2007

Reynes JM, Counor D, Ong S, Faure C, Seng V, Molia S, Walston J, Georges-Courbot MC, Deubel V, Sarthou JL (2005) Nipah virus in Lyle's flying foxes Cambodia. Emerg Infect Dis 11:1042–1047

Richmond JY, Nesby-O'Dell SL (2002) Laboratory security and emergency response guidance for laboratories working with select agents. Centers for Disease Control and Prevention. MMWR Recomm. Rep 51(RR-19):1–6

Roden JA, Nieuwhof GJ, Bishop SC, Jones DA, Haresign W, Gubbins S (2006) Breeding programmes for TSE resistance in British sheep I Assessing the impact on prion protein (PrP) genotype frequencies. Prev Vet Med 73:1–16

Rouquet P, Froment J-M, Bermejo M, Yaba P, Delicat A, Rollin PE, Leroy EM (2005) Wild animal mortality monitoring and human Ebola outbreaks Gabon and Republic of Congo, 2001–(2003) Emerg Infect Dis11:283–290

Seimenis A (1998) Zoonoses: a social and economic burden. East Med Health J 4:220–222

Sergeev N, Distler M, Courtney S, Al-Khaldi SF, Volokhov D, Chizhikov V, Rasooly A (2004) Multipathogen oligonucleotide microarray for environmental and biodefense applications. Biosens. Bioelectron 20:684–698

Smolinksi MS, Hamburg MA, Lederberg J (eds) (2003) Microbial threats to health: emergence, detection and response. The National Academies Press, Washington, DC

Sturm-Ramirez KM, Ellis T, Bousfield B, Bissett L, Dyrting K, Rehg JE, Poon L, Guan Y, Peiris JS, Webster RG (2004) Reemerging H5N1 influenza viruses in Hong Kong in 2002 are highly pathogenic to ducks. J Virol 78:4892–4901

Tam JS (2002) Influenza A (H5N1) in Hong Kong: an overview. Vaccine 20 [Suppl 2]: S77–S81

Tambyah PA, Tan JH, Ong BK, Ho KH, Chan KP (2001) First case of Nipah virus encephalitis in Singapore. Int Med J 31:132–133

Taylor LH, Latham SM, Woolhouse MEJ (2001) Risk factors for human disease emergence. Philos Trans R Soc B Biol Sci 356:983–989

Thanawongnuwech R, Amonsin A, Tantilertcharoen R, Damrongwatanapokin S, Theamboonlers A, Payungporn S, Nanthapornphiphat K, Ratanamungklanon S, Tunak E, Songserm T, Vivatthanavanich V, Lekdumrongsak T, Kesdangsakonwut S, Tunhikorn S, Poovorawan Y (2005) Probable tiger-to-tiger transmission of avian influenza H5N1. Emerg Infect Dis 11:699–701

Tran TH, Nguyen TL, Nguyen TD, Luong TS, Pham PM, Nguyen VC, Pham TS, Vo CD, Le TQ, Ngo TT, Dao BK, Le PP, Nguyen TT, Hoang TL, Cao VT, Le TG, Nguyen DT, Le HN, Nguyen KT, Le HS, Le VT, Christiane D, Tran TT, de Jong M, Schultsz C, Cheng P, Lim W, Horby P, Farrar J and the World Health Organization International

Avian Influenza Investigative Team (2004) Avian influenza A (H5N1) in 10 patients in Vietnam. N Engl J Med 350:1179–1188

UNAIDS (2005) AIDS epidemic update: December (2005) Joint United Nations Programme on HIV/AIDS (UNAIDS) http://www.unaids.org/Epi2005/doc/EPIupdate2005_pdf_en/epi-update2005_en.pdf. Cited 5 December 2005

Ungchusak K, Auewarakul P, Dowell SF, Kitphati R, Auwanit W, Puthavathana P, Uiprasertkul M, Boonnak K, Pittayawonganon C, Cox NJ, Zaki SR, Thawatsupha P, Chittaganpitch M, Khontong R, Simmerman JM, Chunsutthiwat S (2005) Probable person-to-person transmission of avian influenza A (H5N1) N Engl J Med 352:333–340

University of Sydney FAH Report (2004) Focus on Food Safety. Faculty of Veterinary Sciences. Farm Animal and Veterinary Public Health, 2002-2005, Faculty Workshop, 23 August 2004, University of Sydney. http://www.vetsci.usyd.edu.au/research/FAH_Report(2005)pdf

US General Accounting Office (2004) Emerging infectious diseases Asian SARS outbreak challenged international and national responses. Report to the Chairman Subcommittee on Asia and the Pacific Committee on International Relations House of Representatives April (2004) http://www.gao.gov/new.items/d0(4564)pdf. Cited 5 December 2005

Warrell MJ, and Warrell DA (2004) Rabies and other lyssavirus diseases. Lancet 363:959–969

Wildlife Conservation Society (2005) The Manhattan Principles on "One World – One Health". http://www.wcs.org/5060651. Cited 5 December 2005

Williams ES, Miller MW (2003) Transmissible spongiform encephalopathies in non-domestic animals: origin, transmission and risk factors. Rev. Sci Tech 22:145–156

World Health Organization (2000) WHO/CDS/CSR/(2000)3. Global outbreak alert and response: report of a WHO meeting. Geneva Switzerland April 26–2:000: http://www.who.int/csr/resources/publications/surveillance/whocdscsr(2003)pdf. Cited 1 June 2005

World Health Organization (2001) Fifty-fourth World Health Assembly Resolution WHA54.14 Global health security: epidemic alert and response, 21 May 2001 http://www.who.int/gb/ebwha/pdf_files/WHA54/ea54r14.pdf. Cited 1 June 2005

World Health Organization (2003a) WHO manual for surveillance of human transmissible spongiform encephalopathies including variant Creutzfeldt-Jakob disease. World Health Organization, Geneva

World Health Organization (2003b) Severe acute respiratory syndrome (SARS) in Singapore, 10 September 2003. http://www.who.int/csr/don/2003_09_10/en/. Cited 27 February 2007

World Health Organization (2003c) Severe acute respiratory syndrome (SARS) in Taiwan China, 17 December 2003. http://www.who.int/csr/don/2003_12_17/en/. Cited 27 February 2007

World Health Organization (2004a) Nipah virus outbreak(s) in Bangladesh January–April (2004) WHO Weekly Epidemiol Rec 79:168–171

World Health Organization (2004b) Summary of probable SARS cases with onset of illness from 1 November 2002 to 31 July 2003. http://www.who.int/csr/sars/country/table2004_04_21/en/index.html. Cited 5 December 2005

World Health Organization (2004c) Avian influenza – assessment of the situation as of 30 July (2004) Weekly Epidemiol Rec 32:291–292

World Health Organization (2004d) Laboratory biosafety manual. – 3rd edition. World Health Organization Geneva. http://www.who.int/csr/resources/publications/biosafety/Biosafety7.pdf. Cited 10 January 2006

World Health Organization (2004e) SARS: one suspected case reported in China, 22 April 2004. http://www.who.int/csr/don/2004_04_22/en/. Cited 27 February 2007

World Health Organization (2004f) China's latest SARS outbreak has been contained, but biosafety concerns remain – Update 7, 18 May 2004. http://www.who.int/csr/don/2004_05_18a/en/. Cited 27 February 2007

World Health Organization (2005a) Fifty-eighth World Health Assembly Resolution WHA58.3. Revisions of the International Health Regulations, 23 May 2005 http://www.who.int/gb/ebwha/pdf_files/WHA58/WHA58_3-en.pdf. Cited 10 June 2005

World Health Organization (2005b) Asia-Pacific Strategy for Emerging Diseases. World Health Organization, Geneva

World Health Organization (2007) H5N1 avian influenza: Timeline of major events, 13 April 2007. http://www.who.int/csr/disease/avian_influenza/timeline_2007_04_20.pdf

World Health Organization (WHO), Food and Agriculture Organization of the United Nations (FAO), and World Organisation for Animal Health (OIE), 2004 "Report of the WHO/FAO/OIE joint consultation on emerging zoonotic diseases, 3–5 May 2004 – Geneva, Switzerland" http://whqlibdoc.who.int/hq/2004/WHO_CDS_CPE_ZFK_2004.9.pdf (accessed 2007-05-24)

Yamakawa Y, Hagiwara K, Nohtomi K, Nakamura Y, Nishijima M, Higuchi Y, Sato Y, Sata T, Expert Committee for BSE Diagnosis Ministry of Health Labour and Welfare of Japan. (2003) Atypical proteinase K-resistant prion protein (PrPres) observed in an apparently healthy 23-month-old Holstein steer. Jpn J Infect Dis 56:221–222

Index

Current Topics in Microbiology and Immunology

Volumes published since 1989 (and still available)

Vol. 271: **Koehler, Theresa M. (Ed.):** Anthrax. 2002. 14 figs. X, 169 pp. ISBN 3-540-43497-6

Vol. 272: **Doerfler, Walter; Böhm, Petra (Eds.):** Adenoviruses: Model and Vectors in Virus-Host Interactions. Virion and Structure, Viral Replication, Host Cell Interactions. 2003. 63 figs., approx. 280 pp. ISBN 3-540-00154-9

Vol. 273: **Doerfler, Walter; Böhm, Petra (Eds.):** Adenoviruses: Model and Vectors in VirusHost Interactions. Immune System, Oncogenesis, Gene Therapy. 2004. 35 figs., approx. 280 pp. ISBN 3-540-06851-1

Vol. 274: **Workman, Jerry L. (Ed.):** Protein Complexes that Modify Chromatin. 2003. 38 figs., XII, 296 pp. ISBN 3-540-44208-1

Vol. 275: **Fan, Hung (Ed.):** Jaagsiekte Sheep Retrovirus and Lung Cancer. 2003. 63 figs., XII, 252 pp. ISBN 3-540-44096-3

Vol. 276: **Steinkasserer, Alexander (Ed.):** Dendritic Cells and Virus Infection. 2003. 24 figs., X, 296 pp. ISBN 3-540-44290-1

Vol. 277: **Rethwilm, Axel (Ed.):** Foamy Viruses. 2003. 40 figs., X, 214 pp. ISBN 3-540-44388-6

Vol. 278: **Salomon, Daniel R.; Wilson, Carolyn (Eds.):** Xenotransplantation. 2003. 22 figs., IX, 254 pp. ISBN 3-540-00210-3

Vol. 279: **Thomas, George; Sabatini, David; Hall, Michael N. (Eds.):** TOR. 2004. 49 figs., X, 364 pp. ISBN 3-540-00534X

Vol. 280: **Heber-Katz, Ellen (Ed.):** Regeneration: Stem Cells and Beyond. 2004. 42 figs., XII, 194 pp. ISBN 3-540-02238-4

Vol. 281: **Young, John A. T. (Ed.):** Cellular Factors Involved in Early Steps of Retroviral Replication. 2003. 21 figs., IX, 240 pp. ISBN 3-540-00844-6

Vol. 282: **Stenmark, Harald (Ed.):** Phosphoinositides in Subcellular Targeting and Enzyme Activation. 2003. 20 figs., X, 210 pp. ISBN 3-540-00950-7

Vol. 283: **Kawaoka, Yoshihiro (Ed.):** Biology of Negative Strand RNA Viruses: The Power of Reverse Genetics. 2004. 24 figs., IX, 350 pp. ISBN 3-540-40661-1

Vol. 284: **Harris, David (Ed.):** Mad Cow Disease and Related Spongiform Encephalopathies. 2004. 34 figs., IX, 219 pp. ISBN 3-540-20107-6

Vol. 285: **Marsh, Mark (Ed.):** Membrane Trafficking in Viral Replication. 2004. 19 figs., IX, 259 pp. ISBN 3-540-21430-5

Vol. 286: **Madshus, Inger H. (Ed.):** Signalling from Internalized Growth Factor Receptors. 2004. 19 figs., IX, 187 pp. ISBN 3-540-21038-5

Vol. 287: **Enjuanes, Luis (Ed.):** Coronavirus Replication and Reverse Genetics. 2005. 49 figs., XI, 257 pp. ISBN 3-540- 21494-1

Vol. 288: **Mahy, Brain W. J. (Ed.):** Foot-and-Mouth-Disease Virus. 2005. 16 figs., IX, 178 pp. ISBN 3-540-22419X

Vol. 289: **Griffin, Diane E. (Ed.):** Role of Apoptosis in Infection. 2005. 40 figs., IX, 294 pp. ISBN 3-540-23006-8

Vol. 290: **Singh, Harinder; Grosschedl, Rudolf (Eds.):** Molecular Analysis of B Lymphocyte Development and Activation. 2005. 28 figs., XI, 255 pp. ISBN 3-540-23090-4

Vol. 291: **Boquet, Patrice; Lemichez Emmanuel (Eds.):** Bacterial Virulence Factors and Rho GTPases. 2005. 28 figs., IX, 196 pp. ISBN 3-540-23865-4

Vol. 292: **Fu, Zhen F. (Ed.):** The World of Rhabdoviruses. 2005. 27 figs., X, 210 pp. ISBN 3-540-24011-X

Vol. 293: **Kyewski, Bruno; Suri-Payer, Elisabeth (Eds.):** CD4+CD25+ Regulatory T Cells: Origin, Function and Therapeutic Potential. 2005. 22 figs., XII, 332 pp. ISBN 3-540-24444-1

Vol. 294: **Caligaris-Cappio, Federico, Dalla Favera, Ricardo (Eds.):** Chronic Lymphocytic Leukemia. 2005. 25 figs., VIII, 187 pp. ISBN 3-540-25279-7

Vol. 295: **Sullivan, David J.; Krishna Sanjeew (Eds.):** Malaria: Drugs, Disease and Post-genomic Biology. 2005. 40 figs., XI, 446 pp. ISBN 3-540-25363-7

Vol. 296: **Oldstone, Michael B. A. (Ed.):** Molecular Mimicry: Infection Induced Autoimmune Disease. 2005. 28 figs., VIII, 167 pp. ISBN 3-540-25597-4

Vol. 297: **Langhorne, Jean (Ed.):** Immunology and Immunopathogenesis of Malaria. 2005. 8 figs., XII, 236 pp. ISBN 3-540-25718-7

Vol. 298: **Vivier, Eric; Colonna, Marco (Eds.):** Immunobiology of Natural Killer Cell Receptors. 2005. 27 figs., VIII, 286 pp. ISBN 3-540-26083-8

Vol. 299: **Domingo, Esteban (Ed.):** Quasispecies: Concept and Implications. 2006. 44 figs., XII, 401 pp. ISBN 3-540-26395-0

Vol. 300: **Wiertz, Emmanuel J.H.J.; Kikkert, Marjolein (Eds.):** Dislocation and Degradation of Proteins from the Endoplasmic Reticulum. 2006. 19 figs., VIII, 168 pp. ISBN 3-540-28006-5

Vol. 301: **Doerfler, Walter; Böhm, Petra (Eds.):** DNA Methylation: Basic Mechanisms. 2006. 24 figs., VIII, 324 pp. ISBN 3-540-29114-8

Vol. 302: **Robert N. Eisenman (Ed.):** The Myc/Max/Mad Transcription Factor Network. 2006. 28 figs., XII, 278 pp. ISBN 3-540-23968-5

Vol. 303: **Thomas E. Lane (Ed.):** Chemokines and Viral Infection. 2006. 14 figs. XII, 154 pp. ISBN 3-540-29207-1

Vol. 304: **Stanley A. Plotkin (Ed.):** Mass Vaccination: Global Aspects – Progress and Obstacles. 2006. 40 figs. X, 270 pp. ISBN 3-540-29382-5

Vol. 305: **Radbruch, Andreas; Lipsky, Peter E. (Eds.):** Current Concepts in Autoimmunity. 2006. 29 figs. IIX, 276 pp. ISBN 3-540-29713-8

Vol. 306: **William M. Shafer (Ed.):** Antimicrobial Peptides and Human Disease. 2006. 12 figs. XII, 262 pp. ISBN 3-540-29915-7

Vol. 307: **John L. Casey (Ed.):** Hepatitis Delta Virus. 2006. 22 figs. XII, 228 pp. ISBN 3-540-29801-0

Vol. 308: **Honjo, Tasuku; Melchers, Fritz (Eds.):** Gut-Associated Lymphoid Tissues. 2006. 24 figs. XII, 204 pp. ISBN 3-540-30656-0

Vol. 309: **Polly Roy (Ed.):** Reoviruses: Entry, Assembly and Morphogenesis. 2006. 43 figs. XX, 261 pp. ISBN 3-540-30772-9

Vol. 310: **Doerfler, Walter; Böhm, Petra (Eds.):** DNA Methylation: Development, Genetic Disease and Cancer. 2006. 25 figs. X, 284 pp. ISBN 3-540-31180-7

Vol. 311: **Pulendran, Bali; Ahmed, Rafi (Eds.):** From Innate Immunity to Immunological Memory. 2006. 13 figs. X, 177 pp. ISBN 3-540-32635-9

Vol. 312: **Boshoff, Chris; Weiss, Robin A. (Eds.):** Kaposi Sarcoma Herpesvirus: New Perspectives. 2006. 29 figs. XVI, 330 pp. ISBN 3-540-34343-1

Vol. 313: **Pandolfi, Pier P.; Vogt, Peter K. (Eds.):** Acute Promyelocytic Leukemia. 2007. 16 figs. VIII, 273 pp. ISBN 3-540-34592-2

Vol. 314: **Moody, Branch D, (Ed.):** T Cell Activation by CD1 and Lipid Antigens, 2007, 25 figs. VIII, 348 pp. ISBN 978-3-540-69510-3